ITALIAN PHYSICAL SOCIETY

PROCEEDINGS

OF THE

INTERNATIONAL SCHOOL OF PHYSICS « ENRICO FERMI »

Course LXXII

edited by G. TORALDO DI FRANCIA

Director of the Course

VARENNA ON LAKE COMO

VILLA MONASTERO

25 th JULY - 6 th AUGUST 1977

Problems in the Foundations of Physics

1979

NORTH-HOLLAND PUBLISHING COMPANY, AMSTERDAM - NEW YORK - OXFORD

SOCIETA' ITALIANA DI FISICA

RENDICONTI
DELLA
SCUOLA INTERNAZIONALE DI FISICA «ENRICO FERMI»

LXXII Corso

a cura di G. TORALDO DI FRANCIA
Direttore del Corso

VARENNA SUL LAGO DI COMO
VILLA MONASTERO
25 LUGLIO - 6 AGOSTO 1977

Problemi dei fondamenti della fisica

1979

SOCIETÀ ITALIANA DI FISICA
BOLOGNA - ITALY

Library of Congress Cataloging in Publication Data

Varenna, Italy. Scuola internazionale di fisica.
Problems in the foundations of physics.
(Proceedings of the International School of Physics « Enrico Fermi »; course 72).
At head of title: Italian Physical Society.
Added t.p.: Problemi dei fondamenti della fisica.
Course held July 25-Aug. 6, 1977.
Bibliography: p.
1. Physics—Congresses. I. Toraldo di Francia, Giuliano, 1916. II. Società Italiana di Fisica. III. Title. IV. Title: Problemi dei fondamenti della fisica.
QC1.V29 1979 530.1 79-13069
ISBN 0-444-85285-9

Printed in Italy

INDICE

C. Piron – Galilean and Lorentz particles: a new approach of quantization.

I. Prigogine and A. P. Grecos – Topics in nonequilibrium statistical mechanics.

B. C. van Fraassen – Foundations of probability: a modal frequency interpretation.

J. A. Wheeler – Frontiers of time.

Foreword.

« *Allez en avant, la foi viendra* ». This exhortation attributed to D'ALEMBERT seems to reflect very well the attitude taken by many of the great founders of science. Some of them were perhaps even unaware of the unfirm ground they were treading on; they just went boldly forward, and their success seemed to represent the best justification for the starting assumptions.

Somebody has even remarked that science is not founded upon its foundations. That remark certainly contains a great deal of truth; but it is not the entire story.

It is true that a good many, perhaps the majority, of the great achievements of science were not the result of a brick-by-brick construction, gradually carried out from the foundations to the top, like the construction of an old-time building. Pythagoras' theorem was discovered much before the time when EUCLID settled geometry as a well-founded science. Differential calculus yielded the most brilliant results in celestial mechanics much before CAUCHY, RIEMANN and WEIERSTRASS set out to justify its procedures. Today quantum mechanics can be said to have perfectly fulfilled the task for which it was first introduced: *i.e.* to interpret with great precision all atomic phenomena and radiation processes. Yet the foundations of quantum mechanics still present a number of puzzling problems that are currently challenging the ingenuity of many workers. Ironically, it can be said that in several cases physicists have obtained full success and exciting results before they knew exactly what they were talking about.

It would nevertheless be wrong to derive from these and other countless examples the conclusion that the study of the foundations is « useless », and represents only an academic pastime. The foundational problems of a discipline are likely to arise only when the discipline is fully developed and ripe. But, when finally the scientific community becomes aware of them and subjects them to a critical appraisal, a major breakthrough can occur and a more advanced discipline can arise.

Mechanical clocks of comparatively high precision had existed for centuries, when Einstein's keen examination of how time is measured and simultaneity is ascertained paved the way to special relativity. Euclidean geometry had provided for millennia the firm basis for science, when a critical analysis of its foundations led to a conceptual revolution, and eventually to general relativity. As a third example, let us recall that, if physicists had gone on taking for granted the elementary and « evident » fact that any quantity can in principle be measured with any precision, independently of other quantities, we would never have built quantum mechanics.

There is a sort of feed-back in the construction of physical science. First a new chapter is built on shaky grounds; then, when it is fully developed, it starts to react on its own foundations. It is only at this later stage that workers begin to realize what these foundations are and to pinpoint the more delicate questions of principle. Perhaps, the increasing interest in the foundational problems of physics which we are witnessing today is due to the unprecedented developments which physics has experienced in this century and to the vast amount of knowledge which is at our disposal today. An increasing number of physicists are becoming more and more critical and want to know « what it is all about ».

The present volume, a response to this demand, cannot encompass, for obvious reasons, all the foundations of physics. It contains the lectures given at the LXXII Course of the International School of Physics « Enrico Fermi ». The topics and the speakers were selected so as to give at least the highlights of several critical subjects which are currently debated today, but no attempt was made to reach completeness.

The foundational problems of physics are certainly all connected with one another. But in many cases the connection is subtle and hidden at a profound level. This makes it very difficult to give a logical succession to the different topics. Accordingly, the Editor has decided to publish the contributions in the alphabetical order of the names of the authors.

G. TORALDO DI FRANCIA

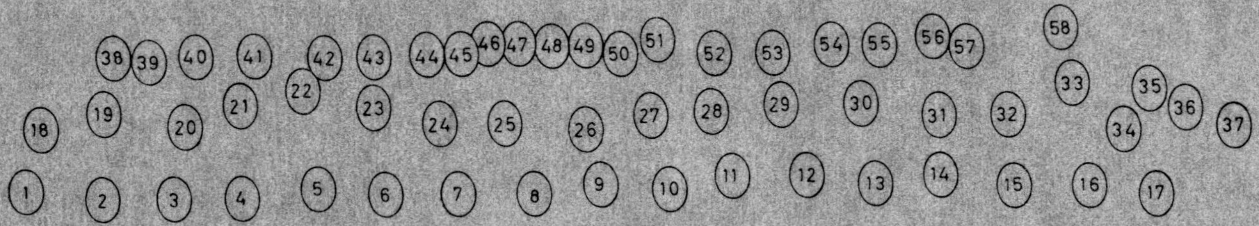

1. M. Bouten
2. R. Serneels
3. H. R. Tschudi
4. F. Balibar
5. L. Carlos Ryff
6. P. Hoyningen
7. I. Prigogine
8. M. Jammer
9. P. Mittelstaedt
10. M. L. Dalla Chiara
11. J. Bub
12. G. Toraldo di Francia
13. J. M. Lévy-Leblond
14. J. A. Wheeler
15. E. Beltrametti
16. A. Amer
17. G. A. Wolzak
18. P. Jasselette
19. C. Garola
20. R. Stella
21. J. Tassart
22. W. Roos
23. O. Ischebeck
24. G. Verrone
25. M. Severi
26. L. Solombrino
27. G. Fonte
28. W. Stachow
29. H. Pfister
30. G. Holdsworth
31. J. Grea
32. J. Kruger
33. D. Dieks
34. E. Mazzi
35. W. M. Honig
36. A. Rodriguez Vargas
37. E. Pittino
38. G. D'Emma
39. M. Battezzati
40. I. de Castro Moreira
41. C. Sa Furtado
42. J. van den Berg
43. F. Dal Fabbro
44. C. Hogan
45. Y. Saillard
46. G. Benettin
47. G. Dixon
48. S. Ruffo
49. A. Ten Ros
50. G. Morandi
51. H. C. Zapp
52. W. de Muynck
53. M. Ricci
54. G. Buffa
55. D. Aerts
56. J. Hamilton
57. J. M. Rodriguez
58. R. Meinhardt

SOCIETÀ ITALIANA DI FISICA

SCUOLA INTERNAZIONALE DI FISICA « E. FERMI »

LXXII CORSO - VARENNA SUL LAGO DI COMO - VILLA MONASTERO - 25 Luglio - 6 Agosto 1977

Radioactivity, a Pragmatic Pillar of Probabilistic Conceptions.

E. AMALDI
Istituto di Fisica dell'Università - Roma

When, years ago, I read the book by Max JAMMER *The Conceptual Development of Quantum Mechanics* [1], I found very interesting and stimulating the presentation he gives of this fundamental subject. I should say, however, that I was slightly disappointed from Chapter 4, devoted to *The transition to quantum mechanics*, since it appeared to me too short, incomplete and in some way one-sided.

The chapter consists of three sections. The first one, entitled *Applications of quantum conceptions to physical optics*, summarizes very effectively the work and conceptual background of A. FRESNEL, J. J. THOMSON, A. H. COMPTON, G. BARKLA, W. H. BRAGG and W. L. BRAGG, P. DEBYE and a few others. The second section, devoted to *The philosophical background of nonclassical interpretations*, can be, in some way, summarized by the following sentence by JAMMER himself: « certain philosophical ideas of the late ninteenth century not only prepared the intellectual climate for, but contributed decisively to the formation of the new conceptions of the modern quantum theory » [2]. The people quoted are C. RENOUVIER, E. BOUTROUX, F. EXNER, S. KIERKEGAARD, H. HØFFDING, and also H. POINCARÉ, L. DE BROGLIE, N. BOHR, C. G. DARWIN and a few others. The new conceptions are probabilistic conceptions which « differ fundamentally from the traditional notions of probability as used, for example, in classical statistical mechanics. In classical physics probability statements were but an expression of human ignorance of the exact details of the individual event, either because of the insufficient resolving power of our measuring instrument or because of the large number of events involved: the individual physical process, however, was always regarded as strictly obeying the law of cause and effect and the result was always considered as uniquely determined.

The new conception of probability, on the other hand, assumed not only that macroscopic determinism is a statistical effect but also that the individual microscopic and submicroscopic event is purely contingent » [3].

The third and last section of Chapter 4, entitled *Nonclassical interpretation of optical dispersion*, deals with the early work by N. BOHR, J. C. SLATER,

H. A. KRAMERS and by M. BORN, A. LANDÉ, R. LADENBURG and a few others.

With this presentation of the first attempts to solve the fundamental problem of optical dispersion, Chapter 4 is closed and the discussion of the transition to the new probabilistic conceptions is practically finished.

My first impression that this presentation is too hastly and too one-sided did not find further support from successive Jammer works such as the book on the *Philosophy of Quantum Mechanics* [4], since, in this case, an emphasis (or perhaps an over-emphasis) of the philosophical aspects of the subject is justified by the same title of the book.

My original impression was, on the contrary, strengthened by the article of P. FORMAN on *Weimar culture, causality and quantum theory, 1918-1927* [5], where the author states: ... « Jammer did not go very far towards demonstrating his propositions ... » the most important of which is « that extrinsic influences led physicists to ardently hope for, actively search for, and willingly embrace an acausal quantum mechanics ».

The aim of Forman's article (according to him) is not « to fill the gap left by Jammer » between « a variety of late ninteenth century philosophers » and « the development of quantum mechanics by German speaking central European physicists circa 1925 ». His aim is « rather to examine closely the lay of the land on the far side ... » with the result of a « overwhelming evidence that in the years after the end of the First World War, but before the development of an acausal quantum mechanics, under the influence of 'currents of thought' large numbers of German physicists, for reasons only incidentally related to developments in their own discipline, distanced themselves from, or explicitly repudiated, causality in physics ».

One of the main conclusions of Forman is that the « extrinsic influences » suggested by JAMMER are demonstrated in his article « but only for the German cultural sphere ».

Certainly I am not here to deny the interest of Jammer's point of view and the suggestiveness of the above propositions nor the cultural value of Forman's article, which deeply analyses the interrelations between the development of physical sciences in those years and their cultural and philosophical craddle and environment. What I would like to express in the following is my impression that, by looking at the problem of « repudiation of causality in physics » from the most general and far away point of view, one can be brought to over-estimate the « extrinsic influences » outlined above and overlook « intrinsic arguments » inherent to two parallel, almost independent developments. The first one starts from the kinetic theory of gases and passes through statistical mechanics, Planck original definition of quantum, the photons conceived as particles and the relations between emission and absorption of photons by atoms.

The other path, also intrinsic to physics, starts with the accidental discovery of radioactive substances, passes through the experimental recognition of their decay properties and quickly finds its natural settlement in a probabilistic conception which can be accused to be acritical, but has certainly a sound pragmatic ground, uncorrelated or at least extremely loosely correlated to contemporary or pre-existing philosophical lines of thought.

1. – Gradual infiltration of probability's laws into physical sciences.

The first one of these two intrinsic lines of development has been discussed on many occasions, in particular by Stephen G. BRUSH in an article appeared a few months ago, entitled *Irreversibility and indeterminism: Fourier to Heisenberg* [6], which according to the author « might be considered a chapter in the history of the changing meaning of the word *statistical* ». I will not summarize the historical succession of scientific steps forming this path nor will I sketch the line traced by BRUSH. It would take too much time for considerations that are very interesting but outside the main scope of my lecture. I will only mention a few points that appeared to me—already before reading Brush article—as milestones of these developments, omitting the very interesting detailed interconnections.

One of the most important aspects of Brush article, in my opinion, is the emphasis he gives to the gradual, continuous infiltration of probability's laws into physical sciences, to which I will add at the end, as one of my conclusions, the multiplicity of paths that all led to the same final winning-post: quantum mechanics.

Starting from Fourier's theory of heat conduction and a comparison of time-reversible mechanics (in the absence of dissipative forces) and time-oriented heat flow, BRUSH passes in review the first and second laws of thermodynamics with special regard to the work of Sadi CARNOT, Rudolf CLAUSIUS and William THOMSON (Lord KELVIN) and arrives at the kinetic theory of gases, in particular at the work of J. C. MAXWELL and L. BOLTZMANN. The words of these authors are ambiguous so that « in the absence of any explicit statements one might legitimately infer that they tacitly accepted the views of their contemporaries », *i.e.* molecular determinism. « But their equations (points out BRUSH) pushed physical theory very definitively in the direction of indeterminism. As in other transformations of physical sciences ... mathematical calculation led to results that forced the acceptance of qualitatively different concepts. »

From BOLTZMANN, BRUSH arrives to PLANCK, to his work on Boltzmann's statistical interpretation of entropy, and to his successive discovery, in 1900, of the spectral distribution of the black-body radiation.

The two laws of thermodynamics gave rise to well-known debates, which were strongly reinforced in 1865 when CLAUSIUS put them in the simple verbal form: « The energy of the Universe is constant. The entropy of the Universe tends towards a maximum ». These two statements, along with various logical and illogical extensions, and the use of probability methods in kinetic theory were matters of widespread debates, in which various philosophers and even theologians [7] were deeply involved. The fact that the Universe was moving towards a state of maximum probability, a configuration of maximum randomness, a condition of minimum available energy, was deemed to led to a degradation of the energy sources and ultimately to that pessimistic state of affairs that was called « heat-death » of the Universe. Themodynamics accordingly predicted an end of everything as a function of time.

Thus the laws of thermodynamics and their statistical interpretation had, directly or indirectly, an influence on the way of thinking of many philosophers between the end of the last century and the beginning of our century, and these, through a kind of feedback process, sometime had an influence on the general world outlook of physicists of successive generations.

The next milestone, as everybody knows, is the photon theory of light presented by EINSTEIN in sections 8 and 9 of the rather long paper published in 1905 and devoted mainly to the *Entropie der Strahlung* [8].

Max BORN wrote at the beginning of his 1926 paper [9] where he suggests the interpretation of the wave function as probability amplitude: « Dabei knüpfe ich an eine Bemerkung Einsteins über das Verhältnis von Wellenfeld und Lichtquanten an; er sagte etwa, das die Wellen nur dazu da seien, um der korpuskularen Lichtquanten den Weg zu weisen, und er sprach in diesem Sinne von einem *Gespensterfeld.* Dieses bestimmt die Wahrscheinlichkeit dafür, das ein Lichtquant, der Träger von Energie und Impuls, einen bestimmten Weg einschlägt; dem Felde selbst aber gehort keine Energie und kein Impuls zu ». (To this end I pick up from a remark by EINSTEIN on the relationship between wave field and light quanta; he said, more or less, that the waves are present only to show the corpuscular light quanta the way and he spoke in this sense of a *ghost field.* This determines the probability that a light quantum, the bearer of energy and momentum, takes a certain path: no energy nor momentum belongs to the wave field itself.)

In his Nobel lecture delivered in 1954 BORN was even more explicit in affirming a direct connection between his statistical interpretation of the wave function and the interpretation suggested by EINSTEIN of the « square of the optical wave amplitudes as probability density for the occurrence of photons » [10].

The last milestone of this line of development is, of course, the group of papers published by EINSTEIN between 1916 [11] and 1918 [12] on the absorption and emission of radiation by molecules [13]. In the 1918 paper [12] the basic assumptions of his theory are summarized in section 2.

The transition from the higher energy level E_m of a molecule to the lower E_n, during the time interval dt, can take place through two different mechanisms that EINSTEIN calls Ausstrahlung and Einstrahlung. The second one is the stimulated transition. Its probability of occurrence is proportional to the radiation density $\varrho(\nu)$ at the frequency of the emitted photon:

$$\text{(B)} \qquad \mathrm{d}W = B^n_m\, \varrho(\nu)\, \mathrm{d}t\,,$$

where $\nu = (E_m - E_n)/h$ and B^n_m is a constant. In the first process, the Ausstrahlung, the transition is not excited externally. The probability of occurrence in the time interval dt is given by

$$\text{(A)} \qquad \mathrm{d}W = A^n_m\, \mathrm{d}t\,,$$

where A^n_m is a constant. EINSTEIN comments this assumption as follows: « Das angenommene statistische Gesetz entspricht dem einer radioaktiven Reaktion, der vorausgesetzte Elementarprocess einer derartigen Reaktion, bei welcher nur γ-Strahlen emittiert werden ... ». (The adopted statistical law corresponds to that of radioactive reactions, the elementary process of which is such that only γ-rays are emitted.)

This final remark is of considerable interest from my point of view, since it establishes a connection between the line of development sketched above and the other line of development, that in my opinion is very often forgotten, in some case mentioned [6], but never discussed with sufficient attention and detail

Before passing to this new subject that constitutes the main scope of my seminar, I should recall that the Einstein's papers on the absorption and emission of light by molecules are discussed at length in sect. 3.2 of Jammer book [1], devoted to *The Correspondence Principle.* A few remarks are added by JAMMER that can be summarized as follows: 1) EINSTEIN did not define his conception of probability. Therefore, we are not authorized to consider him, as some authors do, a precursor of the new probabilistic conceptions. 2) Einstein's comparison of his statistical law with that of radioactive disintegration cannot be used as an argument, since at that time radioactive disintegration was generally considered, for example by PLANCK, as a process involving as yet unknown parameters. 3) The absence of a declaration in favour of causality cannot be interpreted as a declaration in favour of the absence of causality. 4) Referring to the related problem of other reactions of an atom or molecule during emission, EINSTEIN stated expressly that « in the present state of the theory the direction of recoil is only statistically determined », alluding thereby to the merely preliminary character of such an approach.

I frankly should say that it seems to me that the remarks 1), 3) and 4) refer only to the personal position of Einstein, but they are not very significant in the historical development of the new probabilistic conceptions. The remark 2) regards the personal position of Planck, who certainly did never work on radioactive decay. Jammer's remarks only underline the personal position of these two great scientists, who were both irreducible determinists, in spite of the fact that a great part of their fundamental contributions was essential for the developments that brought to quantum mechanics.

2. – The discovery of the law of radioactive decay.

A first indication of an activity decreasing with the passing of time was found in Vienna in 1898 by G. C. SCHMIDT, who observed that thorium compounds continuously emit radioactive particles of some kind, which retain their radioactive power for several minutes [14]. Shortly afterwards in 1899, Marie and Pierre CURIE [15, 16] observed a similar effect in the case of radium (and polonium [15]) and called it « radioactivitè induite » (induced radioactivity), while RUTHERFORD [17, 18], in 1900, went on with Schmidt experiments calling « emanation of thorium » « the radioactive particles given out from the mass of thorium compounds in addition to ordinary radiation ».

The expressions « induced activity » and « emanation of thorium » were due to primitive interpretations of essentially the same phenomenon. A few years later these were both replaced by the expression « active deposit », suggested by RUTHERFORD and still used today, for the radioactive bodies (RaA + + RaB + RaC + RaC'+ RaC'' and ThA + ThB + ThC + ThC'+ ThC'') produced in succession by the decay of the emanations of Ra and Th. These are the noble gases belonging to the radium and thorium families, later called radon (Rn) and thoron (Tn) [19].

What the CURIES and RUTHERFORD had observed?

The CURIES had found that all substances placed in the vicinity of a sample of radium acquired a radioactivity which, after the removal of radium, decreases according to an exponential law.

In the experiment of Rutherford a slow current of air passed over a sample of thorium oxide and « carried away the radioactive particles with it and these were gradually conveyed » into an ionization chamber connected to an electrometer arranged to measure the saturation current (fig. 1). Thus RUTHERFORD measured the decay curve (curve A of fig. 2) of a certain amount of emanation brought from its source (thorium oxide) to the measuring instrument and found an exponential curve, for which he wrote the formula

$$n(t) = n_0 \exp[-\lambda t]\,, \tag{1}$$

where $n(t)$ is the number of ions produced per second by the radioactive particles between the plates. This is connected to the measured current $i(t)$ by the relation $i(t) = en(t)$. RUTHERFORD measured also « the rise of current » (curve B of fig. 2), *i.e.* the rise of production of emanation from thorium as a function of time starting from the instant in which all emanation pre-

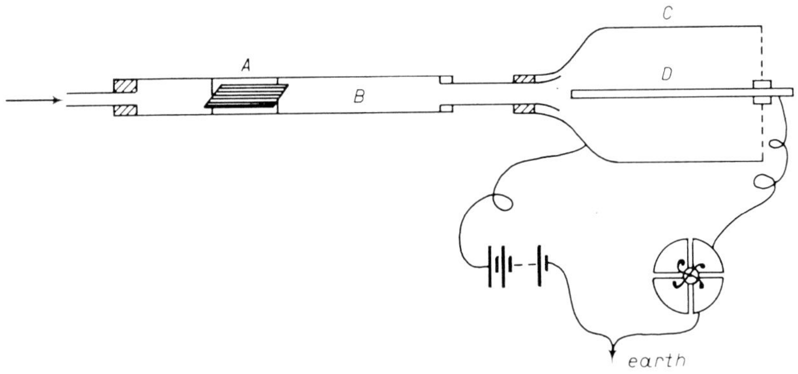

Fig. 1. – Rutherford experimental set-up: a thick layer of thorium oxide was enclosed in a paper vessel A placed inside the long metal tube B. One end of B was connected to a large insulated cylindrical vessel C connected to one terminal of a 100 V battery. Inside C was fixed an insulated electrode D connected to a pair of quadrants of the electrometer. The other pair of quadrants was connected to the other terminal of the battery. A slow current of purified air was passed through the apparatus and carried the emanation of thorium from A to C and then went out through a number of small holes opened in the bottom of C.

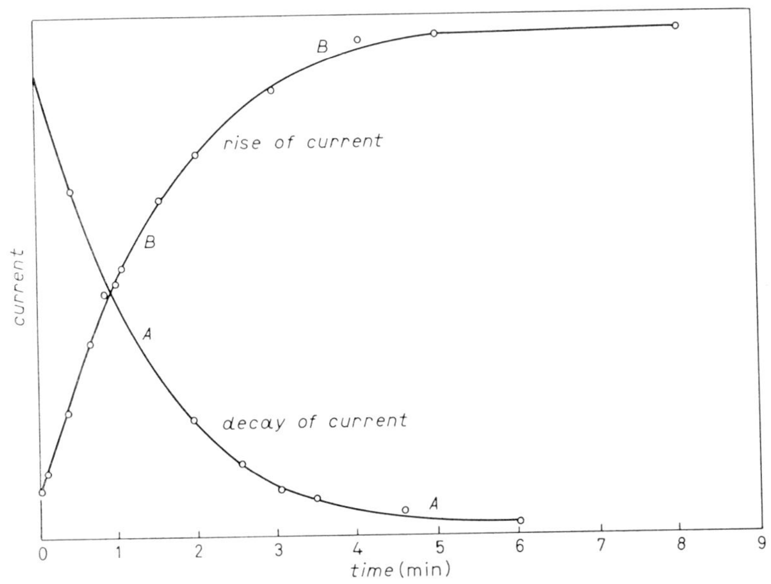

Fig. 2. – Experimental results obtained by RUTHERFORD with the experimental set-up shown in fig. 1.

viously produced was pumped away. For discussing these experimental results RUTHERFORD writes the differential equation

$$\frac{dn(t)}{dt} = q - \lambda n(t) \tag{2}$$

(q is the number of ions supplied per second by the emanation diffusing from thorium) and derives the solutions corresponding to the two cases mentioned above:

$$\left|\begin{array}{lll} \text{for} \begin{cases} q = 0 \\ n(0) = n_0 \end{cases} & \text{(decay of current): } n(t) = n_0 \exp[-\lambda t], \\ \text{for} \begin{cases} q \neq 0 \\ n(0) = 0 \end{cases} & \text{(rise of current): } n(t) = \frac{q}{\lambda}(1 - \exp[-\lambda t]) . \end{array}\right. \tag{3}$$

Thus the decay law was already well established, but the ideas about the nature of the « emanation » were still confused. Two possible interpretations were considered by RUTHERFORD, *viz.*

« 1) That the emanation may be due to fine dust particles of the radioactive substance emitted by the thorium compounds.

2) That the emanation may be a vapor given off from thorium compounds ».

One of the main conclusions of the paper was that the interpretation 1) had to be excluded.

These results of Schmidt, the Curies and Rutherford were the first steps towards the establishment, that came gradually only from systematic experiments, of the general law that any amount of a chemically defined radioactive body, isolated from its source, decreases with the passing of time according to an exponential law.

The experiments described above were made only a few years after the discovery of the radioactivity of uranium made by Henri BECQUEREL [20] in February 1896. The discovery was accidental but made during a systematic investigation of the fluorescence of uranium salts inspired by the discovery made by W. C. RÖNTGEN, in 1895, of X-rays and their remarkable properties.

Becquerel discovery was followed by the discovery of the radioactivity of thorium, made independently in Vienna by SCHMIDT [21] and in Paris by Marie CURIE [22], of polonium [23] and radium [24], both separated chemically from uranium minerals in 1898 by Marie and Pierre CURIE, with the collaboration of G. BÈMONT in the last piece of work.

In none of these cases was observed a time dependence of the activity. Only in the case of polonium such a dependence could have been observed, and it actually was, but only in 1906 [25].

Not only it took a few years to establish the correct interpretation of the active deposit; even the nature of radioactivity remained for years an unsolved puzzle. I may recall, as an example, an interpretation suggested (rather softly) by the CURIES in 1900 [26].

The radiation emitted from radioactive bodies could be a secondary radiation emitted by heavy elements under the action of a primary radiation (similar to X-rays) pervading the whole space and all bodies.

A clear understanding of the nature of radioactivity came by steps. In 1902 the CURIES [27] as well as RUTHERFORD and SODDY [28] express clearly the idea that radioactivity is an atomic phenomenon, more precisely that « radioactivity is a manifestation of sub-atomic chemical change » [28, 29]. RUTHERFORD and SODDY go on and a few lines below say: « The idea of the chemical atom in certain cases spontaneously breaking up with evolution of energy is not of itself contrary to anything that is known of the properties of atoms ... ».

The idea of spontaneous breaking of a chemical atom which thus is transformed into a chemical atom of a different type is further elaborated in various successive papers by RUTHERFORD and SODDY, who start to consider already in 1903 [30] families of radioactive bodies. In this same paper, the authors notice that the exponential law (1) should hold not only for the number of ions produced, but also for the rays emitted by radioactive bodies [31]. Since « the rays emitted must be an accompaniment of the change of the radiating system into the next one produced », the same law should hold for the number $N(t)$ of radioactive atoms of a given type, present at the time t. By differentiating the empirical exponential law, they deduce

$$\frac{\mathrm{d}N}{\mathrm{d}t} = -\lambda N(t) \tag{4}$$

as the general law regulating the behaviour of any radioactive body: « The change of the system at any time is always proportional to the amount remaining unchanged », and « the proportion of the whole which changes in unit time is represented by the constant λ, which possesses for each type of active matter a fixed and characteristic value. λ may therefore be suitably called the *radioactive constant* ».

At the beginning the « activity » measured by the discharge of an electroscope was the quantity decaying exponentially; later it was the current, or the number of ions, still later the number of observed emitted particles and finally, in 1903, the number of decaying radioactive atoms.

In the disintegration theory advanced by RUTHERFORD and SODDY, it is supposed that a definite small portion of the radium atoms (about 1 in every 10^{10} will suffice) breaks up per second. The disintegration of each atom is accompanied by the expulsion of an α-ray or β-particle with great velocity [32]. In the meantime it was shown by RUTHERFORD that the α-rays of radium consist of positively charged bodies of mass about twice that of the hydrogen atom [33].

In the Bakerian lecture, delivered in front of the Royal Society on May 19, 1904, RUTHERFORD reviewed once more the whole subject [34]. The theory of the transformation of a whole succession of radioactive bodies is discussed in general, and is applied to a number of particular cases. In the discussion it is clearly stated that the constant λ « is the fraction of atoms disintegrating per second, and that its inverse is equal to the *average life* of the corresponding radioactive body ».

Furthermore, the value of λ for any substance is independent of all physical and chemical conditions: many authors, already from early times, had attempted, without success, to observe some change of λ as consequence of changes of temperature, concentration, chemical binding, age of the atoms, or due to the presence of intensive electric and magnetic fields [35].

I have dwelt upon certain detail of interest from the point of view of the present discussion, while I have avoided to quote a number of other important results obtained during those years by the CURIES, RUTHERFORD and their collaborators as well as by many other scientists who gave substantial contributions to the study of radioactivity during the first 8 years after Becquerel discovery. A much more complete picture can be found in Rutherford book, published in 1904 [36].

3. – Statistical fluctuations.

The next important step was made in Vienna in 1905 by Egon VON SCHWEIDLER [37], who noticed that, by treating the constant λ multiplied by dt as the probability for one atom to decay during the time interval dt, the exponential decay law can be theoretically deduced without any special assumption about the dynamic processes taking place within the atom, but simply applying the elementary rules of probability calculus. In this paper, presented to the First International Conference for the Study of Radiology and Ionization held in Liège in September 1905, SCHWEIDLER begins by summarizing the situation with the following sentence (fig. 3): « Nach der Vorstellungen, zu denen die Zerfallstheorie der radioaktiven Erscheinungen führt, sind die Atome einer aktiven Substanz instabile Gebilde, denen eine von ihrer Struktur abhangige 'mittlere Lebensdauer' zukommt. Ist $\lambda\, dt$ die Wahrscheinlichkeit, dass ein Atom innerhalb der Zeit dt eine Umwandlung erfährt, so ist die Warscheinlichkeit, dass es eine Zeit t überdauere, gleich $\exp[-\lambda t]$ und $\tau = 1/\lambda$ seine mittlere Lebensdauer ». (According to the picture provided by the disintegration theory of the radioactive phenomena, the atoms of an active substance are unstable systems with a mean life determined by their structure. If $\lambda\, dt$ is the probability that one atom undergoes a transformation in the time interval dt, then the probability that it survives the time interval t is equal to $\exp[-\lambda t]$ and $\tau = 1/\lambda$ is its mean life.)

Uber Schwankungen der radioaktiven Umwandlung

von E. v. SCHWEIDLER

—

Nach den Vorstellungen, zu denen die Zerfallstheorie der radioaktiven Erscheinungen führt, sind die Atome einer aktiven Substanz instabile Gebilde, denen eine von ihrer Struktur abhängige « mittlere Lebensdauer » zukommt. Ist λdt die Wahrscheinlichkeit, dass ein Atom innerhalb der Zeit dt eine Umwandlung erfährt, so ist die Wahrscheinlichkeit, dass es eine Zeit t überdauere, gleich $e^{-\lambda t}$ und $\tau = \frac{1}{\lambda}$ seine mittlere Lebensdauer. Bei einer sehr grossen Anzahl N gleichartiger solcher Atome wird daher, entsprechend dem Gesetz der grossen Zahlen, die Anzahl der nach der Zeit t noch vorhandenen Atome gegeben sein durch $n = Ne^{-\lambda t}$. Es ist selbstverständlich, dass bei einer geringen Anzahl von Atomen der tatsächliche Verlauf ihrer Verminderung von diesem idealen Gesetze abweichen wird, und es soll im Folgenden untersucht werden, ob die durch die Wahrscheinlichkeitsrechnung zu ermittelnde « Streuung » die Grenzen empirischer Nachweisbarkeit erreichen kann.

Es seien N Atome einer Substanz mit der Abklingungskonstante λ gegeben; nach einer gewissen Zeit δ ist für ein bestimmtes einzelnes Atom die Wahrscheinlichkeit noch zu existieren gleich $e^{-\lambda\delta}$, die, inzwischen eine Umwandlung erfahren zu haben, gleich $1 - e^{-\lambda\delta} = \alpha$. Die Wahrscheinlichkeit, dass von den N Atomen die Anzahl x eine Umwandlung erfahren habe, die Anzahl $N-x$ unverwandelt erhalten geblieben sei, ist dann

$$W_x = \alpha^x (1-\alpha)^{N-x} \binom{N}{x}$$

Wie eine einfache Differentiation ergibt, ist

$$W_x = \text{Maximum für } x = \alpha N,$$

also der dem Abklingungsgesetz $n = Ne^{-\lambda t}$ entsprechende Wert ist der wahrscheinlichste. Es lässt sich aber auch die Wahrscheinlichkeit bestimmen, dass x von dem wahrscheinlichsten Werte αN um eine vorgegebene Grösse abweiche.

Fig. 3. – First page of E. von Schweidler theoretical paper on fluctuations of radioactive decay.

Without any knowledge of the causes which determine in single cases the disintegration of a specific atom, we can understand this process as a purely accidental event in the sense of the probability calculus. SCHWEIDLER notices that the exponential law can only be an approximation, because the function $N(t)$ can only take integer values. From the interpretation of this law as a statistical law, it follows that the observable values should fluctuate around the mean value which is provided by the exponential law. SCHWEIDLER in his paper derives the expression of the fluctuations and shows that the variance of the distribution of the actual number of disintegrations approaches $\sqrt{N(t)}$ in the limit of N very large.

Different types of observations can be made on the disintegration process, each corresponding to different treatments of the statistical problem. These, however, are all based on the physical assumption that, for any radioactive atom of an assigned type, the probability that it decays in the time interval dt is given by $\lambda\, dt$, with λ independent of the age of the atom as well as of any external condition.

In the final part of his paper SCHWEIDLER shows that the fluctuations that he has calculated should be easily observable.

The experimental evidence that fluctuations of the order of those computed from the probability calculus do actually take place was provided by K. W. F. KOHLRAUSCH in 1906 [38] (fig. 4) balancing the ionization currents due to two sources of alpha-particles against each other and measuring the fluctuations from the balance by means of an electrometer. Similar observations were made, in 1908, by MEYER and REGENER [39] by balancing the saturation ionization current due to the source of alpha-particles against a Bronson resistance. Experiments were made independently by GEIGER [40] at the same time by balancing one source of particles against another.

I will not try to summarize here all the successive experiments, nor the theoretical considerations mainly of statistical nature that were published for more than twenty years. An excellent presentation can be found in Chapter II of the second edition of the classical book by St. MEYER and E. VON SCHWEIDLER appeared in 1927 [41]. The last of the 29 papers listed in their bibliography was published in 1924.

Some of these papers treat corrections originating from the inertia of the used instruments (electrometers), or from fluctuations of the ionization produced by a single alpha or beta ray, others refer to the fluctuations inherent to the observation of scintillation produced by alpha-particles, others to the fluctuations of the ionization produced by secondary electrons of gamma-rays.

Among the many pubblications I may recall a book appeared in 1913 by L. VON BORTKIEWICH [42] (a.o. Professor an der Universität, Berlin; fig. 5), who treats in some 80 odd pages all possible aspects of the fluctuations and, in particular, *a*) the fluctuations of the number of scintillations observed in preassigned time intervals and *b*) the statistical distribution of the length of time intervals, between two successive decays. Experiments of this second type had been already made by MARSDEN and BARRAT in 1911 [43], and by Mme CURIE, the same year, and in an improved version in 1920 [44].

A different problem was studied experimentally in 1908 and 1910 by RUTHERFORD and GEIGER [45], who measured, by the scintillation method,

Über Schwankungen der radioaktiven Umwandlung

von

K. W. Fritz Kohlrausch.

Aus dem II. physikalischen Institut der Universität in Wien.

(Mit 3 Textfiguren.)

(Vorgelegt in der Sitzung am 21. Juni 1906.)

Die von Rutherford und Soddy begründete Zerfallstheorie geht von der Annahme aus, daß die Atome radioaktiver Substanzen einer Umwandlung unterworfen sind. Der Zerfall erfolgt nach dem Exponentialgesetz $n = Ne^{-\lambda t}$, wo N die Anzahl der Atome zur Zeit $t_0 = 0$, λ eine Konstante und n die Zahl der nach der Zeit t noch vorhandenen Atome bedeutet. Durch Differentiation erhält man $\frac{dn}{dt} = -\lambda n$, das heißt, der Teil der vorhandenen Atome, der in der Zeiteinheit einer Veränderung unterliegt, ist gegeben durch λ; $\frac{1}{\lambda}$ wird als »mittlere Lebensdauer« der betreffenden Atome bezeichnet.

Man gelangt zu diesem Gesetz auch auf anderem Wege. Die Wahrscheinlichkeit, daß ein Atom innerhalb einer gegebenen Zeit eine Umwandlung erfährt, wird desto größer, je größer die Zeit ist. Bezeichnet man die Wahrscheinlichkeit, daß ein Atom in der Zeit Δt, wobei Δt sehr klein sein möge, eine Umwandlung erfährt, mit $\lambda \Delta t$, so wird die Wahrscheinlichkeit w_1, daß dieses Atom in der gleichen Zeit nicht verändert wird: $w_1 = 1 - \lambda \Delta t$; für die Zeiten $2\Delta t$, $3\Delta t, \ldots k\Delta t$ erhält man:

Fig. 4. – First page of K. W. F. Kohlrausch experimental paper on fluctuations of the ionization produced by alpha-particles.

the fluctuations of the number of alpha-particles emitted in a given solid angle. The theory for this case had been developed by Bateman (1910-1911) [46].

Other theoretical papers were published by Tatiana Ehrenfest [47] in 1913 and Schrödinger in 1918-1919 [48].

What I said, although incomplete, is sufficient to show that the problem of the statistical nature of the decay process of radioactive atoms was one of

Die radioaktive Strahlung als Gegenstand wahrscheinlichkeits-theoretischer Untersuchungen

Von

L. v. Bortkiewicz

a. o. Professor an der Universität Berlin

Mit 5 Textfiguren

Berlin

Verlag von Julius Springer

1913

Fig. 5. – Frontispiece of L. von Bortkiewicz book.

those at the centre of the attention of many experimental and theoretical physicists, starting from 1905 until 1928 when GAMOW [49] and CONDON and GURNEY [50] applied quantum mechanics to the decay process of nuclei and derived the connection between the mean life and the energy of the emitted alpha-particles which had been established experimentally by GEIGER and NUTTAL [51] seventeen years before.

This whole problem is not mentioned, of course, in the first book by RUTHERFORD appeared in 1904 [36], but is treated in detail in section 75 (pages 186-191) of his 1913 book [52], where he underlines: « It is important to settle whether the emission of alpha-particles follows a simple probability law, *i.e.* whether the alpha-particles are emitted at random in time and space ».

4. – Early models of the nucleus.

The law of radioactive decay and the fluctuations of the number of disintegrations observed in a preassigned interval of time were matters of serious concern for many physicists of the first quarter of our century. As an example of the prevailing attitude I will quote the beginning of an important paper by SODDY [53] appeared in 1909:

« The cause of atomic disintegration remains unknown. It is difficult to construct any model of the disintegrating mechanism, chiefly on account of certain features in connection with the process. In particular may be mentioned the fact that the period of average life of the atoms disintegrating is the same whether newly formed atoms or those which have already survived many times the average period are considered. What may be termed the inevitableness of the process, and its entire independence of all known conditions, suggest that the cause of disintegration is apart from the atom. It is difficult to believe that the cause is resident in space external to the atom. It seems more probable that it exists within the atom and at the same time is uninfluenced by it. The question about to be discussed is whether necessarily only one mode of instability can exist within the atom at the same time ».

The paper goes on treating and solving the theoretical problem of the decay when two or more alternative disintegration processes are allowed.

Most of the physicists working in the field accepted the well-established fact that the clicks of a counter or the little lights observed in the dark on a fluorescent screen are distributed at random and the majority of them did not dare to propose models.

One of the very few exceptions was André DEBIERNE, who had published in part alone in part in collaboration with Mme CURIE, a number of important papers on radioactivity. On various occasions, in particular at the end of a lecture he gave in Paris the 26 January 1912 in front of the Societè Française de Physique [54], DEBIERNE discussed various possible mechanisms and arrived at the following conclusion:

« On est ainsi conduit a faire l'hypothèse qu'il existe réellement un élément de desordre faisant passer les atomes par un grand numbre d'états differents dans un temp très court, mai que cet élément du desordre est distincte de l'agitation thermique ». (Thus one is brought to the assumption that really there is an element of disorder which causes the atoms to pass through a great number of different states, in a very short time interval, but that such an element of disorder is different from thermal agitation.)

The simplest hypothesis, goes on DEBIERNE, consists in assuming that this element of disorder is shut up in the atom. « For example, one could imagine that inside the atom there are infinitely small elements [today we would say « constituents », but why *infinitely* small?] endowed with disordered movements similar to those of the molecules of a gas inside a container. Such an agitation taking place inside the atom could determine, in certain cases, a state of instability followed by the explosion of the atom » [55].

Debierne lecture was one of a series of seminars [54] in the last of which Henri POINCARÉ [56] summarized the fundamental points mentioned by the various speakers in the presentation of their subjects and added a number of stimulating remarks. Some of these are expressed in the form of questions about the many problems still open in 1912, which sound particularly effective in Poincaré's elegant French.

The part of Poincaré lecture concerning Debierne's ideas begins at the end of p. 361 and finishes at about one third of p. 364. POINCARÉ starts by pointing out that Debierne's ideas are the most suitable « in order to measure the complexity of the atom ». He accepts the fact that the law of radioactive decay is a statistical law by saying: « on y reconnait la marque du hazard » (one there recognizes the stamp of chance). A few lines below he underlines that « the chance that presides the radioactive transformations is an internal chance », *i.e.* the atom of radioactive bodies is a world and a world submitted to chance, « mais qu'on y prenne garde, qui dit hazard, dit grand nombres; un monde formé de peu d'éléments obéira à des lois plus ou moins compliquées, mai qui ne seront pas des lois statistiques ...; puis qu'il y a une statistique et par conséquent une thermodynamique interne de l'atome, nous pouvons parler de la température interne de cet atome; eh bien! elle n'a aucune tendence à se mettre en équilibre avec la temperature extérieure, comme si l'atome était enfermé dans une enveloppe parfaitement adiathermique » (but one should be careful, who says chance says large numbers; a world consisting of a few elements will obey more or less complicated laws, which, however, will not be statistical laws ...; since there is a statistics and therefore a thermodynamics internal to the atom, we can talk of a temperature internal to this atom; eh, well! this temperature does not have any tendency to equilibrium with the external temperature as if the atom were shut in a perfectly insulating thermal container).

This is the first time that I found mentioned explicitly the internal temperature of an atom, an idea very interesting, indeed: about 25 years later it became an essential concept of many nuclear models. It was clearly implied by the whole reasoning of Debierne, although I did not find the words « internal temperature » in any of his papers published before 1915 [57].

The ideas of Debierne were reported by Mme CURIE at the second Solvay Conference held in Bruxelles on 27-31 October 1913 [58]. She insisted on the idea of a disorder inside the central part of the atom (the nucleus) where its constituents should move with very high velocity, judging from that of the emitted particles, and on the possibility even to define a kind of internal temperature much higher than the external temperature.

In the discussion that followed the report by Mme CURIE, NERNST, RUBENS, BRILLOUIN, WIEN and LINDEMANN commented in various forms on the possibility of succeeding perhaps in modifying the mean life of radioactive bodies by going to high temperature (perhaps inside the Sun).

RUTHERFORD and LANGEVIN touched the essence of Debierne model. RUTHERFORD said more or less that « the law of the radioactive substances appears to find a possible explanation only as a consequence of the fortuitous troubles taking place according to the laws of probability inside the nucleus. But in the present state of our knowledge it is not possible to formulate clear ideas neither on the constitution of the atomic nucleus nor on the causes of its distintegration ».

Langevin remark runs more or less as follows: « Debierne ideas require a complex structure for each single atom and imply the necessity of a great numbers of parameters for fixing the configuration of the atom. In the case of three bodies, however, the trajectories are already very complicated.

As it has been shown by POINCARÉ—goes on LANGEVIN—, apart from a few exceptional cases of zero probability, any trajectory turns indefinitely approaching asymptotic solutions and then going away from them so that they have the aspects of wires made into balls .... In the case of radioactive atoms one could ask which is the minimum number of degrees of freedom necessary for obtaining the law of probability within the precision of the experimental results. It would be sufficient, for example, to assume that the distribution of the initial phases of the different atoms is sensibly uniform on the surface of constant energy if all atoms have initially the same energy ».

A number of remarks could be made about Debierne ideas and Langevin remarks. These mix two possible types of assumptions. One introduces inside the nucleus a sufficiently large number of degrees of freedom, to produce a random effect. The other transfers the stochastic nature of the decay process to the initial conditions under which the nucleus is produced by its mother. In this second case some machinery should be introduced in the model for assuring the validity of such an assumption for all members of a radioactive family the mean lives of which are spread over an interval of values greater than 20 powers of 10.

A thorough and complete presentation of his ideas was finally given by DEBIERNE in a 22-page long paper published in 1915 [59], shortly after LINDEMANN [60, 61] had developed a quantitative model along similar lines of thought.

In this model the nucleus of a radioactive element contains particles in movement and becomes instable when N independent particles all pass through some unknown critical interval of positions within a short time τ. These particles are taken to be alpha-particles rotating or oscillating with the frequency ν. Their mean energy $E = H\nu$ is equal to the energy of the emitted alpha-particles. Each of them passes through the critical region $\nu = T^{-1}$ times per second so that the probability for one of them to be in the critical region during one period is τ/T. The probability for N of them to be simultaneously in τ in one period is $(\tau/T)^N$, which multiplied by νdt gives the probability that such an event takes place in the time interval dt. If $\mathcal{N}$ atoms are considered, the number of those which become unstable and explode in the time dt is

$$\mathrm{d}\mathcal{N} = -\mathcal{N}(\tau\nu)^N \nu\,\mathrm{d}t\,,$$

from which it follows that

$$\lambda = \nu(\tau\nu)^N\,. \tag{5}$$

Since the range-energy relation is $R = kE^{\frac{3}{2}}$, LINDEMANN obtains

$$\lambda = \frac{E}{H}\left(\frac{\tau E}{H}\right)^N = \frac{\tau^N}{(Hk^{\frac{2}{3}})^{N+1}} R^{\frac{2}{3}(N+1)}$$

and taking the logarithm

$$\log\lambda = \log\frac{\tau^N}{(Hk^{\frac{2}{3}})^{N+1}} + \frac{2}{3}(N+1)\log R\,,$$

similar to the Geiger-Nuttal relation [51]

$$\log\lambda = a + b\log R\,.$$

Since the experimental value of b is about 53, it follows on comparing the constants that N is about 80.

With the further assumption that the time τ is to be regarded as the time taken by an elastic wave to cross the nucleus, LINDEMANN computed the constant a in terms of known quantities and the unknown radius of the nucleus. By comparison with the experimental value of a he deduced a value of the radius of the radium nucleus of about $4\cdot10^{-13}$ cm, which is too small by about a factor 2.

The theory provided an explanation of the law of radioactive decay as well as of the Geiger-Nuttal law (fig. 6) for reasonable values of the parameters.

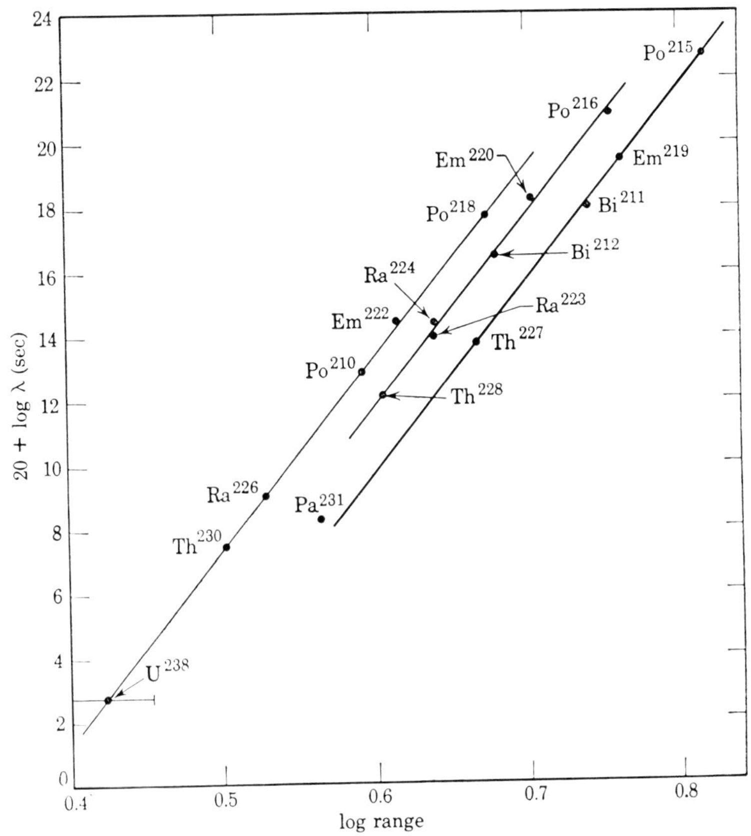

Fig. 6. – Comparison with Geiger-Nuttal rule of the experimental data available to GEIGER in 1921 [62]. This figure is taken from ref. [63].

A few papers were published in the following years [64] attempting to introduce the quantization of the movement of alpha-particles and electrons within the nucleus by applying the same rules used by BOHR for the atomic electrons.

More or less in the same period of time other authors tried to explain the decay law along the lines suggested long before by the CURIE, *i.e.* as due to the action of an external radiation [65]. The most important of these papers is that of Jean PERRIN who devoted about 10 pages of his 103-page paper, trying to extend to the nuclei of radioactive substances his considerations on the chemical processes.

Another type of model was proposed by RUTHERFORD [66] in 1927 for explaining some results obtained by him and CHADWICK [67] and by CHADWICK [61] in 1925, in the study of the elastic scattering at large angles ($\sim 155°$) of the alpha-particles of Ra C' (7.68 MeV) from the uranium nucleus

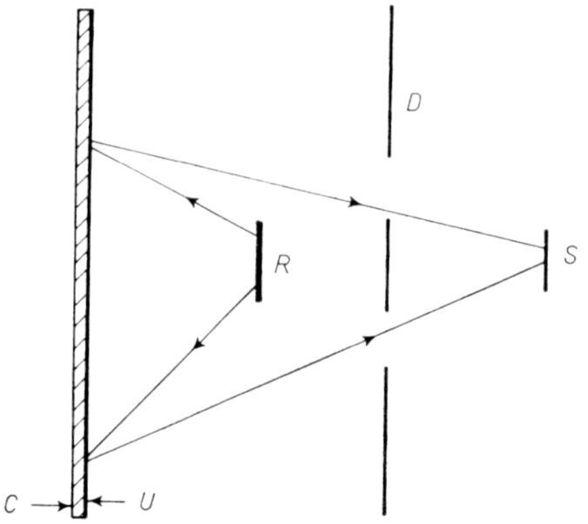

Fig. 7. – Schematic representation of the experimental set-up used in 1925 by CHADWICK to observe the scattering at $\sim 155°$ of the alpha-particles of Ra C' from uranium atom: $R \equiv$ Ra $(B + C)$; $U \equiv$ uranium oxide deposited on $C \equiv$ graphite support; $D \equiv$ diaphragm; $S \equiv$ fluorescent screen used for observing the scintillation by means of a microscope.

(fig. 7). They had found that the observed angular distribution followed very closely that compluted from pure Coulomb potential. The alpha-particles emitted by the decay of the same nucleus have only about one half this energy (4.049 MeV). Thus, it appeared that during the emission process the alpha-particles had to pass through a region of negative kinetic energy (fig. 8). Such a difficulty was avoided in the following model.

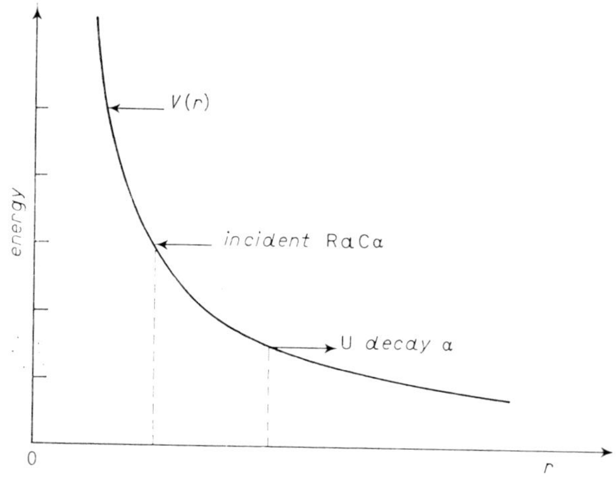

Fig. 8. – The uranium decay alpha-particles are emitted with an energy equal to about one half the energy of the Ra C' alpha-particles used in the scattering experiment of fig. 7. This had shown that the potential $V(r)$ followed the Coulomb law down to distances equal to about one half the distance from which the decay alpha-particles are emitted.

The nucleus of a radioactive atom consists of a very small core of radius less than $1 \cdot 10^{-12}$ cm, surrounded at a distance by a number of neutral satellites describing quantum orbits in the field of the central core. At least some of these satellites are neutral alphas in which the two neutralizing electrons are more closely bound than in the free helium atom but not so closely as in the alpha-particle. A neutral alpha is regarded as stable only in very strong electric fields and is held on its orbit by attractive forces due to its polarization in the field of the central core. The quantized orbit of radius between $1.5 \cdot 10^{-12}$ and $6 \cdot 10^{-12}$ cm is essential to provide the well-defined energy of the emitted alpha.

The disintegration process consists in the *acausal escape* of a neutral alpha satellite from its orbit. When the satellite reaches a distance from the centre of the core where the electric field has a certain critical value, it breaks into two electrons and an alpha-particle: the two electrons are sucked into the core, while the alpha-particle runs away with the observed energy [68].

The model is inconsistent, as it was immediately shown [69], but here it serves to show that the idea of explaining the radioactive decay in term of a spontaneous (or better an acausal) process still in 1927 was alive at the Cavendish Laboratory, in Cambridge, that had been, and still was, the major centre for the study of radioactivity.

Apart from that, the scattering experiments of Rutherford and Chadwick [67] and those (unpublished) by Chadwick alone [61] were made too late for having any influence on the development of quantum mechanics.

They provide, however, one of the most striking examples of a direct observation which imposes quantum mechanics. Apparently they have been forgotten, since they are never quoted in books devoted to the introduction of this fundamental subject.

5. – Final remarks.

A few remarks appear in order at the end of this story, which may appear too long but in reality already constitutes a very shortened version of what happened in the field of radioactivity before the advent of quantum mechanics.

First of all one should keep in mind that the early development of radioactivity took place at the same time as the first steps in the theory of the atom. The existence of the electron was established almost at the same time as the discovery of radioactivity; the quantum of action was introduced at about the same time that the exponential decay law was established, the theoretical prediction and the experimental observation of the fluctuations of the radioactive decay started to be made at the same time of Einstein's theory of photons and of Smolukowski and Einstein theory of Brownian motion.

The people involved in these two developments were always very close to each other, in many cases they worked in the same laboratories. All this went on for more than 15 years before the experiments of Geiger and Marsden established the validity of the Rutherford model, which allowed a separation between the physics of the atom and the physics of the nucleus, and opened the door to the theoretical work of Niels Bohr.

The young BOHR went to Cambridge at the end of September 1911; there he had contacts with LARMOR and J. J. THOMSON and there he met for the first time RUTHERFORD [70]. In the middle of March 1912 BOHR went to Manchester to work in the laboratory of Rutherford, who had been given a professorship there in 1907. RUTHERFORD had collected together in his laboratory a number of the best physicists of those days. Among them H. GEIGER, W. MAKOVER, E. MARSDEN, E. J. EVANS, A. S. RUSSEL, K. FAJANS, H. G. J. MOSELEY, G. HEVESY, J. CHADWICK and C. G. DARWIN.

Before Summer 1912 BOHR started to work on the model of the atom that had been proposed by RUTHERFORD not long before [71], while the second theoretical paper by BATEMAN [46] on the fluctuations and the second paper by GEIGER [72] on the scattering of alpha-particles by nuclei had been received by the *Philosophical Magazine* in January of the same year, and while MARSDEN and BARRAT [43] were just making their experiment on the fluctuations observed in the radioactive decay, as a development of those made by GEIGER in 1908 and by RUTHERFORD and GEIGER in 1910. (Incidentally I may recall that the famous final paper of Geiger and Marsden on the scattering of alpha-particles appeared only in 1913 [73].)

Although the attention of Bohr was mainly concentrated on the model of the atom, it is impossible that the problem of the fluctuations escaped his attention and that the pragmatic attitude of Rutherford and his school did not have a strong influence on him, also on this problem.

The existence of a nucleus at the centre of the atom was just becoming an accepted notion, but its properties, in particular its decay in the case of radioactive elements, were still an « enigma », as Mme CURIE said on many occasions.

A few models for explaining such a behaviour were, of course, proposed. Some of them based on the existence of a hypothetic radiation pervading all space and bodies and provoking the decay of heavy elements, some of a statistical nature such as that proposed by LINDEMANN.

These were perhaps a bit more popular, but not universally accepted. All of them, irrespective of being formally developed or only quantitatively conceived, contained some element of indeterminacy. In Lindemann model this is introduced with the small but nonzero interval of time τ, during which the system becomes unstable. In the case considered by LANGEVIN this element is provided by the initial conditions of the internal motion of the constituents of the nucleus, which are supposed to be distributed at random over a surface of constant energy in the space of phases.

My claim of a historical influence of the study of radioactivity on the general acceptance of probabilistic conceptions in physics clearly does not regard the distinction between statistical and probabilistic points of view, which only years later became a focus of physicists' attention.

It does not concern also the most striking feature of quantum mechanics, *i.e.* the concept of « probability amplitude » and its systematic use as an essential intermediate step for deriving the values of measurable quantities; this concept did not get any inspiration from the investigation of radioactivity. Only the results obtained in 1925 by RUTHERFORD and CHADWICK on the scattering of alphas from uranium were a message of new phenomena requiring this aspect of the theory. The message, however, was not yet construable and in any case it arrived too late for having an influence on the development of quantum mechanics. The only clear experimental indication in favour of the use of « probability amplitude » came from experiments on diffraction of electrons by crystals or molecules.

Max BORN says to have found inspiration from Einstein's theory of photons, but EINSTEIN, in 1916, adopted for the Ausstrahlung the law of radioactive decay.

The value of the development of radioactivity, from the point of view of the acceptance of probabilistic conceptions, lies essentially in the observed phenomena and in the pragmatic attitude that most of the people working in the field had in front of them.

Many other phenomena could have, in principle, played the same role, but radioactivity had two fundamental points of advantage over any other chapter of physics. The mean lives of many radioactive substances were in a range of values very convenient for their observation. Furthermore, the existence of experimental techniques, such as the scintillations, discovered in 1903 by CROOKS and ELSTER and GEITEL [74], and of various types of counters, developed starting from 1908 on by GEIGER and RUTHERFORD and others [75], allowed the observation of the decay of a single nucleus.

I insist on these observational aspects because they provide the basis for the pragmatic attitude of the people working in this field in those days. The situation is not exceptional, as can be seen from many examples, two of which may be mentioned here. The first example is the problem of the value of the velocity of light as it appeared toward the end of the last century. All experiments made during more than half a century [76] were pointing towards the recognition of the invariance of this quantity. Various models were proposed, but were unsatisfactory, so that the invariance of the velocity of light with respect to the reference frame was accepted as an experimental result, still under critical scrutiny. It remained an enigma, repeatedly tested, until EINSTEIN promoted it to the rank of fundamental law.

In more recent years, the θ-τ decay puzzle was based on the equality (within small experimental errors) of the masses and the mean lives of two types of particles: one particle (the θ), decaying into two pions, the other particle (the τ), decaying into three pions, *i.e.* into two final states of opposite parity. Many physicists (especially among the experimentalists working in the field) were ready to accept that the θ and the τ were simply two different decay modes of the same particle, if they had *not* firmly believed in the general validity of the conservation of parity. The puzzle was solved by T. D. LEE and C. N. YANG who proposed a new fundamental law: « parity is violated only by weak interactions », and they indicated the experiments for testing their assumption.

What appears to be the most interesting aspect of the whole story is the convergence of so many lines of thought toward a probabilistic, or, at least, a statistical interpretation as that involved by quantum mechanics.

The line originating from thermodinamics and statistical mechanics, characterized by a sophisticated analysis of the properties of matter and radiation, the line through radioactivity, characterized by an incredible succession of unexpected experimental discoveries, as well as certain lines of philosophical thinking, all appear to converge toward the same target.

These lines, of course, were not independent. Clear reciprocal influences existed between them in all possible combinations. In particular, there were influences of the philosophical thinking on the attitude of the scientists, but there were also strong influences in the opposite direction.

Newton mechanics at the end of 1600, electricity and magnetism in the first part of 1700, thermodynamics and statistical mechanics during the second half of 1800, radioactivity and relativity at the beginning of our century, later quantum mechanics, all became subjects of discussion in much wider circles than those of the people directly involved. They became fashionable matters of conversation, in king's court as well in layman circles.

It is difficult, perhaps impossible, to establish which of the three lines of thought sketched above had more importance for the advent of quantum mechanics. The best is probably to keep in mind all of them.

REFERENCES

[1] M. JAMMER: *The Conceptual Development of Quantum Mechanics* (New York, N. Y., 1966).
[2] M. JAMMER: *The Conceptual Development of Quantum Mechanics* (New York, N. Y., 1966), p. 166-167.
[3] M. JAMMER: *The Conceptual Development of Quantum Mechanics* (New York, N. Y., 1966), p. 170.
[4] M. JAMMER: *The Philosophy of Quantum Mechanics* (New York, N. Y., 1974).
[5] *Historical Studies in the Physical Sciences*, edited by R. MC CORMACH (Philadelphia, Penn., 1971).
[6] S. G. BRUSH: *Journ. Hist. Ideas*, **37**, 603, No. 4 (October-December 1976).
[7] See, for example, E. N. HIEBERT: *The use and abuse of thermodinamics in religion*, *Daedalus*, p. 1046 (Fall, 1966).

[8] A. EINSTEIN: *Über einen die Erzeugung und Verwandlung des Lichtes betreffenden heuristischen Gesichtspunct der Physik, Ann. der Phys.*, **17**, 132 (1905).
[9] M. BORN: *Quantenmechanik der Stossvorgänge, Zeit. für Phys.*, **38**, 803 (1926). An English translation of this article is given at p. 207 of G. LUDWIG: *Wave Mechanics* (London, 1968).
[10] M. BORN: *Die statistische Deutung der Quantenmechanik, Nobelvortrag 1954, Nobel Lectures, Physics, 1942-1962* (Amsterdam and New York, N.Y., 1964), p. 262.
[11] A. EINSTEIN: *Zur Quantentheorie der Strahlung, Mitt. Phys. Ges. Zürich*, **18**, 47 (1916); *Strahlung Emission und Absorption nach der Quantentheorie, Verh. Dtsch. Phys. Ges.* **18**, 318 (1916).
[12] A. EINSTEIN: *Zur Quantentheorie der Strahlung, Phys. Zeits.*, **18**, 121 (1918).
[13] A thorough discussion of this paper and more in general of *Einstein Statistical Theories* has been made by M. BORN at p. 163 of *Albert Einstein: Philosopher-Scientist*, edited by P. A. SCHILPP (New York, N. Y., 1951).
[14] G. C. SCHMIDT: *Wied. Annalen, May 1898*, according to Rutherford quotation (ref. [17]). Most probably it should be identified with ref. [21].
[15] M. CURIE and P. CURIE: *Compt. Rend.*, **129**, 714 (1899), and p. 77, ref. [16]. The effect was observed with radium and polonium. In this second case it was clearly due to incomplete separation from other elements.
[16] *Ouvres de Marie Sklodowska Curie*, recuillies par I. JOLIOT CURIE (Varsovie, 1954).
[17] E. RUTHERFORD: *Phil. Mag.*, **49**, 1 (1900), and p. 220 of ref. [18*a*].
[18] *The Collected Papers of Lord Rutherford of Nelson*, published under the scientific direction of Sir JAMES CHADWICK, F.R.S., Vol. **1** (London, 1962) (*a*); Vol. **2** (London, 1963) (*b*); Vol. **3** (London, 1965) (*c*).
[19] I do not discuss the similar case of actinon (An), the noble gas of the actinium family because discovered later.
[20] H. BECQUEREL: *Compt. Rend.*, **122**, 420, 501, 559, 689, 762, 1086 (1896); **123**, 855 (1896); **124**, 438, 800, 984 (1897).
[21] G. C. SCHMIDT: *Wied. Annalen*, **65**, 141 (1898), paper received on March 24.
[22] M. CURIE: *Compt. Rend.*, **126**, 1101 (1898), paper received on April 12. See also p. 43 of ref. [16].
[23] M. CURIE and P. CURIE: *Compt. Rend.*, **127**, 175 (1898). See also p. 57 of ref. [16].
[24] M. CURIE, P. CURIE and G. BÈMONT: *Compt. Rend.*, **127**, 1215 (1898). See also p. 57 of ref. [16].
[25] M. CURIE: *Compt. Rend.*, **142**, 273 (1906), and p. 314 of ref. [16].
[26] M. CURIE and P. CURIE: *Rapport présentés au Congrès International de Physique*, 1900, Vol. **3**, p. 79; see also p. 106 of ref. [16] (last page).
[27] M. CURIE and P. CURIE: *Compt. Rend.*, **134**, 85 (1902). See also p. 134 of ref. [16].
[28] R. RUTHERFORD and F. SODDY: *Trans. Chem. Soc.*, **81**, 837 (1902), and p. 435 of ref. [18*a*].
[29] This sentence is in Section XII (*General Theoretical Considerations*) of ref. [18*a*], towards the end of p. 455.
[30] E. RUTHERFORD and F. SODDY: *Phil. Mag.*, **5**, 576 (1903), and p. 596 of ref. [18*a*].
[31] These are indicated with a special name (*metabolon*) in order to distinguish them from stable atoms constituting ordinary matter: p. 605 of ref. [18*a*].
[32] E. RUTHERFORD: *Transactions of the Australasian Association for the Advancement of Science* (Dudenin, January 1904), p. 87, and p. 620 of ref. [18*a*].
[33] Only some time later the α-particle was recognized to have a double charge and a mass equal to about twice that of the hydrogen molecule.

[34] E. RUTHERFORD: *Phil. Trans. Roy. Soc.*, **204**, 169 (1904): Bakerian Lecture delivered May 19, 1904; p. 621 of ref. [18*a*].
[35] For an exstensive bibliography of old papers on this problem see at the end of section II.3 (p. 41) of ref. [41].
[36] E. RUTHERFORD: *Radioactivity* (Cambridge, 1904).
[37] E. VON SCHWEIDLER: *Comptes Rendus du Premier Congrès International pour l'Etude de la Radiologie et de l'Jonisation, tenue à Liège du 12 au 14 Septembre 1905, Bruxelles.*
[38] K. W. F. KOHLRAUSCH: *Wien. Ber.*, **115**, 673 (1906).
[39] E. MEYER and F. REGENER: *Verh. Dtsch. Phys. Ges.*, **10**, 1 (1908); *Ann. der Phys.*, **15**, 757 (1908).
[40] H. GEIGER: *Phil. Mag.*, **15**, 539 (1908).
[41] ST. MEYER and E. VON SCHWEIDLER: *Radioactivität* (Leipzig, 1927).
[42] L. VON BORTKIEWICZ: *Die Radioactive Strahlung als Gegenstand Warscheilichkeits-theoretischer Untersuchungen* (Berlin, 1913). Through the kind interest of Prof. W. PAUL of the University of Bonn I had at my disposal, for a few days, a copy of this interesting, and today almost rare, book.
[43] E. MARSDEN and T. BARRATT: *Proc. Phys. Soc.*, **23**, 367 (1911); **24**, 50 (1911).
[44] M. CURIE: *Journ. Phys. Radium*, **8**, 354 (1911), and p. 395 of ref. [16]; **1**, 12 (1920). See also p. 494 of ref. [16].
[45] E. RUTHERFORD and H. GEIGER: *Proc. Roy. Soc.*, **81**, 141 (1908), and p. 109 of ref. [18*b*]; *Phil. Mag.*, **20**, 691 (1910), and p. 196 of ref. [18*b*].
[46] H. BATEMAN: *Phil. Mag.*, **20**, 704 (1910); **21**, 745 (1911). The second paper was made during the two years spent at Bryn Mawr College by Harry BATEMAN (b. Manchester 1882, d. Pasadena 1946), whose contributions to mathematics and theoretical physics refer to solution of potential and wave equations, partial differential equations and integral equations in general, electrodynamics, seismology, relativity, radioactivity, etc.
[47] T. EHRENFEST: *Phys. Zeits.*, **14**, 675 (1913).
[48] E. SCHRÖDINGER: *Wien. Ber.*, **127**, 237 (1918); **128**, 177 (1919).
[49] G. GAMOW: *Zeits. Phys.*, **51**, 204 (1928).
[50] E. U. CONDON and R. W. GURNEY: *Nature*, **122**, 439 (1928).
[51] H. GEIGER and J. M. NUTTAL: *Phil. Mag.*, **22**, 613 (1911); **23**, 439 (1912); **24**, 647 (1912).
[52] E. RUTHERFORD: *Radioactive Substances and Their Radiations* (Cambridge, 1913).
[53] F. SODDY: *Phil. Mag.*, **18**, 739 (1909).
[54] A. DEBIERNE: *Sur les transformations radioactives*, at p. 304 of the volume: *Les idèes modernes sur la constitution de la matière* (Paris, 1913).
[55] Through the kind interest of Prof. J. PRENTKI of CERN and Mme Geneviéve DELVOYE de l'Institut National de Physique Nucléaire et de Physique des Particules, I received the *Notice sur le travaux de M. André Debierne*, prepared by DEBIERNE himself, from which I learned that he already had discussed this same problem in two previous lectures, one in front of La Société Chimique, in 1907, the other at Clark University (USA) in 1909.
[56] H. POINCARÉ: *De la matière et de l'éther*, at p. 357 of the volume quoted in ref. [54].
[57] Perhaps it was mentioned in one of the two lectures of 1907 and 1909 quoted in ref. [55].
[58] M. CURIE: *Sur la loi fondamental des transformations radioactives*, at p. 66 of: *La structure de la matière, Rapports et Discussions du Conseil de Physique tenu a Bruxelles du 27 au 31 Octobre 1913*, Institut International de Physique Solvay (Paris, 1921). See also at p. 507 of ref. [16].

[59] A. DEBIERNE: *Considerations sur le mécanisme des transformations radioactives et la constitution des atomes*, *Ann. de Phys.*, (9) **4**, 323 (1915).
[60] F. A. LINDEMANN: *Verh. Dtsch. Phys. Ges.*, **16**, 281 (1914); *Phil. Mag.*, **30**, 560 (1915); see also p. 324 of ref. [61]. Because of a minor error my expression of λ differs from that of Lindemann by a factor ν.
[61] E. RUTHERFORD, J. CHADWICK and C. D. ELLIS: *Radiation from Radioactive Substances* (Cambridge, 1930).
[62] H. GEIGER: *Zeits. Phys.*, **8**, 45 (1921).
[63] G. C. HANNA: *Alpha radioactivity*, p. 55 of Vol. **3**, *Experimental Nuclear Physics*, edited by E. SEGRÈ (New York, N. Y., 1954). In the quantum-mechanical approach the expression of the decay constant λ can be split into three factors: *a*) the frequency at which the alpha-particle moving inside the nucleus strikes the side of the potential well, *b*) the reflection coefficient of the wall of the well due to the rapid increase of the potential and *c*) the penetrability of the potential barrier. The first term is an exponential and determines the main features of the phenomenon. The factor *a*) corresponds to the factor ν in eq. (5), while the two other factors represent single-particle effects which have nothing to do with the N-fold coincidence factor $(\tau\nu)^N$. Some of the features of the diagram of fig. 6 are not jet fully understood (see p. 98-105 of Hanna's article).
[64] H. TH. WOLFF: *Ann. der Phys.*, (4) **52**, 631 (1917); *Ann. der Phys.*, **60**, 685 (1919); *Phys. Zeits.*, **21**, 175 (1920); A. SMEKAL: *Naturwiss.*, **8**, 206 (1920); *Zeits. Phys.*, **10**, 275 (1922); S. ROSSELAND: *Zeits. Phys.*, **14**, 173 (1923); *Nature*, **11**, 357 (1923); G. KIRSCH: *Naturwiss.*, **8**, 207 (1920); *Phys. Zeits.*, **22**, 20 (1921); W. D. HARKINS: *Journ. Amer. Chem. Soc.*, **42**, 1956 (1920); K. FEHRLE: *Zeits. Phys.*, **16**, 397 (1923).
[65] J. PERRIN: *Ann. de Phys.*, (9) **11**, 5 (1919); E. BRINER: *Compt. Rend.*, **180**, 1586 (1925); A. W. MENZIES and C. A. SLOAT: *Science* (*N.S.*), **63**, 44 (1926).
[66] E. RUTHERFORD: *Phil. Mag.*, **4**, 580 (1927), and p. 181 of ref. [18*c*].
[67] E. RUTHERFORD and J. CHADWICK: *Phil. Mag.*, **50**, 885 (1925); see also p. 143 of ref. [18*c*] and p. 322 of ref. [57].
[68] A similar model was proposed later by D. ENSKOG: *Zeits. Phys.*, **45**, 852 (1927); **52**, 203 (1928).
[69] G. GENTILE jr.: *Rend. Acc. Lincei*, **7**, 346 (1928).
[70] L. ROSENFELD and E. RÜDINGER: *The decisive years: 1911-1918*, at p. 38 of *Niels Bohr*, edited by S. ROZENTAL (Amsterdam, 1967).
[71] E. RUTHERFORD: *Phil. Mag.*, **21**, 669 (1911), received in April 1911. See also at p. 238 of ref. [18*b*].
[72] H. GEIGER: *Proc. Roy. Soc. A*, **83**, 492 (1910); **86**, 235 (1912), received 25 January 1912.
[73] H. GEIGER and E. MARSDEN: *Phil. Mag.*, **25**, 604 (1913).

Note added in proofs. – N. FEATHER has called my attention to the paper by T. J. TRENN (*ISIS*, **65**, 74 (1974)) where the author shows convincingly that the paper quoted above sent by RUTHERFORD (from Manchester) and published in April 1913 is in fact less of a final version than that sent by GEIGER (from Berlin) to Vienna and published in December 1912 (*Wien. Ber.*, **121**, 2361 (1912)). GEIGER, apparently, took a first draft with him when he left Manchester in September 1912. This was amended in correspondence with RUTHERFORD in October but surprisingly the *Phil. Mag.* paper is essentially based on the first draft. I like also to mention that in the paper read by N. FEATHER at a session on *The work of Frederick Soddy* at the *XV International Congress of the History of Science, Edimburgh, August 1977*, the author uses essentially the same quotations from VON SCHWEIDLER (1905) and SODDY (1909) in a discussion of *Isotopes, isomers and* *the fundamental law of radioactive change* (*Notes and Records, Roy. Soc. London*, **32**, No. 2, 225 (March 1978)).

[74] W. CROOKES and also J. ELSTER and H. GEITEL discovered in 1903 the scintillation method but REGENER was the first in 1908 (*Verh. Dtsch. Phys. Ges.*, **19**, 78, 351 (1908) and p. 54 of ref. [61]) to devise methods of counting scintillations in order to determine the number of alpha-particles incident on the screen.

[75] A kind of wire counter was first developed by RUTHERFORD and GEIGER in 1908 (*Proc. Roy. Soc. A*, **81**, 141 (1908)). These were improved by the same authors and became proportional counters a few years later (*Phil. Mag.*, **24**, 618 (1912)). The point counter was developed by GEIGER the successive year (*Verh. Dtsch. Phys. Ges.*, **15**, 524 (1913); *Phys. Zeits.*, **14**, 1129 (1913)).

[76] See, for example, G. C. WICK: *Introduzione alla teoria della relatività, Lezioni di Fisica Teorica, A.A. 1944-1945* (Roma, 1945).

Properties of States in Quantum Logic.

E. G. BELTRAMETTI and G. CASSINELLI

Istituto di Scienze Fisiche dell'Università - Genova, Italia
Istituto Nazionale di Fisica Nucleare - Sezione di Genova, Italia

1. – Introduction.

The various approaches to the foundations of quantum mechanics have the common trend of dismissing from the set of primitive concepts those sophisticated structures, like the Hilbert-space structure, which are assumed in the usual formulation of quantum theory and find their empirical justification only indirectly from the success of the whole theory. Thus, those approaches manifest the aim of starting with a minimal set of primitive concepts, and of qualifying themselves as general approaches to the description of a physical (quantum) system. This attitude will, according to the title, be pursued also in this paper, but, before going on, it is worth remarking that the very notion of physical system still deserves some amount of idealization, a fact which puts limitations on the scopes of the approaches to the foundations of quantum mechanics hitherto worked out.

The concept of physical system corresponds to the separation between the observer and what he observes, a separation which has deep and controversial roots in the philosophy of science. By physical system it is meant a portion of the physical universe whose interaction with the rest of the universe, and with the observing devices in particular, causes no trouble about the identity of the separated portion. Though this notion has unambiguous borders in the cases which form the domain of elementary quantum mechanics, it becomes blurred in situations involving high-energy elementary particles, where, in particular, creation and destruction of particles is a dominant feature.

Also the concepts normally used for the description of physical systems, such as states of the system, observables on the system, and so on, deserve some amount of idealization. Take for instance the notion of state: it is demanded to take care of those attributes of the system which may be different in different situations and may change with time, whereas the permanent attributes are embodied in, and actually form the body of, the physical system itself; thus the information on an actual situation is split in a part pertaining to the system and a part pertaining to its state, a distinction corresponding to the essential and accidental qualities of scholastic philosophy. Besides the fact that the point where the definition of system ends and the definition of state begins is, to a good extent, a matter of taste, some idealization emerges when the notion of state is thought of in terms of the procedure to be used to prepare the system, for such a preparation procedure should fulfil the strong requirement of eliminating any memory of what happened before it was started on. The existence of preparation procedures satisfying this condition looks doubtful whenever the space-time confinement of the system is not ensured, as, for instance, in the domain of relativistic quantum field theory.

Though these remarks may rise the feeling that quantum mechanics will find, a day, place in some general scheme not adopting, from the very beginning, the notion of physical system, the abandonement of this notion looks today unpractical in the study of the foundations of quantum mechanics. The search of the more general schemes alluded to is still in its infancy even if interesting ideas have already been put forward: we refer, for instance, to Toller's recent works [1].

Though the suspect that conventional space-time structure might be untenable at the microscopic level, and though the suspect that the departure from classical to quantum theory might be related to a revised interpretation of space-time at the microscopic level, the various approaches to the foundations of quantum mechanics adopt, explicitly or, more often, implicitly, the conventional space-time structure as a background.

With the conservative trends entailed by the preceding remarks, the approaches to the foundations of quantum mechanics, and mainly those amenable to the area of quantum logic, adhere to the program of decomposing the general structure of elementary quantum mechanics into its basic sub-structures, of reducing them to what may be rested on as direct as possible empirical evidence, of controlling how single assumptions contribute to shape the theory.

As basic ingredients of elementary quantum mechanics we list the following: states, observables, propositions and operations. The first two terms are self-explanatory. The term « propositions » is derived from the logical set-up of the theory and is less universal: it is often replaced by « events », « tests », « questions », « yes-no experiments »; propositions can be viewed as particular observables, namely observables having only two outcomes, and it seems almost evident that the information on the physical system carried by a set of observables can also be carried by a suitable set of propositions. The term operations is, admittedly, somewhat vague and will be specified in the sequel; roughly, we can think of it as instructions to interact with the system and recorder information, or as procedures of measuring observables, or, even more specifically, as state transformations induced by measuring procedures.

In the usual formulation of quantum mechanics, to every physical system corresponds a separable Hilbert space $\mathscr{H}$ over the complex field, and the ingredients under discussion sound as follows. The states of the system are all the bounded, positive, self-adjoint, trace-class operators of unit trace on $\mathscr{H}$; they form a convex set, and the extremal points, the pure states, are in one-to-one correspondence with the unit rays of $\mathscr{H}$ in the absence of superselection rules. Given a family of pure states, say $\varphi_1, \ldots, \varphi_n$, the closed subspace spanned by them is usually referred to as the set of all superpositions of the φ_i's. The observables of the system correspond to self-adjoint operators in $\mathscr{H}$ (in the absence of superselection rules the correspondence is one-to-one); they form an involutive algebra, with involution given by adjoint operation. The propositions are represented by the projection operators in $\mathscr{H}$, *i.e.* by the self-adjoint operators whose spectra consist of the values 0 and 1; they form an ordered structure, technically an orthomodular complete lattice. The operations do not play a very explicit role, but can be associated to the measurement of an observable with given result, and can be identified with the mapping which carries a state vector to its renormed projection onto the eigenspace of the observable corresponding to the obtained result.

Different approaches to the foundations of quantum mechanics make different choices of the ingredients to be considered as (more) primitive and those to be considered as derived. Let us quote a few significant examples.

i) In the historical approach initiated by BIRKHOFF and VON NEUMANN [2], and then pursued by MACKEY [3], PIRON [4], VARADARAJAN [5] and others, the mathematical structure mainly referred to can be expressed by assigning to propositions the role of primitive concepts: they form, at least, an orthomodular partially ordered set $\mathscr{L}$. States can be viewed, *a posteriori*, as probability measures on $\mathscr{L}$, and have thus a status of derived concepts, though they are generally used as undefined quantities to provide the physical interpretation to the order structure of $\mathscr{L}$ (but the Jauch-Piron approach does not follow this attitude). Also the observables have the role of derived concepts, being introduced as $\mathscr{L}$-valued measures on the set (Boolean algebra) of all Borel sets on the real line. As is well known, within this approach it has been developed a number of plausible additional assumptions which are sufficient to ensure for the states the Hilbert-space picture, and for the observables the associated operator algebra. The notion of operation has not received special attention, within this approach, until rather recently, as we shall mention in the sequel.

ii) The status of primitive concepts accorded to observables characterizes the so-called algebraic approach to quantum mechanics; observables are here endowed, from the very beginning, with a rich algebraic structure (von Neumann algebras, C^*-algebras, Segal algebras are examples) which can hardly be justified on direct empirical evidence, a fact that puts the algebraic approach rather aside the area of quantum logics. States are defined as normed, positive linear functionals on the algebra of the observables; propositions are defined as projections in the algebra, and are thus naturally equipped with the familiar structure of orthomodular poset, since it is known that the projections of any involutive ring form an orthomodular poset.

iii) The status of primitive ingredients assigned to the states is characteristic of other approaches particularly developed in the last decade. We refer, in particular, to the so-called convex-set approach [6], which, as the name says, starts from the assumption that the states form a convex set, and then follows the idea that all information on the physical system is embodied in the geometry of that convex set, particularly in the shape of its boundary (pure states). The ambitious program of this approach, a complete geometrization of quantum mechanics on the ground of the state space, is not yet fully accomplished, in particular it is not clear how the conventional quantum mechanics could be recovered from hypotheses, on the geometry of the convex set, having a direct empirical support; however, this approach looks particularly flexible and appropriate for a study of generalizations of quantum mechanics, and the mathematics of convex sets will probably deserve further progress [7]. In this approach the propositions appear as derived concepts, and may be identified with the facets of the convex set; observables can then be derived with the help of propositions. Operations have, here, a major role: they are viewed as filtering procedures on the states, and represented by conditioning maps. Another notion which finds a privileged role within the convex-set approach, and within other approaches using states as primitive concepts, is the one of transition probability between two pure states [6, 8], an abstract version of the inner product in the Hilbert-space formulation of quantum mechanics.

iv) A strictly operational approach to quantum mechanics, namely an approach which accords only to operations the status of primitive concepts, putting states, propositions and observables on the level of derived concepts, has been developed by Foulis and Randall [9]. Operations are thought of as instructions that describe a well-defined, physically realizable, reproducible procedure and that specify what must be observed and what can be recorded; the formal structure they determine is called a manual, which is intended to be an exhaustive collection of physical operations pertaining to quantum phenomena. However, the term « operational approach » is often used in a looser sense and denotes schemes which deserve to operations a fundamental role but admit, as primitive concepts, also other ingredients. Thus, these schemes partially fall into the attitudes quoted in items i)-iii); for instance, Mielnik [6], Davies and Lewis [10], Edwards [11] worked out operational approaches connected with the convex-set approach, Pool [12] has connected operations with the proposition-lattice approach.

Of course, the various points of view alluded to in i) to iv) are, at least partially, complementary, but they exibit also contrasting features, as in the case of Mielnik's convex-set scheme which has rebelling aspects in regard of the traditional proposition-lattice approach.

In the sequel we shall examine the role and the properties of states from the point of view of various approaches amenable to quantum logic. In doing that, propositions, states and operations will alternatively, and sometime jointly, be used as basic ingredients: operations will mostly be thought of as mappings of states, not in the more primitive and radical sense proposed in the Foulis-Randall approach, which, thus, will remain aside from our discussion. Also the algebraic approach will remain aside, based, as it is, on observables.

The mathematical structure dominant in quantum logic is a particular kind of partially ordered set, technically an atomic orthomodular lattice with the covering property [13]. Indeed, such a mathematical structure admits an analytical representation which establishes the contact with the usual Hilbert-space formulation of quantum mechanics. This analytical representation is generally referred to as Piron's representation theorem [14], and constitutes a milestone of quantum logics. Being a rather old and widely quoted result [14], we shall give up an exposition of this representation theorem; rather we shall outline how the basic mathematical structure, the atomic orthomodular lattice with the covering property, can be motivated on different, and alternative, grounds. Moreover, the lattice structure we are referring to is natural support for the logical analysis of the theory.

To recover, from the lattice approach, the Hilbert-space quantum mechanics, one needs, besides Piron's representation theorem, Gleason's theorem, another crucial result which allows identifying states, abstractly defined as probability measures on the proposition lattice, with von Neumann density operators [14]. Both theorems cannot be used for physical systems with two-dimensional space of states; in sect. **6** we examine some pathologies, related to Gleason's theorem, of spin-$\frac{1}{2}$ systems, and comment the concomitant appearance of « exceptional states » which have no place in the Hilbert-space formulation.

2. – States as probability measures on propositions.

Intuitively, but naively, a proposition is thought of as any experimental arrangement which, after interaction with the physical system, produces one of only two possible outcomes, say the mark « yes » or the mark « no »; *vice versa*, any such experimental arrangement is thought of as a proposition. Actually, in order to build up any scheme able to contain, or to fit with, quantum theory, one needs axioms which are not fulfilled by the generality of the experimental arrangements with two outcomes; thus, if we want propositions to remain basic ingredients of the scheme, we cannot use « an experimental arrangement which, after interaction with the physical system, produces one of only two possible outcomes » as the definition of « proposition ».

In this section, we discuss this point using the notion of state, but leaving aside the notion of operation. We start with the set E of all experimental arrangements having dichotomic outcomes (of course we are referring to arrangements pertaining to the physical system under discussion, and E reflects the knowledge reached on the system), and consider the set Π of all different ways of preparing the system. Then call equivalent two preparations, say π_1, π_2, if, for every $e \in E$, the probability of getting the yes outcome of e after interaction with the system prepared according to π_1 is the same as after interaction with the system prepared according to π_2. Finally, call states of the system the equivalence classes of Π; let $\mathscr{S}$ be the set of all states, and denote by $\alpha, \beta, \gamma, \ldots$ single states. Not only E partitions Π in equivalence classes, but also Π partitions E in equivalent classes: we say $e_1, e_2 \in E$ equivalent if, for every $\pi \in \Pi$, the probability of the yes outcome of e_1 equals the one of e_2. Call $\mathscr{E}$ the set of all equivalence classes of E, and denote $a, b, c, \ldots$ its elements. It should be noted that one is considering as equivalent experimental apparata which can be quite different as regards their physical realization; indeed our equivalence does not make requirements on what will be the state of the system after the interaction with the experimental apparatus. Thus the same element of $\mathscr{E}$ can be representative of experimental arrangements transforming in different ways the state of the physical system. For completeness, it will be assumed that $\mathscr{E}$ contains also two trivial elements, denoted by $\underline{0}$ and $\underline{1}$, such that the yes outcome of $\underline{0}$ is excluded (the no outcome is certain) while the yes outcome of $\underline{1}$ is certain (the no outcome is excluded), no matter what the state of the system is. Let $\alpha(a)$ denote the probability of the yes outcome of $a \in \mathscr{E}$ after interaction with the system in the state α; for fixed $a \in \mathscr{E}$, the quantity $\alpha(a)$ determines a functional on $\mathscr{S}$, while, for fixed $\alpha \in \mathscr{S}$, it determines a functional on $\mathscr{E}$. With the help of this functional, it is immediate to verify that $\mathscr{S}$ orders $\mathscr{E}$ through the definition

$$a \leqslant b \text{ whenever } \alpha(a) \leqslant \alpha(b) \text{ for every } \alpha \in \mathscr{S}. \tag{2.1}$$

Thus $\mathscr{E}$ is a poset, with least and greatest elements given, respectively, by $\underline{0}$ and $\underline{1}$. (Mathematically, one might get a reversed definition saying that $\mathscr{E}$ orders $\mathscr{S}$ by means of $\alpha \leqslant \beta$ whenever $\alpha(a) \leqslant \beta(a)$ for every $a \in \mathscr{E}$. We take as an empirical fact the impossibility of finding in $\mathscr{S}$ two nonidentical physical states ordered by such a relation.)

It is natural to assume that, if a is an arbitrary element of $\mathscr{E}$, there is in $\mathscr{E}$ also another element, to be denoted by $a^\perp$, such that

$$\alpha(a^\perp) = 1 - \alpha(a) \text{ for every } \alpha \in \mathscr{S}; \tag{2.2}$$

indeed $a^\perp$ can be interpreted as the arrangement corresponding to a with opposite convention about what is « yes » and what is « no ». Moreover, $\mathscr{S}$ being separating on $\mathscr{E}$, the mapping $a \mapsto a^\perp$ is one-to-one, and obviously satisfies

i) $$a^{\perp\perp} = a\,,$$

ii) $$a \leqslant b \Rightarrow a^\perp \geqslant b^\perp\,.$$

At this stage, $\mathscr{E}$ is naturally endowed with the notion of disjointness: $a, b \in \mathscr{E}$ are called disjoint, and we write $a \perp b$, if and only if $a \leqslant b^\perp$, or, equivalently, if and only if $\alpha(a) + \alpha(b) \leqslant 1$ for all $\alpha \in \mathscr{S}$. Leaving aside a discussion on motivations, we insert now a law, to be considered of empirical nature:

(2.3) There exists in $\mathscr{E}$ the join of disjoint elements, and every state is additive on disjoint elements, *i.e.* $\alpha(a \vee b) = \alpha(a) + \alpha(b)$ when $a \perp b$.

With this law the mapping $a \mapsto a^\perp$ becomes an orthocomplementation in $\mathscr{E}$: indeed, besides items i) and ii) quoted above, we have also

iii) $$a \vee a^\perp = \underline{1}\,.$$

Thus $(\mathscr{E}, \leqslant, \perp)$ is an orthocomplemented poset, and $\mathscr{S}$ takes the status of order-determining set of probability measures on it.

We deduce, in particular, that the fundamental orthomodular (even « quasi-modular » or « weak-modular ») identity holds true:

$$a \leqslant b \Rightarrow b = a \vee (b \wedge a^\perp)\,;$$

indeed, for every $\alpha \in \mathscr{S}$, we have $\alpha\big(a \vee (b \wedge a^\perp)\big) = \alpha(a) + \alpha(b \wedge a^\perp) = \alpha(a) + 1 - \alpha(a \vee b^\perp) = \alpha(a) + 1 - \alpha(a) - \alpha(b^\perp) = \alpha(b)$. Hence $(\mathscr{E}, \leqslant, \perp)$ is an orthomodular poset.

This mathematical structure is the one obtained by MACKEY [3] in his historical approach to quantum mechanics; there, the point of departure consists of the notions of states, observables, and a probability function which assigns the probability that the measured value of an arbitrary observable lie in any given number set when the system has been prepared in any given state: the elements of $\mathscr{E}$ come out as pairs formed by an observable and a number set. We refer to Mackey's book [3] and to Maczynski's papers [15] for a deeper motivation of the orthomodular poset structure of $(\mathscr{E}, \leqslant, \perp)$.

But the structure $(\mathscr{E}, \leqslant, \perp)$ discussed above is not the one on which quantum logics are generally based. Indeed, as the name « quantum logic » anticipates, it is demanded that the ordering relation holding in the set of propositions (dichotomic experiments) be strictly related to a logical entailment: the assertion « a is less than b » is expected, viewing a and b as sentences of a language, to mean the truth of the sentence « a implies b », or, in other words, that « the truth of a entails the truth of b ». Thus, it is clear that the candidate for the order relation of quantum logics is the following relation:

$$a \leqslant b \text{ whenever } \mathscr{S}_1(a) \subseteq \mathscr{S}_1(b)\,, \tag{2.4}$$

where

$$\mathscr{S}_1(a) = \{\alpha \in \mathscr{S} : \alpha(a) = 1\}\,.$$

$\mathscr{S}_1(a)$ represents the « certainly-yes » domain of a, and $a \leqslant b$ says that all situations (states) which make certain the yes outcome of a make also certain the yes outcome of b.

The relation $\leqslant$ is obviously reflexive and transitive, but the third requirement of order relations, the antisymmetry, is, in general, not satisfied: in fact, the implication

$$a \leqslant b \text{ and } b \leqslant a \Rightarrow a = b$$

reads

$$\mathscr{S}_1(a) = \mathscr{S}_1(b) \Rightarrow a = b\,, \tag{2.5}$$

a condition which is not verified by all the elements of $\mathscr{E}$. Thus, if one wants $\leqslant$ to be an order relation, one must eliminate a number of elements of $\mathscr{E}$, restricting to some subset of it.

Let us quote some examples which are inconsistent with condition (2.5). We first remark that all nontrivial experimental arrangements having empty « certainly-yes » domain contradict this condition, for $\mathscr{S}_1(a) = \emptyset$ would imply $a = \underline{0}$; think, *e.g.*, of a semi-transparent mirror which, interacting with a photon, assigns equal probability to reflection and transmission: it has an empty « certainly-yes » domain, but, sure, it is not trivial from the physical point of view. The elimination, from the family of basic ingredients of the theory, of such experimental arrangements is particularly unpleasant, owing to their relevance in describing quantum interference phenomena. Following an elegant analysis of Mielnik [16], consider now two hypothetical devices a and b acting on mixtures of red, yellow and violet light: a transmits the red photons and absorbs the yellow and the violet ones, but re-emits, on the average, 50 % of the absorbed yellow photons in form of red photons; b is similarly transparent to the red photons and absorbs the yellow and violet ones, but re-emits 50 % of the violet photons in form of red photons. Schematically:

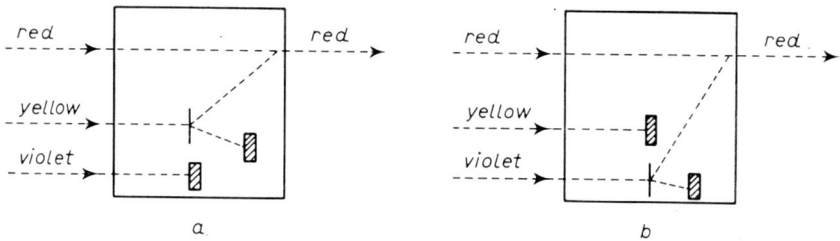

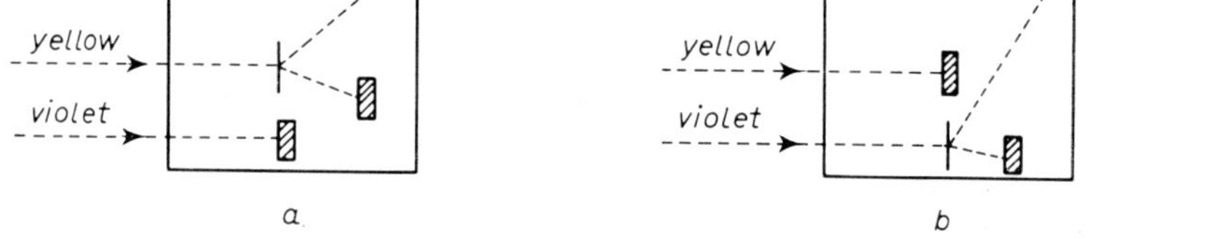

Fig. 1.

If the yes outcome of these devices is acted by the detection of the transmitted (or re-emitted) photon, then both a and b have the same « certainly-yes » domain, which consists in the red state, but do not look physically equal, as condition (2.5) would claim. This example has also another interesting feature: in the language of quantum theory of measurement, these devices are of first kind (idempotent) and pure (map pure states into pure states), an aspect which will be considered in the sequel.

Thus we see that the use of the relation $\leqslant$ as order relation needs the abandonement of significant yes-no arrangements which, on the other hand, found a natural place in the structure $(\mathscr{E}, \leqslant, \perp)$.

Come now to the problem of the orthocomplementation which reflects the negation operation of logic. We can still refer to the mapping $a \mapsto a^{\perp}$ defined by (2.2), assuming that it makes sense also within the set of yes-no arrangements surviving to the selection due to condition (2.5), but an innocent law like (2.3) is no longer sufficient to make this mapping an orthocomplementation with respect to the order relation $\leqslant$. Indeed, the property

$$a \leqslant b \Rightarrow a^{\perp} \geqslant b^{\perp}$$

reads

$$\mathscr{S}_1(a) \subseteq \mathscr{S}_1(b) \Rightarrow \mathscr{S}_0(a) \supseteq \mathscr{S}_0(b)\,, \tag{2.6}$$

where $\mathscr{S}_0(a)$ stands for the « certainly-no » domain of a, *i.e.*

$$\mathscr{S}_0(a) = \{\alpha \in \mathscr{S} : \alpha(a) = 0\}\,,$$

and, again, this is a by no means obvious requirement. The devices of fig. 1 violate it, for $\mathscr{S}_1(a) = \mathscr{S}_1(b)$ but $\mathscr{S}_0(a) \neq \mathscr{S}_0(b)$. Consider, as another counterexample to condition (2.6), the hypothetical devices a and b which act on a four-state system according to the definition: a is transparent to α_1, has transmission coefficient $\frac{1}{2}$ for α_2 and α_3, absorbs α_4; b is transparent to α_1 and α_2, but absorbs α_3 and α_4. Schematically,

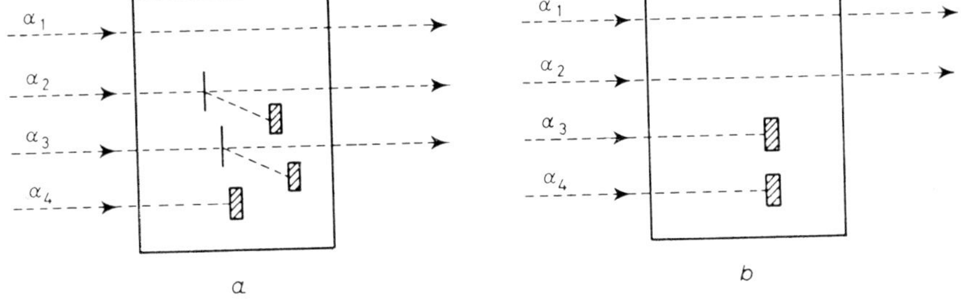

Fig. 2.

(it is assumed that the effective transmission of the system causes the appearance of the yes oucome). Now, $\mathscr{S}_1(a) = \{\alpha_1\}$ is contained into $\mathscr{S}_1(b) = \{\alpha_1, \alpha_2\}$, but $\mathscr{S}_0(a) = \{\alpha_4\}$ is contained into $\mathscr{S}_0(b) = \{\alpha_3, \alpha_4\}$, contrary to what is demanded by (2.6). Intuitively, condition (2.6) is met by those yes-no arrangements which, for given « certainly-yes » domain, have a maximal « certainly-no » domain.

Concluding these remarks, we can say that the aim of building up a structure ordered by $\leqslant$ and with the mapping $a \mapsto a^\perp$ behaving as orthocomplementation, relative to $\leqslant$, can be pursued at the cost of considering only a subclass, say $\mathscr{L}$, of the set $\mathscr{E}$ previously considered. We denote by $(\mathscr{L}, \leqslant, \perp)$ the orthocomplemented poset so constructed. A notion of disjointness can still be introduced in $\mathscr{L}$: $a, b \in \mathscr{L}$ are said to be disjoint, and we write $a)(b$, whenever $a \leqslant b^\perp$ (but we do not require, now, $\alpha(a) + \alpha(b) \leq 1$). Then it is natural to adopt a law analogous to (2.3), namely

(2.3′) There exists in $\mathscr{L}$ the join of disjoint elements, and every state is additive on disjoint elements, *i.e.* $\alpha(a \vee b) = \alpha(a) + \alpha(b)$ when $a)(b$.

Also for the structure $(\mathscr{L}, \leqslant, \perp)$ the orthomodular identity holds true: indeed, by the same argument used for $(\mathscr{E}, \leq, \perp)$, we get

$$a \leqslant b \Rightarrow b = a \vee (b \wedge a^\perp) .$$

Therefore $(\mathscr{L}, \leqslant, \perp)$ is an orthomodular poset, and $\mathscr{S}$ takes the status of a strongly order-determining set of probability measures on it.

It should be clear that most of the axioms supporting this structure have not an empirical meaning, but merely serve to define $\mathscr{L}$; however, we consider the following of empirical nature: i) the requirement that $\mathscr{L}$ is separating on $\mathscr{S}$, ii) the law (2.3′), iii) the assumption that $\mathscr{L}$ is rich enough to be the vehicle of a complete description of the physical system, *i.e.* that every significant property of the system admits a description in terms of the rather idealized elements of $\mathscr{L}$.

We have considered in so far two different notions of order relation: the one defined in (2.1) and the one defined in (2.4). By restriction of the former we can construct another orthomodular poset, namely $(\mathscr{L}, \leq, \perp)$, which should now be compared with $(\mathscr{L}, \leqslant, \perp)$.

Of course, if $a, b \in \mathscr{L}$ and $a \leq b$ then $a \leqslant b$, but, conversely, $a \leqslant b$ does not imply $a \leq b$. In the ordinary Hilbert-space formulation of quantum mechanics $\mathscr{L}$ becomes the set of all projection operators and the structures $(\mathscr{L}, \leqslant, \perp)$ and $(\mathscr{L}, \leq, \perp)$ coincide. Thus, the question arises of whether this coincidence is a basic requirement of quantum mechanics, or it is only a circumstance of the ordinary Hilbert-space formulation. In favour of the first alternative is the fact that there are examples of proposition-state structures which meet almost all the axioms of quantum logics but lack the equivalence of the two order relations under discussion, and, indeed, are of unphysical nature. We shall find in the sequel another strong argument in favour of the equivalence between $(\mathscr{L}, \preccurlyeq, \perp)$ and $(\mathscr{L}, \leqslant, \perp)$: this equivalence will appear necessary in order to make possible the introduction of conditional probabilities.

On account of these remarks we shall assume, hereinafter, the full equivalence, in $\mathscr{L}$, of the two orderings $\preccurlyeq$ and $\leqslant$. We write $\leqslant$ for the order relation, $\perp$ for the disjointness relation; with some abuse of notation, we shall denote by $\mathscr{L}$ the orthomodular poset under discussion, in place of the correct, but needlessly pedantic $(\mathscr{L}, \leqslant, \perp)$.

The mathematics of orthomodular posets is rich, and represents a remarkable example of theory developed under intertwined motivations both of physical and of purely mathematical nature; from an historical perspective, the theory of orthomodular posets (and lattices) can be seen as third step of the sequence which has, as first steps, von Neumann algebras and continuous geometries. A review of this matter goes beyond our purposes [14, 17]; however, it is useful to recall that

i) the orthomodular poset structure can naturally accommodate the notion of commutativity,

ii) meet and join of commuting elements exist in the poset,

iii) if (a, b) is a commuting pair then also $(a^\perp, b)$, $(a, b^\perp)$, $(a^\perp, b^\perp)$ are commuting pairs,

iv) two elements commute if and only if they generate a Boolean subalgebra,

v) the centre of the poset (lattice) is a Boolean subalgebra, containing the least and greatest elements $\underline{0}$, $\underline{1}$, and the poset (lattice) is irreducible when the centre is the two-element set $\{\underline{0}, \underline{1}\}$.

On the basis of the orthomodular poset structure of $\mathscr{L}$ it is possible to continue, in a purely lattice-theoretic framework, the construction of a more elaborated structure which finally admits an analytical representation coinciding with the ordinary Hilbert-space formulation of quantum mechanics: we refer, of course, to the historical lattice approach [2-5, 18], and to its variants. Let us only mention the main assumptions which are added about $\mathscr{L}$. The first assumption is that $\mathscr{L}$ is not only a poset but even a lattice: the nontrivial content of this assumption is the existence of join and meet of noncommuting propositions. After recalling that, in a lattice, a is said to cover b if $a > b$ and $a \geqslant c \geqslant b$ implies either $c = a$ or $c = b$, and after recalling that the elements which cover $\underline{0}$ are called atoms (or points), the second assumption asserts that every nonzero element of $\mathscr{L}$ majorizes at least one atom, in short, that $\mathscr{L}$ is atomic. Actually, when an orthomodular lattice is atomic, then it is atomic in a very strong way, for every nonzero element is the join of all the atoms it majorizes, in short, when it is atomic it is also atomistic. The third assumption is that $\mathscr{L}$ satisfies the famous Birkhoff « covering property » [17], namely that, for every a and every atom p not contained in a, $a \vee p$ covers a; alternatively, one can express the same property by saying that, if p and q are atoms and $a \wedge p = \underline{0}$, then $p \leqslant a \vee q$ implies $q \leqslant a \vee p$, a formulation which reminds the geometric Steinitz-Mac Lane exchange property [19, 20].

If $\mathscr{L}$ is all that, then it is possible to show [4, 14], through a rather heavy itinerary (the « representation theorem »), that $\mathscr{L}$ admits a representation in terms of subspaces of a linear vector space with coefficients from a division ring whose nature seems to have a definite propension towards the complex field (or even the reals, or the quaternions) but, nevertheless, still deserves open questions. If $\mathscr{L}$ is also irreducible, then the representation alluded to involves all the subspaces of the vector space, otherwise the pattern of superselection rules appears.

In an approach of this kind, detailed properties of states come out as a result of the representation procedure, and strong results are known, first of all Gleason's theorem [21] (see sect. **6**). Properties like the superposition principle are found to have roots in the combined effect of atomicity, covering property and irreducibility: more precisely in the assertion that, for every pair of (nonidentical) atoms p, q, the element $p \vee q$ covers some other atom different from p and q. But, in order to view a property of this kind as a property of states, indeed as the core of the superposition principle, a further assumption is required. To formulate it we need first a definition: we say that a proposition is the support of $\alpha \in \mathscr{S}$, and denote it $\sigma(\alpha)$, if and only if $\{\alpha \in \mathscr{L} : \alpha(a) = 1\} = \{a \in \mathscr{L} : a \geqslant \sigma(\alpha)\}$. The further assumption now reads:

(2.7) Every $\alpha \in \mathscr{S}$ has a unique support in $\mathscr{L}$; conversely, every $a \in \mathscr{L}$, $a \neq \underline{0}, \underline{1}$, is the support of some state, and each atom is the support of only one state.

States having an atom as support are identified with the pure states, and the set of pure states is thus in an one-to-one correspondence with the set of the atoms of $\mathscr{L}$. Let us also remark that, as a consequence of (2.7), we have the following property, often called the Jauch-Piron property:

(2.8) $$\text{if} \quad \alpha(a) = \alpha(b) = 1 \quad \text{then} \quad \alpha(a \wedge b) = 1 .$$

Viewing a and b as propositions in logical sense (sentences of a language), and viewing the lattice operation « $\wedge$ » as the counterpart of the « and » connective, condition (2.8) gives the truth table of « and »: the proposition « a and b » is true (in the state α) if and only if both a and b are true (in the same state α).

Recently, RÜTTIMANN [22] has argued that property (2.8), together with the properties of $\mathscr{L}$ and $\mathscr{S}$ adopted before, is incompatible with an assumption of finiteness of $\mathscr{L}$, unless $\mathscr{L}$ is a Boolean algebra. In other words, the combined effect of condition (2.8) and of the finiteness of $\mathscr{L}$ is the one of collapsing the orthomodularity of the lattice $\mathscr{L}$ into the stronger distributivity. Thus, the possibility of a really quantal (*i.e.* nonclassical) proposition lattice formed by a finite number of elements is ruled out by condition (2.8): quantum structure needs an infinite number of propositions and of states. True, this result is familiar in any example of quantum system described in terms of Hilbert space over the complex field; nevertheless, the need of infinity seems to point at some idealized feature and rigidity of the present-day quantum theory.

3. – Propositions as closed sets of states.

In the previous section states have been considered in strict connection with the set $\mathscr{L}$ of propositions; properties like the superposition principle found a place only on the basis of rather elaborated features of $\mathscr{L}$. In this section we examine properties of states introducing from the very beginning a minimal notion of state superposition, and postponing to a later level the connection with propositions.

We shall be mainly concerned with pure states, thus we recall first a few comments on this notion. The distinction between pure and nonpure state is naturally accounted for by assigning to $\mathscr{S}$ the structure of a convex set: a state is pure if it is an extreme point of $\mathscr{S}$, otherwise it is nonpure, or, in other words, it is a mixture. Explicitly, and with reference to propositions, α is a mixture if there exist two positive numbers t_1, t_2 with $t_1 + t_2 = 1$ and two nonidentical states α_1, α_2 such that $\alpha(a) = t_1\alpha_1(a) + t_2\alpha_2(a)$ for every $a \in \mathscr{L}$, in which case α is said to be mixture of α_1, α_2 and we write $\alpha = t_1\alpha_1 + t_2\alpha_2$; otherwise α is pure. It may happen that pure states are so many to allow every nonpure state to be written as a mixture of pure states: this happens for both classical and quantum physical systems.

Classical mixtures admit the ignorance interpretation: $\alpha = \sum t_i \alpha_i$ is taken to mean that the system actually is in one, and only one, of the states α_i, with probability t_i; more selective preparations of the system would allow deciding which one of the α_i's is the real state so that the t_i's represent our initial degree of ignorance. This ignorance interpretation rests on the fact that classical mixtures have a unique decomposition into pure states. Things are more involved in the quantum case, for the decomposition of mixtures into pure states is no longer unique. Think, *e.g.*, to an unpolarized light beam: it can be described by a mixture of linearly polarized states, as well as a mixture of circularly polarized states. This lack of unicity is a crucial branching point between quantum and classical physics: we refer to MIELNIK [6] for an analysis of how this fact is coded into the geometrical properties of the state set $\mathscr{S}$.

We refer also to VAN FRAASSEN [23] for an interpretation of quantum mixtures in the framework of modal logic.

We definitely assume that the set $\mathscr{P}$ of pure states is nonempty, and that every mixed state may be written as convex combination of pure states. Thus, instead of looking at the whole $\mathscr{S}$, we restrict ourselves now to $\mathscr{P}$.

From the usual notion of superposition of states we extract, to begin with, a very general feature, namely that $\mathscr{P}$ is endowed with a closure relation. By this we mean that there is a function $A \mapsto \bar{A}$ defined for all subsets $A \subseteq \mathscr{P}$, satisfying

i) $\bar{A} \subseteq \mathscr{P}$,

ii) $A \subseteq \bar{A}$,

iii) $A \subseteq \bar{B}$ implies $\bar{A} \subseteq \bar{B}$

for all subsets A, B of $\mathscr{P}$. It follows that this function (the closure relation) is order preserving

$$A \subseteq B \quad \text{implies} \quad \bar{A} \subseteq \bar{B}, \tag{3.1}$$

is idempotent

$$\bar{A} = \bar{\bar{A}}, \tag{3.2}$$

and satisfies

$$\overline{A \cap B} = \bar{A} \cap \bar{B}. \tag{3.3}$$

Though the $A \mapsto \bar{A}$ function is quite general, we shall think of $\bar{A}$ as the set obtained by adding to A all the superpositions of its states, thus helping the physical interpretation of what follows.

$\mathscr{P}$ endowed with the closure relation is called a closure space. We shall be particularly concerned with closed subsets of $\mathscr{P}$: $A \subseteq \mathscr{P}$ is called a closed subset, or a « flat », if and only if $A = \bar{A}$.

We add a restriction on the closure relation: that the empty set $\emptyset$ be closed. Then we can prove that the closed subsets (flats) of $\mathscr{P}$ form a lattice $L(\mathscr{P})$. The order relation in $L(\mathscr{P})$ is set-theoretic inclusion, $\emptyset$ and the whole $\mathscr{P}$ are the least and greatest elements $\underline{0}$, $\underline{1}$ of $L(\mathscr{P})$, meet and join are as follows:

$$A \wedge B = A \cap B, \qquad A, B \in L(\mathscr{P}),$$

$$A \vee B = \overline{A \cup B}, \qquad A, B \in L(\mathscr{P}).$$

According to (3.3), $A \cap B$ is closed when A and B are closed, moreover, $A \cap B$ is obviously the greatest flat contained in both A and B: this motivates the meet. According to (3.2), $\overline{A \cup B}$ is closed, and, for every flat C such that $C \supseteq A$ and $C \supseteq B$, we get $C \supseteq A \cup B$, hence by (3.1) $C \supseteq \overline{A \cup B}$: this motivates the join.

Thus we see how intimately a closure space is associated with an order structure.

If we consider this order structure as a candidate to represent a logical structure, we can already remark a departure from classical logic. Indeed, viewing A, B as representative of propositions in logical sense, and interpreting, as usual, the join as the « or » connective, we see that the truth table of our « or » is nonclassical: $A \vee B$ may contain states which are neither in A nor in B, so that « A or B » may be true even if neither A nor B are true.

We proceed now with further restrictions on the closure operation. The next is

iv) $$\{\alpha\} = \overline{\{\alpha\}} \qquad \text{for every } \alpha \in \mathscr{P},$$

i.e. all singletons are closed. Physically, this turns out to assert that there are not nontrivial superpositions of only one state.

The effect of condition iv) on $L(\mathscr{P})$ is the one of ensuring the existence of elements which cover the empty set, namely the existence of atoms: each element of $L(\mathscr{P})$ majorizes one atom at least, and, actually, is the join of the atoms it majorizes, so that $L(\mathscr{P})$ is atomic and atomistic. Moreover, the atoms of $L(\mathscr{P})$ are in one-to-one correspondence with the pure states of the system, the elements of $\mathscr{P}$.

We come now to another restriction on the closure operation: we assume that it satisfies the Steinitz-Mac Lane exchange property [19, 20]. This amounts to saying that, for any states α, $\beta \in \mathscr{P}$ and for any subset $A \subseteq \mathscr{P}$,

v) $$\text{if} \quad \alpha \in \overline{A \cup \{\beta\}} \quad \text{and} \quad \alpha \notin \overline{A} \quad \text{then} \quad \beta \in \overline{A \cup \{\alpha\}}\,.$$

Physically, this says, in particular, that, if γ is a superposition of α and β, then α is a superposition of β and γ (and β is a superposition of α and γ).

The effect of condition v) on the lattice $L(\mathscr{P})$ is the one of ensuring the covering property. Indeed, in terms of $L(\mathscr{P})$, condition v) sound as follows: if $A \in L(\mathscr{P})$, if p, q are atoms of $L(\mathscr{P})$ and $A \wedge p = \underline{0}$, then $p \leqslant A \vee q$ implies $q \leqslant A \vee p$ (and hence implies $A \vee p = A \vee q$). This is one of the formulations of the lattice-theoretic covering property [24].

The structure consisting of

1) the set $\mathscr{P}$,

2) a closure relation with the exchange property and such that the empty set $\emptyset$ and the singletons $\{\alpha\}$ $(\alpha \in \mathscr{P})$ are closed,

will be called geometry of $\mathscr{P}$ and denoted $G(\mathscr{P})$. This use of the word geometry is somewhat loose: only in case we would add a « finite basis » hypothesis we would fall into the notion of combinatorial geometry [20], but an hypothesis of this kind will not be adopted since ordinary quantum mechanics has often to do with infinite-dimensional Hilbert spaces.

We come now to the notion of basis of a subset of states, and the notion of rank, or dimension function. Consider a flat A, and think of it as an element of the lattice $L(\mathscr{P})$. A « chain » in a lattice is a completely ordered subset of the lattice. If $A, B \in L(\mathscr{P})$ and $A > B$, a maximal chain connecting A with B has the form $B < s_1 < \ldots < s_i < \ldots < A$, where each term covers the antecedent. We say that A is a finite element if it can be connected with $\underline{0}$ (the empty set $\emptyset$) by a maximal chain of finite length. Now we have [20]:

(3.4) If A is a finite element of $L(\mathscr{P})$, all maximal chains from $\underline{0}$ to A have the same length.

Proof. We prove, more generally, that, if $B \leqslant A$, then all maximal chains from B to A have the same length. Let $B = s_0 < s_1 < \ldots < s_n = A$ and $B = t_0 < t_1 < \ldots < t_m = A$ be two maximal chains from B to A. If $n = 0$ or $n = 1$ the assertion is obvious; thus, we proceed by induction assuming the truth of the statement for all pairs A', B' between which there exists a maximal chain of length less than n. By the covering property, $s_1 \vee t_1$ covers both s_1 and t_1. Select a maximal chain $s_1 \vee t_1 = u_2 < u_3 < \ldots < u_p = A$ from $s_1 \vee t_1$ to A. Comparing the two paths from s_1 to A, we have $p = n$ by the induction hypothesis. Thus there is a maximal chain from t_1 to A having length $n - 1$, and $m = p = n$ by the induction hypothesis. *q.e.d.*

According to this result, we define the dimension, or rank, of a finite element A of $L(\mathscr{P})$ as the common length of all maximal chains from $\underline{0}$ to A, and denote it $D(A)$; if A is not a finite element, we define $D(A) = \infty$. Thus, the concept of dimension is well defined for every element of the lattice $L(\mathscr{P})$; we have in particular

$$D(A) = 0 \quad \text{if and only if} \quad A = \underline{0}\,,$$

$$A \leqslant B \quad \text{implies} \quad D(A) \leqslant D(B)\,.$$

It becomes justified to call « points » the rank-1 flats, « lines » the rank-2 flats, « planes » the rank-3 flats, etc.

Each maximal chain from $\underline{0}$ to A determines a family of atoms contained in A and having A as join: indeed, if s_{i-1} and s_i are adjacent elements of the chain, s_i is the join of s_{i-1} and an atom not contained in s_{i-1}, so that each step in the chain is produced by the « addition » of one atom. Of course, the family of atoms associated to a maximal chain from $\underline{0}$ to A is formed by a number of atoms equal to the number of steps in the chain, and hence equal to $D(A)$. We call such a family a « basis » of A: each maximal chain from $\underline{0}$ to A determines a basis of A. It is evident that the bases of A can be viewed as the minimal families of atoms contained in A and having A as join.

In a more geometric language, we can say that, given a flat A, it is always possible to select in it a basis, *i.e.* a minimal family of states whose closure is A: all bases of A have the same cardinality, the dimension $D(A)$ of A.

But we can do more: not only the closed subsets of states (flats) have a rank, but we can assign a rank to every subset of states, even if it is not closed. Indeed, if A is any subset of the geometry $G(\mathscr{P})$, we define the rank $r(A)$ of A as the dimension of the smallest flat which contains A, *i.e.* as the dimension of the closure of A. Symbolically,

$$r(A) = D(\bar{A}) , \qquad A \subseteq G(\mathscr{P}) .$$

The use of a symbol different from D stresses the fact that the D-function is defined on the flats, hence on the lattice $L(\mathscr{P})$, while the r-function is defined on every subset of states, hence on the geometry $G(\mathscr{P})$. The number $r(\mathscr{P}) = D(\mathscr{P})$ is the rank of the whole geometry and the dimension of the lattice $L(\mathscr{P})$.

Consider now how the r-function behaves under set-theoretic union and intersection, and how the D-function behaves under lattice-theoretic join and meet. We have at once, for any A, $B \subseteq G(\mathscr{P})$,

$$(3.5) \qquad \begin{cases} r(A \cup B) = D(\overline{A \cup B}) = D(\overline{\bar{A} \cup \bar{B}}) = r(\bar{A} \cup \bar{B}) = D(\bar{A} \vee \bar{B}) , \\ r(A \cap B) = D(\overline{A \cap B}) \leqslant D(\bar{A} \cap \bar{B}) = D(\overline{\bar{A} \cap \bar{B}}) = D(\bar{A} \wedge \bar{B}) . \end{cases}$$

The dimension function on $L(\mathscr{P})$ satisfies a characteristic inequality: if A, B are finite elements of $L(\mathscr{P})$, then [20]

$$(3.6) \qquad D(A \vee B) + D(A \wedge B) \leqslant D(A) + D(B) .$$

In fact, choose a maximal chain from $A \wedge B$ to A, say $A \wedge B = s_0 < s_1 < \dots < s_n = A$. Put $t_i = s_i \vee B$, and observe that t_i covers or equals t_{i-1}, for we know that s_i covers s_{i-1}, whence $s_i \vee B = (s_{i-1} \vee p) \vee B = (s_{i-1} \vee B) \vee p = t_{i-1} \vee p$, where p is an atom (not contained in s_{i-1}, but perhaps contained in t_{i-1}). Thus, except for a possible repetition of some elements, $t_0 < t_1 < \dots < t_n$ is a maximal chain from B to $A \vee B$, and $D(A) - D(A \wedge B) \geqslant D(A \vee B) - D(B)$.

Combining (3.6) with (3.5), we immediately get an analogous inequality for the rank function on $G(\mathscr{P})$: if A, B are subsets of $G(\mathscr{P})$ with finite bases, then

$$(3.7) \qquad r(A \cup B) + r(A \cap B) \leqslant r(A) + r(B) .$$

The inequalities (3.6) and (3.7) represent a relaxation of the conditions holding in the cases of modular lattices and, respectively, of projective geometries, where these inequalities are replaced by equalities. Recall that in ordinary quantum mechanics, where $L(\mathscr{P})$ becomes the lattice of the closed subspaces of a separable Hilbert space, nonmodularity occurs only in the infinite-dimensional case.

Having in mind a logical approach to quantum mechanics, we can observe that the lattice structure so far obtained, the lattice $L(\mathscr{P})$, is, in principle, idoneous to carry logical operations like conjunction, disjunction and implication; however we do not have, at the moment, any operation in $L(\mathscr{P})$ able to carry the logical negation. Thus, let us shortly comment on how $L(\mathscr{P})$ could be made orthocomplemented.

In $L(\mathscr{P})$ a flat A' is a complement of a flat A if and only if $A \wedge A' = \underline{0}$ and $A \vee A' = \underline{1}$ (recall that $\underline{0}$ and $\underline{1}$ stand for the empty subset $\emptyset$ and the whole $\mathscr{P}$, respectively). Every A in $L(\mathscr{P})$ admits complements. Of particular relevance are however the minimal complements: A' is a minimal complement of A if there is no proper subflat of A' which is complement of A. Minimal complements are natural candidates to play the role of orthocomplements; indeed, besides the definitory properties $A \wedge A' = \underline{0}$ and $A \vee A' = \underline{1}$, we have that, if $A \leqslant B$ and A' is a minimal complement of A, then there exists a minimal complement B' of B such that $B' \leqslant A'$. However, to ensure orthocomplementation, we need a further assumption:

(3.8) to each $A \in L(\mathscr{P})$ it is possible to associate a unique minimal complement $A^{\perp}$ such that $A^{\perp\perp} = A$.

If this requirement is satisfied, the mapping $A \mapsto A^{\perp}$ of $L(\mathscr{P})$ onto itself is indeed an orthocomplementation, and $L(\mathscr{P})$ becomes an atomistic orthocomplemented lattice with the covering property. Orthogonality could then be introduced in the usual manner: A is orthogonal to B if $A \leqslant B^{\perp}$; in terms of states, α and β are orthogonal if they belong to orthogonal flats. In order to meet the conditions which allow, for $L(\mathscr{P})$, the use of Piron's representation theorem, one attribute of $L(\mathscr{P})$ has still to be ensured: the orthomodularity (we leave open the suspect that orthomodularity might be proved on the basis of the preceding hypotheses: a proof exists when $L(\mathscr{P})$ is finite-dimensional). There is a relevant characterization of orthomodularity for atomistic orthocomplemented lattices with the covering property: we express it in terms of $L(\mathscr{P})$. $L(\mathscr{P})$ is orthomodular if and only if it satisfies the following condition [24]: if p is an atom of $L(\mathscr{P})$, then there exist an $A \in L(\mathscr{P})$ and two atoms q and r such that $p \leqslant q \vee r$ with $q \leqslant A$ and $r \leqslant A^{\perp}$. In terms of states, and thinking of the closure relation as state superposition, we can say: $L(\mathscr{P})$ is orthomodular if and only if every state can be viewed as superpositions of two states, one of them belonging to a flat A, the other belonging to the flat $A^{\perp}$. This characterization of orthomodularity shows how an important property of Hilbert spaces can be traced back to a purely lattice framework, or, alternatively, to a very general geometric structure.

Once one has made $L(\mathscr{P})$ an atomistic orthomodular lattice with the covering property, it is natural, by analogy with the lattice $\mathscr{L}$ of sect. **2**, to interpret the elements of $L(\mathscr{P})$ as the propositions of the physical system. We would also need the states to be probability measures on $L(\mathscr{P})$, a fact which, we believe, cannot be proved in the present context.

The assumptions needed to ensure the orthomodularity, in particular assumption (3.8) needed to ensure the orthocomplementation of $L(\mathscr{P})$, have, in our opinion, a weaker and less direct physical motivation than the other assumptions adopted on the closure relation in $G(\mathscr{P})$. It is thus interesting to investigate an alternative procedure based on these points: 1) the geometry $G(\mathscr{P})$ is still adopted, together with the associated lattice $L(\mathscr{P})$ which is atomistic with the covering property; 2) the propositions of the system are independently introduced and assumed to form an orthomodular poset $\mathscr{L}$; 3) the closure relation operating in $G(\mathscr{P})$ is given an explicit form looking at the states as probability measures on $\mathscr{L}$. Since in this section we are restricting ourselves to pure states, we take, as order relation in $\mathscr{L}$, $a \leqslant b$ whenever $\mathscr{P}_1(a) \subseteq \mathscr{P}_1(b)$ with $\mathscr{P}_1(a)$ standing for $\{\alpha \in \mathscr{P} : \alpha(a) = 1\}$.

The explicit form of the closure relation we are alluding to is the notion of superposition of states as proposed by VARADARAJAN [5]. Apart from minor technical modifications, it consists of the following. Given $\alpha \in \mathscr{P}$, let $\mathscr{L}_1(\alpha)$ be the set of the propositions whose outcome is certainly yes for the state α:

$$\mathscr{L}_1(\alpha) = \{a \in \mathscr{L} : \alpha(a) = 1\} .$$

Let A be a nonempty subset of $\mathscr{P}$: we say that a state β is a superposition of the states of A if

$$\mathscr{L}_1(\beta) \supseteq \bigcap_{\alpha \in A} \mathscr{L}_1(\alpha) ;$$

then define $\bar{A}$ to be the set of all the states of A and of all superpositions of them:

$$\bar{A} = \Big\{\beta \in \mathscr{P} : \mathscr{L}_1(\beta) \supseteq \bigcap_{\alpha \in A} \mathscr{L}_1(\alpha)\Big\} . \tag{3.9}$$

If A is the empty subset $\emptyset$, put $\bar{\emptyset} = \emptyset$. The mapping $A \mapsto \bar{A}$ is obviously a closure relation. Condition iv) reads

$$\mathscr{L}_1(\alpha) \subseteq \mathscr{L}_1(\beta) \qquad \text{implies} \qquad \alpha = \beta ,$$

a condition ensuring that $\mathscr{L}_1(\alpha)$ determines uniquely the state α (actually, JAUCH and PIRON [25] adopt, as definition of states, the subsets of $\mathscr{L}$ having the properties of the $\mathscr{L}_1(\alpha)$'s). The exchange property v) takes the form

$$\text{if} \quad \mathscr{L}_1(\alpha) \supseteq \mathscr{L}_1(\beta) \cap \Big(\bigcap_{\gamma \in A} \mathscr{L}_1(\gamma)\Big) \quad \text{and} \quad \mathscr{L}_1(\alpha) \not\supseteq \bigcap_{\gamma \in A} \mathscr{L}_1(\gamma) ,$$

$$\text{then} \quad \mathscr{L}_1(\beta) \supseteq \mathscr{L}_1(\alpha) \cap \Big(\bigcap_{\gamma \in A} \mathscr{L}_1(\gamma)\Big) .$$

The question now arises of comparing the two ordered structures we are dealing with: $\mathscr{L}$ and $L(\mathscr{P})$. We have

(3.10) the mapping $a \mapsto \mathscr{P}_1(a)$ is an order-preserving injection of $\mathscr{L}$ into $L(\mathscr{P})$.

Proof. Given $\beta \in \overline{\mathscr{P}_1(a)}$ we get, by (3.9), $a \in \mathscr{L}_1(\beta)$, whence $\beta \in \mathscr{P}_1(a)$: this implies $\overline{\mathscr{P}_1(a)} \subseteq \mathscr{P}_1(a)$, therefore $\mathscr{P}_1(a)$ is closed and is an element of $L(\mathscr{P})$. The mapping $a \mapsto \mathscr{P}_1(a)$ is obviously order preserving. *q.e.d.*

In view of this result, and taking into account basic demands of the superposition principle of quantum mechanics (with Dirac's words [26]: « The intermediate character of the state formed by superposition thus expresses itself through the probability of a particular result for an observation being intermediate between the corresponding probabilities for the original states ... »), we make the next step and assume that

(3.11) every $A \in L(\mathscr{P})$ admits the form $\mathscr{P}_1(a)$ for some $a \in \mathscr{L}$.

Then, combining theorem (3.10) and assumption (3.11), the mapping $a \mapsto \mathscr{P}_1(a)$ provides a one-to-one, order-perserving correspondence between $\mathscr{L}$ and $L(\mathscr{P})$. Pictorially speaking, the orthomodular poset $\mathscr{L}$ inherits from $L(\mathscr{P})$ the lattice structure, the atomicity and the covering property, while $L(\mathscr{P})$ inherits from $\mathscr{L}$ the orthocomplementation and the orthomodularity. This isomorphism between $\mathscr{L}$ and $L(\mathscr{P})$ can be viewed as a general version of the superposition principle [5, 14].

At this stage, one is again in a position to start the machinery of the so-called representation theorem, thus approaching the conventional Hilbert-space formulation of quantum mechanics.

Before closing this section, let us remark that the classical case is recovered by assuming, as closure relation in the state space $\mathscr{P}$, the identical mapping $A \mapsto A$, *i.e.* by assuming that every subset of $\mathscr{P}$ is closed. Then the geometry $G(\mathscr{P})$ becomes (isomorphic to) the « phase space » of the physical system, $L(\mathscr{P})$ becomes a Boolean algebra, the states become « dispersion-free » probability measures on the propositions.

4. – Propositions as mappings of states.

In this section, the relation between states and propositions will be considered from the point of view of the transformations induced on the state of the system by the action of the experimental arrangements which realize the proposition: this will associate to the measurement of a proposition an operation on the states.

The relevant mathematical fact we will use is the connection between involutive semi-groups and ordered structures: indeed, the operations on the states will naturally be endowed with an involutive semi-group structure. Thus, let us first sketch, without proofs, these mathematical facts [24].

Let T be a semi-group, and let $\cdot$ be its binary operation. Suppose T to possess a zero and a unit element, 0_T and 1_T: this means that, for every $x \in T$, $0_T \cdot x = x \cdot 0_T = 0_T$ and $1_T \cdot x = x \cdot 1_T = x$. T is called involutive if there exists a mapping $x \mapsto x^*$ of T onto T such that $x^{**} = x$ and $(x \cdot y)^* = x^* \cdot y^*$. In an involutive semi-group the elements which are idempotent and invariant under the involution are called projections: symbolically, $e \in T$ is a projection if and only if

$$e = e \cdot e = e^* .$$

Let $P(T)$ be the set of all projection of T: it is a partially ordered set with respect to the order

$$e \leqslant f \Leftrightarrow e \cdot f = e (= f \cdot e) , \qquad e, f \in P(T), \tag{4.1}$$

for this relation is reflexive ($f \cdot f = f$), transitive ($e \cdot f = e$, $f \cdot g = f \Rightarrow e \cdot g = e \cdot f \cdot g = e \cdot f = e$) and antisymmetric ($e \cdot f = e$, $f \cdot e = f \Rightarrow e = (e \cdot f)^{**} = (f^* \cdot e^*)^* = (f \cdot e)^* = f^* = f$). Clearly, 0_T and 1_T belong to $P(T)$ and behave as least and greatest elements. In the family of semi-groups with involution, of particular importance for our purposes are the so-called Baer *–semi-groups because of their relation with orthomodular lattices. A Baer *–semi-group is an involutive semi-group T with zero element and equipped with a further mapping $x \mapsto x'$ of T into $P(T)$, such that

$$\{y \in T : x \cdot y = 0_T\} = x' \cdot T , \qquad \forall x \in T . \tag{4.2}$$

In a Baer *–semi-group a projection e is called a closed projection whenever $e'' = e$. Now, the remarkable fact is that the set $P'(T)$ of the closed projections of a Baer *–semi-group T is an orthomodular lattice. More specifically,

i) the order relation in $P'(T)$ is the restriction of the order in $P(T)$, *i.e.* $e \leqslant f \Leftrightarrow e \cdot f = e$;

ii) the orthocomplementation in $P'(T)$ is the restriction of the mapping $x \mapsto x'$ defined in (4.2);

iii) the meet in $P'(T)$ has the form $e \wedge f = e \cdot (f' \cdot e)'$, and the join is obtained through De Morgan's laws;

iv) $e \cdot f = f \cdot e$ if and only if $e = (e \wedge f) \vee (e \wedge f')$; in other words, the commutativity in the sense of the semi-group structure coincides with the commutativity in the sense of the orthomodular lattice structure. We have also that $e \cdot f = f \cdot e$ if and only if the sublattice generated by $\{e, f, e', f'\}$ is distributive. When e, f commute their meet takes the form $e \wedge f = e \cdot f$.

Return now to the problem of the changes in the state of the physical system caused by the measurement of propositions. In general, every proposition can be measured by different experimental devices: our first assumption is that it be always possible to restrict oneself to experimental devices which do not destroy or annihilate the system, but change the initial state of the system into a well-defined final state, at least when the yes outcome occurs. More precisely, and having in mind some general trends of quantum theory of measurement [27], we postulate the possibility of restricting ourselves to the so-called first-kind, pure measurements, *i.e.* to those measurements which do not affect the state of the system if the yes outcome is certain, give with certainty the yes outcome in a replica of the same measurement if the outcome happened to be yes, and leave the system in a pure final state if the initial state was pure. It is worth-while to stress that an assumption of this kind, though already present in von Neumann's formalization of quantum mechanics and implicit in all interpretations of elementary quantum mechanics, is a very strong one and determines an idealized situation: indeed, it presupposes that physical system and measuring apparatus, initially separated, will become again separated after the interaction has taken place, a requirement which rules out all measuring apparata which form a bound state with the physical system.

The restriction to pure, first-kind measurements does not guarantee unicity: given $a \in \mathscr{L}$, we can imagine several pure, first-kind measurements, each of them defining a mapping of $\mathscr{S}$ into $\mathscr{S}$. Let $\{\Omega_a^i\}$ denote this family of mappings, i running in some index set: if the physical system approaches the i-th measuring apparatus in the state α, and produces the yes outcome, then $\Omega_a^i \alpha$ is meant to be the final state of the system.

We are now in a position to introduce a notion of commutativity between propositions: we say that a and b commute, and write aCb, if it is possible to select, in the families $\{\Omega_a^i\}$ and $\{\Omega_b^j\}$, two mappings, say Ω_a^k and Ω_b^l, such that

$$\Omega_a^k \cdot \Omega_b^l = \Omega_b^l \cdot \Omega_a^k ,$$

where $\cdot$ stands for the usual map composition.

Now we implement the previous hypotheses by assuming the possibility of selecting for each $a \in \mathscr{L}$ one mapping, to be simply denoted by Ω_a, in such a way that in the resulting set $\{\Omega_a : a \in \mathscr{L}\}$ *every* pair of commuting propositions is represented by a pair of commuting mappings. We shall refer to this property by saying that the set $\{\Omega_a : a \in \mathscr{L}\}$ is formed by « ideal » mappings. The extreme propositions $\underline{0}$ and $\underline{1}$ are so represented: $\Omega_{\underline{0}}$ is the improper map whose domain is empty, $\Omega_{\underline{1}}$ is the identical map whose domain is the whole $\mathscr{S}$. Symbolically, the fact that the Ω_a's are of first kind and pure is expressed by the following items, where $\mathscr{D}[\Omega_a]$ stands for the domain of Ω_a:

(4.3) $$\mathscr{D}[\Omega_a] = \{\alpha \in \mathscr{S} : \alpha(a) \neq 0\} ;$$

(4.4) $$\text{if} \quad \alpha \in \mathscr{D}[\Omega_a] \quad \text{and} \quad \alpha(a) = 1 \text{, then } \Omega_a \alpha = \alpha;$$

(4.5) $$\Omega_a \alpha(a) = 1 \text{ for every } \alpha \in \mathscr{D}[\Omega_a];$$

(4.6) $$\Omega_a \alpha \in \mathscr{P} \text{ if } \alpha \in \mathscr{P} \text{ and } \alpha \in \mathscr{D}[\Omega_a].$$

Moreover, the following property holds true: for any given pair of commuting propositions, say a and b, and for every state belonging to the intersection of the domains $\mathscr{D}[\Omega_a]$ and $\mathscr{D}[\Omega_b]$, $\alpha(b) = 1$ implies $\Omega_a \alpha(b) = 1$. In fact, from $\alpha(b) = 1$ and from the assumption that Ω_b is of first kind, we have $\alpha = \Omega_b \alpha$, hence $\Omega_a \alpha = \Omega_a \cdot \Omega_b \alpha = \Omega_b \cdot \Omega_a \alpha$ and $\Omega_a \alpha(b) = 1$: if the yes outcome of b is certain, then it remains so after the measurement (with yes response) of any proposition a commuting with b. This property motivates the adjective « ideal » previously attached to the Ω_a's. Let us remark that the ideality requirement can be used to eliminate some exotic ways of measuring propositions as, for instance, the ones of fig. 1.

Let us now make use of the previous mathematical digression on involutive semi-groups and ordered structures, following the approach proposed by Pool [28]. We do not assume, *a priori*, any ordered structure in $\mathscr{L}$. The elements of $\mathscr{S}$ are simply functions from $\mathscr{L}$ to the real interval $[0, 1]$, $\alpha(a)$ being interpreted as probability of the yes outcome of a when the initial state is α. $\mathscr{L}$ is in one-to-one correspondence with $\{\Omega_a : a \in \mathscr{L}\}$ and the Ω_a's satisfy (4.3)-(4.6). There is in $\mathscr{L}$ a mapping $a \mapsto a^\perp$ with $\alpha(a^\perp) = 1 - \alpha(a)$, $\forall\, \alpha \in \mathscr{S}$ (we remark, however, that the experimental device corresponding to $\Omega_{a^\perp}$ is not, in general, obtained by interchanging the « yes » and « no » readings on the device relative to Ω_a, for this would not lead to a first-kind device).

Let T_Ω be the set of the mappings of $\mathscr{S}$ into $\mathscr{S}$ obtained by composing the Ω_a's:

$$T_\Omega = \{\Omega_{a_1} \cdot \ldots \cdot \Omega_{a_n} : a_1, \ldots, a_n \in \mathscr{L},\ n \text{ finite}\} .$$

The elements of T_Ω are called operations. Note that $x \in T_\Omega$ is not, in general, a first-kind map (think of the composition of Nicol analysers).

T_Ω is a semi-group with respect to composition of maps; $\Omega_{\underline{0}}$ and $\Omega_{\underline{1}}$ are its zero and unity elements:

$$\Omega_{\underline{0}} \cdot x = x \cdot \Omega_{\underline{0}} = \Omega_{\underline{0}} \quad \text{and} \quad \Omega_{\underline{1}} \cdot x = x \cdot \Omega_{\underline{1}} = x \quad \text{for all } x \in T_\Omega .$$

Suppose now that the following rule holds in T_Ω [14, 28]:

(4.7) if $\quad \Omega_{a_1}\cdot\ldots\cdot\Omega_{a_n} = \Omega_{b_1}\cdot\ldots\cdot\Omega_{b_m} \quad$ then $\quad \Omega_{a_n}\cdot\ldots\cdot\Omega_{a_1} = \Omega_{b_m}\cdot\ldots\cdot\Omega_{b_1}$,

and note that it becomes obvious in the commutative case. Its impact on T_Ω is to make it a semi-group with involution. In fact, given $x \in T_\Omega$, choose any one of its realizations, say $\Omega_{a_1}\cdot\ldots\cdot\Omega_{a_n}$, and let $x^* = \Omega_{a_n}\cdot\ldots\cdot\Omega_{a_1}$. The mapping $x \mapsto x^*$ of T_Ω onto T_Ω is well defined and is involutive ($x^{**} = x$ and $(x\cdot y)^* = y^*\cdot x^*$).

As we know, the set

$$P(T_\Omega) = \{e \in T_\Omega : e\cdot e = e = e^*\}$$

of all projections of T_Ω is a poset relative to the order

$$e \leqslant f \Leftrightarrow e\cdot f = e (= f\cdot e), \qquad e, f \in P(T_\Omega).$$

Ω_0 and Ω_1 are the least and, respectively, the greatest element of $P(T_\Omega)$. The Ω_a's belong to $P(T_\Omega)$, but, in general, do not exhaust it. By restriction, the order in $P(T_\Omega)$ determines and order in $\{\Omega_a : a \in \mathscr{L}\}$:

(4.8) $$\Omega_a \leqslant \Omega_b \quad \text{whenever} \quad \Omega_a\cdot\Omega_b = \Omega_a,$$

and it may be proved that [14, 29]

(4.9) $$\Omega_a\cdot\Omega_b = \Omega_a \quad \text{if and only if} \quad \mathscr{S}_1(a) \subseteq \mathscr{S}_1(b).$$

By the bijection between $\{\Omega_a : a \in \mathscr{L}\}$ and $\mathscr{L}$ we are led to endow $\mathscr{L}$ with the order

(4.10) $$a \leqslant b \quad \text{whenever} \quad \mathscr{S}_1(a) \subseteq \mathscr{S}_1(b),$$

and thus obtain the order defined in (2.2) and therein commented.

We are now in a position to add a hint in favour of the equivalence between the orders (2.2) and (2.4). Consider the quantity $\alpha(b)\Omega_b\alpha(a)$: we want it to represent the probability of the yes outcome of a conditioned by the yes outcome of b, with initial state α; but to make consistent this interpretation we would require $\alpha(b)\Omega_b\alpha(a) = \alpha(a)$ when $a \leqslant b$ and this has the consequence [14] that $\mathscr{S}_1(a) \subseteq \mathscr{S}_1(b)$ implies $\alpha(a) \leqslant \alpha(b)$, $\forall\, \alpha \in \mathscr{S}$.

We ask now what additional hypothesis would make T_Ω a Baer *–semi-group. According to Pool [28], suppose that

(4.11) for every $x \in T_\Omega$ there exists an $a_x \in \mathscr{L}$ such that $\mathscr{D}[x] = \mathscr{D}[\Omega_{a_x}]$.

The unicity of a_x follows from the fact that $\mathscr{D}[\Omega_a] = \mathscr{D}[\Omega_b]$ implies $\mathscr{S}_1(a^\perp) = \mathscr{S}_1(b^\perp)$, hence $a^\perp = b^\perp$ and $a = b$. In the classical case the property (4.11) can be proved: thus it represents, together with (4.7), a relaxation of commutativity.

Due to (4.11) define the mapping

(4.12) $$x \mapsto x' = \Omega_{a_x^\perp}, \qquad x \in T_\Omega,\ a_x \in \mathscr{L},$$

of T_Ω into $P(T_\Omega)$: it is proved [14, 27] to make T_Ω a Baer *–semi-group. Moreover, the set $P'(T_\Omega)$ of all closed projections of T_Ω is found [14, 28] to coincide with $\{\Omega_a : a \in \mathscr{L}\}$. Recalling the properties of the closed projections of Baer *–semi-groups, and noticing that the one-to-one correspondence between $\{\Omega_a : a \in \mathscr{L}\}$ and $\mathscr{L}$ is order preserving (by (4.8)-(4.10)), we have:

1) $\{\Omega_a : a \in \mathscr{L}\}$ and $\mathscr{L}$ are isomorphic complete lattices.

2) By restriction of the mapping $x \mapsto x'$ defined in (4.12) we get an orthocomplementation in $\{\Omega_a : a \in \mathscr{L}\}$. Since $\Omega_a' = \Omega_{a^\perp}$, the mapping $a \mapsto a^\perp$ is an orthocomplementation in $\mathscr{L}$.

3) $\{\Omega_a : a \in \mathscr{L}\}$ and $\mathscr{L}$ satisfy the orthomodular identity; in $\mathscr{L}$ it reads $a \leqslant b \Rightarrow b = a \vee (b \wedge a^\perp)$.

4) The meet in $\{\Omega_a : a \in \mathscr{L}\}$ is given in terms of the operations in T_Ω by

$$\Omega_a \wedge \Omega_b = \Omega_a \cdot (\Omega_b' \cdot \Omega_a)'.$$

This formula cannot be translated in $\mathscr{L}$ because $\Omega_b' \cdot \Omega_a$ does not belong, in general, to $\{\Omega_a : a \in \mathscr{L}\}$.

5) Ω_a and Ω_b commute as maps ($\Omega_a \cdot \Omega_b = \Omega_b \cdot \Omega_a$) if and only if a and b commute in the sense of orthomodular lattices $\big(a = (a \wedge b) \vee (a \wedge b^\perp)\big)$. If $\Omega_a \cdot \Omega_b = \Omega_b \cdot \Omega_a$, then $\Omega_a \wedge \Omega_b = \Omega_a \cdot \Omega_b$.

The preceding arguments show how the notion of first-kind, ideal, pure measurements can lead to the structure of orthomodular lattice for the set $\mathscr{L}$ of propositions. As we know, two more properties of $\mathscr{L}$ are generally requested to account for the Hilbert-space formulation of quantum mechanics: $\mathscr{L}$ must be atomic and have the covering property. Thus, we shall briefly comment on how these properties fit with the operational approach outlined in this section. For simplicity, let us restrict to pure states, and still denote by $\mathscr{P}$ the set of pure states: this restriction does not produce a real loss of generality if we assume that nonpure states can be thought of as mixtures of pure states. We have seen in sect. **3** that, if $\mathscr{P}$ is endowed with a closure relation (physically interpreted as superposition of states), the set $L(\mathscr{P})$ of all closed subsets of $\mathscr{P}$ is a lattice which becomes atomistic under the quite natural hypothesis that the subsets formed by only one state be closed. Giving to the closure operation the form proposed by VARADARAJAN (see (3.9)), and with the help of assumption (3.11), we were led to recognize the isomorphism between $L(\mathscr{P})$ and $\mathscr{L}$: in this way $\mathscr{L}$ inherited the atomicity. In the context of the operational approach of this section, the statement contained in (3.11) is no longer required as an additional assumption, for it can be proved [14, 29]. Then, recalling (3.10), we get the isomorphism between $L(\mathscr{P})$ and $\mathscr{L}$, and we deduce, in particular, the atomicity of $\mathscr{L}$. As said in sect. **3**, this isomorphism is generally taken as representative of the superposition principle of quantum mechanics: the operational approach outlined in the present section embodies, without *ad hoc* assumptions, that form of the superposition principle.

A few words on the covering property. Assume again, as at the end of sect. **2**, that every $\alpha \in \mathscr{S}$ has a support in $\mathscr{L}$, that every $a \in \mathscr{L}$, $a \neq \underline{0}, \underline{1}$, is the support of some state, and that pure states are mapped, one-to-one, into the atoms of $\mathscr{L}$ by the support operation. Then it is natural to ask how the support of $\Omega_a \alpha$ can be expressed in terms of a and α; a partial answer is provided by the following inequality

$$\sigma(\Omega_a \alpha) \leqslant (\sigma(\alpha) \vee a^\perp) \wedge a, \qquad \forall\, \alpha \in \mathscr{D}[\Omega_a],$$

of which we omit the proof [14, 28]. Now, since the Ω_a's are assumed to map pure states into pure states, both $\sigma(\alpha)$ and $\sigma(\Omega_a \alpha)$ are atoms, so that the covering property is equivalent to the replacement of the inequality by the equality

$$\sigma(\Omega_a \alpha) = (\sigma(\alpha) \vee a^\perp) \wedge a \tag{4.13}$$

(we are referring to a formulation of the covering property which reads: $(p \vee a^\perp) \wedge a$ is an atom if p is an atom not contained in $a^\perp$).

Summing up, we have seen how the structure of orthomodular atomic (actually, atomistic) lattice with the covering property can be motivated, for the set $\mathscr{L}$, in the framework based on the state transformations caused by the measurement of propositions. Let us remark that we have not used, nor deduced, all basic properties of states: in particular, the property of being additive on disjoint elements

$$\alpha(a \vee b) = \alpha(a) + \alpha(b) \qquad \text{whenever} \qquad a \leqslant b^\perp$$

has never been encountered. As far as we know, it is not possible, in the scheme outlined in this section, to identify the elements of $\mathscr{S}$ with probability measures on $\mathscr{L}$, in the technical sense.

Now we reverse the problem so far considered in this section and ask whether the various requirements demanded to the Ω_a's, in order to make $\mathscr{L}$ an orthomodular, atomic lattice with the covering property, can be deduced as consequences of the structure of $\mathscr{L}$ and $\mathscr{S}$. The question has been answered affirmatively [29, 30]: here we shall briefly sum up the results. We adopt, as a starting mathematical structure, a pair $(\mathscr{L}, \mathscr{S})$ endowed with the following structure:

i) $\mathscr{L}$ is an orthomodular atomic lattice with the covering property;

ii) $\mathscr{S}$ is a σ-convex set of probability measures on $\mathscr{L}$, and is connected with the order relation of $\mathscr{L}$ through

$$a \leqslant b \quad \text{if and only if} \quad \mathscr{S}_1(a) \subseteq \mathscr{S}_1(b)$$

(*i.e.* $\mathscr{S}$ is strongly ordering on $\mathscr{L}$);

iii) $\mathscr{S}$ contains a nonempty subset $\mathscr{P}$ of pure states and every nonpure state is a convex combination of pure states;

iv) each $\alpha \in \mathscr{S}$ has support in $\mathscr{L}$, each $a \in \mathscr{L}$, $a \neq \underline{0}, \underline{1}$, is the support of some $\alpha \in \mathscr{S}$, and the support operation maps bijectively $\mathscr{P}$ onto the set of all atoms of $\mathscr{L}$.

Then, again by use of the theory of Baer *–semi-groups, it is possible to prove that, for every $a \in \mathscr{L}$, there exists a unique mapping $\Omega_a : \mathscr{S} \to \mathscr{S}$ which satisfies the properties (4.3)-(4.7), (4.12), (4.13). Moreover, these Ω_a's are proved to satisfy also the property

$$\text{if} \quad \Omega_a \cdot \Omega_b = \Omega_b \cdot \Omega_a , \quad \text{then} \quad \Omega_a \alpha(b) = \Omega_a \alpha(a \wedge b) ,$$

which is necessary to get, out of the Ω_a's, conditional probabilities in the technical sense.

Thus we see that all the properties of the Ω_a's that we assumed in the first part of this section, to motivate the structure of $\mathscr{L}$ and $\mathscr{S}$, can be recovered in the inverse problem, *i.e.* they are deduced from the detailed structure of the pair $(\mathscr{L}, \mathscr{S})$.

5. – Transition probability spaces.

We have neglected, in so far, an important notion: the one of transition probability. In the Hilbert-space formulation of quantum mechanics it finds place in the inner product of the Hilbert space $\mathscr{H}$, and thus takes the form of a particular mapping of pairs of states into the interval $[0, 1]$. This inner product is the key of the probabilistic interpretation of the theory. In the traditional quantum logic approach, in which the structure of $\mathscr{S}$, thought of as the set of the probability measures on $\mathscr{L}$, is mainly determined by the properties of the lattice $\mathscr{L}$, we can individuate the transition probability between two states α and β in the quantity $\alpha(\sigma(\beta))$, where $\sigma(\beta)$ is the support of β. In fact, this quantity admits the desired physical interpretation: if we restrict ourselves, for simplicity, to pure states, $\alpha(\sigma(\beta))$ is the probability of getting the yes outcome of the atomic proposition representing β when the incoming state is α. We can also say that $\alpha(\sigma(\beta))$ is the attenuation produced by the « filter » associated to β on a beam emerging from the « filter » associated to α.

A question can now be raised [6]: does the representation theorem (which represents the elements of $\mathscr{P}$ by projectors in a separable Hilbert space) guarantee the transition probability to be numerically equal to the modulus squared of the inner product between the two vectors representing the given states? This question can be answered affirmatively by recourse to Gleason's theorem, but this goes beyond our purposes.

Our aim, in this section, is to introduce in the set of states, without any reference to Hilbert-space structures, a minimal notion of transition probability, and explore its impact on the geometry of states. We shall first do that leaving aside $\mathscr{L}$ and its properties, thus using states as primitive undefined concepts.

We restrict ourselves to pure states and still denote by $\mathscr{P}$ the (nonempty) set they form. Consider then a mapping $\langle\cdot|\cdot\rangle$ of $\mathscr{P}\times\mathscr{P}$ into $[0, 1]$ satisfying the following properties [6, 8]:

i) $\langle\cdot|\cdot\rangle$ is separating, *i.e.* $\langle\alpha|\beta\rangle = 1$ if and only if $\alpha = \beta$;

ii) $\langle\cdot|\cdot\rangle$ is symmetric, *i.e.*, for every α, $\beta\in\mathscr{P}$, $\langle\alpha|\beta\rangle = \langle\beta|\alpha\rangle$;

iii) if we call α, β orthogonal when $\langle\alpha|\beta\rangle = 0$, every maximally pairwise orthogonal subset B of $\mathscr{P}$ is a basis, *i.e.*

$$\sum_{\beta\in B}\langle\alpha|\beta\rangle = 1 \quad \text{for every } \alpha\in\mathscr{P}.$$

The set $\mathscr{P}$ equipped with a mapping $\langle\cdot|\cdot\rangle$ satisfying i)-iii) is called a transition probability space.

As a trivial example of transition probability space consider the mapping $\langle\cdot|\cdot\rangle$ defined by $\langle\alpha|\beta\rangle = 0$ if $\alpha \neq \beta$, and $\langle\alpha|\alpha\rangle = 1$; this example corresponds to the classical case, with $\mathscr{P}$ standing for the phase space of the classical mechanical system.

We have at our disposal a notion of orthogonality between states: α is orthogonal to β if and only if $\langle\alpha|\beta\rangle = 0$. Thus, we define the orthogonal complement of any subset A of $\mathscr{P}$ as the set of all states orthogonal to every state of A; we write A' for this orthogonal complement, so that, symbolically,

$$A' = \{\beta\in\mathscr{P} : \langle\alpha|\beta\rangle = 0 \text{ for every } \alpha\in A\}.$$

The following assertions are obvious:

(5.1) $$A \cap A' = \emptyset\,, \qquad (A \cup A')' = \emptyset\,;$$

(5.2) $$\text{if} \quad A \subseteq B \quad \text{then} \quad B' \subseteq A'\,;$$

(5.3) $$A \subseteq A''\,, \qquad A' = A'''\,.$$

We consider now the mapping $A \mapsto A''$ and remark that, owing to the properties just mentioned, it is a closure relation in $\mathscr{P}$; indeed, besides (5.3), we need $A \subseteq B'' \Rightarrow A'' \subseteq B''$, which is an immediate consequence of (5.2) and (5.3).

Therefore, once a transition probability space is given, a parallel structure of closure space is naturally generated.

As seen in sect. **3**, a closure space is, in turn, naturally associated to lattice structures. To exploit this fact we consider those subsets of $\mathscr{P}$ which are equal to their own closures, namely those subsets which satisfy the condition $A = A''$, and call them orthoclosed subsets, to remind the particular closure operation we are dealing with. The empty subset $\emptyset$ and the whole $\mathscr{P}$ are obviously orthoclosed. It is then possible to transplant into the present situation what we have seen in sect. **3** about the lattice of the closed subsets of a closure space: the set $M(\mathscr{P})$ of all orthoclosed subsets of $\mathscr{P}$ is a lattice with order relation given by set-theoretic inclusion, the empty set $\emptyset$ and the whole $\mathscr{P}$ are the least and, respectively, the greatest elements of $M(\mathscr{P})$, meet and join are given by

(5.4) $$A \wedge B = A \cap B\,, \qquad A, B \in M(\mathscr{P}),$$

(5.5) $$A \vee B = (A \cup B)''\,, \qquad A, B \in M(\mathscr{P}).$$

In $M(\mathscr{P})$ the mapping $A \mapsto A'$ is an orthocomplementation in the sense of poset theory: indeed, $A'' = A$, $A \leqslant B \Rightarrow B' \leqslant A'$, and $A \vee A' = (A \cup A')'' = \mathscr{P}$ by use of (5.1). Therefore, $M(\mathscr{P})$ is an orthocomplemented lattice.

However, $M(\mathscr{P})$ is not the only ordered structure naturally attached to the transition probability space $(\mathscr{P}, \langle\cdot|\cdot\rangle)$. To show what the other one is, we need first a definition [6, 8]. A subset A of a transition probability space is said to be a subspace if one obtains a transition probability on A by restricting $\langle\cdot|\cdot\rangle$ to $A \times A$. Since $\langle\cdot|\cdot\rangle$ is obviously separating and symmetric when restricted to $A \times A$, the only nontrivial requirement contained in the definition is

(5.6) A is a subspace if and only if every maximally pairwise orthogonal subset B of A is a basis for A, *i.e.*

$$\sum_{\beta \in B} \langle \alpha | \beta \rangle = 1 \text{ for every } \alpha \in A\,.$$

Of course, if A is a subspace of $(\mathscr{P}, \langle\cdot|\cdot\rangle)$, then $(A, \langle\cdot|\cdot\rangle)$ is a transition probability space of its own. The following is a relevant property of subspaces [6, 8]:

(5.7) if $(A, \langle\cdot|\cdot\rangle)$ is a transition probability space, then all bases of A have the same cardinality.

We take this common cardinality as the « dimension », or « rank », of A. Now, following BELINFANTE [8], we restrict ourselves to those subspaces of $(\mathscr{P}, \langle\cdot|\cdot\rangle)$ which are orthoclosed: this is, *a priori*, an effective restriction, for subspaces do not need to be orthoclosed, nor orthoclosed subsets need to be subspaces.

If B is a basis for the subspace A, then $A' = B'$, *i.e.* A and B have the same orthogonal complement [8]. Indeed, if $\gamma \in B'$, then $B \cup \{\gamma\}$ is a pairwise orthogonal subset, so that $\langle\alpha|\gamma\rangle + \sum_{\beta\in B} \langle\alpha|\beta\rangle \leqslant 1$ for every $\alpha \in \mathscr{P}$; taking $\alpha \in A$ we have $\sum_{\beta\in B} \langle\alpha|\beta\rangle = 1$, hence $\langle\alpha|\gamma\rangle = 0$, and $\gamma \in A'$. Thus $B' \subseteq A'$; but we have also $A' \subseteq B'$ because $B \subseteq A$, therefore $A' = B'$.

We prove now that, if A is a subspace, then also A' is a subspace [8]. According to the previous result, it suffices to prove that B' is a subspace if B is pairwise orthogonal. Let C be a maximal pairwise orthogonal subset of B': then $B \cup C$ is a basis for the whole $\mathscr{P}$ so that, for every $\alpha \in \mathscr{P}$, $\sum_{\beta\in B} \langle\alpha|\beta\rangle + \sum_{\gamma\in C} \langle\alpha|\gamma\rangle = 1$. Take now any state α in B'; we have $\sum_{\beta\in B} \langle\alpha|\beta\rangle = 0$, hence $\sum_{\gamma\in C} \langle\alpha|\gamma\rangle = 1$, showing that B' is a subspace.

We denote by $N(\mathscr{P})$ the set of all orthoclosed subspaces of $(\mathscr{P}, \langle\cdot|\cdot\rangle)$. The empty subset and the whole $\mathscr{P}$ are trivial examples of orthoclosed subspaces; as less trivial example, take, for any given $\alpha \in \mathscr{P}$, the singleton $\{\alpha\}$: it is evidently a subspace and is also orthoclosed, indeed $\{\alpha\}$ is a basis for $\{\alpha\}''$, so that $\langle\alpha|\beta\rangle = 1$ for all $\beta \in \{\alpha\}''$, hence $\alpha = \beta$.

We make $N(\mathscr{P})$ a poset endowing it with the order relation given by set-theoretic inclusion. The empty subset $\emptyset$ and the whole $\mathscr{P}$ behave as least and, respectively, greatest elements of $N(\mathscr{P})$. The singletons $\{\alpha\}$, $\alpha \in \mathscr{P}$, exhaust the atoms of $N(\mathscr{P})$, and every element of $N(\mathscr{P})$ majorizes one atom at least: therefore $N(\mathscr{P})$ is an atomic poset. Moreover, according to the facts proved before, the function $A \mapsto A'$ maps $N(\mathscr{P})$ into itself. As in the case of $M(\mathscr{P})$, we are thus allowed to assert that the mapping $A \mapsto A'$ is an orthocomplementation in $N(\mathscr{P})$, in the sense of poset theory. Then, we can introduce in $N(\mathscr{P})$ the notion of disjointness: $A, B \in N(\mathscr{P})$ are said to be disjoint, and we write $A \perp B$, if and only if $A \leqslant B'$.

Meet and join, when they exist, are still given by (5.4) and (5.5); the caution on their existence being due to the fact that the right-hand side of (5.4) and (5.5) is certainly orthoclosed but not necessarily a subspace. BELINFANTE [8] proves a further property of $N(\mathscr{P})$:

(5.8) $N(\mathscr{P})$ is an orthomodular poset.

Proof [8]. Since $N(\mathscr{P})$ is orthocomplemented, we have only to prove that the join of disjoint elements exists, and that the orthomodular identity

$$A \leqslant B \Rightarrow B = A \vee (B \wedge A') = \big(A \cup (B \cap A')\big)''$$

holds true. Let A_1, A_2 be disjoint elements of $N(\mathscr{P})$, and let B be a maximally pairwise orthogonal subset of $A_1 \cup A_2$: we have

$$B' = \big((B \cap A_1) \cup (B \cap A_2)\big)' = (B \cap A_1)' \cap (B \cap A_2)' = A_1' \cap A_2' = (A_1 \cup A_2)'$$

and we know that B' is a subspace, hence $B'' = (A_1 \cup A_2)''$ is again a subspace and is, indeed, the join of A_1 and A_2. Come now to the orthomodular identity. Let A, $B \in N(\mathscr{P})$ with $A \leqslant B$, and prove first that $B \cap A' \in N(\mathscr{P})$. If C is a maximally pairwise orthogonal subset of $B \cap A'$, and if D is a basis for A, we write

$$B \cap (C \cup D)' = B \cap D' \cap C' = (B \cap A') \cap C' = \emptyset$$

and conclude that $C \cup D$ is a basis for B. If $\alpha \in B \cap A'$, then

$$1 = \sum_{\beta \in C \cup D} \langle \alpha | \beta \rangle = \sum_{\beta \in C} \langle \alpha | \beta \rangle ,$$

so that C is a basis for $B \cap A'$, and $B \cap A' \in N(\mathscr{P})$. We remark now that $A \leqslant B$ implies the disjointness of A and $B \cap A'$, so that $C \cup D$ is a basis for $A \cup (B \cap A')$; recalling that $C \cup D$ is also a basis for B, we have $B = (C \cup D)'' = \big(A \cup (B \cap A')\big)''$. *q.e.d.*

Since we have obtained, for $N(\mathscr{P})$, the structure of orthomodular poset, it becomes meaningful to ask if the states can be viewed as probability measures on $N(\mathscr{P})$. The answer is affirmative:

(5.9) $\mathscr{P}$ is a strongly ordering set of probability measures on $N(\mathscr{P})$.

Proof. Let $A \in N(\mathscr{P})$, $A \neq \emptyset$, and let B be a basis for A. For given $\alpha \in \mathscr{P}$, put, with some abuse of notation,

$$\alpha(A) = \sum_{\beta \in B} \langle \alpha | \beta \rangle . \tag{5.10}$$

To prove that the r.h.s. is independent of the particular choice of the basis, let B_1 be another basis for A, and let C be a basis for A': then remark that

$B \cup C$ as well as $B_1 \cup C$ are bases for $\mathscr{P}$ so that, for every $\alpha \in \mathscr{P}$,

$$\sum_{\beta \in B} \langle\alpha|\beta\rangle = 1 - \sum_{\beta \in C} \langle\alpha|\beta\rangle = \sum_{\beta \in B_1} \langle\alpha|\beta\rangle .$$

Implementing (5.10) with $\alpha(\emptyset) = 0$, we see that α determines a function from $N(\mathscr{P})$ into the real interval $[0, 1]$. Now, let A_1, A_2 be disjoint elements of $N(\mathscr{P})$, and let B_1 and B_2 be bases for A_1 and A_2, respectively. $B_1 \cup B_2$ is a pairwise orthogonal subset, thus it is a basis for $(B_1 \cup B_2)''$, since $(B_1 \cup B_2)' \cap \cap (B_1 \cup B_2)'' = \emptyset$. But

$$(B_1 \cup B_2)'' = (B_1' \cap B_2')' = (A_1' \cap A_2')' = (A_1 \cup A_2)'' = A_1 \vee A_2 ,$$

therefore $B_1 \cup B_2$ is a basis for $A_1 \vee A_2$, and $\alpha(A_1 \vee A_2) = \sum_{\beta \in B_1 \cup B_2} \langle\alpha|\beta\rangle = \alpha(A_1) + \alpha(A_2)$. Thus $\mathscr{P}$ is a set of probability measures on $N(\mathscr{P})$. To prove that $\mathscr{P}$ is strongly ordering on $N(\mathscr{P})$, it suffices to notice that $\{\alpha \in \mathscr{P} : \alpha(A) = 1\} = A$. *q.e.d.*

As the preceding analysis shows, to a transition probability space $(\mathscr{P}, \langle\cdot|\cdot\rangle)$ we may associate, in a natural way, two ordered structures: the orthomodular atomistic poset $N(\mathscr{P})$ and the orthocomplemented lattice $M(\mathscr{P})$. In the Hilbert-space model of quantum mechanics $\mathscr{P}$ becomes the set of rays of a separable Hilbert space $\mathscr{H}$, $\langle\cdot|\cdot\rangle$ becomes the scalar product (modulus squared), and both $M(\mathscr{P})$ and $N(\mathscr{P})$ become the lattice $\mathscr{L}(\mathscr{H})$ of all closed (in the sense of Hilbert spaces) subspaces of $\mathscr{H}$. Physically, the coincidence $M(\mathscr{P}) = N(\mathscr{P}) = \mathscr{L}(\mathscr{H})$ can be interpreted as follows: a subset A of states is closed in the sense of Hilbert spaces if an only if it is closed in the sense of superpositions (*i.e.* if an only if it contains all the superpositions of its states), but a superposition of the states of A can equivalently be characterized as an element of the subspace generated by A, or as an element of the double orthogonal complement A''. Returning to our transition probability space, and viewing the closure relation leading to $M(\mathscr{P})$ and the subspace condition leading to $N(\mathscr{P})$ as two aspects of the physical notion of state superposition, we have a hint to advance the following additional axiom:

(5.11) $M(\mathscr{P})$ and $N(\mathscr{P})$ have the same elements.

The effect of this axiom is, of course, of identifying the orthocomplemented lattice $M(\mathscr{P})$ and the atomic orthomodular poset $N(\mathscr{P})$ with a unique ordered structure, say $\mathscr{L}(\mathscr{P})$, which is an atomistic orthomodular lattice, and $\mathscr{P}$ is a strongly ordering set of probability measures on $\mathscr{L}(\mathscr{P})$. We have constructed a quantum logic using only a transition probability space and (5.11) (to have a lattice rather than simply a poset). Actually, comparing with the usual structure adopted as basis for the representation theorem, we notice that the covering property is missing for $\mathscr{L}(\mathscr{P})$. Remark, however, that the main effect of the covering property is to allow a definition of dimension function and of basis for the elements of the orthomodular lattice [14], and these notions are already present in $\mathscr{L}(\mathscr{P})$. We leave open the question of whether a representation theorem might be worked out for $\mathscr{L}(\mathscr{P})$ even in the absence of the covering property. Anyhow, the natural way of ensuring that property would be to assume that the closure operation $A \mapsto A''$, $A \in \mathscr{L}(\mathscr{P})$, satisfy the exchange property

$$\text{if} \quad \alpha \in (A \cup \{\beta\})'' \text{ and } \quad \alpha \notin A'', \quad \text{then} \quad \beta \in (A \cup \{\alpha\})''$$

(notice that, in the Hilbert-space model of quantum mechanics, the exchange property asserts the commutativity of the addition operation of vectors).

Also the notion of support of a (pure) state finds a natural place in the framework of transition probability spaces. Indeed, if $\alpha \in \mathscr{P}$, it is clear that the corresponding atom $\{\alpha\}$ of $\mathscr{L}(\mathscr{P})$ is the support $\sigma(\alpha)$ of α, for $\{A \in \mathscr{L}(\mathscr{P}): :\alpha(A) = 1\} = \{A \in \mathscr{L}(\mathscr{P}): A \geqslant \{\alpha\}\}$. Moreover, we have, by (5.10),

$$\alpha(\sigma(\beta)) = \langle \alpha | \beta \rangle . \tag{5.12}$$

Remark that the symmetry property $\alpha(\sigma(\beta)) = \beta(\sigma(\alpha))$ is here ensured by the properties of the transition probability, while it is not, generally, included in the properties of the conventional propositions-states structure (the symmetry is however untenable if nonpure states are considered). Remark also that, using (5.12) as definition of the transition probability, and postulating the symmetry property, one can reverse the itinerary of this section and deduce, from the conventional $(\mathscr{L}, \mathscr{P})$ structure, a transition probability space structure.

Now, we can use (5.12) to express with different words the orthoclosure relation $A \mapsto A''$ holding in $\mathscr{P}$. Given $A \subseteq \mathscr{P}$, we have, noticing that $\alpha(\sigma(\beta)) = 0$ is equivalent to $\sigma(\beta) \leqslant \sigma(\alpha)'$,

$$A' = \Big\{\beta \in \mathscr{P} : \sigma(\beta) \leqslant \bigwedge_{\alpha \in A} \sigma(\alpha)'\Big\},$$

$$A'' = \Big\{\gamma \in \mathscr{P} : \sigma(\gamma) \leqslant \bigwedge_{\beta \in A'} \sigma(\beta)'\Big\}.$$

Put $a = \bigwedge_{\alpha \in A} \sigma(\alpha)'$, $b = \bigvee_{\beta \in A'} \sigma(\beta)$, and observe that $a = b$: indeed $b \leqslant a$, but the strict inequality would imply the existence of $c \in \mathscr{L}(\mathscr{P})$ such that $c \perp b$ and $b \vee c = a$, so that any atom in c would be majorized by a, *i.e.* it would be support of a state $\bar{\beta}$ with $\sigma(\bar{\beta}) < a$, hence $\sigma(\bar{\beta}) < b$, contrary to the hypothesis $c \perp b$. Then we rewrite A'' in the form

$$A'' = \Big\{\gamma \in \mathscr{P} : \sigma(\gamma) \leqslant \Big(\bigvee_{\beta \in A'} \sigma(\beta)'\Big)\Big\} = \Big\{\gamma \in \mathscr{P} : \sigma(\gamma) \leqslant \bigvee_{\alpha \in A} \sigma(\alpha)\Big\}.$$

Since $\sigma(\gamma) \leqslant \bigvee_{\alpha \in A} \sigma(\alpha)$ is fully equivalent, by definition of support, to $\mathscr{L}_1(\gamma) \supseteq$

$\supseteq \bigcap_{\alpha \in A} \mathscr{L}_1(\alpha)$, we conclude that the orthoclosed subsets are precisely the subsets which are closed with respect to Varadarajan's notion of state superposition (as in sect. **3**, $\mathscr{L}_1(\alpha)$ stands for $\{A \in \mathscr{L}(\mathscr{P}) : \alpha(A) = 1\}$).

DELIYANNIS [31] has shown how to characterize, in the framework of transitions probability spaces, those mappings of $\mathscr{P}$ into $\mathscr{P}$ which can be interpreted as state transformations accompanying the measurement of propositions. These mappings organize themselves into an ordered structure, similarly to what happened to the Ω_a's discussed in sect. **4**. One could also approach this problem starting with the ordered structure of $\mathscr{L}(\mathscr{P})$, and use the arguments sketched at the end of sect. **4** to find out the mappings of states one is looking for [32].

Transition probability spaces admit a natural topology associated to the following definition of « distance » between two (pure) states α, β [8]:

$$d(\alpha, \beta) = \sup \{|\langle\alpha|\gamma\rangle - \langle\beta|\gamma\rangle| : \gamma \in \mathscr{P}\} . \tag{5.13}$$

We remark, however, that there are several possibilities of endowing the set of states with a notion of distance. For instance, in the conventional $(\mathscr{L}, \mathscr{S})$ structure, one can define [14, 33, 34]

$$d(\alpha, \beta) = \sup \{|\alpha(a) - \beta(a)| : a \in \mathscr{L}\} ,$$

of which (5.13) represents a restriction to pure states and to the atoms of $\mathscr{L}$. Another possibility, based on the convex nature of the set $\mathscr{S}$ of states, is the following [34, 35]:

$$d(\alpha, \beta) = \inf \{0 \leqslant t \leqslant 1 : (1-t)\alpha + t\alpha_1 = (1-t)\beta + t\beta_1,\ \alpha_1, \beta_1 \in \mathscr{S}\} .$$

We refer also to CANTONI [36] for the analysis of other possibilities.

In Mielnik's convex scheme of quantum mechanics [6, 16] a generalized notion of transition probability (the symmetry requirement is missing) appears to be derivable from more primitive concepts. In that scheme the ordered structures one naturally generates are of unconventional type, mainly because they are not necessarily orthocomplemented; thus there are drastic departures from what is usually meant to be a « quantum logic » approach to quantum mechanics. A discussion of the convex scheme goes beyond our present purposes.

6. – Gleason's theorem and exceptional states.

6·1. – In the previous sections the notion of state has repeatedly taken the status of probability measure on the orthomodular poset (lattice) $\mathscr{L}$ formed by the propositions. The question of whether this notion of state is fully corresponding to the one encountered in ordinary quantum mechanics has found an almost complete answer in Gleason's theorem, one of the cornerstones of the mathematical foundations of quantum mechanics. Indeed, this theorem asserts that, except for particular cases, the states defined as probability measures on the orthomodular lattice $\mathscr{L}$ coincide with the von Neumann density operators of ordinary quantum mechanics. The exceptional cases, on which Gleason's theorem has nothing to say, are the systems with two-dimensional space of states, say the spin-$\frac{1}{2}$ particle without translational motion. We shall now take this simple physical system to show explicitly that, for it, the states defined as probability measures on its proposition lattice are indeed « much more » than the states considered by Hilbert-space quantum mechanics (that is Pauli spinors and density matrices). These states, which have no image in the Hilbert-space picture, will be called exceptional, and their appearance is found to be responsible of some unusual mathematical features of these systems. Of course, due to the very content of Gleason's theorem, exceptional states can occur only for the two-dimensional case, say for the spin-$\frac{1}{2}$ system.

We shall also show that the models of hidden-variable (h.v.) theories for the spin-$\frac{1}{2}$ system considered by BELL [37] and by KOCHEN and SPECKER [38] make precisely use of the exceptional states. Admittedly, the discussion of this last point will make the present section somewhat aside from the spirit of the previous ones.

It seems that the subject of h.v. theories is an intriguing one and that opinions on it have been influenced by ideological or philosophical prejudices. We do not claim to have new arguments neither in favour nor against h.v.: our aim is simply to show that some models of h.v. theories can describe only spin-$\frac{1}{2}$ systems and that they cannot be generalized, maintaining all the properties they have in this case.

6'2. – As seen in the previous sections, there are strong motivations to describe a physical system by means of an ordered structure $\mathscr{L}$ (at least an orthomodular poset), whose elements are the propositions, and a set $\mathscr{S}$ of probability measures defined on it, representing the set of states. A scheme of this kind looks general enough to include both classical and quantum theories.

Let us briefly summarize how the notions of observable and distribution of an observable in a state can fit into this $(\mathscr{L}, \mathscr{S})$ framework.

An observable of a physical system is defined as a mapping x from $\mathscr{B}(\mathbf{R})$ (the Boolean σ-algebra of real Borel sets) to $\mathscr{L}$, such that

$$x(\emptyset) = \underline{0}\,, \qquad x(\mathbf{R}) = \underline{1}\,,$$

$$\text{if } E_1 \text{ and } E_2 \text{ are disjoint Borel sets, then } x(E_1)\perp x(E_2)\,,$$

$$\text{if } \{E_i\} \text{ is a sequence of disjoint Borel sets then } x(\cup E_i) = \vee x(E_i)\,.$$

Recall that a state is defined as a function α from $\mathscr{L}$ into $[0, 1]$ such that

$$\alpha(\underline{0}) = 0\,, \qquad \alpha(\underline{1}) = 1\,,$$

$$\alpha(\vee a_i) = \sum \alpha(a_i)\,.$$

If the observable x is measured in the state α, then the probability that the observed value of x is an element of the (Borel) set E is given by $\alpha(x(E))$; this rule is the link between the abstract mathematical structures above defined and the empirical properties of physical systems. The probability measure on the reals $\alpha(x(E))$ is called the distribution of x in the state α.

As already remarked, the usual quantum description of a physical system is recovered identifying $\mathscr{L}$ with the orthomodular lattice $\mathscr{L}(\mathscr{H})$ of projection operators on a complex separable Hilbert space $\mathscr{H}$. In this case the definition of an observable on $\mathscr{L}$ becomes the definition of a spectral measure [3] and, owing to the spectral theorem [3], there is a one-to-one correspondence between self-adjoint operators in $\mathscr{H}$ and spectral measures.

The states become probability measures on $\mathscr{L}(\mathscr{H})$ and Gleason's theorem [21] ensures us that every such probability measure α, if $\dim \mathscr{H} \geqslant 3$, has the form

$$\alpha(P) = \operatorname{tr}(D_\alpha P) \qquad \text{for all} \qquad P \in \mathscr{L}(\mathscr{H})\,,$$

where D_α is a trace class operator of unit trace called the density of the state α. In the case that α is a pure state, the preceding formula is greatly simplified; in fact, it is easy to see that, if α is a pure probability measure on $\mathscr{L}(\mathscr{H})$, it has the form

$$\alpha(P) = (P\varphi, \varphi) \qquad \text{for all} \qquad P \in \mathscr{L}(\mathscr{H})\,,$$

where $(.\,,.)$ is the scalar product in $\mathscr{H}$ and φ is an element of $\mathscr{H}$ such that $\|\varphi\| = 1$. $\psi \in \mathscr{H}$, such that $\|\psi\| = 1$, defines the same state as φ if and only if $\varphi = c\psi$, where c is a complex number and $|c| = 1$. Every nonpure state can be written as a convex linear combination of pure ones, that is

$$\alpha(P) = \sum_i t_i(P\varphi_i, \varphi_i) \qquad \text{for all} \qquad P \in \mathscr{L}(\mathscr{H})\,,$$

where $\{\varphi_i\}$ is a sequence of unit vectors in $\mathscr{H}$ and $\{t_i\}$ is a sequence of positive real numbers such that $\sum_i t_i = 1$.

Due to this fact, we can restrict our attention to pure states: by means of the spectral theorem the functional $P \mapsto (P\varphi, \varphi)$ can be extended uniquely to a linear positive functional on the self-adjoint operators on $\mathscr{H}$, defined by $A \mapsto (A\varphi, \varphi)$ which is called the expectation of A, and whose physical meaning is the mean value of the possible outcomes of measurements of A in the state represented by φ. Every bounded operator T in $\mathscr{H}$ can be written uniquely as $T = T_1 + iT_2$, where

$$T_1 = \frac{T + T^*}{2} \qquad \text{and} \qquad T_2 = i\,\frac{T - T^*}{2}$$

are self-adjoint, then a linear functional on self-adjoint operators can be extended in a unique way to a linear functional on the von Neumann algebra $\mathscr{B}(\mathscr{H})$. Concluding, Gleason's theorem ensures us that we can think of quantum-mechanical states either as probability measures on $\mathscr{L}(\mathscr{H})$ or as normal positive linear functionals on $\mathscr{B}(\mathscr{H})$ if $\dim \mathscr{H} \geqslant 3$; we want to stress that this highly nontrivial result is obtained only by supposing that states are probability measures.

Of course, when $\dim \mathscr{H} = 2$, Gleason's theorem cannot be used, and we cannot say that probability measures on $\mathscr{L}(\mathscr{H}_2)$ are necessarily restrictions of linear functionals; in fact, in the sequel we shall explicitly construct examples of probability measures which do not extend to linear functionals on $\mathscr{B}(\mathscr{H}_2)$.

6·3. – The self-adjoint elements in $\mathscr{B}(\mathscr{H}_2)$, that is the Hermitean two-by-two complex matrices, have the form

$$A = aI + \boldsymbol{b}\cdot\boldsymbol{\sigma}\,,$$

where I is the unit two-by-two matrix, $a \in \mathbf{R}$, $\boldsymbol{b} \in \mathbf{R}^3$ and $\boldsymbol{\sigma}$ is a vector whose components are the Pauli spin matrices

$$\sigma_1 = \begin{pmatrix} 0 & 1 \\ 1 & 0 \end{pmatrix}, \qquad \sigma_2 = \begin{pmatrix} 0 & -i \\ i & 0 \end{pmatrix}, \qquad \sigma_3 = \begin{pmatrix} 1 & 0 \\ 0 & -1 \end{pmatrix}.$$

It is easy to see that projections (that is Hermitean matrices such that $A^2 = A$) are the matrices of the form $\frac{1}{2}I + \frac{1}{2}\boldsymbol{u}\cdot\boldsymbol{\sigma}$ (where $|\boldsymbol{u}| = 1$), besides I and the zero matrix. Then nontrivial projectors in $\mathscr{H}_2$ are in one-to-one correspondence with the points of the surface of a sphere of radius 1, this correspondence being defined by

$$\tfrac{1}{2}I + \tfrac{1}{2}\boldsymbol{u}\cdot\boldsymbol{\sigma} \mapsto (u_1, u_2, u_3)\,.$$

By a straightforward, but somewhat tedious, calculation we can find that the necessary and sufficient condition for two projectors $\frac{1}{2}I + \frac{1}{2}\boldsymbol{u}_1\cdot\boldsymbol{\sigma}$ and $\frac{1}{2}I + \frac{1}{2}\boldsymbol{u}_2\cdot\boldsymbol{\sigma}$ being orthogonal (that is $(\frac{1}{2}I + \frac{1}{2}\boldsymbol{u}_1\cdot\boldsymbol{\sigma})(\frac{1}{2}I + \frac{1}{2}\boldsymbol{u}_2\cdot\boldsymbol{\sigma}) = 0$) is $\boldsymbol{u}_1 = -\boldsymbol{u}_2$.

The eigenvalues of the matrix $aI + \boldsymbol{b}\cdot\boldsymbol{\sigma}$ are given by $a \pm |\boldsymbol{b}|$ and it is not difficult to show that its spectral decomposition is

$$aI + \boldsymbol{b}\cdot\boldsymbol{\sigma} = (a + |\boldsymbol{b}|)\left(\frac{1}{2}I + \frac{1}{2}\frac{\boldsymbol{b}}{|\boldsymbol{b}|}\cdot\boldsymbol{\sigma}\right) + (a - |\boldsymbol{b}|)\left(\frac{1}{2}I - \frac{1}{2}\frac{\boldsymbol{b}}{|\boldsymbol{b}|}\cdot\boldsymbol{\sigma}\right).$$

Let ψ be any vector in $\mathscr{H}_2$, $\|\psi\| = 1$, and P any projection, by a change of base we can put ψ in the form $\begin{pmatrix}1\\0\end{pmatrix}$; let the projector P in this new base have the form $\frac{1}{2}I + \frac{1}{2}\boldsymbol{u}\cdot\boldsymbol{\sigma}$, then we can define a functional α_ψ from the set of projection operators into the real interval $[0, 1]$, in the following way

$$\alpha_\psi(P) = \tfrac{1}{2} + \tfrac{1}{2}\,\mathrm{sign}\, x\,,$$

where $\mathrm{sign}\, x = 1$ if $x \geqslant 0$, $\mathrm{sign}\, x = -1$ if $x < 0$, and $x = u_3$ if $u_3 \neq 0$, $x = u_1$ if $u_3 = 0$ and $u_1 \neq 0$, $x = u_2$ if $u_3 = 0$ and $u_1 = 0$. Moreover, on the projectors 0 and I we define

$$\alpha_\psi(I) = 1 \qquad \text{and} \qquad \alpha_\psi(0) = 0\,.$$

The functional so defined is quite evidently a probability measure on the set of projectors: of course, if P and P' are orthogonal, in any base of $\mathscr{H}_2$ they have the form

$$P = \tfrac{1}{2}I + \tfrac{1}{2}\boldsymbol{u}\cdot\boldsymbol{\sigma} \qquad \text{and} \qquad P' = \tfrac{1}{2}I - \tfrac{1}{2}\boldsymbol{u}\cdot\boldsymbol{\sigma}\,,$$

then $P + P' = I$ and $\alpha_\psi(P + P') = 1$.

By means of the spectral decomposition, α_ψ can be extended to a functional (which we shall denote again by α_ψ) on the set of Hermitean matrices, defining

$$\alpha_\psi(A) = \lambda_1\alpha_\psi(P_1) + \lambda_2\alpha_\psi(P_2)\,,$$

where A must be written in the base of $\mathscr{H}_2$ where ψ has the form $\begin{pmatrix}1\\0\end{pmatrix}$, λ_1 and λ_2 are the eigenvalues of A, and $\lambda_1 P_1 + \lambda_2 P_2$ is the spectral decomposition of A; then we have

$$\alpha_\psi(aI + \boldsymbol{b}\cdot\boldsymbol{\sigma}) = (a + |\boldsymbol{b}|)\,\alpha_\psi\left(\frac{1}{2}I + \frac{1}{2}\frac{\boldsymbol{b}}{|\boldsymbol{b}|}\cdot\boldsymbol{\sigma}\right) + (a - |\boldsymbol{b}|)\,\alpha_\psi\left(\frac{1}{2}I - \frac{1}{2}\frac{\boldsymbol{b}}{|\boldsymbol{b}|}\cdot\boldsymbol{\sigma}\right) =$$

$$= (a + |\boldsymbol{b}|)\left(\frac{1}{2} + \frac{1}{2}\,\mathrm{sign}\, x\right) + (a - |\boldsymbol{b}|)\left(\frac{1}{2} - \frac{1}{2}\,\mathrm{sign}\, x\right) = a + |\boldsymbol{b}|\,\mathrm{sign}\, x\,.$$

We stress that the value of α_ψ coincides with one of the eigenvalues of the matrix A, we refer to this property by saying that α_ψ is dispersion free.

The functional α_ψ is linear on commuting matrices, in fact $aI + \boldsymbol{b}\cdot\boldsymbol{\sigma}$ and $a'I + \boldsymbol{b}'\cdot\boldsymbol{\sigma}$ commute if and only if $\boldsymbol{b}\times\boldsymbol{b}' = \boldsymbol{0}$, and two such vectors have the property that $|\boldsymbol{b}+\boldsymbol{b}'| = |\boldsymbol{b}| + |\boldsymbol{b}'|$.

On noncommuting matrices α_ψ is not linear, for example

$$\alpha_\psi(\sigma_1+\sigma_2) = \sqrt{2}\,, \qquad \alpha_\psi(\sigma_1) = 1\,, \qquad \alpha_\psi(\sigma_2) = 1\,.$$

In this way we have seen that there exist probability measures on the set $\mathscr{L}(\mathscr{H}_2)$ of projectors on a 2-dimensional complex Hilbert space $\mathscr{H}_2$ which are not the restrictions of linear functionals on the set of Hermitean operators on $\mathscr{H}_2$ and which are dispersion free; we shall call them exceptional states. Their existence is not in contradiction with Gleason's theorem, which does not hold for 2-dimensional Hilbert spaces.

In ordinary quantum mechanics of spin-$\frac{1}{2}$ systems, exceptional states do not appear, but from the point of view of the axiomatization of quantum theory they are perfectly « legal » as spinors and density matrices are.

Of course the problem of deciding whether exceptional states have some physical meaning and whether there exist physical effects of spin-$\frac{1}{2}$ systems which could be explained by them can be solved only by experimental testings.

6'4. – The h.v. models of ref. [37, 38] use in an essential way the exceptional states described in the preceding section: let us briefly describe Bell's model [37].

Consider a spin-$\frac{1}{2}$ system without translational degrees of freedom. The hidden states are labelled by a vector ψ $(\|\psi\| = 1)$ and a real parameter λ (the hidden variable). The observable A in the state specified by ψ and λ has with certainty the value given by

$$\alpha_{\psi,\lambda}(A) = a + |\boldsymbol{b}|\,\mathrm{sign}\,(\lambda|\boldsymbol{b}| + \tfrac{1}{2}|b_3|)\,\mathrm{sign}\,x\,,$$

where A is written in the basis where ψ has the form $\begin{pmatrix}1\\0\end{pmatrix}$ and $x = b_3$ if $b_3 \neq 0$, $x = b_1$ if $b_3 = 0$ and $b_1 \neq 0$, $x = b_2$ if $b_3 = 0$ and $b_1 = 0$, and with the usual definition of sign x. The ordinary quantum expectation value of A in the state ψ is obtained by uniform averaging of the values $\alpha_{\psi,\lambda}(A)$ for $-\frac{1}{2} \leqslant \lambda \leqslant \frac{1}{2}$. Indeed such expectation is given by

$$\int\limits_{-\frac{1}{2}}^{\frac{1}{2}} \mathrm{d}\lambda\,\{a + |\boldsymbol{b}|\,\mathrm{sign}\,(\lambda|\boldsymbol{b}| + \tfrac{1}{2}|b_3|)\,\mathrm{sign}\,x\}\,.$$

To compute this integral, we note that $\lambda|\boldsymbol{b}| + \frac{1}{2}|b_3| \geqslant 0$ when

$$\lambda \geqslant -\frac{1}{2}\frac{|b_3|}{|\boldsymbol{b}|} = \bar{\lambda} \qquad \left(-\frac{1}{2} \leqslant \bar{\lambda} < 0\right).$$

Then

$$\int_{-\frac{1}{2}}^{\frac{1}{2}} \mathrm{d}\lambda \left\{ a + |\boldsymbol{b}| \operatorname{sign}\left(\lambda|\boldsymbol{b}| + \frac{1}{2}|b_3|\right) \operatorname{sign} x \right\} =$$

$$= a \int_{-\frac{1}{2}}^{\frac{1}{2}} \mathrm{d}\lambda + |\boldsymbol{b}| \operatorname{sign} x \int_{-\frac{1}{2}}^{\frac{1}{2}} \operatorname{sign}\left(\lambda|\boldsymbol{b}| + \frac{1}{2}|b_3|\right) \mathrm{d}\lambda =$$

$$= a + |\boldsymbol{b}| \operatorname{sign} x \left(-\int_{-\frac{1}{2}}^{\bar{\lambda}} \mathrm{d}\lambda + \int_{\bar{\lambda}}^{\frac{1}{2}} \mathrm{d}\lambda \right) = a + |\boldsymbol{b}| \operatorname{sign} x(-2\bar{\lambda}) =$$

$$= a + |\boldsymbol{b}| \operatorname{sign} x \frac{|b_3|}{|\boldsymbol{b}|} = a + b_3 ,$$

as quantum mechanics requires.

One can verify at once that the hidden dispersion-free states $\alpha_{\psi,\lambda}$ are exceptional states as defined in the previous section.

A quite similar analysis can be carried over also for the h.v. model proposed by KOCHEN and SPECKER [38]; we refer to [39], where this analysis is done in a very clear way (this reference contains a discussion of the topics dealt with in the present section from a slightly different point of view). Also for this case it is shown that the hidden states used are exceptional states on $\mathscr{L}(\mathscr{H}_2)$.

These remarks seem to suggest the conclusion that models of quantum systems which admit hidden variables are possible only for Hilbert spaces of dimension 2. In fact, if the « hidden states » have to be considered as probability measures on the logic $\mathscr{L}$ of the system, like in the models of ref. [37, 38], then they can be defined only for spin-$\frac{1}{2}$ systems and cannot be generalized to more complex ones, because in this case Gleason's theorem does not allow for the existence of exceptional states.

Of course there is the possibility of defining h.v. theories whose hidden states are not probability measures on $\mathscr{L}$, but this fact would constitute a radical departure from the usual conception of state (hidden or not) as used in all physical theories. In this last sense a hidden state must be defined as a probability measure only on a sub σ-algebra of $\mathscr{L}$. A definition of such a h.v. theory, which is similar in the spirit to the one proposed by BOHM and BUB [40], has been given in the axiomatic framework of proposition-state systems by GUDDER [41], who has shown that it always exists. On the other hand, such a general theory would imply experimentally testable consequences (as pointed out in [41]) that are not exhibited (to our knowledge) by any physical system. A discussion of this further topic falls beyond the aims of the present paper.

REFERENCES

[1] M. TOLLER: *Int. Journ. Theor. Phys.*, **12**, 349 (1975); *An operational analysis of the space-time structure*, preprint (1976).
[2] G. D. BIRKHOFF and J. VON NEUMANN: *Ann. Math.*, **37**, 823 (1936).
[3] G. W. MACKEY: *Mathematical Foundations of Quantum Mechanics* (New York, N. Y., 1963).
[4] C. PIRON: *Helv. Phys. Acta*, **42**, 330 (1969); *Found. Phys.*, **2**, 287 (1972); *Foundations of Quantum Theory* (New York, N. Y., 1976).
[5] V. S. VARADARAJAN: *The Geometry of Quantum Theory* (Princeton, N. J., 1968).
[6] B. MIELNIK: *Comm. Math. Phys.*, **9**, 55 (1968); **15**, 1 (1969); **37**, 221 (1974).
[7] See, *e.g.*, E. M. ALFSEN and F. W. SHULTZ: *Mem. Ann. Math. Soc.*, **6**, No. 172 (1976); *State space of Jordan algebras*, preprint (1976).
[8] J. G. F. BELINFANTE: *Journ. Math. Phys.*, **17**, 285 (1976).
[9] See, also for further bibliography, C. H. RANDALL and D. J. FOULIS: *The operational approach to quantum mechanics*, in *The Logico-Algebraic Approach to Quantum Mechanics*, edited by C. A. HOOKER (Dordrecht, 1977).
[10] E. B. DAVIES and J. T. LEWIS: *Comm. Math. Phys.*, **17**, 239 (1970).
[11] C. M. EDWARDS: *Comm. Math. Phys.*, **16**, 207 (1970).
[12] J. C. T. POOL: *Comm. Math. Phys.*, **8**, 118, 212 (1968).
[13] We shall freely use basic notions of lattice theory, however the terms « atomic », « orthomodular », « covering property » will find explanation in the sequel. The following is an exhaustive reference on lattice theory: F. MAEDA and S. MAEDA: *Theory of Symmetric Lattices* (Berlin, 1970).
[14] See, also for further bibliography, E. G. BELTRAMETTI and G. CASSINELLI: *Riv. Nuovo Cimento*, **6**, 321 (1976).
[15] M. J. MACZYNSKI: *Rep. Math. Phys.*, **3**, 209 (1972); *Journ. Math. Phys.*, **14**, 1469 (1973).
[16] B. MIELNIK: *Quantum logic: is it necessarily orthocomplemented?*, in *Quantum Mechanics, Determinism, Causality, and Particles*, edited by M. FLATO *et al.* (Dordrecht, 1976).
[17] See, also for further bibliography, S. S. HOLLAND jr.: *Current interest in orthomodular lattices*, in *Trends in Lattice Theory*, edited by J. C. ABBOTT (Princeton, N. J., 1969).
[18] See, also for further bibliography, J. M. JAUCH: *Foundations of Quantum Mechanics* (Reading, Mass., 1968).
[19] S. MAC LANE: *Duke Math. Journ.*, **4**, 455 (1938).
[20] H. CRAPO and G. C. ROTA: *Studies Appl. Math.*, **49**, 109 (1970).
[21] A. M. GLEASON: *Journ. Math. Mech.*, **6**, 885 (1957).
[22] G. T. RÜTTIMANN: *Journ. Math. Phys.*, **18**, 189 (1977).
[23] B. C. VAN FRAASSEN: *Semantic analysis of quantum logic*, in *Contemporary Research in the Foundations and Philosophy of Quantum Theory*, edited by C. HOOKER (Dordrecht, 1973).
[24] See the book quoted in ref. [13].
[25] J. M. JAUCH and C. PIRON: *Helv. Phys. Acta*, **43**, 842 (1969).
[26] P. A. M. DIRAC: *The Principles of Quantum Mechanics*, III edition (Oxford, 1947), p. 13.
[27] We refer to Bub's paper in this volume, p. 71.
[28] J. C. T. POOL: *Comm. Math. Phys.*, **8**, 118, 212 (1968).

[29] E. G. Beltrametti and G. Cassinelli: in *Colloquio internazionale sulle teorie combinatorie* (Roma, 1976), p. 481.
[30] G. Cassinelli and E. G. Beltrametti: *Comm. Math. Phys.*, **40**, 7 (1975); *Problems on the proposition-state structure of quantum mechanics*, preprint (1977).
[31] P. C. Deliyannis: *Journ. Math. Phys.*, **17**, 653 (1976).
[32] E. G. Beltrametti and G. Cassinelli: *Journ. Phil. Logic*, **6**, 369 (1977).
[33] N. S. Kronfly: *Journ. Theor. Phys.*, **3**, 191 (1970).
[34] S. P. Gudder: *Int. Journ. Theor. Phys.*, **7**, 205 (1973).
[35] S. P. Gudder: *Comm. Math.*, **29**, 249 (1973).
[36] V. Cantoni: *Comm. Math. Phys.*, **44**, 125 (1975).
[37] J. S. Bell: *Rev. Mod. Phys.*, **38**, 447 (1966).
[38] S. Kochen and E. P. Specker: *Journ. Math. Mech.*, **17**, 59 (1967).
[39] K. Bugajska and S. Bugajski: *Ann. Inst. H. Poincaré*, **16** A, 93 (1972).
[40] D. Bohm and J. Bub: *Rev. Mod. Phys.*, **38**, 453 (1966).
[41] S. P. Gudder: *Journ. Math. Phys.*, **11**, 431 (1970).

The Measurement Problem of Quantum Mechanics.

J. Bub

University of Western Ontario - London, Ont., Canada

Introductory remarks.

The measurement problem of quantum mechanics is, roughly, the problem of reconciling the change of state of a system described by the dynamical group of the theory, *i.e.* unitary transformations of the state vector in Hilbert space, with the stochastic transitions of the measurement process, characterized by von Neumann's projection postulate. More fundamentally, the problem concerns all aspects of the general problem of interpretation of the theory. It is perhaps for this reason that the problem is often dismissed as a pseudoproblem.

There is, of course, a way of looking at quantum mechanics which immediately resolves all questions of interpretation, including the measurement problem. If we adopt a dogmatic antirealist position and regard quantum mechanics—or, indeed, any physical theory—as no more than an algorithm for generating statistical predictions concerning a certain class of measurement operations, then there can be no sense in which the statistics of the theory is conceptually puzzling. For problems of interpretation only arise insofar as the statistical relations of the theory incorporate features excluding a certain representation, which is characteristic of other « standard » statistical theories, and is regarded as relevant in an essential way to the status of these theories as fundamental theories of the structure of physical processes. In other words, it is only by comparing quantum mechanics with its predecessors, in particular with classical mechanics or classical statistical mechanics, as generically different structural theories that a problem of interpretation is introduced. A solution to such a problem would then involve clarifying the significance of the transition from classical to quantum mechanics, specifically in this case the significance of the transition from a commutative to a noncommutative algebra of physical magnitudes.

In the literature, reference is often made to a « statistical interpretation » of the quantum state, attributed to Einstein, or perhaps Ballentine [1]. It is suggested that conceptual problems such as the measurement problem are avoided on this interpretation. Now, a « statistical interpretation » of the state vector ψ can only mean an interpretation according to which the statistics generated by ψ is understood as referring to an ensemble of systems with a distribution of quantum-mechanical properties, *i.e.* a distribution of values of the quantum-mechanical physical magnitudes or observables. From the point of view of orthodox quantum mechanics, the differences between individual systems in the ensemble are represented by values of variables which are « hidden ». In other words, the « statistical interpretation » is simply the proposal that « hidden variables » underly the quantum statistics.

I shall deal with the difficulties in the way of the hidden-variable interpretation in part I. The Copenhagen interpretation, that the transition from a commutative to a noncommutative algebra of physical magnitudes has to do with the impossibility of simultaneously measuring certain magnitudes, because of an irreducible and uncontrollable disturbance of the measured system by the measuring instrument, should be understood as originating with the rejection of the hidden-variable thesis. Similarly, von Neumann's interpretation, in terms of the restricted validity of the distributive law of logic, is proposed in the light of the nonexistence of hidden variables underlying the quantum statistics.

I shall subsequently consider both these interpretations and attempt to bring out the issue between BOHR and VON NEUMANN. In 1938, at a conference in Warsaw organized by the International Union of Physics and the Polish Intellectual Co-operation Committee [2], BOHR presented a paper on *The Causality Problem in Atomic Physics*. VON NEUMANN commented extensively, with an analysis of the hidden-variable problem and an exposition of quantum logic, in which he contrasted his own view with that of Bohr. This is the closest thing to a debate between BOHR and VON NEUMANN that I have seen in print. I shall argue below that there is an aspect of the measurement problem—I shall refer to this as « the statistical problem of measurement »—which is properly resolved by interpreting the projection postulate as the probability conditionalization rule appropriate to a non-Boolean possibility structure of events, or noncommutative algebra of random variables. This is the thesis of part I. Nevertheless, there is a remaining core problem which is intimately bound up with the « irreducibility » of the quantum statistics, and the nonexistence of hidden variables. I shall refer to this as « the semantic problem of measurement », since it concerns the totality of properties that can be ascribed to a quantum-mechanical system at a particular time, or the totality of propositions that can be asserted to be true of a quantum-mechanical system at a particular time. I shall discuss the Bohr-von Neumann debate and abstract two opposing directions for a solution to this problem in part II.

PART I

The statistical problem of measurement.

1. – von Neumann's formulation of the projection postulate.

von Neumann's argument for the projection postulate is introduced in § 3 of Chapter III of his book, *Mathematical Foundations of Quantum Mechanics* [3]. Here VON NEUMANN points out an important feature of quantum mechanics as a statistical theory. In general, a statistical theory, which assigns a probability distribution with nonzero dispersion to each physical magnitude, might either predict the same, or similar, distribution with nonzero dispersion for the results of an immediate re-measurement of a magnitude A on systems which have all been found to have a particular value a of A, or predict the value a of A with no dispersion. The first possibility corresponds to the Bohr-Kramers-Slater theory, which is refuted by considering the manner in which light is scattered by electrons. VON NEUMANN cites the Compton-Simon experiment which, he argues, shows that, if a system is initially in a state in which the value of a magnitude A cannot be predicted with certainty, then this state is transformed by a measurement into another state in which the value of A is uniquely determined.

The details of the argument are not relevant here. What I am concerned with is von Neumann's conclusion, which he formulated as a postulate labelled M, « indispensable for the conceptual structure of quantum mechanics » [4]. The postulate states that, if a physical magnitude A is measured twice in succession on a system S, then we get the same value each time, even though A has a dispersion in the original state of S, and the A measurement can change the state of S.

Applied to a maximal (*i.e.* nondegenerate) magnitude A with eigenvalues $a_1, a_2, \ldots$ and corresponding eigenvalues $\alpha_1, \alpha_2, \ldots$, the postulate states that, if a measurement yields the result a_i, then the initial quantum state of the system is transformed to the state α_i. For a nonmaximal magnitude B, VON NEUMANN states that, if a measurement yields a result corresponding to a degenerate eigenvalue b_i, then « the state φ after the measurement is not uniquely determined by the knowledge of the result of the measurement » [5]. If b_i has multiplicity k_i, then the corresponding eigenvectors span a k_i-dimensional subspace $\mathscr{H}_{b_i}$, the range of the projection operator P_{b_i}. VON NEUMANN argues that φ can be any vector in the subspace $\mathscr{H}_{b_i}$. This suggests that, after a measurement yielding the result b_i, the system is represented by the statistical operator

$$\frac{P_{b_i}}{\mathrm{Tr}\,(P_{b_i})} = \frac{P_{b_i}}{k_i},$$

which represents a mixture, not a pure state. In the case of a magnitude C represented by an operator with a continuous spectrum, VON NEUMANN concludes that, after a measurement which localizes the value c of C to the set S, the state vector of the system can be anywhere in the subspace $\mathscr{K}_c(S)$, the range of the projection operator $P_c(S)$ in the spectral measure of C (where $C = \int r\,\mathrm{d}P_c(r)$). This suggests that, after a measurement of C gielding the result $c \in S$, the relative probabilities of properties of the system are generated by the projection operator $P_c(S)$. The operator $P_c(S)$ cannot, of course, be normalized by the factor $1/\mathrm{Tr}\,(P_c(S))$. von Neumann's use of the term « statistical operator » covers both the normalized operators generating probabilities for the system, and the unnormalized (and perhaps unnormalizable) operators generating relative probabilities [6].

Now, this form of the projection postulate for nonmaximal magnitudes and magnitudes with continuous spectra cannot be correct in general, for fairly obvious reasons. (Consider, for example, a « measurement » of the magnitude represented by the unit operator I on a system in the pure state ψ. The projection postulate requires the transition $P_\psi \to I$ for the statistical operator, *i.e.* the final state of the system can be any vector in the Hilbert space of the system with equal relative probability.) A subsequent discussion in chapter V shows that this was not von Neumann's intention. The question posed here is: What happens to a mixture represented by the statistical operator W after a measurement of a magnitude A? In the case of a nonmaximal magnitude B, VON NEUMANN argues that an initial statistical operator W is always transformed to a mixture W', even if W represents a pure state. However, W' is not in general uniquely determined by the measured value of b alone, but depends on the choice of basis vectors in the subspace $\mathscr{K}_b$. This is fixed by regarding the B-measurement as the measurement of a particular maximal magnitude A, with loss of information, *i.e.* the measurement procedure for B is not defined unless B is specified as a function of some maximal magnitude A. Each choice of A—involving a different measurement procedure for B—fixes a particular set of k_i basis vectors (eigenvalues of A) in each k_i-dimensional subspace $\mathscr{K}_{b_i}$ corresponding to the degenerate eigenvalue b_i of B. In other words, if $B = f(A)$, the measurement result b_i of B is regarded as an A-measurement restricting the value of A to the range of values $a_{i1}, a_{i2}, \ldots, a_{ik_i}$, so that the projection postulate yields the transition

$$W \to W' = \frac{\sum_{j=1}^{k_i} \mathrm{Tr}\,(WP_{\alpha_{ij}})\, P_{\alpha_{ij}}}{\sum_{j=1}^{k_i} \mathrm{Tr}\,(WP_{\alpha_{ij}})}$$

(where I have labelled the eigenvalues and corresponding eigenvectors of A $\{a_{ij}\}$, $\{\alpha_{ij}\}$). W' clearly depends on the initial statistical operator W, on the measured value b_i of B and on the maximal magnitude A through which B is measured.

There is a special case in which W' is independent of the maximal magnitude A, *i.e.* the precise eigenbasis in each subspace $\mathscr{K}_{b_i}$. If $\mathrm{Tr}\,(WP_{\alpha_{ij}}) = c_i$ for each j in the above expression for W', then

$$W' = \frac{c_i}{k_i c_i} \sum_{j=1}^{k_i} P_{\alpha_{ij}} = \frac{P_{b_i}}{\mathrm{Tr}\,(P_{b_i})}.$$

If this holds for each eigenvalue b_i, then the transition $W \to W'$ is the same for each choice of A. Thus, the projection postulate yields the transition

$$W \to W' = \frac{P_{b_i}}{\mathrm{Tr}\,(P_{b_i})}$$

after a measurement of B with the result b_i if and only if W assigns equal probability p_i to each eigenvalue a_{ij}, $j = 1, \ldots, k_i$, of A. If this holds for some set of basis vectors in the subspace $\mathscr{K}_{b_i}$, it holds for every set, so that $\mathrm{Tr}\,(WP_\varphi) = c_i\|\varphi\|$ for every vector $\varphi \in \mathscr{K}_{b_i}$. An equivalent condition on W is

$$P_{b_i} W P_{b_i} = c_i P_{b_i}.$$

Clearly,

$$c_i = \frac{\mathrm{Tr}\,(WP_{b_i})}{\mathrm{Tr}\,(P_{b_i})},$$

i.e. the condition is

$$P_{b_i} W P_{b_i} = \frac{\mathrm{Tr}\,(WP_{b_i})}{\mathrm{Tr}\,(P_{b_i})} \cdot P_{b_i}.$$

Notice that this condition can never be satisfied if W represents a pure state, *i.e.* $W = P_\psi$, since

$$\begin{aligned} P_{b_i} P_\psi P_{b_i} &= P_{\theta_i}, \qquad \text{where } \theta_i = P_{b_i}\psi, \\ &= \|P_{b_i}\|^2 P_{\theta'_i} = \mathrm{Tr}\,(P_\psi P_{b_i}) P_{\theta'_i}, \end{aligned}$$

where θ'_i is the normalized projection of ψ onto the subspace $\mathscr{K}_{b_i}$. The condition reduces to

$$P_{\theta'_i} = \frac{P_{b_i}}{\mathrm{Tr}\,(P_{b_i})},$$

which cannot be satisfied unless the range of P_{b_i} is 1-dimensional.

The general conclusion, then, is this: For a maximal magnitude A with a pure discrete spectrum of eigenvalues $a_1, a_2, \ldots$, a measurement of A with the result a_i yields the transition

$$W \to W' = P_{\alpha_i} .$$

For a nonmaximal magnitude B with a pure discrete spectrum $b_1, b_2, \ldots$ a measurement of B with the result b_i yields the transition

$$W \to W' = \frac{P_{b_i}}{\mathrm{Tr}\,(P_{b_i})}$$

in the special case of an initial mixture represented by a statistical operator W satisfying the condition $P_{b_i} W P_{b_i} = c_i P_{b_i}$ or, equivalently, $\mathrm{Tr}\,(WP_\varphi) = c_i$ for every unit vector $\varphi \in \mathscr{K}_{b_i}$. If this condition is not satisfied, for example if $W = P_\psi$, representing a pure state, then the resulting mixture W' depends on the measured eigenvalue b_i of B as well as the choice of basis in the subspace $\mathscr{K}_{b_i}$. In effect, a measurement of B is regarded as a measurement of some maximal magnitude A, such that $B = f(A)$, with loss of information. Thus, the measurement procedure for B is only uniquely defined in general by specifying B as a function of some maximal magnitude. The case of a magnitude C with a continuous spectrum is treated by regarding a measurement which localizes the value c of C to a set S_i in a partition $\{S_i\}$ of the range of values of C, as a measurement of a magnitude $G = g(C)$ with a pure discrete spectrum, each distinct eigenvalue g_i of G corresponding to a set S_i in the partition. As far as I can see, this means that an approximate measurement of C with the result $c \in S_i$ yields the transition

$$W \to W' = P_C(S_i) .$$

It was first pointed out by LÜDERS [7] that these considerations are not completely correct. Consider a nonmaximal magnitude B such that

$$B = f(A) = g(D) ,$$

where A and D are two noncommuting maximal magnitudes. Suppose the subspace $\mathscr{K}_{b_i}$ corresponding to the eigenvalue b_i of B is spanned by the eigenvectors $\alpha_{i1}, \alpha_{i2}, \ldots, \alpha_{ik_i}$ of A, and by the eigenvectors $\delta_{i1}, \delta_{i2}, \ldots, \delta_{ik_i}$ of D. We assume that none of the α_{ij} vectors are the same as the δ_{ij} vectors. Then certainly an ensemble prepared by measuring A on the systems in an initial ensemble represented by the statistical operator W, and collecting those systems with eigenvalues in the set $a_{i1}, a_{i2}, \ldots, a_{ik_i}$, is different from the ensemble prepared by measuring D on the systems in the same initial ensemble, and collecting those systems with eigenvalues in the set $d_{i1}, d_{i2}, \ldots, d_{ik_i}$. In the first case, the resulting ensemble would be correctly represented by the statistical operator

$$W'_A = \sum_{j=1}^{k_i} \mathrm{Tr}\,(WP_{\alpha_{ij}})P_{\alpha_{ij}} \Big/ \sum_{j=1}^{k_i} \mathrm{Tr}\,(WP_{\alpha_{ij}})\,.$$

In the second case, the resulting ensemble would be correctly represented by the statistical operator

$$W'_D = \sum_{j=1}^{k_i} \mathrm{Tr}\,(WP_{\delta_{ij}})P_{\delta_{ij}} \Big/ \sum_{j=1}^{k_i} \mathrm{Tr}\,(WP_{\alpha_{ij}})\,.$$

And $W'_A \neq W'_D$, unless the initial W satisfies von Neumann's condition $\mathrm{Tr}\,(WP_{\alpha_{ij}}) = \mathrm{Tr}\,(WP_{\delta_{ij}}) = c_i$ for all j. Both the transitions $W \to W'_A$ and $W \to W'_D$ may be regarded as characterizing a B-measurement with the result b_i on an initial ensemble represented by W, in the first case via a measurement of A with loss of information, in the second case via a measurement of B with loss of information. Nevertheless, there is a clear sense in which the notion of measuring the nonmaximal magnitude B is unambiguous and does not depend on the representation of B as $f(A)$ or $g(D)$. For example, if the Hilbert space is a tensor product space $\mathscr{H}_1 \otimes \mathscr{H}_2$, representing a pair of particles S_1 and S_2, then a magnitude represented by an operator of the form $A_1 \otimes I_2$ (where A_1 is maximal in $\mathscr{H}_1$ and I_2 is the unit operator in $\mathscr{H}_2$) is nonmaximal in the tensor product space. But a measurement of $A_1 \otimes I_2$, which represents a maximal magnitude of S_1, surely makes sense without specifying any S_2-magnitude at all.

The transition rule that LÜDERS proposes for such an « autonomous » measurement of a nonmaximal magnitude B is

$$W \to W' = \frac{P_{b_i} W P_{b_i}}{\mathrm{Tr}\,(P_{b_i} W P_{b_i})}\,.$$

If W represents a pure state, *i.e.* $W = P_\psi$, then W' represents a pure state, *the normalized projection of ψ onto the subspace P_{b_i}*. In the case of a magnitude C with a continuous spectrum, the Lüders rule is

$$W \to W' = \frac{P_C(S)\, W P_C(S)}{\mathrm{Tr}\,\big(P_C(S)\, W P_C(S)\big)}$$

for a measurement which localizes the value of C to the range S.

Since my purpose here is not to present a critique of von Neumann, but to propose an interpretation of the projection postulate as a probability conditionalization rule, it is irrelevant whether we take the Lüders version of the projection postulate as an extension of von Neumann's rule, or as a correction to von Neumann's analysis. What is important is that the projection postulate in the form

$$W \to W' = \frac{PWP}{\mathrm{Tr}\,(PWP)},$$

where P is the projection operator corresponding to the subspace determined by the measurement result, can be taken as a general rule, applying to all cases. An A-measurement with loss of information is different from an autonomous B-measurement, even though the subspace corresponding to the set of eivenvalues $a_{i1}, a_{i2}, \ldots, a_{ik_i}$ is identical with the subspace corresponding to the eigenvalue b_i of B. For an autonomous B-measurement with the result b_i, the Lüders rule yields

$$W \to W' = \frac{P_{b_i} W P_{b_i}}{\mathrm{Tr}\,(P_{b_i} W P_{b_i})}.$$

For an A-measurement with the result a_{ij}, the Lüders rule yields

$$W \to W' = \frac{P_{\alpha_{ij}} W P_{\alpha_{ij}}}{\mathrm{Tr}\,(P_{\alpha_{ij}} W P_{\alpha_{ij}})}.$$

If we form a new ensemble by combining the ensembles corresponding to the different results a_{ij}, $j = 1, \ldots, k_i$, with relative weights determined by W, *i.e.* $\mathrm{Tr}\,(WP_{\alpha_{ij}})$, then the resulting mixture is represented by the statistical operator

$$W' = \frac{\sum_{j=1}^{k_i} P_{\alpha_{ij}} W P_{\alpha_{ij}}}{\sum_{j=1}^{k_i} \mathrm{Tr}\,(P_{\alpha_{ij}} W P_{\alpha_{ij}})} = \frac{\sum_{i=1}^{k_i} \mathrm{Tr}\,(WP_{\alpha_{ij}}) P_{\alpha_{ij}}}{\sum_{j=1}^{k_i} \mathrm{Tr}\,(WP_{\alpha_{ij}})}.$$

Thus the Lüders formulation of the projection postulate yields von Neumann's rule as a special case, if we understand the measurement of a nonmaximal magnitude B as a function of a maximal magnitude A as an A-measurement with loss of information, and distinguish this from an autonomous B-measurement. An autonomous measurement of a nonmaximal magnitude B always transforms a pure state ψ into a new pure state, the normalized projection of ψ onto the subspace corresponding to the measurement result. An A-measurement with loss of information which determines the same subspace as the autonomous B-measurement in general leads to a mixture of A-eigenstates in the subspace, with weights specified by the squares of the lengths of the projections of ψ onto the A-eigenvectors.

2. – von Neumann's measurement problem.

With the projection postulate for state transitions on measurement, the theory is apparently characterized by two fundamentally different modes in which the state of a system can change. If an A-measurement is performed on all the systems in an ensemble defined by the statistical operator W, without selecting those systems for which the measurement yields a particular eigenvalue a_i, then the systems in the ensemble undergo changes of state in such a way as to lead to the transition

$$W \to W' = \sum P_{a_i} W P_{a_i} .$$

This is von Neumann's « process 1 » (in the amended Lüders version). If no measurement is performed on the ensemble in the time interval $t - t_0$, the systems evolve in time according to the equations of motion of the theory, so that W undergoes a unitary transformation:

$$W \to W' = U^{-1}(t - t_0)\, W U(t - t_0) .$$

von Neumann refers to this as « process 2 » [8].

The measurement problem for von Neumann, is to clarify the relation between process 1 and process 2, and to provide an explanation for the statistical character of process 1. von Neumann's explanation for the duality is this [9]:

« We must always divide the world into two parts, the one being the observed system, the other the observer. In the former, we can follow up all physical processes (in principle at least) arbitrarily precisely. In the latter, this is meaningless. The boundary between the two is arbitrary to a very large extent ... . Now quantum mechanics describes the events which occur in the observed portions of the world, so long as they do not interact with the observing portion, with the aid of the process 2, but as soon as such an interaction occurs, *i.e.* a measurement, it requires the application of process 1. The dual form is therefore justified. »

The remaining problem is a consistency problem [10]:

« Let I be the system actually observed, II the measuring instrument, and III the actual observer. It is to be shown that the boundary can just as well be drawn between I and II+III as between I+II and III ... . That is, in one case 2 is to be applied to I, and 1 to the interaction between I and II+III; and, in the other case, 2 to I+II, and 1 to the interaction between I+II and III. (In each case, III itself remains outside of the calculation.) The proof of this assertion, that both procedures give the same results regarding I (this and only this belongs to the observed portion of the world in both cases), is then our problem. »

von Neumann resolves this problem by a detailed examination of the statistics of composite systems. Consider two systems, S_1 and S_2, with associated Hilbert spaces $\mathscr{H}_1$ and $\mathscr{H}_2$. The composite system $S_1 + S_2$ is represented in the Hilbert space $\mathscr{H}_1 \otimes \mathscr{H}_2$. Let $\{\psi_m\}$ be a complete orthonormal set of basis vectors in $\mathscr{H}_1$, and $\{\varphi_n\}$ a complete orthonormal basis in $\mathscr{H}_2$. Then $\{\Phi_{mn} = \psi_m \otimes \varphi_n\}$ is a complete orthonormal basis in $\mathscr{H}_1 \otimes \mathscr{H}_2$. S_1-magnitudes, S_2-magnitudes and $(S_1 + S_2)$-magnitudes are represented by self-adjoint operators in $\mathscr{H}_1$, $\mathscr{H}_2$ and $\mathscr{H}_1 \otimes \mathscr{H}_2$, respectively.

A statistical operator W in $\mathscr{H}_1 \otimes \mathscr{H}_2$ determines a unique statistical operator W^1 in $\mathscr{H}_1$, which defines the statistics of all S_1-magnitudes, *i.e.* of all magnitudes represented by operators of the form $A^1 \otimes I^2$ [11]. Similarly, W defines a unique statistical operator W^2 of S_2, which specifies the statistics of all magnitudes represented by operators of the form $I^1 \otimes A^2$. Let the matrix of W with respect to the basis $\{\Phi_{mn}\}$ be

$$W_{mn,m'n'} .$$

von Neumann shows that W^1 and W^2 are represented by the matrices

$$W^1_{mm'} = \sum_{n=1}^{\infty} W_{mn,m'n} ,$$

$$W^2_{nn'} = \sum_{m=1}^{\infty} W_{mn,mn'} .$$

He considers the general problem of finding a statistical operator $W \in \mathscr{H}_1 \otimes \mathscr{H}_2$ that reduces to two given statistical operators $W^1 \in \mathscr{H}_1$ and $W^2 \in \mathscr{H}_2$, and shows that this problem has a unique solution if and only if at least one of the two statistical operators represents a pure state. In this case the solution is

$$W = W^1 \otimes W^2 ,$$

or, in terms of the representative matrices,

$$W_{mn,m'n'} = W^1_{mm'} W^2_{nn'} .$$

von Neumann also shows that for any vector $\Phi \in \mathscr{H}_1 \otimes \mathscr{H}_2$ there exists a maximal magnitude A^1 defining a complete orthonormal basis $\{\alpha_m\}$ in $\mathscr{H}_1$, and a maximal magnitude B^2 defining a complete orthonormal basis $\{\beta_n\}$ in $\mathscr{H}_2$, such that

$$\Phi = \sum_{k=1}^{M} \sqrt{w_k}\, \alpha_{r_k} \otimes \beta_{s_k}$$

with M finite or infinite, *i.e.* the values of the magnitudes $A^1 \in \mathscr{H}_1$ and $B^2 \in \mathscr{H}_2$ are correlated in the state Φ.

Now, in general, for a state represented by a vector

$$\Phi = \sum_{mn} f_{mn} \alpha_m \otimes \beta_n$$

the matrix of the statistical operator $W = P_\Phi$ with respect to the basis $\alpha_m \otimes \beta_n$ is

$$W_{mn,m'n'} = f_{mn} f^*_{m'n'}$$

and so

$$W^1_{mm'} = \sum_n f_{mn} f^*_{m'n}$$

and

$$W^2_{nn'} = \sum_m f_{mn} f^*_{mn'} \,.$$

Since in this case $f_{mn} = \sqrt{w_k}$ for $m = r_k$, $n = s_k$ and $f_{mn} = 0$ otherwise, it follows that $\sum_n f_{mn} f^*_{m'n} = 0$ unless $m = m' = r_k$, in which case $\sum_n f_{mn} f^*_{m'n} = w_k$, and $\sum_m f_{mn} f^*_{mn'} = 0$ unless $n = n' = s_k$, in which case $\sum_m f_{mn} f^*_{mn'} = w_k$, and so

$$W^1 = \sum_{k=1}^{M} w_k P_{\alpha_{r_k}} \,,$$

$$W^2 = \sum_{k=1}^{M} w_k P_{\beta_{s_k}} \,.$$

Thus, a *pure* statistical operator $W = P_\Psi$ in $\mathscr{H}_1 \otimes \mathscr{H}_2$ in general determines a *mixture* in the factor space $\mathscr{H}_1$ or $\mathscr{H}_2$. (The exception is the case $M = 1$.)

To sum up: If the states of the systems S_1 and S_2 are represented by the vectors $\psi \in \mathscr{H}_1$ and $\varphi \in \mathscr{H}_2$, respectively, then the state of the system $S_1 + S_2$ is represented by the vector $\Psi = \psi \otimes \varphi$ in $\mathscr{H}_1 \otimes \mathscr{H}_2$. If the statistical operator of $S_1 + S_2$ is $W = P_\Psi$, where Ψ is not a tensor product $\psi \otimes \varphi$, then the statistical operators W^1 of S_1 and W^2 of S_2 represent mixtures. There exists a S_1-magnitude and a S_2-magnitude such that the statistical correlations determined by Ψ establish a one-to-one correspondence between the values of these magnitudes.

The consistency problem is resolved in the following way: Suppose we have a system S whose statistics is represented by the statistical operator $W = P_\psi$, where

$$\psi = \sum_i (\alpha_i, \psi) \alpha_i$$

and the α_i are the eigenvectors of a maximal magnitude A [12]. By the projection postulate, measurement of the magnitude A results in the transition

$$\psi \rightarrow \alpha_i$$

with probability $|(\alpha_i, \psi)|^2$. Now consider the measurement as an interaction between the system S and a measuring instrument M suitable for measuring A. This means that S and M interact in such a way that the state vector of the composite system $S + M$ evolves during the interaction, by a unitary transformation, to the vector

$$\sum_i (\alpha_i, \psi)\alpha_i \otimes \varphi_i .$$

The φ_i are eigenvectors of some M-magnitude, say R, and this representation of the statistical state of $S + M$ correlates the eigenvalues of R (representing the « pointer-readings » of the instrument M) with the eigenvalues of the S-magnitude A:

$$\text{prob}\,(\alpha_i \,\&\, r_i) = |(\alpha_i, \psi)|^2 ,$$

$$\text{prob}\,(a_i \,\&\, r_j) = 0 \qquad (i \neq j) .$$

In other words, if the measuring instrument is represented initially by the vector φ, then the measurement process is an interaction governed by the equation of motion of the theory which results in the transition

$$\psi \otimes \varphi \rightarrow \sum_i (\alpha_i, \psi)\alpha_i \otimes \varphi_i .$$

What von Neumann shows is that there exist measuring instruments in this sense: Given a complete orthonormal set of vectors $\{\alpha_i\}$ in $\mathscr{H}_1$ (the Hilbert space of the system S) and any vector ψ representing the initial pure state of S, there exists a complete orthonormal set $\{\varphi_j\}$ in $\mathscr{H}_2$ (the Hilbert space of M) and a vector $\varphi \in \mathscr{H}_2$ (representing the initial state of the instrument M), such that the vector

$$\Psi(t) = \sum_i (\alpha_i, \psi)\alpha_i \otimes \varphi_i$$

(representing the state of $S + M$ in $\mathscr{H}_1 \otimes \mathscr{H}_2$) is a solution of the equation of motion for the composite system $S + M$, if

$$\Psi(t_0) = \psi \otimes \varphi$$

represents the initial state of $S + M$. Equivalently: There exists a unitary transformation

$$\psi \otimes \varphi \to \sum_i (\alpha_i, \psi)\alpha_i \otimes \varphi_i .$$

Now, by the projection postulate, a measurement of the magnitudes A and R on the composite system $S + M$ by a second measuring instrument results in the transition

$$\Psi(t) \to \alpha_i \otimes \varphi_i$$

with probability $|(\alpha_i, \psi)|^2$, from which it follows that the state of the system S after the interaction is α_i with probability $|(\alpha_i, \psi)|^2$. If $W = P_\Psi$, with $\Psi = \alpha_i \otimes \varphi_i$, then

$$W^1 = P_{\alpha_i} ,$$

$$W^2 = P_{\varphi_i} .$$

We have consistency in the following sense. The application of the projection postulate directly to the system S is consistent with its application to the system $S + M$ after a suitable interaction between S and M governed by the equation of motion of the theory.

Now, von Neumann's original problem concerned the status of the projection postulate as a statistical state transition rule, and the reconciliation of this mode of transformation of the state (process 1) with the unitary evolutions in time described by the equation of motion of the theory (process 2). Showing that the postulation of process 1 avoids a certain kind of inconsistency with the unitary measurement interactions does not explain why a fundamental theory of mechanics should incorporate a special postulate applicable only to state transitions on measurement, nor why these transitions should be only statistically determined. von Neumann's explanation that process 2 can be expected to apply only to the observed portions of the world, and that a special process may be expected to apply to the « interaction » between the observed portions of the world and the observer, *i.e.* to measurements, is hardly reassuring, if only because classical mechanics requires no such process.

Recall that VON NEUMANN introduced process 1, *i.e.* the projection postulate, in order to characterize a special feature of quantum mechanics as a statistical theory, not shared, for example, by the Bohr-Kramers-Slater theory. Immediate re-measurement of a magnitude A with initial result a yields the value a with probability 1, even though the initial state shows a dispersion in A. Actually it is precisely this feature of the theory which VON NEUMANN sees as the reason underlying the impossibility of extending postulate P [13]— « the most general probability assertion possible »—to magnitudes represented by noncommuting operations. Postulate P states that, in state φ, the probability that the magnitudes $A_1, \ldots, A_n$ take on values from the respective intervals $S_1, \ldots, S_n$ is $\|P_{A_1}(S_1)\, P_{A_2}(S_2) \ldots P_{A_n}(S_n)\|^2$, where $P_{A_1}(S_1), P_{A_2}(S_2), \ldots, P_{A_n}(S_n)$ are the operators in the spectral measures of $A_1, A_2, \ldots, A_n$ corresponding to the sets $S_1, S_2, \ldots, S_n$. P holds only for pairwise commuting operators $A_1, A_2, \ldots, A_n$. VON NEUMANN remarks [14]:

« The most obvious step would be to assume that this is an incompleteness in P, and that there must exist a more general formula, containing this as a special case. Because even if quantum mechanics furnishes only statistical information regarding nature, the least we can expect from it is that it describe not only the statistics of individual quantities, but also the relations among several such quantities.

But, contrary to this concept, which appears reasonable at first glance, we shall soon see that such a generalization of P is not possible, and that in addition to the formal reasons (intrinsic in the structure of the mathematical tools of the theory), weighty physical grounds also suggest this type of a limitation. »

There follows a discussion of the Compton-Simon experiment motivating the formulation of the projection postulate as a state transition rule on measurement. A subsequent analysis establishes that the condition for the simultaneous measurability of an arbitrary (finite) number of magnitudes is the pairwise commutativity of their operators, for this is a necessary and sufficient condition for the existence of a common magnitude of which they are all representable as functions. VON NEUMANN concludes [15]:

« We have now produced the mathematical proof that P makes the most extensive statement which is in general possible in this theory (*i.e.* in one that includes P). This is due to the fact that it preserves only the commutativity of the operators [$A_1, A_2, \ldots, A_n$]. Without this condition, nothing can be said regarding the results of simultaneous measurements for the [quantities corresponding to these operators], since simultaneous measurements of these quantities are then not in general possible. »

Thus, process 1 described by the projection postulate is introduced by VON NEUMANN as a mode of state transition peculiar to quantum mechanics, that is intimately related to the impossibility of simultaneously measuring certain quantities, reflected by their noncommutativity in the theory. The question concerning the reason for a special stochastic state transition on measurement in quantum mechanics, and the relation between process 1 and process 2, surely concerns the role played by the noncommutative algebra of physical magnitudes of the theory in securing the irreducible character of the statistics, and has nothing to do with general observations about the relation between the observer and the observed which, even if plausible, would apply to all theories. It is for this reason that von Neumann's solution to the measurement problem is unsatisfactory. Solving the consistency problem does not answer the original question: Why does a mechanical theory, which is distinguished by incorporating a noncommutative algebra for the physical magnitudes of mechanical systems, involve stochastic state transitions on measurement, and what is the relation between these measurement transitions and the non-stochastic, unitary transitions governed by the equation of motion of the theory?

In the following section, I shall present an aswer to this question.

3. – The projection postulate as a probability conditionalization rule.

I want to sketch a representation of classical probability theory as an operator calculus, analogous to the operator calculus of quantum mechanics, and show that the *classical* conditionalization rule in this calculus is just the Lüders version of von Neumann's projection postulate.

Consider, for simplicity, a countable classical probability space $(X, \mathscr{F}, \mu)$. I shall label the atomic events or elementary possibilities by $x_1, x_2, \ldots$. These are associated with singleton subsets $X_1, X_2, \ldots$, or indicator functions (characteristic functions) $I_1, I_2, \ldots$. I shall label other, possibly nonatomic, events by $a_1, a_2, \ldots$. Thus, the set $a_1, a_2, \ldots$ might denote a set of nonatomic, mutually exclusive and collectively exhaustive events $\left(\sum_i I_{a_i} = I, I_{a_i} I_{a_j} = 0, \ i \neq j\right)$.

Now, for any probability measure μ, it is possible to introduce a « statistical operator » $W = \sum_i w_i I_i$, where $\sum_i w_i = 1$, $w_i \geqslant 0$ for all i, in terms of which the probability of an event a may be represented as

$$p_\mu(a) = \mu(X_a) = \sum_j \left(\sum_i w_i I_i(x_j)\right) I_a(x_j) .$$

I shall write $p_W(a)$ for $p_\mu(a)$, where W corresponds to μ, *i.e.*

$$p_W(a) = \sum_j W(x_j) I_a(x_j) .$$

To simplify notation, I shall abbreviate this expression as

$$p_W(a) = \sum W I_a ,$$

where a summation sign without an index is understood as summing over all the atomic events x_j. This convention will be used below.

In terms of the statistical operator, the conditional probability (relative to an initial measure μ associated with the statistical operator W) of an event b

given an event a_i may be represented as

$$p_W(b|a_i) = \frac{\sum WI_{a_i}I_b}{\sum WI_{a_i}}$$

$$\left(i.e.\ p_W(b|a_i) = \frac{\sum_j W(x_j)I_a(x_j)I_b(x_j)}{\sum_j W(x_j)I_{a_i}(x_j)}\right).$$

To see this, simply notice that

$$p_\mu(b|a_i) = \frac{\mu(X_{a_i} \cap X_b)}{\mu(X_{a_i})} = \frac{\sum WI_{a_i}I_b}{\sum WI_{a_i}}.$$

Thus, the transition

$$\mu \to \mu'$$

on conditionalization with respect to a_i (where μ' is defined by

$$\mu'(X_e) = \frac{\mu(X_{a_i} \cap X_e)}{\mu(X_{a_i})} = p_\mu(e|a_i)$$

for an event e) may be represented as the transition

$$W \to W' = \frac{WI_{a_i}}{\sum WI_{a_i}}$$

so that $p_W(b|a_i) = \sum W'I_b$.

The statistical operator construction allows the replacement of the measure function μ, which is a set function whose domain is the field of measurable subsets of X, by a corresponding random variable W, a point function whose domain is X. If we regard the probability space associated with a classical system as defined by the algebra of physical magnitudes A, B, etc., whose values $a_1, a_2, \ldots, b_1, b_2, \ldots$, etc., correspond to the possible events represented by the field $\mathscr{F}$, then the statistics of this system is now represented by a physical magnitude W belonging to the algebra of magnitudes of the system. In fact, W is a linear sum of atomic idempotent magnitudes of the system. The advantage of this construction is that it provides a purely algebraic way of representing the statistics of a system, which is appropriate whether or not a representation of the algebra of magnitudes as real-valued functions on a space is possible. I want to suggest that we take W as representative of the statistics in a primary sense—the measure function exists only if the algebra of magnitudes is commutative. In this special case (a classical probability space), the subalgebra of idempotent magnitudes forms a Boolean algebra, which has a representation as a field of subsets of a set, by Stone's theorem [16].

The measure function defined as a set function on this field is essentially the « Stone representative » of the statistical operator W, which is the element in the algebra of magnitudes incorporating the statistics.

If we bear in mind the possibility of noncommutative algebras of magnitudes as in quantum mechanics (*i.e.* non-Boolean possibility structures of events represented by the associated algebra of idempotent magnitudes), it seems appropriate to represent the transition corresponding to conditionalization with respect to an event a_i by the symmetrical expression

$$W \to W' = \frac{I_{a_i} W I_{a_i}}{\sum I_{a_i} W I_{a_i}},$$

which is just the Lüders version of the projection postulate.

In quantum mechanics, the statistics of a system is specified by a statistical operator W which may be represented as

$$W = \sum_i w_i P_{\alpha_i},$$

where the P_{α_i} are projection operators onto atomic events (*i.e.* projection operators onto 1-dimensional subspaces spanned by the vectors α_i), and correspond to the commutative functions I_i in the classical case. Thus, the quantum-mechanical statistical operator is a linear sum of atomic idempotent magnitudes of the system. In terms of this operator, the probability of an event b is represented as

$$p_W(b) = \mathrm{Tr}\,(WP_b)\,.$$

Now, the trace of an operator O is just the sum of the eigenvalues of O, *i.e.* the sum of the possible values of O at each atom in the maximal Boolean subalgebra defined by O (*i.e.* the maximal Boolean subalgebra generated by the projection operators in the spectral measure of O, in the non-Boolean possibility structure of the system defined by the algebra of all projection operators). Thus, the operation Tr in the noncommutative algebra of magnitudes of a quantum-mechanical system is a noncommutative extension of the operation $\sum$ in the commutative algebra of magnitudes of a classical mechanical system. It follows that the transition

$$W \to W' = \frac{P_{c_i} W P_{c_i}}{\mathrm{Tr}\,(P_{c_i} W P_{c_i})},$$

specifying the change in the statistical operator W of a system S on measurement of a magnitude C with result c_i, is just the conditionalization of W with respect to the event c_i. In other words, the conditional probability (relative to an initial measure associated with the statistical operator W) of an event e

given an event c_i is

$$p_W(e|c_i) = \mathrm{Tr}\,(W'P_e)\,,$$

where

$$W' = \frac{P_{c_i} W P_{c_i}}{\mathrm{Tr}\,(P_{c_i} W P_{c_i})}\,.$$

The relation between von Neumann's process 1 and process 2 is now clarified. Process 1 does not describe a special model of state transition applicable on measurement of a quantum-mechanical system, in addition to the unitary transitions governed by the equation of motion of the theory. Moreover, the stochastic character of process 1 does not reflect the irreducible and uncontrollable disturbance of the system measured by the measuring instrument. Rather, the projection postulate describes the conditionalization of the statistics of the system.

In the classical case, the system is associated with a possibility structure of events which is a Boolean algebra, equivalently by a commutative algebra of random variables. The possibility structure is defined by the subalgebra of idempotent variables. The rule

$$W \to W' = \frac{I_{c_i} W I_{c_i}}{\sum I_{c_i} W I_{c_i}}$$

represents conditionalization with respect to the event c_i, in the sense that *W' retains all initial statistical information specified by W concerning the system, consistent with the event c_i.* That is, the transition $W \to W'$ preserves the relative probabilities of events a represented by subsets $X_e \subseteq X_{c_i}$. By contrast, the rule $W \to W' = I_{c_i}/\sum I_{c_i}$ represents conditionalization with respect to the event c_i and *randomization* of the initial measure corresponding to W on the subset X_{c_i}, so that all initial information specified by W or μ concerning the relative probabilities of the events e is eliminated. Such a conditionalization-cum-randomization might characterize a situation in which the information c_i is derived by a measurement process that disturbs the system in a random way. In effect, the condition is now not just c_i, but c_i together with the information that the system has been disturbed in a certain way. An ideal measurement of the magnitude C with result c_i corresponds to the condition c_i and the conditionalization rule

$$W \to W' = \frac{I_{c_i} W I_{c_i}}{\sum I_{c_i} W I_{c_i}}\,.$$

A nonideal (disturbing) measurement of C yielding the result c_i corresponds to the condition c_i and a *further condition* specifying the extent to which information concerning the relative probabilities of events e represented by subsets $X_e \subseteq X_{c_i}$ has been altered by the measurement process. The extreme case, in which all information concerning the relative probabilities of events e has been destroyed by the measurement, is represented by the conditionalization rule

$$W \to W' = \frac{I_{c_i}}{\sum I_{c_i}}.$$

A quantum-mechanical system is associated with a noncommutative algebra of physical magnitudes or random variables, *i.e.* with a non-Boolean possibility structure, represented by the algebra of idempotent magnitudes. The rule

$$W \to W' = \frac{P_{c_i} W P_{c_i}}{\mathrm{Tr}\,(P_{c_i} W P_{c_i})}$$

represents conditionalization with respect to the event c_i, in a precisely analogous sense to the corresponding rule in the classical Boolean or commutative case. That is, *the transition $W \to W'$ preserves the relative probabilities of events e represented by subspaces $\mathscr{K}_e \subseteq \mathscr{K}_{c_i}$, and so W' retains all initial statistical information specified by W concerning the system, consistent with the event c_i.* For

$$p_{W'}(e) = \mathrm{Tr}\,(W' P_e) = \frac{1}{\mathrm{Tr}\,(P_{c_i} W P_{c_i})} \cdot \mathrm{Tr}\,(P_{c_i} W P_{c_i} P_e) = \frac{1}{\mathrm{Tr}\,(P_{c_i} W P_{c_i})} \cdot \mathrm{Tr}\,(W P_e).$$

Hence

$$\frac{p_{W'}(e_1)}{p_{W'}(e_2)} = \frac{\mathrm{Tr}\,(W P_{e_1})}{\mathrm{Tr}\,(W P_{e_2})} = \frac{p_W(e_1)}{p_W(e_2)}.$$

It follows that the transition $W \to W'$ does not represent conditionalization of the statistics with respect to a condition specifying an irreducible and uncontrollable measurement disturbance *in addition* to the event c_i, for there is no randomization of the statistics specified by W for events e consistent with c_i and inconsistent with c_j, for all $j \neq i$. Such a conditionalization-cum-randomization would be characterized by the rule $W \to W' = P_{c_i}/\mathrm{Tr}\,(P_{c_i})$ (in the extreme case of a maximal disturbance, in which all initial information concerning relative probabilities of events e, represented by subspaces $\mathscr{K}_e \subseteq \mathscr{K}_{c_i}$, is destroyed).

Of course, the projection postulate for a quantum-mechanical system S can yield transitions $W \to W'$ on conditionalization which lead to changes in the statistical specification of S that are quite impossible for systems in classical mechanics. For example, if $W = P_{\alpha_1}$, so that $p_W(a_1) = 1$, where a_1 is the eigenvalue of a quantum-mechanical magnitude A corresponding to the eigenvector α_1, then conditionalization of W to W' can yield $0 < p_{W'}(a_1) < 1$. In the classical case, if $p_W(e) = 1$, then $p_{W'}(e) = 1$, for every proper conditionalization $W \to W'$. But this difference between the quantum-mechanical and classical-mechanical case cannot be regarded as grounds for interpreting the quantum-mechanical conditionalization rule as reflecting the irreducible and uncontrollable disturbance of the system measured by any measuring instrument suitable for obtaining information concerning the system. Rather, the peculiar features of the quantum-mechanical conditionalization rule relative to the classical rule reflect the non-Boolean character of the possibility structures of quantum-mechanical systems.

An extension of classical probability theory to noncommutative algebras of random variables, *i.e.* to non-Boolean possibility structures, has been developed by UMEGAKI [17] and generalized in certain ways by DAVIES and LEWIS [18]. Umegaki's theory exploits an idea first proposed by MOY [19], that the classical notion of conditional expectation with respect to a Boolean algebra can be understood as a linear map on the algebra of random variables.

The expectation value of a random variable on a classical probability space $(X, \mathscr{F}, \mu)$, *i.e.* a real-valued function on X measurable with respect to $\mathscr{F}$, is defined by

$$\mathrm{Exp}_\mu(A) = \int_X a\, d\mu\,.$$

If we assume that A is bounded, *i.e.*

$$\mathrm{Exp}_\mu(|A|) = \int_X |a|\, d\mu < \infty\,,$$

the conditional expectation of A with respect to a σ-field $\mathscr{F}' \subseteq \mathscr{F}$ is defined as any random variable A' satisfying the conditions

i) A' is measurable with respect to $\mathscr{F}'$, *i.e.* $\{x\colon A(x) \leqslant k\} \in \mathscr{F}'$ for all k,

ii) $\mathrm{Exp}_\mu(AI_Y) = \mathrm{Exp}_\mu(A'I_Y)$, for all $Y \in \mathscr{F}'$ or, alternatively,

ii′) $\mathrm{Exp}_\mu(AZ) = \mathrm{Exp}_\mu(A'Z)$ for every bounded random variable Z that is $\mathscr{F}'$-measurable.

Such a random variable always exists, but is not unique. If A' and A'' are two versions of the conditional expectation with respect to $\mathscr{F}'$, then

$$\mu\{x : A'(x) = A''(x)\} = 1\,,$$

i.e. A' and A'' take the same value at all points, except for a set of measure zero. Loosely, A' is any random variable that has the same average value as A on the sets $Y \in \mathscr{F}'$.

As an example, consider a countable probability space, where $X = \{x_i, i = 1, 2, \ldots\}$ and the points or atomic events x_i are assigned probabilities p_i. Then $\mathrm{Exp}(A) = \sum A(x_i)p_i = \sum a_i p_i$.

Let $\mathscr{F}'$ be generated by a countable partition $\mathscr{F}'_0 = \{X_{e_1}, X_{e_2}, \ldots\}$ of X, *i.e.* the X_{e_i} represent the new atomic events of the field $\mathscr{F}'$, which may be understood as labelled by the possible values of a magnitude E. The conditional expectation of the random variable A on the event e_i is

$$\operatorname{Exp}(A|X_{e_i}) = \sum_{x_j \in X_{e_i}} A(x_j) \frac{p_j}{\mu(X_{e_i})},$$

where $\mu(X_{e_i})$ is the measure of the set X_{e_i}, *i.e.* $\sum_{x_j \in X_{e_i}} p_j$. Now A', the conditional expectation of A with respect to the field $\mathscr{F}'$, may be constructed as the random variable defined by the condition

$$A'(x_j) = \operatorname{Exp}(A|X_{e_i}), \qquad \text{if } x_j \in X_{e_i},$$

i.e.

$$A'(x_j) = \sum_{i=1}^{\infty} \operatorname{Exp}(A|X_{e_i}) I_{e_i}(x_j).$$

In other words, A' is constant on each of the sets X_{e_i} representing the atomic events of the field $\mathscr{F}'$, and takes the value $\operatorname{Exp}(A|X_{e_i})$ on X_{e_i}.

The concept of conditionalization with respect to a σ-field generalizes the classical notion of the conditional probability of an event a, given the event b. Just as

$$\operatorname{prob}_\mu(a) = \operatorname{Exp}_\mu(I_a),$$

so the conditional probability of an event a with respect to the field $\mathscr{F}$ is defined by

$$\operatorname{prob}_\mu(a|\mathscr{F}) = \operatorname{Exp}_\mu(I_a|\mathscr{F}).$$

Notice that the expectation value of a random variable A may be understood as the conditional expectation of A with respect to the field generated by the partition $\{0, X\}$. The conditional probability of an event a given the event e_i is the value of the random variable I'_a, representing the conditional expectation of I_a with respect to a field containing the set X_{e_i}, on the set X_{e_i}. In the above example of a countable probability space, taking $A = I_a$, we get

$$I'_a(x_j) = \sum_{i=1}^{\infty} \operatorname{Exp}(I_a|X_{e_i}) I_{e_i}(x_j),$$

so that

$$\operatorname{prob}(a|e_i) = \operatorname{Exp}(I_a|X_{e_i}) = \sum_{x_j \in X_{e_i}} I_a(x_j) \frac{p_j}{\mu(X_{e_i})} = \frac{\sum_{x_j \in X_{e_i} \cap X_a} p_j}{\mu(X_{e_i})} = \frac{\mu(X_a \cap X_{e_i})}{\mu(X_{e_i})}.$$

Moy pointed out [20] that, if we fix the field $\mathscr{F}'$, then classical conditionalization with respect to $\mathscr{F}'$ defines a map on the commutative algebra of random variables, such that each random variable A is mapped onto A', defined as above. Notice that the map depends on the measure μ.

From this standpoint, classical probability theory can be looked at in the following way: We have a commutative algebra of random variables. The subalgebra of idempotent random variables defines a Boolean algebra $\mathscr{B}$ (corresponding to the field $\mathscr{F}$), which represents the classical possibility structure. Conditionalization with respect to a Boolean subalgebra $\mathscr{B}' \subseteq \mathscr{B}$ (corresponding to the field $\mathscr{F}'$) is a map on the algebra of random variables satisfying certain conditions. With $\mathscr{B}'$ fixed, there is a one-to-one correspondence between such maps and a particular set of random variables, the « statistical operators », so that the condition satisfied by the conditionalization map $A \to A'$ corresponding to the statistical operator W is essentially

$$\sum WAI_e = \sum WA'I_e$$

for all events $e \in \mathscr{B}'$, where $\sum O$ represents the operation of summing the values of the random variables O over all the atomic events in $\mathscr{B}$. It follows that the unconditioned expectation value of a random variable A is defined, for each statistical operator W, by

$$\mathrm{Exp}_W(A) = \sum WA$$

and the conditional probability of an event a, given the event $e \in \mathscr{B}'$, is defined by

$$\mathrm{prob}_W(a|e) = \frac{\sum WI_aI_e}{\sum WI_e}.$$

The notion of conditionalization as a map on an algebraic structure is particularly suited to the generalization of classical probability theory from a commutative to a noncommutative algebra of random variables. In fact the probabilities $\mathrm{prob}_W(a|e)$ can be regarded as induced by the conditionalization map, so that we can think of a family of probabilities as numbers representing the « filtering » of a given possibility structure through a (Boolean) subalgebra of the structure. Classically, the underlying possibility structure is taken as Boolean. In quantum mechanics, the underlying possibility structure is a non-Boolean lattice or partial Boolean algebra $\mathscr{L}$. If we fix a Boolean subalgebra $\mathscr{B} \subset \mathscr{L}$, generated by the projection operators P_i, $i = 1, 2, \ldots$, then, for each statistical operator W, a generalized conditional expectation is defined by the map $A \to A'$ on the noncommutative algebra of random variables represented by self-adjoint Hilbert-space operators, satisfying the conditions

i) A' is compatible with $\mathscr{B}$, *i.e.* the projection operators in the spectral measure of A' commute pairwise with the projection operators P_i which generate $\mathscr{B}$;

ii) $$\mathrm{Exp}_W(AP_i) = \mathrm{Exp}_W(A'P_i) \qquad \text{for all } P_i,$$

i.e.

$$\mathrm{Tr}_W(AP_i) = \mathrm{Tr}_W(A'P_i) \qquad \text{for all } P_i,$$

or, alternatively,

ii') $\mathrm{Tr}_W(AZ) = \mathrm{Tr}_W(A'Z)$ for all Z compatible with the projection operators P_i.

UMEGAKI shows [21] that the map

$$A \to A^{|P_1|P_2|\dots},$$

where $A^{|P}$ is von Neumann's operator [22] $PAP + (1-P)A(1-P)$, defines the conditional expectation of A with respect to the Boolean subalgebra $\mathscr{B}$ generated by the P_i. (This may be understood as the conditional expectation relative to an initial random or uniform measure defined by the statistical operator $W = I$, *i.e.* $A^{|P_1|P_2|\dots} = \mathrm{Exp}_W(A|\mathscr{B})$, where $W = I$.)

For example, let $\mathscr{B}_A$ be the maximal Boolean subalgebra generated by the projection operators $P_{\alpha_1}, P_{\alpha_2}, P_{\alpha_3}$ of the orthogonal basis $\alpha_1, \alpha_2, \alpha_3$ in a 3-dimensional Hilbert space. Then

$$P_\beta^{|P_{\alpha_1}|P_{\alpha_2}|P_{\alpha_3}} = P_{\alpha_1}P_\beta P_{\alpha_1} + P_{\alpha_2}P_\beta P_{\alpha_2} + P_{\alpha_3}P_\beta P_{\alpha_3} = \\ = |(\alpha_1, \beta)|^2 P_{\alpha_1} + |(\alpha_2, \beta)|^2 P_{\alpha_2} + |(\alpha_3, \beta)|^2 P_{\alpha_3}.$$

This is the conditional expectation of the random variable P_β with respect to the maximal Boolean subalgebra $\mathscr{B}_A$. It follows that $|(\alpha_i, \beta)|^2$ is the conditional probability of the atomic event b (corresponding to β) given the atomic event a_i (corresponding to α_i).

Since every statistical operator can be represented as a weighted sum of orthogonal projection operators, every statistical operator can be generated by conditionalizing a 1-dimensional projection operator representing an atomic property or event with respect to some Boolean subalgebra. Thus, the statistical operator

$$W = \sum_i |(\alpha_i, \psi)|^2 P_{\alpha_i}$$

represents the conditionalization of the projection operator P_ψ, where $\psi = \sum_i (\alpha_i, \psi)\alpha_i$, with respect to the maximal Boolean subalgebra $\mathscr{B}_A$ generated by the 1-dimensional projection operators P_{α_i}, for

$$P_\psi^{|P_{\alpha_1}|P_{\alpha_2}|\dots} = \sum_i P_{\alpha_i} P_\psi P_{\alpha_i} = \sum_i |(\alpha_i, \psi)|^2 P_{\alpha_i}.$$

Similarly, if a system S initially in the state ψ interacts with a system M initially in the state φ, yielding the correlated state

$$\Psi = \sum_i (\alpha_i, \psi)\alpha_i \otimes \varphi_i$$

via a unitary transformation of the product state $\psi \otimes \varphi$, then conditionalizing P_Ψ with respect to the Boolean subalgebra generated by $P_1 = I \otimes P_{\varphi_1}$, $P_2 = I \otimes P_{\varphi_2}$, ... yields the mixture represented by

$$W = P_\Psi^{|P_1|P_2|\ldots} = \sum_i |(\alpha_i, \psi)|^2 P_{\alpha_i} \otimes P_{\varphi_i} .$$

If the interaction is regarded as establishing correlations between the eigenvalues of a magnitude A of S and the « pointer-readings » of M (*i.e.* eigenvalues of some M-magnitude R), then W represents the conditionalization of P_Ψ with respect to the Boolean algebra generated by elements in $\mathscr{L}$ representing the possible pointer-readings.

The transition defined by the Lüders rule

$$W \to W' = \sum_i P_{a_i} W P_{a_i}$$

is just the conditionalization of W with respect to the Boolean subalgebra generated by the P_{a_i}, *i.e.*

$$\sum_i P_{a_i} W P_{a_i} = W^{|P_{a_1}|P_{a_2}|\ldots} .$$

The particular transition

$$W \to W' = \frac{P_{a_i} W P_{a_i}}{\mathrm{Tr}\,(P_{a_i} W P_{a_i})}$$

is the conditionalization of W with respect to the particular event a_i in the subalgebra.

4. – Hidden variables.

I have shown that von Neumann's process 1 in quantum mechanics is to be understood as the conditionalization of an initial statical operator W representing the statistics of a system with a non-Boolean possibility structure. For example, conditionalization with respect to an atomic property c_i of S yields the transition

$$W \to W' = P_{c_i} ,$$

where P_{c_i} is the projection operator onto the 1-dimensional subspace spanned by the eigenvector γ_i, say, corresponding to c_i. This means that the probability of a property b_j conditional on c_i (where the magnitude B may be incompatible with C) is to be computed according to the rule

$$p_W(b_j|c_i) = p_{W'}(b_j) = \mathrm{Tr}\,(P_{c_i}P_{b_j}) \,.$$

If b_j is an atomic property, corresponding to the vector β_j, we have

$$p_W(b_j|c_i) = |(\beta_j, \gamma_i)|^2 \,.$$

Thus, the probability assigned by the state vector γ_i (an eigenvector of the magnitude C, representing the association of the atomic property c_i with the system) to an incompatible atomic property b_j may be interpreted as *the conditional probability of the property b_j given the property c_i*.

Of course, the state vector cannot be interpreted in this way if the possible values of the maximal quantum-mechanical magnitudes are represented as generating a classical probability space. It is precisely in this sense that « hidden variables » are excluded in quantum mechanics.

For simplicity, consider a quantum-mechanical system associated with a 2-dimensional Hilbert space. The question at issue is whether probabilities like $|(\beta_2, \alpha_1)|^2$ associated with pairs of atomic properties b_2, a_1 of a quantum-mechanical system (*e.g.* the probability of spin down in the direction $\boldsymbol{b}$, given spin up in the direction $\boldsymbol{a}$, etc.) can be represented as conditional probabilities on a classical probability space $(X, \mathscr{F}, \mu)$.

Let X_{a_1}, X_{a_2} be two mutually exclusive and collectively exhaustive subsets of X, which partition X into two regions associated respectively with the two possible values a_1 and a_2 of the magnitude A. Similarly, partition X into X_{b_1} and X_{b_2} for the magnitude B, and so on. The sets X_{a_1}, X_{a_2}, X_{b_1}, X_{b_2}, etc. generate the field $\mathscr{F}$. Now, for any sets X_s, X_t, X_u in $\mathscr{F}$ and any measure μ, we have

$$\mu(X_s \cap X_t) = \mu(X_s \cap X_t \cap X_u) + \mu(X_s \cap X_t \cap X'_u) \,,$$
$$\mu(X_s \cap X_u) = \mu(X_s \cap X_u \cap X_t) + \mu(X_s \cap X_u \cap X'_t) \,,$$
$$\mu(X_t \cap X'_u) = \mu(X_t \cap X'_u \cap X_s) + \mu(X_t \cap X'_u \cap X'_s) \,,$$

and so

$$\mu(X_s \cap X_t) \leqslant \mu(X_s \cap X_u) + \mu(X_t \cap X'_u) \,.$$

(The $'$ here denotes the set complement.)

Hence, if we take $s = a_2$, $t = c_1$, $u = b_1$, it follows that

$$\mu(X_{a_2} \cap X_{c_1}) \leqslant \mu(X_{a_2} \cap X_{b_1}) + \mu(X_{c_1} \cap X'_{b_1}) \,,$$

i.e.

$$\mu(X_{a_2} \cap X_{c_1}) \leqslant \mu(X_{a_2} \cap X_{b_1}) + \mu(X_{c_1} \cap X_{b_2}) .$$

I shall refer to this inequality as a coherence condition on the probabilities. If the quantum-mechanical probabilities $p_{\alpha_2}(c_1)$, $p_{\alpha_2}(b_1)$, $p_{\beta_2}(c_1)$ can be represented as conditional probabilities with respect to the initial measure μ, then

$$p_{\alpha_2}(c_1) = \frac{\mu(X_{a_2} \cap X_{c_1})}{\mu(X_{a_2})} ,$$

$$p_{\alpha_2}(b_1) = \frac{\mu(X_{a_2} \cap X_{b_1})}{p(X_{a_2})} ,$$

$$p_{\beta_2}(c_1) = \frac{\mu(X_{b_2} \cap X_{c_1})}{\mu(X_{b_2})} .$$

It follows that the probabilities $p_{\alpha_2}(c_1)$, $p_{\alpha_2}(b_1)$, $p_{\beta_2}(c_1)$ are required to satisfy the coherence condition, provided that $\mu(X_{a_2}) \geqslant \mu(X_{b_2})$, since then

$$\frac{\mu(X_{a_2} \cap X_{c_1})}{\mu(X_{a_2})} \leqslant \frac{\mu(X_{a_2} \cap X_{b_1})}{\mu(X_{a_2})} + \frac{\mu(X_{c_1} \cap X_{b_2})}{\mu(X_{b_2})} ,$$

and so

$$p_{\alpha_2}(c_1) \leqslant p_{\alpha_2}(b_1) + p_{\beta_2}(c_1) .$$

But this condition cannot be satisfied in general for all triples of quantum-mechanical magnitudes represented by operators A, B, C in $\mathscr{H}_2$. There exist unit vectors $\alpha_2, \beta_1, \beta_2, \gamma_1 \in \mathscr{H}_2$ such that

$$|(\gamma_1, \alpha_2)|^2 \nleqslant |(\beta_1, \alpha_2)|^2 + |(\gamma_1, \beta_2)|^2 ,$$

or, alternatively, there exist directions $\boldsymbol{a}, \boldsymbol{b}, \boldsymbol{c}$ in real space (corresponding to the directions in the spin magnitudes A, B, C) such that

$$\sin^2 \tfrac{1}{2}\theta_{ac} \nleqslant \sin^2 \tfrac{1}{2}\theta_{ab} + \sin^2 \tfrac{1}{2}\theta_{bc}$$

(where θ_{ac} is the angle between $\boldsymbol{a}$ and $\boldsymbol{c}$, etc.) [23].

The proviso that $\mu(X_{a_2}) \geqslant \mu(X_{b_2})$ involves no loss of generality here. If $\mu(X_{a_2}) < \mu(X_{b_2})$, take $s = b_2$, $t = c_1$, $u = a_2$ and derive the inequality

$$\frac{\mu(X_{b_2} \cap X_{c_1})}{\mu(X_{b_2})} \leqslant \frac{\mu(X_{b_2} \cap X_{a_1})}{\mu(X_{b_2})} + \frac{\mu(X_{c_1} \cap X_{a_2})}{\mu(X_{a_2})} ,$$

i.e.

$$p_{\beta_2}(c_1) \leqslant p_{\beta_2}(a_1) + p_{\alpha_2}(c_1) ,$$

or

$$|(\gamma_1, \beta_2)|^2 \leqslant |(\alpha_1, \beta_2)|^2 + |(\gamma_1, \alpha_2)|^2 .$$

This is the same inequality, with α and β interchanged.

In a classical probability space, if a and b represent different atomic properties of a system, we always have $p_\mu(a|b) = 0$. In quantum mechanics, if A and B represent incompatible maximal magnitudes, the probability of the atomic property b_j given a_i is $|(\beta_j, \alpha_i)|^2$, which is nonzero (for example, the probability that a photon will pass a polaroid with its axis in direction $\boldsymbol{b}$, given that it has passed a polaroid with its axis in direction $\boldsymbol{a}$). The obvious reaction to this feature of the quantum theory is to suppose that a_i and b_j do not *really* correspond to *atomic* properties of the system (even though a_i and b_j represent atoms in the non-Boolean algebra of idempotent quantum-mechanical magnitudes). If a_i and b_j could be represented by different subsets with a nonempty intersection on a classical probability space, then it would be possible to interpret the nonzero probability $|(\beta_j, \alpha_i)|^2$ as a standard conditional probability. This, of course, means introducing additional «hidden variables» which label the different points in the subsets representing a_i and b_j. The above argument shows that we cannot regard the probabilities assigned by the state vector α_i to atomic properties represented by eigenvalues of the magnitudes B, C, etc. as conditional probabilities derived by conditionalizing an initial measure μ to the property a_i, assuming a Boolean possibility structure, *i.e.* a representation of the probabilities on a classical probability space.

The only way around this argument is to suppose that the measurement of quantum-mechanical magnitudes always involves a disturbance of the system measured by the measuring instrument, so that conditionalization is always properly with respect to a condition characterizing the measurement disturbance as well as the measurement result. An analysis of the nature of this disturbance shows that it must be «maximal», *i.e.* any initial information concerning the relative probabilities of subsets in the set representing a property that is atomic in the quantum-mechanical algebra of idempotents is completely destroyed by a measurement of the property [24]. In other words, a measurement of a quantum-mechanical magnitude must be accompanied by an irreducible and uncontrollable disturbance of the system, so that the hidden variables are «randomized» in the measurement process.

Now, it can be shown that a measurement disturbance of this sort is excluded by the correlations of the Einstein-Podolsky-Rosen experiment [25]. If we consider a pair of spin-$\frac{1}{2}$ particles, S and S', in the singlet spin state ψ, and represent the atomic properties of S and S' by subsets in a classical probability space $(X, \mathscr{F}, \mu)$, so that the atomic properties $a_i, b_j, \ldots$ of S are represented by *unique* subsets $X_{a_i}, X_{b_j}, \ldots$ in $\mathscr{F}$, and similarly for the atomic properties $a'_i, b'_j, \ldots$ of S', then it follows that

$$p_\psi(a_1 \,\&\, c'_1) = \mu(X_{a_1} \cap X_{c'_1}) ,$$

where a_1 represents the S-property « spin up in the $\boldsymbol{a}$-direction » and c_1' represents the S'-property « spin up in the $\boldsymbol{c}$-direction ». According to quantum mechanics, the joint probability $p_\psi(a_1 \,\&\, c_1')$ is equal to the sequential probability $p_\psi(a_2' - c_1')$, the probability that a measurement of spin in the $\boldsymbol{a}$-direction on S' followed by a measurement of spin in the $\boldsymbol{c}$-direction on S' yields the sequence: $\boldsymbol{a}$-spin down–$\boldsymbol{c}$-spin up. A purely combinatorial argument [26] exploiting the statistical correlations of the singlet spin state shows that

$$\mu(X_{a_1} \cap X_{c_1'}) = \mu(X_{a_2'} \cap X_{c_1'})$$

and hence

$$p_\psi(a_2' - c_1') = \mu(X_{a_2'} \cap X_{c_1'}) = \mu(X_{a_2'}) \frac{\mu(X_{a_2'} \cap X_{c_1'})}{p(X_{a_2'})} = \frac{1}{2} \frac{\mu(X_{a_2'} \cap X_{c_1'})}{\mu(X_{a_2'})} .$$

Since the sequential probability is just one half the conditional probability of c_1' given a_2', it follows that the conditionalization with respect to a_2' is derived by conditionalizing the measure μ in the usual way *with respect to the property* a_2' *alone* (*i.e.* without an additional condition characterizing a disturbance by the instrument which measured the magnitude « spin in the direction $\boldsymbol{a}$ » on the system S').

It seems to me that this is the significance of Bell's result [27]. Bell's inequality for « local » hidden-variable theories is, in effect, a coherence condition on conditional probabilities, analogous to the condition formulated above, where conditionalization is understood in the « ideal » sense, excluding any measurement disturbance or randomization of the hidden variables. The rejection of hidden-variable theories on the basis that the quantum statistics violates the inequality is equivalent to a demonstration that the statistical correlations of the Einstein-Podolsky-Rosen experiment exclude the assumption of measurement disturbances which randomize the hidden variables. Since I have discussed this issue elsewhere, I shall not repeat the details of the argument here [28].

The general conclusion is this: The statistics of quantum mechanics cannot be attributed to distributions over hidden variables, *i.e.* properties which correspond to atoms in the non-Boolean algebra of idempotent magnitudes of a quantum-mechanical system cannot be represented by nonatomic elements in a classical probability space, in such a way that the quantum-mechanical probabilities are derivable as conditional probabilities, even if the conditions are required to specify a randomizing measurement disturbance in addition to the measurement result. The thesis that the statistical character of quantum mechanics has its origin in the irreducible and uncontrollable disturbance of the system measured by the measuring instrument is therefore excluded by the statistical correlations of the Einstein-Podolsky-Rosen experiment.

This argument for the nonexistence of hidden variables underlying the quantum statistics—that a randomizing measurement disturbance is required to account for the interference properties of the statistics, but that such a disturbance is excluded by the statistical correlations of the Einstein-Podolsky-Rosen experiment—is quite different from the Kochen and Specker argument [29], or the argument from Gleason's theorem [30]. KOCHEN and SPECKER show that the functional relations between quantum-mechanical magnitudes cannot be preserved by their representative random variables in a proposed representation of the quantum statistics on a classical probability space, irrespective of any considerations concerning the statistical relations. This conclusion holds for Hilbert spaces of three or more dimensions. Preservation of the functional relations means that, if A and B are any two quantum-mechanical magnitudes represented by operators in a Hilbert space, such that $B = g(A)$, then the corresponding random variables f_A, f_B on the classical probability space in a hidden-variable reconstruction of the quantum statistics are required to satisfy the relation

$$f_B = g(f_A)$$

or

$$f_{g(A)} = g(f_A) .$$

It follows that the correspondence between operators in Hilbert space and random variables on the classical probability space defines a 2-valued homomorphism on the non-Boolean algebra of idempotent magnitudes of a quantum-mechanical system, for each point in the classical probability space. Loosely, fixing the hidden variables determines a value for each magnitude in such a way that functional relationships between the magnitudes are preserved by their values, *i.e.* 1 or 0 is assigned to each property of the system in such a way as to define a homomorphism on the algebra of properties. It can be shown that no such homomorphism exists, except in the case of a 2-dimensional Hilbert space [31].

As BELL pointed out [32], one might legitimately consider a hidden-variable theory in which the functional relationships between quantum-mechanical magnitudes are preserved by their representative random variables only for measures on the classical probability space corresponding to the states of quantum mechanics, and not pointwise, *i.e.* not for general measures on the space. The exclusion of measures in which the quantum-statistical relations are violated requires the assumption of a randomizing effect of the measurement process, in virtue of which such nonquantal measures are transformed to measures corresponding to quantum states in any measurement of quantum-mechanical magnitude. The rejection of such hidden-variable theories requires the statistical argument from the Einstein-Podolsky-Rosen experiment. Notice that the existence of 2-valued homomorphisms on the algebra of idempotent magnitudes in the 2-dimensional case does not enable the representation of the quantum statistics on a classical probability space, without introducing the assumption of a randomizing measurement disturbance. Such an assumption is required in order to generate the correct quantum-mechanical properties for conditionalization on the results of measurement [33].

From this standpoint, the central interpretative problem of quantum mechanics may be expressed as follows: The probabilities associated with pairs of properties in quantum mechanics (*e.g.* the probability of spin up in the direction $\boldsymbol{b}$, given spin up in the direction $\boldsymbol{a}$, etc.) cannot be represented as conditional probabilities on a classical probability space. Yet, in some sense, a probability like $p_{\alpha_i}(b_j) = |(\beta_j, \alpha_i)|^2$ is to be understood as the conditional probability of b_j given a_i. How do we make sense of this probability as a conditional probability? We cannot simply say that $p_{\alpha_1}(b_2)$ represents the probability that a *measurement* of spin in direction $\boldsymbol{b}$ will yield a result corresponding to spin down, immediately after a *measurement* of spin in the direction $\boldsymbol{a}$ has yielded a result corresponding to spin up. For this resolves the problem only if we refuse to consider the further question of how these probabilities are to be represented. No representation of these probabilities on a classical probability space $(X, \mathscr{F}, \mu)$ is possible, even on the assumption that an irreducible randomization of the probability measure accompanies any measurement of a property represented by an element in the field. In other words, the problem is just this: *Since the quantum statistics cannot be derived from distributions over hidden variables, how do we interpret the probabilities?*

The problem involves two components. Firstly, the problem concerns the impossibility of generating the quantum-statistical relations by conditionalizing measures on a classical probabilility space, *i.e.* a Boolean possibility structure. The demonstration that the quantum-statistical relations can be generated by conditionalizing generalized probability measures represented by statistical operators on a non-Boolean possibility structure resolves this problem. Secondly, the problem concerns the significance of the notion of « conditionalizing a probability with respect to an event » when there are no 2-valued homomorphisms on the algebra of events. I shall take up this second problem of measurement in part II. I conclude part I by showing how certain initially puzzling features of the quantum statistics are explained by interpreting the projection postulate as a probability conditionalization rule on a non-Boolean possibility structure.

5. – Conditionalization of non-Boolean possibility structures.

5·1. *The 2-slit experiment.* – We have a screen with two slits, A and B, and a second detecting screen or photographic plate. A photon in a pure quantum state represented by a plane wave moves towards the slits. Each slit can be regarded as localizing the photon to a region, Δ_A or Δ_B, in the plane of the slit screen. In other words, there is a magnitude M, representing position in the slit screen plane, and the passage of a particle though a slit is a measurement of the magnitude M, in the sense that a range Δ_A or Δ_B is assigned to M for the photon at the time of passage. We are interested in the probability that the photon will arrive at a certain region on the detecting screen, conditional on localization to a certain range of values of M (Δ_A, Δ_B or $\Delta_B \cup \Delta_B$). Localization to a region Δ on the detecting screen is a measurement of a magnitude N, representing position in the detecting-screen plane. N may be taken as the magnitude M, if the regions Δ are the same size as the slits, or at least as a function of M otherwise.

If we represent the probabilities and events on a classical probability space $(X, \mathscr{F}, \mu)$, so that the ranges Δ of N and Δ_A, Δ_B, $\Delta_A \cup \Delta_B$ of M correspond to sets $X_N(\Delta)$, $X_M(\Delta)$, $X_M(\Delta_A \cup \Delta_B) = X_M(\Delta_A) \cup X_M(\Delta_B)$ in $\mathscr{F}$, then conditionalizing any initial probability measure μ on the event $X_M(\Delta_A) \cup X_M(\Delta_B)$ yields a measure μ', defined for any event X_e by

$$\mu'(X_e) = \frac{\mu(X_e \cap (X_M(\Delta_A) \cup X_M(\Delta_B)))}{\mu(X_M(\Delta_A) \cup X_M(\Delta_B))} \propto \mu(X_e \cap X_M(\Delta_A)) + \mu(X_e \cap X_M(\Delta_B)) \, .$$

It follows that the probability of the photon arriving at a region on the detecting screen is proportional to a sum of two terms, the first term representing the probability of arrival conditional on slit A being open, the second representing the probability of arrival conditional on slit B being open, and this conflicts with the observed distribution of hits on the screen predicted by quantum mechanics.

The only way of deriving the correct probability distribution on the assumption of a classical probability space is by postulating that passage through the slit system disturbs the photon, so that the appropriate conditionalization is with respect to a condition specifying both the slit (*i.e.* Δ_A, Δ_B or $\Delta_A \cup \Delta_B$) and the measurement disturbance. As I have pointed out above, it can be shown that it is just this disturbance that is excluded by the correlations of the Eistein-Podolsky-Rosen experiment.

If we assume a non-Boolean possibility structure, the conditionalized statistical operator yielding the photon statistics immediately after the photon has passed through slit A (with slit B closed) is the *pure* statistical operator $W = P_{\psi_A}$, where ψ_A is the normalized projection of ψ onto the subspace which is the range of the projection operator $P_M(\Delta_B)$. Immediately after the photon has passed through slit B (with slit A closed), the conditionalized statistical operator is $W_B = P_{\psi_B}$, where ψ_B is the normalized projection of ψ onto the subspace defined by $P_M(\Delta_B)$. Since ψ represents a plane wave, which assigns equal probability to equal ranges of M, it follows that the projection of ψ onto the subspace which is the range of the projection operator $P_M(\Delta_A) +$

$+ P_M(\Delta_B)$ bisects the angle between ψ_A and ψ_B. Thus, immediately after the photon has passed through the slit system with both slits open, the conditionalized statistical operator is the *pure* statistical operator $W_{AB} = P_\theta$, where

$$\theta = \frac{\psi_A + \psi_B}{\|\psi_A + \psi_B\|}\,.$$

Since ψ_A and ψ_B are orthogonal unit vectors, $\|\psi_A + \psi_B\|^2 = \|\psi_A\|^2 + \|\psi_B\|^2 = 2$, and so

$$\theta = \frac{\psi_A + \psi_B}{\sqrt{2}}\,.$$

With slit A open and B closed, the probability of the photon arriving at a region Δ on the detecting screen after a time t is given by

$$p_{W_A}(n \in \Delta) = \mathrm{Tr}\,(U_t^{-1} P_{\psi_A} U_t P_N(\Delta))\,.$$

With slit B open and A closed, the probability is

$$p_{W_B}(n \in \Delta) = \mathrm{Tr}\,(U_t^{-1} P_{\psi_B} U_t P_N(\Delta))\,.$$

With both slits open, the probability is

$$p_{W_{AB}}(n \in \Delta) = \mathrm{Tr}\,(U_t^{-1} P_\theta U_t P_N(\Delta))\,.$$

If $t \neq 0$,

$$p_{W_{AB}}(n \in \Delta) \neq \tfrac{1}{2} p_{W_A}(n \in \Delta) + \tfrac{1}{2} p_{W_B}(n \in \Delta)\,.$$

To see this, let

$$U_t^{-1} P_{\psi_A} U_t = P_{\psi_{A'}}\,,$$
$$U_t^{-1} P_{\psi_B} U_t = P_{\psi_{B'}}\,,$$
$$U_t^{-1} P_\theta\, U_t = P_{\theta'}\,,$$

and to simplify notation write

$$P_M(\Delta_A) = P_A\,,$$
$$P_M(\Delta_B) = P_B\,,$$
$$P_N(\Delta) \;= Q\,,$$
$$\|P_A \psi\| = \|P_B \psi\| = l\,.$$

Then

$$p_{W_A}(n \in \Delta) = \|Q\psi_{A'}\|^2,$$
$$p_{W_B}(n \in \Delta) = \|Q\psi_{B'}\|^2,$$
$$p_{W_{AB}}(n \in \Delta) = \|Q_{0'}\|^2 = \left\|Q\left(\frac{\psi_{A'}+\psi_{B'}}{\sqrt{2}}\right)\right\|^2 = \tfrac{1}{2}\|Q\psi_{A'}+Q\psi_{B'}\|^2 =$$
$$= \tfrac{1}{2}p_{W_A}(n \in \Delta) + \tfrac{1}{2}p_{W_B}(n \in \Delta) + (Q\psi_{A'}, Q\psi_{B'}) + (Q\psi_{B'}, Q\psi_{A'}).$$

If $t \neq 0$, the « interference terms » $(Q\psi_{A'}, Q\psi_{B'})$, $(Q\psi_{B'}, Q\psi_{A'})$ are nonzero. We have

$$(Q\psi_{A'},\ Q\psi_{B'}) = (\psi_{A'},\ Q\psi_{B'}) = (U_t\psi_A, QU_t\psi_B) =$$
$$= \left(U_t\frac{P_A\psi}{l},\ QU_t\frac{P_B\psi}{l}\right) = \frac{1}{l^2}(\psi, P_A U_t^{-1}QU_tP_B\psi).$$

Although $(\psi, P_AQP_B\psi) = 0$, since P_A and P_B both commute with Q, and $P_AP_B = 0$, it is not the case that $(\psi, P_A(U_t^{-1}QU_t)P_B\psi) = 0$, since P_A does not commute with $Q' = U_t^{-1}QU_t$, and P_B does not commute with Q'. In general, if $P_A\psi \perp P_B\psi$, *i.e.* $(P_A\psi, P_B\psi) = 0$, then

$$QP_A\psi \perp QP_B\psi, \quad i.e.\ (QP_A\psi, QP_B\psi) = 0,$$

if and only if the projection operator Q commutes with both P_A and P_B. This is geometrically obvious if we think of Q as defining a plane $\mathscr{K}$ in a 3-dimensional Hilbert space. If the plane $\mathscr{K}_{AB}$ defined by the orthogonal pair $P_A\psi$ and $P_B\psi$ is orthogonal to $\mathscr{K}$, then the angle between the projection of $P_A\psi$ and $P_B\psi$ onto $\mathscr{K}$ is π. If we rotate the plane about the line defined by the intersection between $\mathscr{K}_{AB}$ and $\mathscr{K}$, then the angle between the projections of $P_A\psi$ and $P_B\psi$ onto $\mathscr{K}$ decreases continuously to $\pi/2$, when the planes $\mathscr{K}$ and $\mathscr{K}_{AB}$ coincide (and Q commutes with P_A and P_B).

This analysis resolves the « paradox » involved in the 2-slit experiment by showing precisely how the assumption of a non-Boolean possibility structure explains the existence of the « anomalous » interference effects. The conclusion

$$p_{W_{AB}}(n \in \Delta) = \tfrac{1}{2}p_{W_A}(n \in \Delta) + \tfrac{1}{2}p_{W_B}(n \in \Delta),$$

or

$$p_\psi(n \in \Delta | m \in \Delta_A \cup \Delta_B) = \tfrac{1}{2}p_\psi(n \in \Delta | m \in \Delta_A) + \tfrac{1}{2}p_\psi(n \in \Delta | m \in \Delta_B),$$

which could only be avoided in a representation of the events on a classical probability space by assuming a special kind of randomizing disturbance of the system on passing through the slit system, does not follow when the probabilities are conditionalized in a representation of the events on a generalized non-Boolean probability space. Thus, the « interference effects » which appear to originate from the irreducible and uncontrollable disturbance by the measuring instrument arise naturally in a non-Boolean possibility structure of events, without the assumption of a randomizing measurement disturbance.

5·2. *The Einstein-Podolsky-Rosen experiment.* – The core of the Einstein-Podolsky-Rosen argument is the following disjunction, which is presented as an *exclusive* and *exhaustive* alternative [34]:

« Either 1) the quantum-mechanical description of reality given by the wave function is not complete or 2) when the operators corresponding to two physical quantities do not commute the two quantities cannot have simultaneous reality ».

The alternative 2), as the only sense in which 1) might be denied, is understood in the following way: If the state vector of a system S is an eigenvector α_i of a magnitude A, then S has the property a_i (corresponding to the eigenvalue a_i of A), but *no* property b_j, for *any* j, corresponding to a value of *any* magnitude B incompatible with A, is a property of S. The *totality* of properties characterizing S (when S is assigned the state vector α_i) comprises a_i and those properties consistent with a_i corresponding to values of magnitudes compatible with A. Thus, the state vector represents the state of a quantum-mechanical system in an analogous sense to the state of a classical mechanical system. The state of a system in classical mechanics is represented by a point in phase space, which assigns a value to *every* magnitude of the system, *i.e.* the state associates a set of properties with the system. More precisely, the state defines a 2-valued homomorphism on the possibility structure of the system—the Boolean algebra of idempotent magnitudes—which selects an ultrafilter of properties, the principal filter generated by the atomic property representing the state [35]. The state of a system in quantum mechanics assigns values only to a certain subset of magnitudes, selecting an ultrafilter of properties in a maximal Boolean subalgebra of the non-Boolean possibility structure. If the system S is in the state α_i, where $\alpha_i = \sum_j (\beta_j, \alpha_i)\beta_j$, then the probabilities $|(\beta_j, \alpha_i)|^2$ refer to the possible results of measurements of the magnitude B on S, where a measurement is understood as an interaction altering the state of the system to one of the states β_j. That is, a measurement does not reveal the value of the magnitude B characterizing S—since the system has no B-property in the state α_i—but rather disturbs or changes S so that the set of properties characterizing S is no longer given by α_i, but by some β_j.

Actually, it would seem more natural to regard the state vector ψ as associating with S the set of properties assigned probability 1 by ψ. This is the principal filter in the non-Boolean lattice of idempotents generated by the atomic property P_ψ, *i.e.* the set of properties represented by lattice elements above P_ψ, and includes elements which are not all mutually compatible. In this case, the filter is not a prime filter in the non-Boolean lattice. If ψ is a common eigenvector of two incompatible magnitudes A and B, there would seem to be no grounds for ascribing reality status to an A-property and not to a B-property, unless ψ were derived from an A-measurement and not a B-measurement. If all measurements were maximal, there would be no problem in maintaining such a distinction. The possibility of nonmaximal measurements would seem to require the more general notion of state. This would not, however, affect the Einstein-Podolsky-Rosen argument, which depends only on the assumption that a state in the sense of a selection of properties characterizing the system is determined by ψ, and that, for some magnitudes M, no M-property at all belongs to the selection.

Now, the Einstein-Podolsky-Rosen argument shows that this conception of the state of a system, *i.e.* alternative 2), is excluded if we consider the quantum statistics of a pair of correlated systems. The simplest case is a pair of spin-$\frac{1}{2}$ particles, S and S', in the singlet spin state:

$$\Psi = \frac{1}{\sqrt{2}}\alpha_1 \otimes \alpha_2' - \frac{1}{\sqrt{2}}\alpha_2 \otimes \alpha_1' ,$$

where α_1, α_2 are eigenvectors of a spin magnitude A of S (spin in the direction $\boldsymbol{a}$), and α_1', α_2' are eigenvectors of the corresponding spin magnitude A' of S'. In the state $\Psi \in \mathscr{H} \otimes \mathscr{H}'$, no S-property belongs to the principal filter of properties generated by the vector Ψ, and no S'-property belongs to the principal filter of properties generated by Ψ. If we measure A' on S', then the measurement induces the transition

$$\Psi \to \alpha_1 \otimes \alpha_2'$$

or

$$\Psi \to \alpha_2 \otimes \alpha_1' .$$

In either case, the new state defines a principal filter of properties containing an A-property of S and an A'-property of S'. Thus, in virtue of the measurement, which can only plausibly be considered as affecting the system S', an A-property has been established for S. Moreover, the state is such that

$$\Psi = \frac{1}{\sqrt{2}}\beta_1 \otimes \beta_2' - \frac{1}{\sqrt{2}}\beta_2 \otimes \beta_1' ,$$

where β_1, β_2 are eigenvectors of a spin magnitude B of S (spin in the direction $\boldsymbol{b}$, for any direction $\boldsymbol{b}$), and β_1', β_2' are eigenvectors of the corresponding spin magnitude B' of S'. Hence, a measurement of B' on S' will establish a B-property for S. Since neither measurement can be regarded as influencing S in any physical way, it would seem proper to ascribe both an A-property and a B-property to S. But this contradicts alternative 2), according to which a complete specification of the state of a system (in the sense of a specification of all properties belonging to the system) never includes *both* an A-property and a B-property, if A and B are incompatible and have no common eigenvectors. It follows that there must be a more complete specification of the state of a system than that given by the state vector.

Now, this conclusion follows from a particular semantic interpretation of the state vector Ψ, *i.e.* the interpretation of Ψ as the analogue of the classical state, in the sense that Ψ picks out a principal filter of properties generated by an atom in the non-Boolean possibility structure of a quantum-mechanical system, just as the classical state picks out a principal filter of properties generated by an atom in the Boolean possibility structure of a classical system. I shall refer to this as the « orthodox » semantic interpretation of the state vector Ψ, although this conception was articulated in an explicit way for the first time by JAUCH and PIRON [36]. The singlet-spin-state projection operator corresponds to a 1-dimensional subspace, which may be regarded as the intersection of a family of 2-dimensional planes in the 4-dimensional Hilbert space $\mathscr{H}\otimes\mathscr{H}'$:

$$[(\mathscr{K}_{a_1}\wedge\mathscr{K}_{a_2'})\vee(\mathscr{K}_{a_2}\wedge\mathscr{K}_{a_1'})]\wedge[(\mathscr{K}_{b_1}\wedge\mathscr{K}_{b_2'})\vee(\mathscr{K}_{b_2}\wedge\mathscr{K}_{b_1'})]\wedge\dots .$$

Here $\mathscr{K}_{a_1}\wedge\mathscr{K}_{a_2'}$ represents the line spanned by the vector $\alpha_1\otimes\alpha_2'$ (*i.e.* the intersection of the planes represented by the projection operators $P_{\alpha_1}\otimes I$ and $I\otimes P_{\alpha_2'}$), and so $[(\mathscr{K}_{a_1}\wedge\mathscr{K}_{a_2'})\vee(\mathscr{K}_{a_2}\wedge\mathscr{K}_{a_1'})]$ represents the plane spanned by the two orthogonal vectors $\alpha_1\otimes\alpha_2'$ and $\alpha_2\otimes\alpha_1'$. This plane represents the following proposition in the non-Boolean possibility structure:

Either the spin of system S in the $\boldsymbol{a}$-direction is up and the spin of system S' in the $\boldsymbol{a}$-direction is down, or the spin of S in the $\boldsymbol{a}$-direction is down and the spin of S' in the $\boldsymbol{a}$-direction is up.

The line P_Ψ, an atom in the non-Boolean possibility structure, represents the conjunction of a family of such propositions, for all spin directions $\boldsymbol{a}$, $\boldsymbol{b}$, $\boldsymbol{c}$ ..., *i.e.*

$$\bigwedge_{x=\boldsymbol{a},\boldsymbol{b},\boldsymbol{c},\dots}(x_1\wedge x_2')\vee(x_2\wedge x_1') .$$

The orthodox semantics specifies this proposition, and all propositions implied by it, as true of the composite system in the singlet spin state, and no other proposition as true of the system in this state. Since neither the proposition a_1' nor the proposition a_2' belongs to this set of propositions, we cannot understand the measurement of spin in the $\boldsymbol{a}$-direction on S' as merely ascertaining whether or not a_1' is true or a_2' is true (*i.e.* which of a_1' or a_2' belongs to this set). Rather a measurement of spin in the $\boldsymbol{a}$-direction on S' must be understood as a physical disturbance of S' in such a way that either a_1' or a_2' (exclusively) becomes true of S'. What is incomprehensible is that, *in virtue of this very same disturbance of S', either a_2 or a_1 (exclusively) becomes true of S*, even though S may be physically separated from S'. It is on this basis that EINSTEIN, PODOLSKY and ROSEN argue that either a_1 or a_2 (exclusively) must be true of S before the disturbance of S' by the measurement, and similarly either b_1 or b_2 must be true of S, and so on, leading to the conclusion that the state vector Ψ cannot be a complete specification of the (semantic) state of the system.

If we interpret Ψ purely statistically, *i.e.* as the non-Boolean analogue of a classical probability measure, then the statistical correlations of the composite system in the Einstein-Podolsky-Rosen experiment arise naturally by conditionalizing the statistical operator $W = P_\Psi$ corresponding to Ψ with respect to the (nonmaximal) Boolean subalgebra generated by a spin magnitude of a subsystem.

For example, consider conditionalization with respect to the Boolean subalgebra $\mathscr{B}_{A'}$ generated by the projection operators $P_1 = I \otimes P_{a_1'}$, $P_2 = I \otimes P_{a_2'}$. The Lüders rule, or von Neumann operation, yields

$$P_\Psi^{|P_1|P_2} = P_1 P_\Psi P_1 + P_2 P_\Psi P_2 = P_{\theta_1} + P_{\theta_2} ,$$

where

$$\theta_1 = P_1 \Psi = I \otimes P_{\alpha_1'}\left(\frac{1}{\sqrt{2}}\alpha_1 \otimes \alpha_2' - \frac{1}{\sqrt{2}}\alpha_2 \otimes \alpha_1'\right) = -\frac{1}{\sqrt{2}}\alpha_2 \otimes \alpha_1'$$

and

$$\theta_2 = P_2 \Psi = \frac{1}{\sqrt{2}}\alpha_1 \otimes \alpha_2' ,$$

i.e.

$$P_\Psi^{|P_1|P_2} = \tfrac{1}{2} P_{\alpha_2} \otimes P_{\alpha_1'} + \tfrac{1}{2} P_{\alpha_1} \otimes P_{\alpha_2'} .$$

Thus, conditionalization with respect to the Boolean subalgebra $\mathscr{B}_{A'}$ yields the transition

$$W = P_\Psi \to W' = \tfrac{1}{2} P_{\alpha_2} \otimes P_{\alpha_1'} + \tfrac{1}{2} P_{\alpha_1} \otimes P_{\alpha_2'} .$$

The conditionalized statistical operator with respect to the event $a_1' \in \mathscr{B}_{A'}$ is

$$W' \cdot I \otimes P_{\alpha_1'} = \tfrac{1}{2} P_{\alpha_2} \otimes P_{\alpha_1'} ,$$

which specifies a probability of 0 for a_1 and 1 for a_2. Similarly, the conditionalized statistical operator with respect to the event $a'_2 \in \mathscr{B}_{A'}$ is

$$W' \cdot I \otimes P_{\alpha'_2} = \tfrac{1}{2} P_{\alpha_1} \otimes P_{\alpha'_2},$$

which specifies a probability of 1 for a_1 and 0 for a_2.

Just as the interference phenomena of the 2-slit experiment are explained by deriving the conditionalized probabilities via a representation of the events on a non-Boolean possibility structure, without requiring the assumption of a disturbance by the measuring instrument, so the statistical correlations of the Einstein-Podolsky-Rosen experiment are explained by showing how the statistical operator of the singlet spin state yields these correlations after conditionalization in an appropriate way on a non-Boolean possibility structure. There is, however, a further question raised by the Einstein-Podolsky-Rosen experiment, which the statistical analysis does not answer. The argument can be read as an objection to the orthodox semantic interpretation of the quantum-mechanical state vector ψ, *i.e.* the interpretation of ψ as selecting the set of properties assigned probability 1 by ψ, this set of properties being understood as the totality of properties possessed by the system in the state ψ. But if the orthodox semantics is incorrect, just what totality of properties is the quantum-mechanical analogue of the classical state? In other words, when a quantum-mechanical system is represented by the statistical operator P_ψ, can we list all the properties possessed by the system, or perhaps specify the possible list of properties consistent with the interpretation of P_ψ as assigning generalized conditional probabilities?

The problem raised by the Einstein-Podolsky-Rosen experiment is how to make sense of the generalized theory of conditional probabilities in terms of the possible totalities of properties possessed by a quantum-mechanical system, given the nonexistence of 2-valued homomorphism on the non-Boolean possibility structure (*i.e.* the impossibility of a classical semantics) and the difficulties in the way of the orthodox semantics. This problem underlies the semantic problem of measurement.

PART II

The semantic problem of measurement.

1. – The standard formulation of the measurement problem.

The measurement problem is usually formulated in the following way. We have a microsystem S and a measuring instrument M. Suppose M measures a magnitude A of S. This means that there exists an interaction between S

and M which transforms the initial state of the composite system by a unitary transformation, in such a way that correlations are established between the eigenvalues a_i of A and the eigenvalues r_i of a M-magnitude R (representing the « pointer-readings » of the instrument). If the initial state of S is represented by the eigenvector $\alpha_i \in \mathscr{H}$ and the initial state of M is $\varrho_0 \in \mathscr{H}'$, the measurement interaction yields the transition

$$\alpha_i \otimes \varrho_0 \xrightarrow{U(t-t_0)} \alpha_i \otimes \varrho_i ,$$

i.e. the instrument has undergone a change of state from ϱ_0 to ϱ_i, indicating the value a_i of A for the system. If the initial state of S is $\psi = \sum_i (\alpha_i, \psi)\, \alpha_i$, then it follows that the measurement interaction yields the transition

$$\Psi_0 = \psi \otimes \varrho_0 \xrightarrow{U(t-t_0)} \Psi(t) = \sum (\alpha_i, \psi) \alpha_i \otimes \varrho_i .$$

Now, the instrument M is a macrosystem. If M has functioned properly, we expect that after the interaction M has registered a particular eigenvalue of A, *i.e.* that M is in one of the states ϱ_i corresponding to a particular pointer-reading, and that S is in the correlated state α_i. We expect, in fact, that, if the measurement is performed on an ensemble of systems, all in the state ψ, the instrument will register the value r_i in a fraction $|(\alpha_i, \psi)|^2$ of the trials, yielding the transition

$$\psi \otimes \varrho_0 \rightarrow \alpha_i \otimes \varrho_i .$$

But this means that the statistics of the ensemble of composite systems $S + M$ after the interaction is represented by the mixture

$$W = \sum_i |(\alpha_i, \psi)|^2 P_{\alpha_i} \otimes P_{\varrho_i} .$$

The problem is how to account for the mixture W, in the light of the fact that the theoretical account of measurement in quantum mechanics yields the pure state $\Psi(t)$ as the final state of $S + M$ after the measurement. Now, this problem arises only if we implicitly invoke the orthodox semantic interpretation of the state vector. After a particular measurement interaction, *i.e.* after the instrument has registered a particular value of the magnitude A by exhibiting the reading r_i, we suppose that M has the property r_i and S has the property a_i, for some i. It follows that the statistical operator specifying the statistics of $S + M$ after the interaction should represent a probability distribution over the various possible totalities of properties of S and M, each possible totality of properties containing a_i and r_i, for some i. On the orthodox semantic interpretation of the state vector, each such possible set is defined by the state $\alpha_i \otimes \varrho_i$, for some i. The vector $\Psi(t)$ picks out a set of properties of $S+M$, the properties assigned probability 1 by this vector, or the principal filter of elements in the lattice generated by the atom corresponding to $\Psi(t)$, but this set of properties is distinct from any of the sets defined by $\alpha_i \otimes \varrho_i$. In fact, no properties a_i, r_i, for any i, belong to the principal filter generated by $\Psi(t)$. So, on the orthodox semantics, the final state of $S+M$ after the measurement should be represented by the mixture $W = \sum_i |(\alpha_i, \psi)|^2 P_{\alpha_i} \otimes P_{\varrho_i}$ and not by the pure state $\Psi(t) = \sum_i (\alpha_i, \psi)\alpha_i \otimes \varrho_i$.

JAUCH [37] has argued that the measurement problem is resolved by noticing that $\Psi(t)$ and W are equivalent in a certain sense. The system $S+M$ is classical, *i.e.* the observable or macroscopic magnitudes of $S+M$ form a commutative algebra. Every such magnitude has the spectral representation $\sum k_i \Pi_i$, where $\Pi_i = P_{\alpha_i} \otimes P_{\varrho_i}$. It follows that the two statistical operators, W and $P_{\Psi(t)}$, are equivalent for all macroscopic magnitudes, since

$$\mathrm{Tr}\,(P_{\Psi(t)} \sum k_i \Pi_i) = \sum k_i\, \mathrm{Tr}\,(P_{\Psi(t)} \Pi_i) =$$
$$= \sum k_i |(P_{\alpha_i} \otimes P_{\varrho_i}, \Psi(t))|^2 = \sum k_i |(\alpha_i, \psi)|^2 = \mathrm{Tr}\,(W \sum k_i \Pi_i)\,.$$

JAUCH remarks [38]:

« Thus, if the system $S+M$ is truly classical, the two states $\Psi(t)$ and W cannot be distinguished from one another, and one of the most vexing problems of quantum mechanics dissolves into a pseudoproblem.

« A final question: does this result mean that the two states $\Psi(t)$ and W can, under all circumstances, never be distinguished by measurement? It does not mean this. It means, this can never be accomplished with measurements from the Abelian set $\mathfrak{C}$ which contains the [projections Π_i]. In order to distinguish them it is necessary to have at one's disposal an observable which is not in this set. An observation of such a quantity will no doubt reveal that it is indeed $\Psi(t)$ which is the final state after the interaction, in agreement with the Schrödinger equation.

« This conclusion does not invalidate the statement that the measurement produces the state W (or rather the equivalence class containing W) because, as Bohr has always emphasized, the very possibility of measurement implies a classical apparatus. »

Now, the semantic problem of measurement is not solved by showing that $P_{\Psi(t)}$ and W belong to the same equivalence class, for this means only that $P_{\Psi(t)}$ and W are statistically equivalent for all magnitudes in a certain class, *i.e.* that these statistical operators yield the same probabilities for all ranges of values of these magnitudes. But the semantic problem of measurement arises because W is proposed as the statistical operator representing the final state of $S+M$, and not $P_{\Psi(t)}$, on the basis that W and not $P_{\Psi(t)}$ is consistent with the possible totalities of properties possessed by $S+M$ after the measurement interaction. Showing that W and $P_{\Psi(t)}$ yield the same *statistics* for a certain class of measurements does not resolve the semantic problem, that the totality of properties specified as true by $P_{\Psi(t)}$—the *theoretically derived* statistical operator of $S + M$ after the measurement interaction—does not include the properties a_i, r_i for any i.

A similar objection applies to the Daneri-Loinger-Prosperi theory of measurement [39]. A sophisticated quantum ergodic theory of macrosystems leads to the conclusion that, because of the macroscopic nature of the measuring instrument, the statistical operator $P_{\Psi(t)}$ undergoes an ergodic movement, so that after a sufficiently long time the statistics defined by the transformed statistical operator is equivalent to that defined by a statistical operator representing a corresponding mixture, with respect to all macroscopic quantities of M. It is *not* shown that, because of the macroscopic nature of M, $\Psi(t)$ is transformed to the mixture W, or that $P_{\psi} \otimes W_0$ is transformed to W, where W_0 is a mixture representing the initial macrostate of M. VON NEUMANN [40], and later WIGNER [41], showed that no such analysis, exploting the fact that the measuring instrument is a macrosystem and that macrostates are to be represented by appropriate mixtures, can explain the resulting mixture in a measurement interaction. It should be pointed out that DANERI, LOINGER and PROSPERI do not claim this. Like JAUCH, they emphasize that the desired mixture represents the statistics of the system only as far as the macroscopic quantities of M are concerned—*i.e.* measurements of magnitudes incompatible with macroscopic magnitudes are not excluded. Such measurements would, however, destroy the character of M as a macrosystem.

Again, these considerations do not resolve the semantic problem of measurement. What we want is a micro-theoretical account of a quantum-mechanical measurement process, showing precisely how conditionalizing the probabilities on the basis of a particular result is related to the possible totalities of properties possessed by the systems S and M.

Perhaps this point can be clarified by considering the difference between « coarse graining » the phase space of a classical system and the analogous procedure in the Hilbert space of a quantum-mechanical system. The state of a classical system is represented by a point in phase space. Coarse graining involves imposing a cell structure on the phase space—representing, say, the maximal resolution of our measuring instruments. Now, whatever the state of the system is, whether this state is known or not, it is represented by a phase point in a particular cell, irrespective of how the cells are chosen. The cell structure representing coarse graining does not influence what is true or false of the system, *i.e.* the totality of properties possessed by the system. In quantum mechanics, the state of a system is represented by a vector in Hilbert space. If we coarse grain the Hilbert space, we introduce cells of a minimal number of dimensions corresponding to the maximal resolution of our measuring instruments. But now, it does not follow that any initial state vector is necessarily in some particular cell. A general vector will be a superposition of components belonging to different cells. In the classical case, even if the particular cell containing the state is unknown, we can say that the system has the property corresponding to this cell, and that measurement will in principle reveal this cell. In the quantum-mechanical case, we cannot interpret a general state vector as representing a situation in which the system actually has the property corresponding to some particular cell, which is unknown prior to measurement. In fact, we are forced (by the orthodox semantic interpretation of the state vector) to admit that the system has *no* property corresponding to a particular cell, if the state is represented by a superposition of vectors from different cells. Showing that the superposition is statistically equivalent to a mixture of states from different cells, with respect to properties corresponding to the cell structure, does not in any way show that the statistics defined by the superposition can be interpreted as referring to a distribution of properties corresponding to the different cells, *i.e.* that the system represented by the superposition has *some* property corresponding to a particular cell, even though this property is unknown. The fact that the difference between the superposition and the mixture can only be revealed in the statistics of measurements which are impossible from a practical point of view is irrelevant here. What has to be explained is how coupling a microsystem to a macrosystem with an infinite number of degrees of freedom leads always to a state vector, perhaps unknown but statistically determined, which *is actually in a particular cell of the coarse-graining in the Hilbert space of the combined system imposed by the character of the instrument as a macrosystem.*

The most sophisticated analysis of the measurement problem is probably that of Hepp [42], who discusses some explicitly solvable dynamical models for measurement processes within the C^*-algebra approach to systems with infinitely many degrees of freedom developed by HAAG and KASTLER [43], SEGAL [44] and others, and apparently shows that « probability amplitudes evolve into probabilities ».

HEPP remarks [45]:

« It is very satisfactory that in this quantum-mechanical description of a microsystem coupled to a macroscopic apparatus, the coherent superposition of states and their incoherent mixture can become equivalent with respect to all quasi-local observables, when some classical observable assumes different values in these states. »

Of course, the analysis is vastly more sophisticated than that of Jauch——Jauch's « macrosystem » is associated with a 2-dimensional Hilbert space; HEPP considers in detail passage to the limit for a system with infinitely many degrees of freedom. Nevertheless, there is this similarity: Even Hepp's « realistic » analysis succeeds only in showing the statistical equivalence of the superposition and a corresponding mixture *with respect to a certain class of observables.*

The fact that this vast class of observables contains all the observables we might conceivably be interested in, and more, is irrelevant here. For this statistical equivalence of the superposition and the mixture does not show that the superposition and the mixture select *the same possible totalities of properties* for the system after the measurement interaction.

This is surely the point of the Schrödinger cat paradox [46]. SCHRÖDINGER considered a cat in a closed box containing a small amount of radioactive material. The probability that at least one atom will decay within an hour is $\frac{1}{2}$. If a decay occurs, a Geiger counter is activated, closing a circuit which electrocutes the cat. Let S be the radioactive atom, α_1 the undecayed state and α_2 the decayed state. Let M represent the cat, r_1 the property of being alive, and r_2 the property of being dead. The vector $\alpha_1 \otimes \varrho_1$ represents a state of affairs in which the atom has not decayed and the cat is alive, and the vector $\alpha_2 \otimes \varrho_2$ represents a state of affairs in which the atom has decayed and the cat is dead. The mixture $W = \frac{1}{2}P_{\alpha_1} \otimes P_{\varrho_1} + \frac{1}{2}P_{\alpha_2} \otimes P_{\varrho_2}$ can be understood as a statistical state consistent with either of the two possibilities: (atom undecayed, cat alive), (atom decayed, cat dead), *i.e.* one or the other possibility obtains, the probability of each being $\frac{1}{2}$. By contrast, the statistical state P_Ψ, where $\Psi = (1/\sqrt{2})\alpha_1 \otimes \varrho_1 + (1/\sqrt{2})\alpha_2 \otimes \varrho_2$, assigns the same probabilities to the events (atom undecayed, cat alive) and (atom decayed, cat dead), but, on the orthodox semantic interpretation of Ψ, neither possibility actually obtains. It follows that it is not true that the cat is alive, and it is not true that the cat is dead, when the composite system is in the state Ψ. Since it seems absurd to suppose that the cat becomes dead, or becomes alive, only when we observe it to be so (*i.e.* when we perform a further measurement on $S + M$), the inescapable conclusion seems to be that the state of $S + M$ after the interaction between S and M is *not* Ψ, but W. Yet it is Ψ, and not W, that is theoretically derivable on the basis of quantum mechanics.

To sum up: The semantic problem of measurement is not resolved by showing that P_Ψ and W are in the same equivalence class, in the sense that they yield the same statistics for all macroscopic magnitudes. It would hardly convince the cat, who is prior to observation perhaps happily in a state of suspended animation represented by P_Ψ, that, because P_Ψ and W are statistically equivalent for all sufficiently coarse observations, he should already regard himself as definitely either alive or dead. Similarly, the statistical operators

$$P_\psi, \qquad \text{where } \psi = \sum (\alpha_i, \psi)\alpha_i,$$

$$W = \sum |(\alpha_i, \psi)|^2 P_{\alpha_i}$$

yield the same probabilities for all ranges of A. Yet we cannot argue that, because W represents a situation in which the system possesses a particular property a_i, for some i, and W and P_ψ are statistically equivalent for measurements of A, it follows that P_ψ similarly represents a situation in which the system possesses a particular property a_i, for some i, provided only that we refrain from measuring any magnitudes B incompatible with A.

The proper conclusion to draw from the semantic problem of measurement is that the orthodox semantic interpretation of the state vector conflicts with the interpretation of the statistical operators as defining conditional probabilities on a non-Boolean possibility structure. The solution to this problem requires an alternative semantics consistent with the statistical interpretation of the pure statistical operators. What has to be shown is that P_ψ and W are semantically equivalent, in a certain sense, *i.e.* that they specify the same possible totalities of properties possessed by $S + M$, if M is a macrosystem.

In the remaining sections of this essay, I shall discuss Bohr's complementarity interpretation and von Neumann's quantum-logical interpretation as two opposing proposals for a solution to the semantic problem of measurement. Both these interpretations involve a rejection of what I have termed the « orthodox » semantic interpretation of the state vector.

2. – Bohr.

Bohr's complementarity interpretation begins with the rejection of hidden variables underlying the quantum statistics:

« ... the statistical character of the uncertainty relations in no way originates from any failure of measurements to discriminate within a certain latitude between classically describable states of the object, but rather expresses an essential limitation of the applicability of classical ideas to the analysis of quantum phenomena » [47].

The key concept here is the notion of a « quantum phenomenon ». In the Warsaw lecture, BOHR is quite explicit [48]:

« The essential lesson of the analysis of measurements in quantum theory is thus the emphasis on the necessity, in the account of the phenomena, of taking the whole experimental arrangement into consideration, in complete conformity with the fact that *all unambiguous interpretation of the quantum mechanical formalism involves the fixation of the external conditions, defining the initial state of the atomic system concerned and the character of the possible predictions as regards subsequent observable properties of that system*. Any measurement in quantum theory can in fact only refer either to a *fixation of the initial state or to the test of such predictions*, and it is first *the combination of measurements of both kinds which constitutes a well-defined phenomenon.* » [My italics.]

A *phenomenon*, then, is constituted by *two measurements*. And a *measurement* is [49]

« ... the unambiguous comparison of some property of the object under investigation with a corresponding property of another system, serving as a measuring instrument, and for which this property is directly determinable according to its definition in everyday language or in the terminology of classical physics. »

I suggest that BOHR understands a measuring instrument as classical in the sense that the system is associated with a Boolean algebra of possible properties, the actual atomic property (« pointer-reading ») at any particular time being « directly determinable ». Now, the coupling of the quantum system with a classical or Boolean measuring instrument M selects a unique Boolean subalgebra in the propositional lattice of the composite system $S+M$.

Consider the interaction between S and M yielding the state

$$\Psi = \sum_i (\alpha_i, \psi)\alpha_i \otimes \varrho_i \,.$$

The system M is regarded as classical, in the sense that the projection operators P_{ϱ_i} represent the idempotent macroscopic magnitudes of M and generate a Boolean algebra. The atoms of $\mathscr{B}_M$ are the P_{ϱ_i}; I shall also denote these by r_i. If the measuring instrument is classical, *i.e.* Boolean, and Ψ represents the above correlated state, then the set of vectors $\{\alpha_i\}$ in the Hilbert space $\mathscr{H}$ of S is unique for values of i corresponding to nonzero coefficients (α_i, ψ), in the sense that, if

$$\Psi = \sum_i a_i \alpha_i \otimes \varrho_i = \sum_i b_i \beta_i \otimes \varrho_i \,,$$

then

$$a_i = b_i$$

and

$$\alpha_i = \beta_i \,,$$

unless

$$a_i = b_i = 0 \,.$$

To see this, notice that

$$\Psi = \sum_k I \otimes P_{\varrho_k} \cdot \Psi$$

and

$$I \otimes P_{\varrho_k} \cdot \Psi = a_k \alpha_k = b_k \beta_k \,.$$

Hence $\alpha_k = b_k$ and $\alpha_k = \beta_k$, unless $a_k = b_k = 0$ (since α_k and β_k are unit vectors). This depends on the correlated form of Ψ.

It follows that the coupling between S and a Boolean measuring instrument M selects a *unique* Boolean subalgebra in the non-Boolean propositional or possibility structure of $S + M$, *viz.* the subalgebra $\mathscr{B}_{S+M}$ generated by the projection operators $P_{\alpha_i} \otimes P_{\varrho_i}$, for values of i corresponding to nonzero coefficients (α_i, ψ). If all values of i correspond to nonzero coefficients, then the set of vectors $\{\alpha_i\}$ forms a basis in $\mathscr{H}$, and the Boolean subalgebra $\mathscr{B}_{S+M}$ is maximal [50].

Consider, now, the following modification of the orthodox semantics: The possible totalities of properties possessed by a quantum-mechanical system S whose statistics is represented by a statistical operator W_S, or, equivalently, the possible maximal sets of events which can occur simultaneously, are only defined if a Boolean subalgebra $\mathscr{B}_S$ is selected in the non-Boolean possibility structure through a measurement interaction between S and a classical, *i.e.* Boolean, measuring instrument M. Since W'_{S+M} (the statistical operator of $S + M$ after the measurement interaction) defines a *classical* probability measure on the Boolean subalgebra $\mathscr{B}_{S+M}$, the possible totalities of properties possessed by $S + M$ are just those principal filters generated by the atoms in any maximal extension of $\mathscr{B}_{S+M}$ which are not assigned probability 0 by W'_{S+M}. The possible maximal sets of properties possessed by S are those principal filters generated by the atoms in any maximal extension of $\mathscr{B}_S$ which are assigned nonzero probability by the reduced statistical operator W'_S. Notice that, if $W_S = P_\psi$, where $\psi = \sum_i (\alpha_i, \psi)\alpha_i$ with some $(\alpha_i, \psi) = 0$, so that $\mathscr{B}_S$ (and $\mathscr{B}_{S+M}$) are therefore nonmaximal, then every extension of $\mathscr{B}_S$ (or $\mathscr{B}_{S+M}$) to a maximal Boolean subalgebra involves the addition of atoms which are assigned zero probability by W'_S (or W'_{S+M}).

If the statistical operator of $S + M$ after the measurement interaction is the pure state P_Ψ, where $\Psi = \sum_i (\alpha_i, \psi)\alpha_i \otimes \varrho_i$, then P_Ψ defines a *classical* probability measure on the Boolean subalgebra $\mathscr{B}_{S+M}$ generated by the projection operators $P_{\alpha_i} \otimes P_{\varrho_i}$, for values of i corresponding to nonzero coefficients (α_i, ψ), or on any maximal extension of $\mathscr{B}_{S+M}$. The possible maximal sets of properties are just the principal filters generated by the atoms in any maximal extension of $\mathscr{B}_{S+M}$ which are not assigned probability zero by P_Ψ, *viz.* the atoms (a_i, r_i)—corresponding to the projection operators $P_{\alpha_i} \otimes P_{\varrho_i}$—, since the atoms (a_i, r_j), with $i \neq j$, are assigned probability 0 by P_Ψ. Thus the system $S + M$ has the pair of properties (a_i, r_i), for some i, and the probabilities of these different occurrences are given by P_Ψ as $|(\alpha_i, \psi)|^2$. The conception here is that, when a maximal Boolean subalgebra $\mathscr{B}_{S+M}$ is selected for $S + M$ in a measurement interaction between S and a macroinstrument M, then the statistical state of $S + M$ is given by a *classical* measure on $\mathscr{B}_{S+M}$. The statistical operators P_Ψ and $W = \sum_i |(\alpha_i, \psi)|^2 P_{\alpha_i} \otimes P_{\varrho_i}$ are equivalent in the sense that they define the same *classical* measure on $\mathscr{B}_{S+M}$. Conditionalizing this measure with respect to a particular result yields the statistical state

$$\frac{P_{\alpha_i}\otimes P_{\varrho_i}\cdot P_{\Psi}\cdot P_{\alpha_i}\otimes P_{\varrho_i}}{\mathrm{Tr}\,(P_{\alpha_i}\otimes P_{\varrho_i}\cdot P_{\Psi}\cdot P_{\alpha_i}\otimes P_{\varrho_i})}=\frac{P_{\alpha_i}\otimes P_{\varrho_i}\cdot W\cdot P_{\alpha_i}\otimes P_{\varrho_i}}{\mathrm{Tr}\,(P_{\alpha_i}\otimes P_{\varrho_i}\cdot W\cdot P_{\alpha_i}\otimes P_{\varrho_i})}=P_{\alpha_i}\otimes P_{\varrho_i}\,,$$

and so the conditionalized statistical operator of S after the measurement is P_{α_i}. After an ensemble of measurements, if the different outcomes are not selected, the statistical operator is $\sum_i |(\alpha_i,\, \psi)|^2 P_{\alpha_i}$.

At first sight it might appear that this conclusion is open to the following objection: Since W and P_Ψ are semantically equivalent with respect to the Boolean subalgebra $\mathscr{B}_{S+M}$, and define the same probabilities for the different possibilities, W and P_Ψ ought to generate the same *conditional* probabilities for *any* subsequent measurement on S. But, while it is well known that W and P_Ψ define the same statistics on a particular Boolean sublagebra $\mathscr{B}_{S+M}$, it is just the difference between W and P_Ψ for properties belonging to other Boolean subalgebras incompatible with $\mathscr{B}_{S+M}$ that reveals the impossibility of interpreting P_Ψ as referring to a distribution over certain possibilities, one and only one of which actually obtains.

What has to be shown is that W and P_Ψ actually do define the same conditional probabilities for subsequent measurements of any magnitude B of S incompatible with A—conditional with respect to a particular value r_i of R.

For the pure state P_Ψ, we have

$$\mathrm{prob}_\Psi\,(b_j|r_i) = \mathrm{prob}_{W'}\,(b_j)\,,$$

where

$$W' = \frac{I\otimes P_{\varrho_i}\cdot P_\Psi\cdot I\otimes P_{\varrho_i}}{\mathrm{Tr}\,(I\otimes P_{\varrho_i}\cdot P_\Psi\cdot I\otimes P_{\varrho_i})}\,.$$

Now this operator represents the normalized projection of Ψ onto the subspace defined by the projection operator $I\otimes P_{\varrho_i}$, *i.e.*

$$W' = P_{\Psi'}\,,$$

where

$$\Psi' = \frac{I\otimes P_{\varrho_i}\cdot\Psi}{\|I\otimes P_{\varrho_i}\cdot\Psi\|} = \alpha_i\otimes\varrho_i\,,$$

and so

$$W' = P_{\alpha_i}\otimes P_{\varrho_i}\,.$$

Hence

$$\text{prob}_{W'}(b_j) = \text{prob}_{W'_S}(b_j) = \text{Tr}(P_{\alpha_i} P_{\beta_j}) = |(\alpha_i, \beta_j)|^2 .$$

For the mixture W

$$\text{prob}_W(b_j|r_i) = \text{prob}_{W'}(b_j) ,$$

where

$$W' = \frac{I \otimes P_{\varrho_i} \cdot W \cdot I \otimes P_{\varrho_i}}{\text{Tr}(I \otimes P_{\varrho_i} \cdot W \cdot I \otimes P_{\varrho_i})} = P_{\alpha_i} \otimes P_{\varrho_i} .$$

Hence, the conditional probability for b_j is the same as that specified by the pure state P_Ψ.

Notice that the state vector is always interpreted purely statistically, as the analogue of the classical measure function. The replacement of the state $\Psi = \sum_i (\alpha_i, \psi)\alpha_i \otimes \varrho_i$ by one of the states $\alpha_i \otimes \varrho_i$ is understood as the conditionalization of the classical probability measure defined by Ψ on $\mathscr{B}_{S+M}$ with respect to the « pointer-reading ». On this view *the probabilities of quantum mechanics arise through measurement interactions with classical systems.* The conditional probabilities always refer to the results of a subsequent measurement B, conditional on a prior measurement A, where the measurements of A and B involve the selection of two (incompatible) Boolean subalgebras in the non-Boolean possibility structure.

Here the notion of *measurement* is understood in a very special sense. Part of the function of a measurement is to select a Boolean subalgebra in the non-Boolean possibility structure of a quantum-mechanical system. I suggest that BOHR regards the notion of *truth* as meaningful only in the context of a Boolean possibility structure, *i.e.* to ascribe a property to a system only makes sense with respect to a structure of possible properties which forms a Boolean algebra. In the case of a quantum-mechanical system the possibility structure is non-Boolean. The application of the classical notion of truth, or the attribution of physical properties to such a system, requires reference to a classical measuring instrument which fixes a particular Boolean algebra in the non-Boolean possibility structure.

From this standpoint, we can say that the probability $p_{\alpha_i}(b_j) = |(\beta_j, \alpha_i)|^2$, for example, refers to the probability that the value of the magnitude B *will be found to be* b_j *if a B-measurement is made*, after a state preparation in which the state α_i is prepared. But this has to be understood in the context of Bohr's solution to the semantic problem of measurement, in the light of the non-existence of hidden variables underlying the quantum statistics, and does not involve an instrumentalist rejection of the problem, still less a position of philosophical subjectivism.

3. – von Neumann.

von Neumann's quantum-logical interpretation is very clearly set out in his reply to Bohr's Warsaw lecture [51]. He begins with a review of his proof that the quantum statistics cannot be understood in terms of distributions of hidden variables. On this issue he is in agreement with Bohr. What he rejects is the conclusion, implicit in Bohr's analysis of measurement in quantum mechanics, *that logical conjunction and disjunction are no longer always defined operations of the propositional calculus.* For example, the nonsimultaneous measurability of position and momentum (which are associated with phenomena involving incompatible Boolean subalgebras of the propositional lattice) means that « two perfectly legitimate propositions, like 'x has a co-ordinate between q and $q + \Delta q$' and 'x has a momentum between p and $p + \Delta p$' cannot be used as parts of the same composite proposition without paradoxes (at least not, if $\Delta q \Delta p \ll h$) » [52]. von Neumann asks: « Is this interpretation really inescapable? » [53].

He argues that the uncertainty principle need not be understood in Bohr's sense as a principle of nonsimultaneous measurability of certain quantum-mechanical properties, but can be constructed as showing that the distributive law of logic fails for quantum propositions. If a, b, c, ... represent the propositions $m\Delta p \leqslant p < (m+1)\Delta p$ for all integers $m = 0, \pm 1, \pm 2, \ldots$, and a', b', c', ... represent the proposition $n\Delta q \leqslant q < (n+1)\Delta q$ for all integers $n = \pm 1, \pm 2, \ldots$ (for some fixed choice of Δp, Δq, with $\Delta p \Delta q \ll h$), then

$$a \vee b \vee c \vee \ldots \quad = 1$$

and

$$a' \vee b' \vee c' \vee \ldots = 1\,,$$

where 1 here denotes the unit or maximum element in the algebra (*i.e.* the equivalence class of tautologies). But the uncertainty principle requires that

$$a \wedge a' = a \wedge b' = a \wedge c' = \ldots = b \wedge a' = b \wedge b' = b \wedge c' = \ldots$$
$$\ldots = c \wedge a' = c \wedge b' = c \wedge c' = \ldots = 0\,,$$

where 0 here denotes the null or minimum element in the algebra (*i.e.* the equivalence class of logically false propositions). It follows that

$$(a \vee b \vee c \vee \ldots) \wedge (a' \vee b' \vee c' \vee \ldots) \neq (a \wedge a') \vee (a \wedge b') \vee (a \wedge c') \vee \ldots$$

and the distributive law fails.

von Neumann concludes [54]:

« So we see: The 'principle of indeterminacy' means that the 'distributive law' of logics fails. The current view of quantum mechanics therefore forbids us to form both

$$a \vee b \vee c \vee \ldots$$

and

$$a' \vee b' \vee c' \vee \ldots$$

in the same consideration. *We have freed ourselves of such restrictions*, but had to sacrifice the 'distributive law' of logics instead. » [My italics.]

It appears that von Neumann's view is that position and momentum can both be regarded as (simultaneously) obtaining properties of a quantum-mechanical system, provided we drop the distributive law of logic. But even on the assumption of a non-Boolean possibility structure for the properties of a quantum-mechanical system, maintaining the following view is problematic:

a) For any system S, at every time t, every magnitude has a value, and these values preserve the functional relations between the magnitudes, *i.e.*, if the value of M for S at time t is m, and the value of N for S at time t is n, and $M = f(N)$, then $m = f(n)$.

b) The statistical states of quantum mechanics represent all possible probability distributions over the values of the magnitudes for S, and transformations of probability distributions on measurement are derived by conditionalizing the initial probability distribution in accordance with the measurement result, where a measurement is regarded as in principle ideal (*i.e.* no additional disturbance transformation is invoked).

VON NEUMANN suggests that a new notion of truth and probability is required to make sense of quantum mechanics. He remarks [55]:

« A complete derivation of quantum mechanics is only possible if the propositional calculus of logics is so extended, as to include probabilities, in harmony with the ideas of J. M. Keynes. In the quantum-mechanical terminology: the notion of 'transition probability' from a to b, to be denoted by $P(a, b)$, must be introduced. $P(a, b)$ is the probability of b, if a is known to be true. $P(a, b)$ can be used to define $a \leqslant b$ and $-a$: $P(a, b) = 1$ means $a \leqslant b$, $P(a, b) = 0$ means $a \leqslant -b$. But $P(a, b) = \phi$, with a $\phi > 0$, < 1 is a new 'sui generis' statement, only understandable in terms of probabilities. »

His comments end with a reference to his work on « continuous geometries », in which he hoped to elaborate on these questions. Much of von Neumann's work on this subject was never published. Remarks in the paper entitled *Continuous geometry* [56] suggest that his conception of a continuous geometry as a « geometry without points » was suggested by J. W. ALEXANDER, who developed a similar programme for set theory, *i.e.* « sets without points ».

Now, this notion of a « set without points » seems to be just the right sort of thing one would want to represent a property in a non-Boolean possibility structure, in which distributivity fails in such a way as to exclude the existence of 2-valued homomorphisms on the structure. I suggest that VON NEUMANN is pointing to a generalized theory of properties for quantum-mechanical systems, in which properties are associated with « sets », but no 2-valued homomorphisms exist on the structure of properties, *i.e.* the « sets » have no points.

Conclusion.

The source of the measurement problem is the nonexistence of hidden variables underlying the quantum statistics, or, equivalently, the nonexistence of 2-valued homomorphisms on the possibility structure of a quantum-mechanical system, which correspond in the classical or Boolean case to maximal selections of properties of the system, or assignments of values to all the physical magnitudes of the system. This raises the problem of how to interpret the probabilities generated by the states of the theory, since a Boolean representation of the probabilities is impossible.

I have shown in part I how the probabilities may be interpreted as generalized conditional probabilities on a non-Boolean possibility structure. This resolves an aspect of the measurement problem, the « statistical problem » of measurement, *i.e.* the significance of the stochastic state transitions on measurement described by the projection postulate.

The remaining « semantic problem » of measurement discussed in part II is generated by the « orthodox » semantic interpretation of the state vector ψ, *i.e.* the interpretation of ψ as selecting the set of properties assigned probability 1 by ψ, this set of properties being understood as the totality of properties possessed by the system in the state ψ (regarded as the quantum-mechanical analogue of the classical state, which defines a 2-valued homomorphism on the algebra of properties of a classical system). The theoretical account of measurement in quantum mechanics yields a state for the combined system, object+measuring instrument, after a measurement interaction which, on the orthodox semantic interpretation of the state, is inconsistent with an *actual* property of the instrument being correlated with an *actual* property of the object. Yet, if the orthodox semantics is rejected, the question arises: Just what totality of properties is the quantum-mechanical analogue of the classical state, and how is the process of conditionalization on measurement in the generalized theory of conditional probabilities on a non-Boolean possibility structure related to the possible totalities of properties possessed by a quantum-mechanical system (given the nonexistence of 2-valued homomorphisms on the non-Boolean possibility structure)?

I have discussed Bohr's solution to this problem as a proposal showing how the classical semantics can be maintained. von Neumann's suggestion is to re-interpret the classical notions of truth and probability in a manner appropriate to a nondistributive logic, or a non-Boolean possibility structure. At the present time, this remains a research programme.

While the quantum-logical interpretation must be regarded as programmatic, it might well present the most promising avenue of research as far as a solution to the measurement problem is concerned. Orthodox quantum-mechanical theories of measurement ultimately confuse the semantic problem of measurement with the statistical problem. A classical statistical interpretation, or hidden-variable interpretation, simply does not account for the peculiarities of the quantum statistics. And Bohr's solution to the measurement problem saves the appearance of classical logic and classical probability theory, while the nature of the microworld remains a mystery that transcends rational explanation.

REFERENCES

[1] L. E. BALLENTINE: *Rev. Mod. Phys.*, **42**, 358 (1970).

[2] Proceedings were published in 1939 under the title *New Theories in Physics* by the Polish Intellectual Co-operation Committee and the International Union of Physics (no editor).

[3] J. VON NEUMANN: *Mathematical Foundations of Quantum Mechanics* (Princeton, N. J., 1955).

[4] *Ibid.*, p. 335.

[5] *Ibid.*, p. 218.

[6] If W is a normalized statistical operator, *i.e.* $\mathrm{Tr}\,(W) = 1$, then the probability of a property represented by the projection operator P is $\mathrm{Tr}\,(WP)$. If W is unnormalized, then the ratios of the probabilities of the properties represented by the projection operators P and Q is given by $\mathrm{Tr}\,(WP)/\mathrm{Tr}\,(WQ)$.

[7] G. LÜDERS: *Ann. der Phys.*, **8**, 322 (1951). The Lüders rule is discussed at some length by W. FURRY in *Some aspects of the quantum theory of measurement*, in *Lectures in Theoretical Physics*, Vol. VIII A (*Statistical Physics and Solid State Physics*) (Boulder, Colo., 1966).

[8] J. VON NEUMANN: *op. cit.*, p. 351, 352, 417, 418.

[9] *Ibid.*, p. 420.

[10] *Ibid.*, p. 421.

[11] The superscripts here denote the corresponding Hilbert space, *i.e.* $\mathscr{H}_1$ or $\mathscr{H}_2$. I have used superscripts rather than subscripts for the operators or physical magnitudes, in order to avoid confusion with the subscripts used below to denote matrix elements of the operators.

[12] Here and subsequently I suppress the superscripts on the operators representing physical magnitudes of S or M, except in the case of statistical operators.

[13] J. VON NEUMANN: *op. cit.*, p. 200, 206.

[14] *Ibid.*, p. 211, 212.

[15] *Ibid.*, p. 230.
[16] M. H. STONE: *Trans. Amer. Math. Soc.*, **40**, 37 (1936).
[17] H. UMEGAKI: *Tôhoku Math. Journ.*, I, **6**, 177 (1954); II, **8**, 86 (1956); *Kodai Math. Semi. Rep.*, III, **11**, 51 (1959); IV, **14**, 59 (1962). M. NAKAMURA and H. UMEGAKI: *Math. Jap.*, **7**, 151 (1961-62).
[18] E. B. DAVIES and J. T. LEWIS: *Comm. Math. Phys.*, **17**, 239 (1970).
[19] S. C. MOY: *Pacific Journ. Math.*, **4**, 47 (1954).
[20] *Ibid.*
[21] H. UMEGAKI: *op. cit.*, II, p. 97.
[22] J. VON NEUMANN: *Ann. Math.*, **41**, 94 (1949), p. 118.
[23] This inequality is closely related to Bell's inequality (see below), and the above argument was in fact suggested by Wigner's derivation of Bell's inequality in *Amer. Journ. Phys.*, **38**, 1005 (1970).
[24] J. BUB: *Found. Phys.*, **6**, 511 (1976).
[25] A. EINSTEIN, B. PODOLSKY and N. ROSEN: *Phys. Rev.*, **47**, 777 (1935).
[26] J. BUB: *op. cit.*
[27] J. S. BELL: *Physics*, **1**, 195 (1964); *Proc. S.I.F.*, Course IL, edited by B. D'ESPAGNAT (New York, N. Y., 1967).
[28] J. BUB: *op. cit.*
[29] S. KOCHEN and E. P. SPECKER: *Journ. Math. Mech.*, **17**, 59 (1967).
[30] A. M. GLEASON: *Journ. Math. Mech.*, **6**, 885 (1957).
[31] S. KOCHEN and E. P. SPECKER: *op. cit.*
[32] J. S. BELL: *Rev. Mod. Phys.*, **38**, 453 (1966).
[33] J. BUB: *op. cit.*, p. 521, 522.
[34] A. EINSTEIN, B. PODOLSKY and N. ROSEN: *op. cit.*, p. 778.
[35] In a Boolean algebra, the principal filter generated by an atom is an ultrafilter (a maximal filter) and has the property that for every element a, either a or the complement of a belongs to the filter, *i.e.* the filter is a *prime* filter. The elements belonging to such a filter are mapped onto 1 by a 2-valued homomorphism on the algebra.
[36] J. M. JAUCH and C. PIRON: *Helv. Phys. Acta*, **42**, 842 (1969).
[37] J. M. JAUCH: *Helv. Phys. Acta*, **37**, 293 (1964).
[38] *Ibid.*, p. 314.
[39] A. DANERI, A. LOINGER and G. M. PROSPERI: *Nucl. Phys.*, **33**, 297 (1962).
[40] J. VON NEUMANN: *op. cit.* [3], p. 438, 439.
[41] E. WIGNER: *Amer. Journ. Phys.*, **31**, 6 (1963).
[42] K. HEPP: *Helv. Phys. Acta*, **45**, 237 (1972).
[43] D. HAAG and D. KASTLER: *Journ. Math. Phys.*, **5**, 848 (1964).
[44] I. E. SEGAL: *Ann. Math.*, **48**, 930 (1947).
[45] K. HEPP: *op. cit.*, p. 239.
[46] E. SCHRÖDINGER: *Naturwiss.*, **23**, 807-812, 824-825, 844-849 (1935).
[47] N. BOHR: *The causality problem in atomic physics*, in *New Theories in Physics*, Polish Intellectual Co-operation Committee and International Union of Physics, (1939), p. 19.
[48] *Ibid.*, p. 20.
[49] *Ibid.*, p. 19.
[50] My original argument here has been modified to take account of an objection by VAN FRAASSEN, who correctly pointed out that the coupling between S and M does not select a *unique maximal* Boolean subalgebra.

[51] J. von Neumann: *op. cit.* [2].
[52] *Ibid.*, p. 33.
[53] *Ibid.*, p. 33.
[54] *Ibid.*, p. 37.
[55] *Ibid.*, p. 38.
[56] J. von Neumann: *Continuous geometry*, in *Collected Works*, Vol. IV, p. 126, 127. In 1937, von Neumann delivered a series of four lectures on *Continuous geometry* at a colloquium sponsored by the American Mathematical Society at Pennsylvania State University. In von Neumann's comments on Bohr's paper at the Warsaw meeting, the fourth lecture is cited as relevant to the questions concerning logic and probability discussed here. von Neumann intended publishing these lectures as a monograph, but the war intervened and the proposed work was never published.

Toward a Generalized Probability Theory: Conditional Probabilities.

G. CASSINELLI

Istituto di Scienze Fisiche dell'Università - Genova
Istituto Nazionale di Fisica Nucleare - Sezione di Genova

P. TRUINI

Scuola di Perfezionamento in Fisica dell'Università - Genova

1. – The main mathematical object of interest in the so-called quantum logic approach to the foundations of quantum mechanics is a pair, usually indicated with $(\mathcal{L}, \mathcal{S})$, such that

$\mathcal{L}$ is an orthomodular lattice,

$\mathcal{S}$ is a set of probability measures, or states, defined on $\mathcal{L}$.

The physical (possibly various) meanings of this mathematical structure and its role in the problem of an axiomatic approach to quantum mechanics are well known (see, for example, [1] and references therein). It is meaningful, however, to study this structure « per se », independently from the intuitive or physical motivations of its definition, as a generalized probability theory. In turn the building-up of such a probability theory could eventually throw light on the mathematical structure of Hilbert-space quantum mechanics as a particular concrete model of this generalized theory.

To understand in what sense we can speak of the pair $(\mathcal{L}, \mathcal{S})$ as a generalized probability theory we recall briefly that, in the classical Kolmogorov axiomatization, a probability space is a triple $\langle \Omega, \mathcal{B}, \mu \rangle$, where Ω is a set, $\mathcal{B}$ a Boolean σ-algebra of subsets of Ω and the probability measure μ is a function from $\mathcal{B}$ to $[0, 1]$ such that

$$\mu(\emptyset) = 0\,, \qquad \mu(\Omega) = 1\,,$$

$$\mu\Big(\bigcup_i E_i\Big) = \sum_i \mu(E_i)$$

if E_i is a sequence of disjoint sets.

We stress that in this definition there is no reference at all neither to the Boolean nature of $\mathscr{B}$, nor to the fact that $\mathscr{B}$ is a σ-algebra of subsets of a set. Only the ordered structure of $\mathscr{B}$ (to give meaning to the set-theoretic union) and the notion of disjointness (which also can be given purely in terms of the order) are needed to give meaning to the definition. If we drop out the requirement of distributivity and look for the minimal mathematical structure on which we can define a probability measure, we are led to consider the pair $(\mathscr{L}, \mathscr{S})$ which we talked of at the beginning.

For convenience we recall here some definitions from orthomodular-lattice theory and measure theory on orthomodular lattices. For details we refer to the existing literature, namely for lattice theory to [2] and for measure theory on lattices to [3, 4].

A complete orthomodular lattice $\mathscr{L}$ is

i) a lattice with 0 and I (that is a partially ordered set with least element 0 and greatest element I, in which every subset $\{a_i\}$ of elements has a least upper bound, denoted by $\vee a_i$, and a greatest lower bound, denoted by $\wedge a_i$),

ii) with an orthocomplementation (that is a mapping $a \mapsto a^\perp$ of $\mathscr{L}$ onto itself satisfying $a \vee a^\perp = I$, $a \wedge a^\perp = 0$, $a \leqslant b \Rightarrow b^\perp \leqslant a^\perp$, $(a^\perp)^\perp = a$),

iii) satisfying the orthomodular identity $a \leqslant b \Rightarrow b = a \vee (a^\perp \wedge b)$. We say that a and b are disjoint (written $a \perp b$) in case $a \leqslant b^\perp$; if $a \perp b$, we write $a \oplus b$ instead of $a \vee b$ and, if $a \leqslant b$, we write $b - a$ for $b \wedge a^\perp$. In this notation the orthomodular identity is written as $a \leqslant b \Rightarrow b = a \oplus (b - a)$.

In an orthomodular lattice $\mathscr{L}$ two elements a and b of $\mathscr{L}$ commute, written $a \,\mathbf{C}\, b$, when $a = (b \wedge a) \oplus (a \wedge b^\perp)$ (which is equivalent to $b = (a \wedge b) \oplus$ $\oplus (b \wedge a^\perp)$).

A probability measure on $\mathscr{L}$ is a function α of $\mathscr{L}$ into $[0, 1]$ such that

$$\alpha(0) = 0\,, \qquad \alpha(I) = 1\,, \qquad \alpha(\oplus\, a_i) = \sum \alpha(a_i)\,.$$

In the sequel we shall consider as our main object of interest a probability measure α on an orthomodular lattice $\mathscr{L}$, and we shall refer to the pair $(\mathscr{L}, \alpha)$ as a generalized probability theory.

Before going on to develop our theory, we want to examine briefly the form that random variables have in this generalized theory. Classically a (real) random variable is a real-valued function f, defined on Ω, which has the following property

$$f^{-1}(E) = \{\omega \in \Omega : f(\omega) \in E\} \in \mathscr{B}$$

for all Borel real sets E.

The inverse image x of f defines a function from the set $\mathscr{B}(\mathbf{R})$ of all Borel real sets to $\mathscr{B}$, which has the following properties:

i) $x(\emptyset) = \emptyset$, $x(\mathbf{R}) = \Omega$;

ii) if E_1 and E_2 are disjoint sets, then $x(E_1)$ and $x(E_2)$ are disjoint sets;

iii) $x\left(\bigcup_i E_i\right) = \bigcup_i x(E_i)$ if E_i is a sequence of disjoint sets.

The probability measure on the real line $\mu(x(E))$ $(E \in \mathscr{B}(\mathbf{R}))$ is called the distribution of the random variable f (in the state μ) and the integral

$$\int_{-\infty}^{+\infty} \lambda \mu\left(x(\mathrm{d}\lambda)\right)$$

is called (when it exists) the expectation of f (in the state μ).

Paraphrasing the classical definition, we can define a (real) random variable on the generalized probability space $(\mathscr{L}, \alpha)$ as a mapping x from $\mathscr{B}(\mathbf{R})$ into $\mathscr{L}$ such that

$$x(\emptyset) = 0\,, \qquad x(\mathbf{R}) = I\,,$$

$$x(E_k) \perp x(E_i)$$

if $\{E_i\}$ is a sequence of disjoint Borel sets,

$$x\left(\bigcup_i E_i\right) = \bigoplus_i x(E_i)$$

if $\{E_i\}$ is a sequence of disjoint Borel sets.

The image in $\mathscr{L}$ of $\mathscr{B}(\mathbf{R})$ with respect to x is a Boolean separable sub σ-algebra of $\mathscr{L}$. The probability measure on the real line $\alpha(x(E))$ is called the distribution of x and the integral

$$\int_{-\infty}^{+\infty} \lambda \alpha\left(x(\mathrm{d}\lambda)\right)$$

the expectation (if it exists) of x; it will be denoted also as $\alpha(x)$.

2. – It is possible to define another generalization of ordinary probability theory, from a point of view different from the one adopted in the preceding section. As we shall need this point of view to understand what we are going to develop, we shall briefly outline, without entering into mathematical technicalities, what is meant by the phrase « noncommutative probability theory ».

Think of a classical probability space: the set of all complex random variables has a natural structure of commutative involutive algebra with respect to usual addition and multiplication of functions, and complex conjugation; the expectation is a linear positive functional (an integral) on this algebra. Consider now the characteristic function of an element b of $\mathscr{B}$, call it $\chi_b(\omega)$, it is obviously measurable and its expectation is, by definition, the probability, $\mu(b)$, of b; moreover, the characteristic functions of the elements of $\mathscr{B}$ are the idempotent (that is $f^2 = f$) and self-adjoint (that is $\bar{f} = f$) elements of the involutive algebra of all random variables, they form a Boolean algebra and the expectation induces on them the measure μ.

The noncommutative extension of classical probability theory is obtained by dropping out the commutativity property of the algebra of all random variables, and thus considering a probability algebra as a pair consisting of a $*$-algebra $\mathscr{A}$ with identity 1 and involution $*$ and a linear functional ω, defined on $\mathscr{A}$, which is positive $(\omega(xx^*) \geq 0$ for all $x \in \mathscr{A})$ and normalized $(\omega(1) = 1)$. As in the classical case the set $\mathscr{P}(\mathscr{A})$ of all self-adjoint and idempotent elements of $\mathscr{A}$ (projectors of the algebra) has a distinguished structure: it is an orthomodular lattice (which is not Boolean as long as $\mathscr{A}$ is not commutative) and ω induces on it a finitely additive measure. If $\mathscr{A}$ has some additional topological structure which allows us to define « infinite sums » and ω has some continuity property (we do not enter into mathematical details), then the restriction of ω to $\mathscr{P}(\mathscr{A})$ is indeed a measure in the sense explained in the previous section.

The algebra $\mathscr{A}$ is usually identified with an algebra of bounded operators on some separable complex Hilbert space $\mathscr{H}$, the resulting noncommutative probability theory has been investigated by several authors [5-10].

We shall pay particular attention in the sequel to the relatively simple case in which $\mathscr{A}$ is the von Neumann algebra $\mathscr{B}(\mathscr{H})$ of all bounded operators on $\mathscr{H}$. In this case a further property is available, namely the fundamental theorem of Gleason [11]: if $\dim \mathscr{H} \geq 3$, then every probability measure on the lattice $\mathscr{P}(\mathscr{B}(\mathscr{H}))$ of projectors uniquely induces a normal positive linear normalized functional on $\mathscr{B}(\mathscr{H})$. Then there is a one-to-one correspondence between probability measures on $\mathscr{P}(\mathscr{B}(\mathscr{H}))$ and normal positive linear normalized functionals on $\mathscr{B}(\mathscr{H})$, and every probability measure α on $\mathscr{P}(\mathscr{B}(\mathscr{H}))$ has the form

$$\alpha(P) = \operatorname{tr}(D_\alpha P) \text{ for all } P \in \mathscr{P}(\mathscr{B}(\mathscr{H})), \qquad \dim \mathscr{H} \geq 3,$$

where D_α is a trace class operator of unit trace, fixed for each α, which is called the density of α.

In view of Gleason's theorem the noncommutative probability theory on the von Neumann algebra $\mathscr{B}(\mathscr{H})$ contains the generalized probability theory on the orthomodular lattice $\mathscr{P}(\mathscr{B}(\mathscr{H}))$; this seems to be the case also for every von Neumann algebra which contains no factor of type I_2 as direct summand (Lodkin's theorem [12]). In spite of these considerations it seems to us meaningful to develop the generalized probability theory, defined in sect. **1**, purely lattice theoretically, without reference to noncommutative theory on algebras, for two reasons at least. The first is that an orthomodular lattice in general cannot be thought of as the lattice of projections of a von Neumann algebra, indeed it is an open and important problem to understand which are the abstract characterizations of the lattice of projections of a von Neumann algebra. The second is that some properties of a von Neumann algebra can be traced back to the lattice-theoretical properties of the set of its projections; we refer, for example, to the dimension theory and classification of factors [13], and to some forms of Radon-Nikodym theorem for operator algebras [3].

Hence, in the sequel, we shall develop some aspects of generalized probability theory in a purely lattice-theoretical way, without referring to the lattice of projectors of a von Neumann algebra but to an abstract orthomodular lattice, in this way we obtain a greater generality but weaker results; nevertheless, we shall use the classical theory of probability and noncommutative probability on $\mathscr{B}(\mathscr{H})$ as hints for the theory we are looking for.

3. – In this section and in the following $\mathscr{L}$ will be an orthomodular lattice and α a probability measure on it. For mathematical simplicity we shall always suppose that $\mathscr{L}$ has a property similar to the separability of Hilbert space, that is we assume that every sequence of disjoint elements of $\mathscr{L}$ is at most countable.

Now we examine the definition of conditional probability of an event (an element of $\mathscr{L}$) given another event.

In a classical probability space, given two elements a and b of $\mathscr{B}$ such that $\mu(b) \neq 0$, we define the conditional probability of a given b, denoted by $P(a|b)$, as

$$P(a|b) = \frac{\mu(a \cap b)}{p(b)} \,.$$

We stress that $P(\cdot|b)$ is a probability measure and that

i) if $\mu(b) = 1$, then $P(a|b) = \mu(a \cap b) = \mu(a)$;

ii) $P(a|a) = 1$.

Let us turn to $\mathscr{B}(\mathscr{H})$; according to the usual quantum theory of measurement, after the *a priori* measurement of $Q \in \mathscr{P}(\mathscr{B}(\mathscr{H}))$, the initial state whose density operator is D changes into a state whose density operator is $D' = QDQ/\text{tr}\,(DQ)$, hence the noncommutative conditional probability of P given Q can be defined as

$$\text{tr}\,(PD') = \frac{\text{tr}\,(PQDQ)}{\text{tr}\,(DQ)} \,.$$

We note that when $P \leqslant Q$ (in the sense of projection operators) we have $\text{tr}\,(PD') = \text{tr}\,(PD)/\text{tr}\,(QD)$, in agreement with the commutative case.

Now we want to find the lattice-theoretical analogue of the state $\mathrm{tr}(\cdot D')$, to the interpreted as a conditional probability. Let a be an element of $\mathscr{L}$, we shall formally denote the transformation of the state α analogue to the transformation $\mathrm{tr}(\cdot D) \mapsto \mathrm{tr}(\cdot D')$ as $\Omega_a \alpha$, where the element a of $\mathscr{L}$ plays the role of the element Q of $\mathscr{P}(\mathscr{B}(\mathscr{H}))$. Then we are looking for the existence of a mapping $\alpha \mapsto \Omega_a \alpha$ of the set $\mathscr{S}$ of states of $\mathscr{L}$ into itself which has the properties agreeing with the intuitive concept of conditioning of a state. The proof of the existence of such a mapping can be found elsewhere [14]; we shall give here only the main results. In an orthomodular lattice $\mathscr{L}$ with the properties specified in [14], for every element a of $\mathscr{L}$, different from 0 and I, there exists a mapping Ω_a of $\mathscr{S}$ into $\mathscr{S}$ such that

i) Ω_a is defined on the subset of $\mathscr{S}$ $\{\alpha \in \mathscr{S} : \alpha(a) \neq 0\}$;

ii) if $\alpha(a) = 1$, then $\Omega_a \alpha = \alpha$;

iii) $\Omega_a \alpha(a) = 1$;

iv) if $a \mathbf{C} b$ and $\alpha(b) = 1$, then $\Omega_a \alpha(b) = 1$;

v) if $a \mathbf{C} b$, then $\Omega_a \alpha(a \wedge b) = \Omega_a \alpha(b)$.

All these properties suggest us that $\Omega_a \alpha(b)$ might be interpreted as the conditional probability of b given a in the state α; nevertheless, to ensure that $\Omega_a \alpha(b)$ is the analogue of $P(b|a)$, we should need the further property

vi) if $b \leqslant a$, then $\Omega_a \alpha(b) = \alpha(b)/\alpha(a)$, which is the analogue of $\mathrm{tr}(PD') = \mathrm{tr}(PD)/\mathrm{tr}(QD)$ when $D' = QDQ/\mathrm{tr}(QD)$ and $P \leqslant Q$.

To find conditions on $\mathscr{L}$ or $\mathscr{S}$ that allow us to prove this property seems a difficult task and it is an open problem in the development of a probability theory on orthomodular lattices. In the sequel we shall suppose that our mappings Ω's have also this property. With this assumption we define the conditional probability of b given a in the state α as $\Omega_a \alpha(b)$.

4. – The main purpose of this paper is to define and find the properties of the conditional probability of an event with respect to a sublattice $\mathscr{L}_0$ of $\mathscr{L}$. We shall give only the main results without entering into the technicalities of the proofs; for all mathematical details we refer to [15]. We shall examine particularly the case in which $\mathscr{L}_0$ is a Boolean sub σ-algebra of $\mathscr{L}$, it being possible to find in this situation some meaningful properties and characterizations.

Classically, if $\mathscr{B}_0$ is a sub σ-algebra of $\mathscr{B}$ and a an element of $\mathscr{B}$, we define the conditional probability of a given $\mathscr{B}_0$, denoted by $P(a|\mathscr{B}_0)$, as the unique function defined on Ω, $\mathscr{B}_0$-measurable, finite almost everywhere with respect to the restriction $\mu_{\mathscr{B}_0}$ of μ to $\mathscr{B}_0$, such that

$$\int_b P(a|\mathscr{B}_0)(\omega)\, \mathrm{d}\mu_{\mathscr{B}_0}(\omega) = \mu(a \cap b) \qquad \text{for all } b \in \mathscr{B}_0 .$$

The existence of $P(a|\mathscr{L}_0)$ is ensured by the classical Random-Nikodym theorem.

We have seen in the previous section that $\mu(a \cap b)$ must be identified in the noncommutative case with $\operatorname{tr}(PQDQ)$. If $\mathscr{L}_0$ is a sublattice of $\mathscr{P}(\mathscr{B}(\mathscr{H}))$, the $\mathscr{L}_0$-probability of $Q \in \mathscr{P}(\mathscr{B}(\mathscr{H}))$ in the state α is defined as a bounded self-adjoint operator to be denoted by $P(Q|\mathscr{L}_0)$ whose spectral resolution belongs to $\mathscr{L}_0$, satisfying

$$\operatorname{tr}(QRD_\alpha R) = \operatorname{tr}\big(P(Q|\mathscr{L}_0)RD_\alpha R\big) \qquad \text{for all } R \in \mathscr{L}_0 .$$

Further discussion on this definition and the proofs of the properties of $P(Q|\mathscr{L}_0)$ can be found in ref. [9], where it is examined the more general case of an arbitrary von Neumann algebra.

In the lattice-theoretical case it is possible to find an analogue of $\operatorname{tr}(QRD_\alpha R)$ by means of the mappings Ω's defined in the previous section. Namely, if $\mathscr{L}_0$ is a sublattice of $\mathscr{L}$ and a is an element of $\mathscr{L}$, the defining property of the $\mathscr{L}_0$-probability of a in the state α is written, in analogy with the noncommutative case, as

$$(4.1) \qquad \alpha(b)\,\Omega_b\alpha(a) = \alpha(b)\,\Omega_b\alpha\big(P(a|\mathscr{L}_0)\big) \qquad \text{for all } b \in \mathscr{L}_0 ,$$

where $P(a|\mathscr{L}_0)$ is a bounded random variable with range in $\mathscr{L}_0$, $\Omega_b\alpha(P(a|\mathscr{L}_0))$ is the usual expectation of a random variable, and we adopt the rule that $\alpha(b)\Omega_b\alpha(a) = 0$ if $\alpha(b) = 0$ (note that $\Omega_b\alpha$ is not defined when $\alpha(b) = 0$).

Now we give some properties which can be deduced from our definition (for the proofs see [15]).

Theorem 4.1. If $P(a|\mathscr{L}_0)$ exists such that (4.1) is satisfied, then

i) the spectrum of $P(a|\mathscr{L}_0)$ is a subset of the interval $[0, 1]$,

ii) $P(I|\mathscr{L}_0) = I$.

Now we want to examine the physically meaningful case in which $\mathscr{L}_0$ is a Boolean sub σ-algebra of $\mathscr{L}$; in this case it is possible to prove the existence of the observable $P(a|\mathscr{L}_0)$ by giving an explicit construction of it. Let us suppose that $\mathscr{B}_0$ is an atomic Boolean sub σ-algebra of $\mathscr{L}$, in this case it is well known that $\mathscr{B}_0$ is atomistic [2], and moreover the set of its atoms is countable (the atoms of a Boolean algebra are disjoint, and every set of disjoint elements in $\mathscr{L}$ is at most countable). Hence every element b of $\mathscr{B}_0$ can be written as $b = \bigoplus_i b_i$, where $\{b_i\}$ is a suitable subset of the set of atoms of $\mathscr{B}_0$. To show the existence of $P(a|\mathscr{B}_0)$ we need a strong hypothesis on $\Omega_b\alpha(a)$, which we shall further discuss at the end of the section. Consider $\alpha(b)\Omega_b\,\alpha(a)$, suppose that, for fixed a and α, $\alpha(b)\Omega_b\alpha(a)$ is a measure on $\mathscr{B}_0$; of course, this is a condition on either a or α (or both). The classical analogue of $\alpha(b)\Omega_b\alpha(a)$ is

$\int_b \chi_a(\omega)\,\mathrm{d}\mu(\omega) = \mu(a \cap b)$, which is obviously a measure with respect to the variable b; in the noncommutative theory the analogue of $\alpha(b)\,\Omega_b\alpha(a)$ is $\mathrm{tr}\,(QAQD_\alpha)$ (where D_α is the density operator of the state α, A is an element of $\mathscr{B}(\mathscr{H})$ and Q is an element of a Boolean sub σ-algebra $\mathscr{B}_0$ of $\mathscr{P}(\mathscr{B}(\mathscr{H}))$). It is possible to show [9] that $\mathrm{tr}\,(QAQD_\alpha)$ is a measure with respect to the variable Q for every density operator D_α if and only if $A \in \mathscr{B}_0'$ (the commutant of $\mathscr{B}_0$), and is a measure for every $A \in \mathscr{B}(\mathscr{H})$ if and only if $D_\alpha \in \mathscr{B}_0'$. Our main result is

Theorem 4.2. $P(a|\mathscr{B}_0)$ exists if and only if

$$b \mapsto \alpha(b)\,\Omega_b\alpha(a)$$

is a measure on $\mathscr{B}_0$.

As shown in [15], the random variable $x \equiv P(a|\mathscr{B}_0)$ has the following characterizations:

i) $x(E) = \oplus\,(b_i : \Omega_{b_i}\alpha(a) \in E)$, for all $E \in \mathscr{B}(\mathbf{R})$ such that $0 \notin E$,

$x(\{0\}) = \oplus\,(b_i : \alpha(b_i)\,\Omega_{b_i}\alpha(a) = 0)$.

ii) The spectrum of x is either $\mathrm{Cl}\,\{\Omega_{b_i}\alpha(a)\}$ if $\alpha(b_i) \neq 0$ for all atoms b_i of $\mathscr{B}_0$, or $\mathrm{Cl}\,\{\Omega_{b_i}\alpha(a)\} \cup \{0\}$ if $\alpha(b_i) = 0$ for some atom b_i of $\mathscr{B}_0$.

iii) $x = \sum_{\alpha(b_i)\neq 0} \Omega_{b_i}\alpha(a)\,x_{b_i}$, where x_{b_i} is the random variable with spectrum $\{0, 1\}$, such that $x_{b_i}(\{1\}) = b_i$ and $x_{b_i}(\{0\}) = b_i^\perp$, and sum of random variables must be understood in the sense that $z = x + y$ when $\alpha(z) = \alpha(x) + \alpha(y)$ for all $\alpha \in \mathscr{S}$ [4]. Note that in the noncommutative case it is possible to derive an expression quite analogous to ours, see corollary 3.1 of ref. [9]. To explain the meaning of the preceding formula, let us think of $\mathscr{B}_0$ generated by the two elements $b, b^\perp$, that is $\mathscr{B}_0 = \{0, b, b^\perp, I\}$. Then $x = \Omega_b\alpha(a)\,x_b + \Omega_{b^\perp}\alpha(a)\,x_{b^\perp}$, which agrees with our previous definition of $\Omega_b\alpha(a)$ as the conditional probability of a given b.

The condition that $b \mapsto \alpha(b)\,\Omega_b\alpha(a)$ is a measure on $\mathscr{B}_0$ can be proved in the lattice-theoretical case under hypotheses quite analogous to those employed in the noncommutative case [9]. We have indeed

Theorem 4.3. $b \mapsto \alpha(b)\,\Omega_b\alpha(a)$ is a measure (not normalized) on a sublattice $\mathscr{L}_0$ of $\mathscr{L}$ for all α if a commutes with every element of $\mathscr{L}_0$.

Under further assumptions on the pair $(\mathscr{L}, \mathscr{S})$ it is possible to find a condition on α which ensures us that $b \mapsto \alpha(b)\,\Omega_b\alpha(a)$ is a measure, on a sublattice $\mathscr{L}_0$ of $\mathscr{L}$, for all $a \in \mathscr{L}$. We do not enter into the intriguing details of this point and refer for them to [15], where the analogy with the noncommutative case is fully exploited.

Concluding, we have found the conditions for the existence of the probability of an event of $\mathscr{L}$ with respect to a sub σ-algebra of $\mathscr{L}$. Thinking of $\mathscr{B}_0$ as the range of a random variable, it is easy to understand that this is the first fundamental step in the way of defining the « conditional expectation » of a random variable, given another one, which is an essential problem in any probability theory.

REFERENCES

[1] E. G. Beltrametti and G. Cassinelli: *Riv. Nuovo Cimento*, **6**, 321 (1976).
[2] F. Maeda and S. Maeda: *Theory of Symmetric Lattices* (Berlin, 1970).
[3] S. S. Holland jr.: *Trans. Amer. Math. Soc.*, **108**, 66 (1963).
[4] S. P. Gudder: *Pacific Journ. Math.*, **19**, 81 (1966).
[5] H. A. Dye: *Trans. Amer. Math. Soc.*, **72**, 243 (1952).
[6] I. E. Segal: *Ann. of Math.*, **57**, 401 (1953).
[7] H. Umegaki: *Tohoku Math. Journ.*, **8**, 86 (1956).
[8] N. Nakamura and T. Turumaru: *Tohoku Math. Journ.*, **6**, 182 (1954).
[9] S. P. Gudder and J. P. Marchand: *Journ. Math. Phys.*, **13**, 799 (1972).
[10] L. Accardi: *Funct. Anal. Appl.* (in Russian), **9**, 1 (1975).
[11] A. M. Gleason: *Journ. Math. Mech.*, **6**, 885 (1957).
[12] A. A. Lodkin: *Funct. Anal. Appl.* (in Russian), **8**, 54 (1974).
[13] L. H. Loomis: *Mem. Amer. Math. Soc.*, **18**, 1 (1955).
[14] G. Cassinelli and E. G. Beltrametti: *Comm. Math. Phys.*, **40**, 7 (1975).
[15] G. Cassinelli and P. Truini: preprint University of Genoa (November 1977).

Formal Analysis of Physical Theories.

M. L. Dalla Chiara

Istituto di Filosofia dell'Università - Firenze

G. Toraldo di Francia

Istituto di Fisica Superiore dell'Università - Firenze

Introduction.

The sixth of the celebrated twenty-three problems posed by Hilbert at the International Congress of Mathematicians, held in Paris in 1900, was the axiomatization of physical theories. He said:

« Durch die Untersuchungen über die Grundlagen der Geometrie wird uns die Aufgabe nahe gelegt, *nach diesem Vorbilde diejenigen physikalischen Disziplinen axiomatisch zu behandeln, in denen schon heute die Mathematik eine hervorragende Rolle spielt* ... » [1].

Several more or less successful attempts in this direction have appeared since. One can mention the axiomatizations of classical mechanics by Hamel [2] and others [3], the axiomatization of thermodynamics by Carathéodory [4], the efforts of Hilbert himself, as well as of von Neumann [5] and several others, on quantum mechanics, and so on. As to probability theory, which Hilbert explicitly included in the physical theories, suffice it to name Kolmogorov [6].

If we are not grossly mistaken, this kind of work has never widely appealed to physicists (though the situation seems to have been changing in recent times). The prevalent opinion appears to be that it is futile to spend time and energy in a task, which merely represents an academic exercise, and will never lead to a real advancement of physics. Some workers go as far as to bluntly affirm that axiomatization simply ends up in a sort of fake physics.

There is certainly a grain of truth in all this, and it must be recognized that a certain type of mathematical physics, which is still widely found in scientific journals, does all but give strength to the physicist's critical attitude.

There is however, in our opinion, a widespread misunderstanding which must be dispelled. To *axiomatize* is not to *dogmatize*!

Axioms, in modern science, are certainly not construed as sacred dogmas, engraved in marble, once for all. They are simply exact and formalized expressions of the assumptions, which, at a certain historical epoch, are made in the theory of a certain branch of science. In physics, of course, the axioms and all their consequences must agree with experimental evidence, and this largely determines the *evolution* of physical theories. But, both in order to be sure of this agreement, and to ascertain the self-consistency of the theory, it is necessary 1) to use a rigorous language, 2) to express the assumptions in an inequivocal fashion, and 3) to be aware of the rules of inference that are applied. In other words, one must *formalize* the theory.

It must be recognized that, in physics, different *degrees* of formalization are accepted in different contexts, and that it is quite sensible to do so. It is nonetheless useful for everybody to lay the *foundations* of physics in a reasonably formalized fashion. Only when the general conceptual layout is understood with sufficient precision, can one pass with confidence to develop the more specialized and less formalized fields of science.

It is worth-while to mention a few characteristic features of *the kind of formalization that we present in these lectures* (*). As a general remark, let us stress that our aim throughout is to adhere as much as possible to a *realistic* picture of physics. We want our formalization to mirror a real science, a science that has to do with laboratories, instruments and measurements. We want to describe in formal terms *what physicists actually do*, and not to prescribe what they *ought to do*. As is well known, it is difficult to resist the temptation of dealing all the time with a sort of *ideal* science, of which real physics represents, as it were, an awkward and distorted reproduction. When one takes this view, one can be led to believe that real physics is always missing the pure truth, the crystalline core of that ideal science. We find this attitude fruitless, to say the least, and accordingly try a different approach.

First, it will be emphasized that any reasonable formalization must make clear what are the rules of inference that are made use of. Some people may believe that, to this end, it is sufficient to mention and to make explicit only the *deductive* rules. We think, on the contrary, that also the *inductive* rules of inference used in physics are of essence for a correct formalization. We do not want, of course, to solve Hume's problem and to «justify» induction; our much less ambitious aim is to make clear what rules of inductive inference are used with confidence by physicists, and what other rules have long been discarded, because they have been found unreliable. In this context, we find [12] that essentially *one* single rule of inductive inference is sufficient to build physical science, namely the rule of the space-time invariance of probability. We will maintain that, in contrast with what is believed by a number of inductive logicians, a physical law represents the universal assertion of a probability, and not the assessment of the probability of a universal assertion.

(*) Among the modern authors who have worked on the general problem of the formalization of physics, one can mention SUPPES [7], SNEED [8], PRZELECKI [9], as well as the present authors [10]. Partial analogy with our line can be found in Wójcicki's approach [11].

Secondly, we introduce, as an essential part of the theoretical structure of physics, *the precision* of the apparatus used to collect the experimental evidence. In this way, we avoid the pitfall represented by the assumption of a fake physics, the physics where the result of a measurement is a real number. As is well known, this kind of physics has never existed, and its formalization can even be misleading.

Thirdly, the above approach allows us to define the concept of *truth* in physics, in a satisfactory way. Abandoning the unacceptable notion of *approximate truth*, we are able to show that physics can arrive at true assertions. Instead of speaking of increasingly more approximate theories, we speak of true theories, valid in increasingly larger domains. In the same context, we are able to define the concept of deterministic theory, and to show in what sense its predictions can be true.

Finally, we will show that our approach can shed a considerable amount of light on a number of much debated problems, arising in the logic of quantum mechanics.

We will first discuss the « logician's dilemma of quantum mechanics » (classical logic or quantum logic?). We will stress that the deductive structure of quantum theory, in its present form, seems to be essentially founded on a kind of « mixture » of different logics. Nevertheless, a classical *modal interpretation* of quantum logic permits to solve, to a certain extent, the « logical schizophrenia » of the quantum physicist. This will lead us to distinguish two different concepts of truth within quantum theory: an *empirical* truth and a *quantum-logical* truth, the latter intuitively corresponding to a kind of « *a priori* certainty ». The distinction between these different ideas of truth will have a bearing also on the much debated question concerning the measurement problem of quantum theory.

1. – The inductive inference in physics.

1·1. *What is an inductive inference?* – The whole of science is derived from two main sources: experience and intellectual processing. Experience *per se* can *at most* (if anyway: see subsect. **2·1**) give us knowledge of isolated historical facts, but only a suitable processing can lead us to classifications, laws and theories, without which there is no science.

It has been held for a long time, at least from classical Greece up to our days, that, apart from classification, which is based upon the recognition of common characters in different individuals, two main intellectual procedures are applied in the construction of science, namely *deduction* and *induction*. While the former is commonly believed to be above criticism, in the sense that, if correctly applied to a true assertion, it can only lead to other true assertions, the latter's claim to represent a valid procedure to be used in science has often been challenged on quite serious grounds. In particular, the very clear and drastic criticism set forth by David HUME in his *Enquiry Concerning Human Understanding* seems to have never been convincingly refuted, and scholars largely agree that it is essentially correct.

However, one would wish to know what exactly is meant by « justifying », in this context. When and how would we agree that induction has been « justified »? Perhaps, we would like to *deduce* the soundness of induction. But the question soon arises: to deduce *from what*? Perhaps, from a universally accepted truth; but that *some* form of induction must be valid is precisely a universally accepted truth! No man (or higher animal) could survive without accepting and applying some form of induction.

The validity of induction is sometimes justified by saying that *Nature is regular*. This statement is certainly too vague, but can be made more precise. If we take sense stimuli as our information source, we discover that such a source does not have maximum entropy. In other words, a given collection of sense stimuli is not unconditionally compatible with any other collection of sense stimuli. The *real* sensory world is a member of a proper subset of the set of all « possible » sensory worlds. *Defining this subset is the task of science* [13].

However, all this talk about Nature's regularity is merely *a posteriori* speculation, and cannot be seriously taken as a rigorous justification of the inductive procedure. All we can say *at best* is that, *so far*, Nature has been regular. We cannot by any logical means conclude that Nature *is* (and consequently *will be*) regular. The legitimacy of prediction is still unproven.

Sometimes, a *philogenetic* explanation of induction has been put forward, but that represents no proof either. Briefly, let α denote the sentence « Nature is regular » and β the sentence « natural selection endows the individual with a propensity to use induction ». One can probably assert that $\alpha \rightarrow \beta$; but the assertion $\beta \rightarrow \alpha$ is simply the very old error of inverting material implication.

A sensible attitude to take may be the pragmatic one. If Nature *is* regular, the strategy of inductive inference can be successful, while, if Nature *is not* regular, no strategy can be successful, and science is impossible. No one can predict the next element of a completely random sequence. As a consequence, we have better assume that Nature *is* regular.

A mere denial of the legitimacy of *any* process of induction does not help us very much in analysing the foundations of science, as long as we recognize some objectivity, or at least *intersubjectivity*, to science. It is widely admitted that an observational report cannot serve as a basis for science, unless it can be made intersubjective (dreams and hallucinations do not help us to understand the outer world). But as soon as one assumes that something is intersubjective, one has made full use of an inductive inference. We believe that our observational report: « This paper is white », has an intersubjective validity, because in all (but finite in number) cases that we have reported « white », we have obtained confirmation from all (but finite in number) people.

From this, we jump to the conclusion that in *all* cases, *all* people *will* agree with our statement.

It is largely believed that, unlike the legitimacy of the inductive inference, the legitimacy of the *deductive* inference is simply *evident* and *indisputable*. However, this may be to a large extent a delusion, unduly supported by a long philosophical tradition. We have recently heard a good logician ask: why should one be certain that, if $\alpha \to \beta$ and α, then β? In our opinion, the first and strongest reason is that so far, every time we have asked ourselves that question, we have carried out a rapid inner experiment, and have obtained the answer: because it is evident. And all of our fellow men seem to have obtained the same answer. But who can assure us that tomorrow, on asking the same question, all of us will not obtain a different answer? Nobody can, if HUME is right. As a matter of fact, we can « justify » deduction, only by applying a sort of inductive procedure! Again and again, we hit upon Hume's problem.

Anyway, the *certainty* of deductive logic was connected in the course of history with a sort of faith in the *a priori* nature of logic and in its uniqueness. Today we know very well that this is only a prejudice. The opinion that logic is empirical, in much the same way as physics, is gradually gaining ground among logicians.

Thus, the request of justifying induction seems to be identical with that of justifying the validity of the whole of human knowledge, a probably hopeless task, which certainly cannot concern us here. We have no ambition to solve this old and much debated philosophical problem. We therefore take the scientist's, rather than the philosopher's attitude, and, setting aside the question about justification (*quid iuris?*), we will confine ourselves to the question of how induction is actually applied in physics (*quid facti?*).

Physicists have a long record of using induction as a tool. It is well known that, by doing experiments and applying the inductive procedure, they have learned a lot about the physical world. However, it is not very widely realized that, at the same time, *they have learned something about induction itself*. It stands to reason that, while uncorking bottles, one should learn something about the corkscrew!

Today, the situation is considerably different from what it appeared to Francis BACON, or even to Isaac NEWTON. Perhaps, physicists have not learned from experience what kind of inductive procedure *is* philosophically *justified*. This would be a *petitio principii*, for it would involve induction. But they have definitely learned what procedures are *wrong*, by stumbling upon a number of *counterexamples*. The doctrine that inductive procedures cannot be *verified*, but can be *falsified* by experimental evidence, has been called by one of us a meta-Popper theory [14].

There is a point that we want to emphasize. Physics, by its own nature, strives to arrive at rigorous and fully reliable statements. It is only on this level that physicists have been *forced* to give up some very common and long cherished forms of induction. But they can very well make use of such forms of inference on a *heuristic* level. For instance, in any trial and error procedure, the trials are suggested by some sort of induction, otherwise they would not make sense.

Accordingly, we will only refer to the *reliable* forms of induction, those that can lead to statements accepted with perfect confidence by the physics community.

Even so reduced to the *quid facti* question, the problem still offers some curious difficulties to the formalist. The first and very big difficulty encountered is that it is not at all easy to make a rigorous and clear-cut distinction between deduction and induction! There is probably some analogy with the pairs *analytic-synthetic, fact-law*, for which no rigorous distinctions can be made, as has been argued by QUINE [15, 16]. We shall, therefore, content ourselves with a *semi-formal* treatment, and accept a few pragmatic and conventional notions that are currently used in empirical sciences.

A *rule of inference* is any rule which allows us to derive from a set of sentences (or formulae) $\alpha_1, \ldots, \alpha_n$ a new sentence β, and will be denoted by

$$(1) \qquad \frac{\alpha_1, \ldots, \alpha_n}{\beta}.$$

The rules of deduction of classical logic (CL) are particular examples of this definition.

Let L represent a language, whose only logical constants are the usual connectives ($\wedge, \vee, \rightarrow, \neg$) and the quantifiers ($\forall, \exists$). Let A be a set of sentences α_i of L, which will be taken as the *premises*, and D the system of rules of deduction of CL.

An *inductive rule* of inference IR with respect to $\langle L, A, D\rangle$ must satisfy the two conditions:

$$(2) \qquad \text{IR is compatible with CL},$$

$$(3) \qquad \exists\beta\,(\beta \in L) : A \not\vdash_D \beta, \quad \text{and} \quad A \vdash_{D+\text{IR}} \beta.$$

In some particular cases, the scientist may derive a sentence γ from A by applying an IR, even though $A \vdash_D \gamma$. This can happen either because the scientist is not aware of the latter possibility, or because otherwise the calculations would be too long and difficult. Of course, this will generally require a suitable extension of A, to include the results α_i' of some *ad hoc* experiments, but the assertion $A \vdash_D \gamma$ will remain valid. In such cases we can speak of a nonessential application of an IR.

In order to say something more specific or restrictive than the very general conditions (2), (3), one must examine the concrete forms that an IR can have in science.

1·2. *The many-to-all rule.* – We will now restrict our discussion to those IR's which have the form usually applied by empirical scientists to derive *general laws.*

Let $c_1, c_2, \ldots$ represent a (possibly infinite) set of names for the individuals (cases) of a universe of interest U. We will assume that no two different names c_i, c_j apply to one and the same element of U. Let $\alpha(x)$, $\beta(x)$ represent two *properties*, which can apply to the elements of U. The IR of empirical science can be put in the general form

$$\frac{\alpha(c_1)\wedge\ldots\wedge\alpha(c_n)}{\forall x\,\beta(x)}, \tag{4}$$

where $n<\infty$ and x ranges over U. Usually, $\alpha(c_i)$ represents a conjunction

$$\alpha(c_i)=\alpha_1(c_i)\wedge\alpha_2(c_i)\,, \tag{5}$$

while $\beta(x)$ represents the conditional

$$\beta(x)=\alpha_1(x)\to\alpha_2(x)\,. \tag{6}$$

For instance, α_1 and α_2 may describe the properties « to be an electron » and « to have spin $\frac{1}{2}$ » (in $\hbar$ units), respectively. Every electron that is found by experience to have spin $\frac{1}{2}$, that is every particle that is found to have the property (5), is said to represent a *positive instance* for the universal assertion

$$\forall x[\alpha_1(x)\to\alpha_2(x)] \tag{7}$$

that *all* electrons have spin $\frac{1}{2}$.

We will distinguish three subcases of the rule specified by (4), (5), (6). They will be called the *all-to-all* (AA), the *many-to-all* (MA) and the *one-to-all* (OA) rules, respectively.

In the AA case, the universe has exactly n elements, so that $\alpha(c_1), \ldots, \alpha(c_n)$ exhaust the universe in respect of α. This case is trivial and does not yield an IR, because the result can be *deduced* by CL (provided, of course, the premises include the cardinality of U), contrary to the requirement (3).

The most popular of the three subcases is that of the MA rule, which is defined by $1\ll n<$ (cardinality of U). Many people still believe that *induction* is nothing but the MA rule. Of course, scholars are aware that the MA rule cannot have an absolute validity. But many of them are likely to claim that the more numerous the positive instances (5) are, *in the absence of negative instances*, the better the universal assertion (7) is *confirmed* (Nicod's criterion) [17, 18]. In other words, *positive instantiation* should render the universal statement (7) more « probable », in one or other of the different meanings that the notion of probability can have.

Now, physicists have known for a long time that that criterion is certainly *unsound*. The reason is that they have learned, at the cost of repeated failures, that, no matter how many positive instances (5) have been found, the world where

$$\exists x[\alpha_1(x) \wedge \neg\, \alpha_2(x)] \tag{8}$$

is *not less probable* than the world where (7) is valid. This is because modern physics has taught us that, contrary to common belief, *very rare events are very common*!

An example will clarify this point. Let α_1 represent the property « to be a ^{238}U nucleus » and α_2 the property « to live longer than one hour ». We can observe one thousand, or even one million, ^{238}U nuclei and still be (virtually) certain that all of them have the property α_2. This is because less than one out of ten thousand billion ^{238}U nuclei decay within an hour. Nevertheless, *if we conclude that* (7) *is valid, we are wrong.*

Still, one might be tempted to think that, after all, we are only *very little* wrong in assuming the validity of (7), and that we are thereby missing a negligible detail of the physical world. Nothing could be farther away from the truth. We would instead miss a fact of tremendous importance in nuclear theory and in cosmology!

Would a physicist, after observing one million ^{238}U atoms, be justified in saying that (7) is at least highly *probable*? By no means, for the case of the uranium atom does not represent a rare exception in modern physics. On the contrary, it is rather the rule, due to the fact that any potential barrier can be tunnelled through, and *will* be tunnelled through, in the long run. In particle physics, very rare modes of decay are always to be expected.

For these reasons, a physicist is not authorized to think that a world in which (8) is valid is less probable than a world in which (7) is valid.

The unsoundness of the MA rule could have been suspected even on purely theoretical grounds, owing to the paradoxes it gives rise to. Such *paradoxes of confirmation* were described by HEMPEL [19]. Ironically, the attitude taken on this subject by many scholars is that one should try hard and « solve » the paradoxes, rather than simply recognize that the MA rule and Nicod's criterion are wrong.

Let us briefly recall the paradoxes, by going back to the example of the electron and its spin, and taking for U the universe of elementary particles. The universal assertion (7) is logically equivalent to

$$\forall x[\neg\, \alpha_2(x) \rightarrow \neg\, \alpha_1(x)]\,, \tag{9}$$

that is every particle that does not have spin $\frac{1}{2}$ is not an electron. According to Nicod's criterion, a positive instance of (9) is represented by a c_i for which

$$\neg\, \alpha_2(c_i) \wedge \neg\, \alpha_1(c_i)\,. \tag{10}$$

Therefore, a particle which neither has spin $\frac{1}{2}$ nor is an electron would confirm (9). And, since (9) is logically equivalent to (7), one should conclude that the observation of, say, a π-meson confirms that all electrons have spin $\frac{1}{2}$!

But, by juggling with the formulae, one can also do some other curious tricks. For instance, the universal assertion (7) is also equivalent to

$$(11) \qquad \forall x\{[\alpha_1(x) \vee \neg\, \alpha_1(x)] \rightarrow [\neg\, \alpha_1(x) \vee \alpha_2(x)]\}\ ,$$

which can be confirmed by *any* particle, whether an electron or not, even by a proton, which has spin $\frac{1}{2}$.

No physicist can accept these queer conclusions. HEMPEL maintains that « the startling results are paradoxical only in a psychological, but not in a logical sense ». Be that as it may, we need not *explain away* the paradoxes, once we have recognized that the MA rule (even in its probabilistic version) is unsound.

1·3. *The* OA *rule.* – Have then physicists followed Popper's advice, and given up any attemp to apply induction? The answer is no. But, although this is not very widely realized, they have confined themselves to the application of only a special case of the OA rule. The experiments of physics are essentially unique. When an experiment is repeated several times, it is either because one wishes to eliminate any possible error, or because one wants to improve on the precision of the measurement. But it would be absurd to repeat the experiments in order to test whether or not Nature is uniform and well behaved.

The good behaviour of Nature is taken for granted, but only *in a very restricted sense.* It is not the broad and vague sense described by Hume's words: « From causes which appear similar, we expect similar effects ». In this context, the notion of « similar » does not lend itself to an easy and convincing analysis. The principle on which the OA rule of physics rests is the principle of *space-time invariance* of physical laws. In simple words, it says that Nature acts *here now* in the same way as it acts *there then*, both in the past and in the future.

This principle is taken as a *postulate*, not only in physics or in science in general, but even in everyday life. When we walk, eat or speak, we rely completely on the assumption that Nature (including our body) will behave at the next instant of time in a given place exactly in the same way as we have known her to behave at past times, in different places. If the content of this postulate were false, we could not live at all.

So frequent and general is the application of the postulate, that it seems to have acquired the role of an *a priori* structure of our mind. Probably, one should not ascribe to it the necessary and immutable character of Kantian *a priori*, nor is it an *innate* structure of the human mind. It seems rather to be one of those structures that are dealt with in the theory of *cognitive develop-*

ment, and are acquired very early by children. Since it is a *necessary* condition for carrying out life as we know it, it soon becomes an apparently necessary condition of human thought.

As a matter of fact, for the purposes of everyday life, the postulate is necessary only in a *weak* sense: it need only concern a suitable space-time *neighbourhood* of our present location. The *universal* validity of space-time invariance is not required by any inborn or acquired structure of our mind, but represents rather an estension of the natural structure, brought about by modern science. As is well known, GALILEO had a hard time when he tried to convince his contemporaries that the laws of Nature are the same in the heavens as on the earth.

We will now state in precise terms the rule of inference derived in classical physics from the postulate of space-time invariance. Let $r = VT$ and $r' = V'T'$ represent two regions of space-time, such that the *space* volumes V, V' are both at rest in one and the same inertial frame F (say, as a first approximation, our laboratory frame) and can be superimposed on each other by translation and/or rotation, while the *time* intervals T, T', as measured in F, can be superimposed on each other by translation in time. We will say that r and r' are mutually *congruent*. Let IB (r) represent the assertion that a specified set of initial and boundary conditions (the latter being possibly functions of time) are verified in r, and $\mathrm{P}(r)$ the assertion that a certain physical phenomenon takes place in r. If we take as our universe U the equivalence class of all space-time regions congruent with r, the OA rule of classical physics can be written in the form

$$\frac{\mathrm{IB}(r) \wedge \mathrm{P}(r)}{\forall x[\mathrm{IB}(x) \rightarrow \mathrm{P}(x)]}\,. \tag{12}$$

The paradox of confirmation is avoided by the particular form that the rule (12) must strictly adhere to. For, it is true that the conclusion of (12) is logically equivalent to

$$\forall x[\neg\, \mathrm{P}(x) \rightarrow \neg\, \mathrm{IB}(x)]\,, \tag{13}$$

but this has not the form that appears in (12). Indeed nobody would deem to derive conclusion (13) from a positive instance of the form $\neg\, \mathrm{P}(r_i) \wedge \neg\, \mathrm{IB}(r_i)$. For instance, consider a guitar string with the initial and boundary conditions set to give the note A, so that it does give the note A. One can say that this represents an instance of the form $\neg\, \mathrm{P}(r_i) \wedge \neg\, \mathrm{IB}(r_i)$, where P stands for the sound of the note G and IB for the stimulation of the note E. We cannot certainly conclude therefrom that, every time that we do not obtain G, we have not stimulated E!

The principles of special and general relativity seem to be successive extensions of the rule (12). Precisely, the principle of special relativity extends the notion of congruence to space-time regions measured in *different* inertial frames, provided that all measurements of IB and P are also made in those frames. General relativity is an attempt to do away with the requirement that the frames be inertial. But it must be emphasized that these generalizations are themselves based, for their experimental verification, upon the simple and prerelativistic rule (12). For instance, we must take for granted that, if a clock prepared in a certain manner and travelling at a certain speed with respect to our laboratory appears to us to slow down by a certain amount, the same will happen to *all* identical clocks in identical conditions.

Of course, the ascertainment that the IB conditions are really identical in each case is vital for a correct application of space-time invariance. As an example, take the case of the 11-year sunspot cycle. Astronomers had too hastily assumed that at the outset of each cycle the Sun is in the same initial conditions, and that consequently sunspots must appear in each cycle, only to discover to their dismay that for 70 years (from 1645 to 1715) there had virtually been no sunspot! No wonder that, from wrong premises, one should have derived a wrong conclusion.

This is as far as the application of the OA rule goes in classical physics, for those processes that do not involve the notion of *probability*. But, as everybody knows, probability is widely present, even at the level of classical physics, and moreover represents the cornerstone of the modern theory of microphysics. Accordingly, we will now turn to the probabilistic case.

We will refrain here from entering a long discussion about the concept of probability. This would lead us astray from our main subject, and we, therefore, refer the reader to the critical analyses to be found elsewhere. We will content ourselves with observing that in the *theory* of physics the interpretation of probability is largely that of a *relative frequency*. But we do not want thereby to deny that, in the *metatheory*, a *subjectivistic* interpretation is sometimes understood, as when one says: it is probable that quarks exist.

Sticking to our *quid facti* attitude, let us first examine the procedure of an astrophysicist who, observing a great number n of stars, finds that all of them share the property $\alpha_2(x)$. Does he conclude that *all* stars have the property $\alpha_2(x)$? By no means. No modern astrophysicist would be so naive as to apply that old-fashioned sort of inductivism. Nor would he assert that it is *probable* that all stars share the property $\alpha_2(x)$. He would instead say that the probability of a star *not* having property $\alpha_2(x)$ is $O(1/n)$ or less.

This conclusion can be justified by a simplified Bayesian argument. Let $p = a/n^k$ represent the probability of a *negative instance*, that is the probability that a star does not have property $\alpha_2(x)$, with $a > 0$ and $k > 0$. The probability of the sequel of n positive instances is then $(1 - a/n^k)^n$. When $n \to \infty$ this expression tends to 1 for $k > 1$, to $\exp[-a]$ for $k = 1$, and to 0 for $k < 1$. Hence, if the astrophysicist discards the hypothesis of having hit upon an extremely rare case, he will infer that p is $O(1/n)$ or less. The same type of argument would be applied by a physicist investigating the possible decay channels of a particle, or of a radioactive nucleus.

In the case of *probabilistic* laws, the specification of the IB conditions is *incomplete*. In classical physics, this means that not all the parameters characterizing the IB conditions have been precisely fixed; some of them can vary at random (within a given range) in an unknown way. In microphysics, besides this kind of incompleteness (which may either be present or not in the specification of the initial state), there are of course the limitations imposed by the uncertainty principle. Conventionally, we will denote the incomplete and/or uncertain IB conditions by $\widetilde{\text{IB}}$.

We will now rely on an extension of the space-time invariance postulate, which may be called the *space-time invariance of probability*.

Let E indicate an *event* or the result of a prescribed set of measurements. Denoting by $p_r(E)$ the probability that the event E takes place in the space-time region r, one may be tempted to write the probabilistic OA rule in the form

$$\frac{\widetilde{\text{IB}}(r)\wedge p_r(E)=p}{\forall x[\widetilde{\text{IB}}(x)\rightarrow p_x(E)=p]}\,. \tag{14}$$

However, this does not make much sense, because the value p of the probability cannot be measured on a single case r. Of course, one must instead consider *a set of repeatable experiments*.

As is well known, there has been a long controversy about the meaning of p, some people having maintained that it represents the limit which the relative frequency f of occurrence of E tends to, when the number n of the experiments tends to infinity. We do not want here to argue in favour or against this definition. But we want to stress that one will never be able to measure an *exact* value p; what one derives from a set of experiments is f.

The sensible thing to do is to recognize that, when $0<f<1$, a physicist will write $p=f\pm\varepsilon$, and say that p can be any real number of the interval $f-\varepsilon$, $f+\varepsilon$, where $\varepsilon=O(n^{-\frac{1}{2}})$. When $f=0$ or 1 the interval reduces to 0, ε or $1-\varepsilon$, 1, respectively, and $\varepsilon=O(n^{-1})$.

Let us now take as our universe U the set of all finite subsets of the equivalence class of all space-time regions congruent with r. If R is one such subset, let $\widetilde{\text{IB}}\,(R)$ indicate the statement that the $\widetilde{\text{IB}}$ conditions apply to all the n elements of R. Further, denote by $f_R(E)$ the relative frequency and by $p_R(E)$ the probability of occurrence of E in R. In the case of $0<f<1$, the OA rule will be written as

$$\frac{\widetilde{\text{IB}}(R)\wedge f_R(E)=f}{\forall X[\widetilde{\text{IB}}(X)\rightarrow p_X(E)=f\pm O(n^{-\frac{1}{2}})]}\,. \tag{15}$$

A similar rule will hold in the case $f=0$ or 1.

The meaning of the botton row of (15) is this: the physicist expects that, in a new set of n' trials, event E will occur $n'p \pm O(n'^{\frac{1}{2}})$ times, where p can have any value in the interval $f \pm O(n^{-\frac{1}{2}})$.

Strictly speaking, the specification of the event E should *precede* the n trials. However, one does not always adhere to this procedure and accepts any specification of E that is not too far-fetched and *could* reasonably be imagined *a priori*.

Rule (15) is very powerful and of very general application. But, because it refers to the probability of an event E, one may have some difficulty to recognize that it can apply also to the derivation of the quantitative laws of physics, expressed by equations of the type $f(q_1, \ldots, q_m) = 0$, where the $q_1, \ldots, q_m$ represent the values of m physical quantities. Even without anticipating the formal discussion about physical quantities (see subsect. **2**·2), we can recall that an equation of this kind has, of course, a domain of validity, specified by the class of phenomena to which it applies, by the allowed range of each quantity q_i, and by the precision with which each quantity is measured. Due to this last factor, the correct form of the equation, when f is a continuous and differentiable function, is

$$|f(q_1, \ldots, q_m)| \leqslant \varepsilon \,, \tag{16}$$

where ε depends on the precisions, or resolving powers, of the instruments used.

When one states the physical law (16), one apparently implies that (16) should be true for *any* m-tuple of values found for the $q_1, \ldots, q_m$, within the allowed ranges. But this cannot be verified experimentally, for it would require an infinite number of experiments. When we assert (16), we assert something even about *unobserved* cases that are not space-time congruent with *observed* cases. Is this a sound procedure? Of course not. Spectral lines, resonances in particle physics, and related phenomena, have taught us that experimental plots cannot be smoothed out without risk. In a totally unknown case, one can come across a fine spike anywhere.

The correct way to describe the procedure is this. One measures the relevant quantities in n cases, selecting the conditions *at random* within the allowed ranges of the q_i (IB conditions), and calculates the expression $f(q_1, \ldots, q_m)$. If, in at least one case, $|f|$ is found to exceed ε, the law (16) is not valid and one should look for a different function. If, on the contrary, (16) is found to be verified in all n cases, one must conclude that *the probability that the result of a new measurement does not satisfy* (16) *is* $O(n^{-1})$ *or less*. It will be emphasized that, strictly speaking, the form of the function f should be selected *before* carrying out the n measurements, for, *a posteriori*, one could choose any wild function that passes through the n points measured. In practice, there is much tolerance on this point. But everyone knows that such laws as, say, Bode's law for the distances from the Sun to the planets are not much more than guess-work. In order to verify to some degree of accuracy Bode's law, one should try it on a number of *new* planets. This has been done once, and has led to the discovery of the asteroid belt. However, one single (and somewhat doubtful) case cannot be considered to represent a sufficient verification.

One may be surprised to note that we are thus converting all physical laws into probabilistic laws. But this is precisely what physical laws are. In this connection, one should take into account two things.

First, in the case of the classical and time-honoured laws of physics, n is so large that people tend to disregard $1/n$ and simply say that (16) is *verified.* There is no harm in this, but one must be aware that such locution is just an abbreviation and a matter of convention.

Second, it is well known from statistical mechanics that the laws of classical (macroscopic) physics *are* probabilistic. There is a finite probability that any one of them be violated in a particular case. However, the probability is generally so small, that one can take it to be exactly zero. No one would normally say that it is only *probable* that heat should pass from a hotter to a colder body. This is why the OA rule of classical physics, as expressed by (12), can generally be applied with perfect confidence. In *theory* it is wrong, and must be replaced by the more correct rule (15), but in *practice* it is nearly always applicable in macrophysics. Nevertheless, it is good to remember that there may be exceptions, as is also stressed by PRIGOGINE in this volume [20]. As for the laws of microphysics, we do know that they *are* essentially probabilistic.

Before ending this section, we note that, although we have confined ourselves, for the sake of simplicity, to the very common case when the function f is continuous and differentiable, one can apply the same concepts also to the case of discontinuous functions, with only slight modifications. In this way one can deal, for instance, with the case of « qualitative » properties, by using step functions, and assigning the value 1 to « having the property » and 0 to « not having the property ». If the property is a function of q, one should be very careful and take the samples at random over the *whole* range of interest of q. We may note, incidentally, that in this way one can avoid the apparently paradoxical result pointed out by GOODMAN for a predicate that changes abruptly, say tomorrow [21]. That predicate is a function of time, and it is certainly true that we cannot conclude anything as to its application in the future, by utilizing uniquely the evidence collected till today.

2. – Physical quantities and physical states.

2·1. *Observation and operation.* – Not many physicists are familiar with a problem, which apparently is still worrying the minds of several science philosophers. It has to do with the nature of the terms that are used in the language of physics. Allegedly, the language of physics contains three different types of terms:

1) mathematical terms;

2) nonobservational, or theoretical terms;

3) observational terms.

There has been a lot of argument on whether or not all theoretical terms can be conveniently « reduced » to the observational terms.

We will not deal with the nature of mathematical terms, and will discuss briefly the relation between observational and theoretical terms, having chiefly in mind the problem of defining physical quantities.

In fact, it has sometimes been claimed that the distinction between the primary and derived quantities of physics is akin to the distinction between observational and theoretical terms. Take, for instance, the case of the specific weight of a body x; if one assumes that the weight $w(x)$ and the volume $v(x)$ of the body can be directly « observed », the specific weight $s(x) = w(x)/v(x)$ is claimed to be a theoretical term [22].

In our opinion, there is a misunderstanding, which arises when one fails to analyse correctly the concept of « directly observable ».

What do we mean, when we affirm that we observe, say, a visible object? Today we know pretty well the chain of processes that take place in vision.

A beam of radiation, say solar light, hits the object O and is scattered thereby. In order to collect the information carried by the scattered light, we could employ a great number of photosensitive and directive devices, placed far apart from O, and thus detect the light scattered in different directions by the different points of O. In the language of ordinary optics, we would be examining the scattered light *ray by ray*. By taking the results of all these measurements and feeding them into a computer, we would be able to derive all the visual properties of the object.

Of course, one could make use of an analogue computer, capable of simulating an *inverse scattering*, *i.e.* a process where the light would retrace its steps and go back to O. A very simple and convenient data processor of this kind is represented by a convergent lens, which collects the scattered radiation and forms a real image of O over a screen. Nature discovered this analogue computer a long time ago, when she endowed us with the eye. When we use this computer, we say that we *directly observe* the object.

But what *fundamental* difference is there between this case and the case when we use a conventional digital computer? Perhaps the fact that our computer is made up of organic substance? Let us for a moment accept this queer point of view and claim that with the eye we see *directly* the object. We will encounter some difficulties.

For one thing, the light from the object does not (ordinarily) travel through a vacuum, but reaches us through the atmosphere. For the sake of the argument, let us forget this fact. But suppose we are looking at a fish in a pond. Do we see it directly, even if the rays come to us through the water? Presumably, we will say yes. What if the fish is in an aquarium and the ray paths are partly inside the glass? And if the aquarium wall is slightly convex, and we see the fish with a magnification? What if the wall is replaced by a lens? And if the lens is replaced by two lenses, forming a microscope? As we see, we have passed with continuity from direct observation to observation through an optical instrument.

It is, therefore, a fallacy to believe that the distinction between direct observation and observation by means of an instrument may have any precise and fundamental meaning. We have also established that an instrument can include (and in most cases *does* include) an analogue computer. Then, why not a digital computer? What conceptual difference is there between using a lens and using a system of electronic circuits? And what basic distinction can be made between the case of an electronic computer and the case of a person using paper and pencil and whatever kind of mathematics is desired?

It seems difficult to avoid the conclusion that no clear-cut distinction can be made between what is generally called *direct observation* and observation plus *data processing*. There is *always* some amount of data processing.

At this point one can even dispense with the restriction to visible light. It is evident that information can be collected from any suitable kind of scattered radiation, such as X-rays, electrons, α-particles, and so on, as is largely done in modern physics. As a consequence, one can very well call such terms as « atom », « nucleus », « proton », *theoretical terms*, if one so wishes; but then « fish » is a theoretical term, too. No harm in this; in our opinion, one could very well maintain that there are no purely observational terms in physics. All terms are, to some extent, theoretical.

If this is true, the distinction between observational and theoretical terms is arbitrary and nonessential, in physics. A logical analysis of physics based on that distinction is only likely to generate noise and confusion.

Going back to the definition of physical quantities, we observe that the usual textbook distinction between primary and derived quantities is largely arbitrary and conventional, even if useful from a tutorial point of view. In particular, it is perfectly possible to build an instrument that directly measures specific weight $s(x)$, in much the same way as other instruments can measure separately weight $w(x)$ and volume $v(x)$. Whether the quotient $w(x)/v(x)$ is carried out by an instrument or by a person with paper and pencil has no foundational significance.

What is essential instead is that the chain of operations to be carried out in order to measure the quantity of interest should be precisely specified. This is what has been called the *operational definition* of physical quantities.

Operationalism has been the object of a lot of criticism, sometimes as a result of an analysis, and sometimes on more or less metaphysical grounds.

It would be pointless for us to enter here this discussion. We will limit ourselves to emphasizing that we are not defending here a general and philosophical point of view; we are only analysing the precise implications of a methodology that is largely accepted by physicists. The purpose of the operational definition of a physical quantity is to ensure, as far as possible, 1) that we know what we are talking about, and 2) that any two physicists are talking about the same things. The fact that, even with the operational definition, the path of the theoretician is strewn with difficulties can only confirm that it is vital to use very clear and unequivocal definitions. We hardly believe that dispensing with the operational definition of physical quantities can help to overcome the difficulties.

2·2. *Generalized operational definition of a physical quantity.* – Having accepted the operational definition of a physical quantity, as being represented by the description of the sequence of operations required to measure it, let us now turn to a more precise and formal analysis of the relevant concepts.

Suppose we have defined a physical quantity by means of an unequivocal sequence of operations (possibily including some operations with paper and pencil). As is well known, this definition is not necessarily *unique*. Many different sequences of operations can define one and the same physical quantity. For instance, length can be measured either by means of a yardstick, or by means of an optical triangulation, or by means of a radar set, and so on. By properly adjusting the units of measurement, one can obtain from all of these procedures the same results. From Bridgman's point of view, each procedure defines a different physical quantity, which, in principle, should have a different name. When we say that any two of these procedures yield the same results, we essentially state an empirical *law*.

Theoretically, this attitude seems to be perfectly correct. However, it is extremely inconvenient, and does not represent the way in which physicists usually talk of physical quantities.

A sensible way out is to identify a physical quantity with the *class* of all the different procedures which give one and the same result for its measurement. However, the issue is more complicated than it might appear, and one must be careful.

We will be talking of (*isolated*) *physical systems* and of *procedures of measurement*. These terms will be largely understood as *primitive terms*, however a little comment will be in order, to make clear for the reader what we are talking about.

A physical system σ will be represented, for us, by all the physical entities existing in a space-time region $r = VT$. Note the difference from the commonly accepted physical jargon, in which our σ would rather be described by the long expression « a physical system in V, considered during the time interval T ».

As a particular case, a physical system can consist of an ensemble (*statistical ensemble*) of identical subsystems, each one existing in a different element $r \in R$ of a class R of congruent regions.

As to the concept of «isolated», it is well known that its precise definition involves a tremendous difficulty. Of course, in our work of formalization we cannot have the ambition to solve the most profound difficulties of physics. We will, therefore, take «isolated» as a primitive term and admit that isolated physical systems exist. In practice, when one suspects that a physical system σ_1 is influenced by another system σ_2, one will have to consider the overall system $\sigma_1 + \sigma_2$.

A procedure of measurement π will be represented by a sequence of operations to be performed on σ. The prescription can possibly include also the specification of which operation should be carried out at what time (with respect to the initial time). When π is applied to σ, it yields a *result* $\pi(\sigma)$. What exactly is a «result» will be defined later on.

More generally, a procedure of measurement π can be applied to several different parts of one and the same physical system σ, to yield a number $n \geqslant 1$ of distinct results, which will be indicated by $\pi_{(1)}(\sigma), \ldots, \pi_{(n)}(\sigma)$. Think, for example, of the measurement of different masses, or charges, on σ.

We shall say that a procedure π is *defined* with respect to σ, when it can be applied to σ, to yield a nonempty set of results $\pi_{(1)}(\sigma), \ldots, \pi_{(n)}(\sigma)$.

Let Σ represent a class of physical systems. We will call $\Pi(\Sigma)$ the class of all procedures that are defined on every element σ of Σ. Let π and π' be two procedures of $\Pi(\Sigma)$. We will say that π and π' are *equivalent* with respect to Σ if and only if,

A) for every $\sigma \in \Sigma$, the application of π and π' to the same parts of σ yields the same number n of results $\pi_{(1)}(\sigma), \ldots, \pi_{(n)}(\sigma)$ and $\pi'_{(1)}(\sigma), \ldots, \pi'_{(n)}(\sigma)$, respectively;

B) for every pair $\sigma_1, \sigma_2 \in \Sigma$ for which $\pi_{(i)}(\sigma_1)$ and $\pi_{(i)}(\sigma_2)$ represent equal results (for all values of i), also $\pi'_{(i)}(\sigma_1)$ and $\pi'_{(i)}(\sigma_2)$ represent equal results, and the other way around.

The notion «equal result» is, for the time being, understood in an intuitive sense, but will be made more precise later. For the sake of simplicity, the distinction among $\pi_{(1)}(\sigma), \ldots, \pi_{(n)}(\sigma)$ will often be omitted in the sequel and we will simply talk of a single result $\pi(\sigma)$.

As an example of equivalent procedures, we may refer to the measurement of temperature between the freezing and boiling points of water, by means of a mercury thermometer (π), or an alcohol thermometer (π'). A water thermometer is instead not equivalent to π (or π') in this range, for the volume of water increases both above and below 4 °C; hence, one and the same result obtained with a water thermometer can correspond to two different results of the mercury (or alcohol) thermometer.

Whenever π and π' are equivalent, one can adjust the units or the scales, so as to obtain equal results from both π and π'.

The equivalence relation above defined is clearly reflexive and symmetric; it is also transitive, because of the transitivity of the logical connective « implication ». Thus, it specifies a set of equivalence classes $C_a, C_b, \ldots \subset \Pi(\Sigma)$. By abstraction, $C_a, C_b, \ldots$ will define the physical quantities $Q_a, Q_b, \ldots$.

Let us now select a pair of quantities Q_1, Q_2, referring to two different classes Σ_1, Σ_2 of physical systems, such that $\Sigma_1 \cap \Sigma_2 = \emptyset$. It is a result of experience that in some case the corresponding equivalence classes C_1, C_2 may have a nonempty intersection.

As an example, let us take the case of the measurement of length. Class C_1 may include all procedures which can be applied and give equal results for the length L_1 of any object in the range 1 m to 1 km (class Σ_1). Class C_2 may include instead all procedures which can be applied and give equal results for the distance L_2 from the Earth to any object of the solar system, say 10^5 to 10^{10} km (class Σ_2). It is evident that, although C_1 and C_2 do not coincide (for, *e.g.*, the Earth-Sun distance cannot be measured with a yardstick), their intersection is not empty (for, *e.g.*, length in both ranges can be measured by optical triangulation). It is this intersection that allows us to match the units (according to some *reasonable* criteria, such as continuity, linearity, etc.) and to consider L_1 and L_2 as one and the same physical quantity. In other words, length can be defined by $C = C_1 \cup C_2$ with respect to $\Sigma = \Sigma_1 \cup \Sigma_2$.

We are now in a position to formulate a *generalized operational definition* of a physical quantity. A physical quantity Q is defined by the union $C = \bigcup \{C_k\}$ over a set $\{C_k\}$ of equivalence classes of measuring procedures, such that

a) set $\{C_k\}$ is connected, namely, if $\{C_k\} = F \cup G$ and $F, G \neq \emptyset$, then $\bigcup F \cap \bigcup G \neq \emptyset$;

b) each C_k is defined over a well-determined class Σ_k of physical systems and $\Sigma_k \neq \Sigma_l$ for $k \neq l$.

For convenience, we will replace the latter condition with the stronger condition that all Σ_k are disjoint, namely $\Sigma_k \cap \Sigma_l = \emptyset$ for $k \neq l$.

Condition *b*) is more important than it might appear. In order to see this, let us suppose that class C, defined over Σ, includes m procedures $\pi_1, \ldots, \pi_n$, $\pi_{n+1}, \ldots, \pi_m$, that are applicable also in the domain Σ'. In this domain, we may find that, while $\pi_1, \ldots, \pi_n$ are mutually equivalent and $\pi_{n+1}, \ldots, \pi_m$ are also mutually equivalent, no one element of the first set is equivalent to an element of the second set. Thus, a number of procedures that belong to one and the same class for Σ are found to belong to two different classes C_1', C_2' for Σ'.

This situation is often likely to arise when Σ refers to ordinary macroscopic bodies, while Σ' refers to microphysics or to cosmic physics. For instance, suppose that Σ consists of a set of macroscopic solid spheres, whose radius r we want to measure. Procedure π_1 may consist of packing closely together a number n^3 of spheres, identical to the one of interest, inside a cube of side l, and taking for r the value $r = l/2n$. Procedure π_{n+1} may consist of shooting at the sphere a collimated beam of low-energy electrons of uniform intensity I (number of electrons across unit surface per unit time), counting the number N of electrons removed from the primary beam per unit time, and taking $r = (N/\pi I)^{\frac{1}{2}}$. As is well known, procedures π_1 and π_{n+1} are equivalent for all spheres in the macroscopic range. But both π_1 and π_{n+1} apply also to a class Σ' of spheres in the domain of atomic sizes. Precisely, π_1 consists of packing together the atoms in a crystal and measuring the overall size, while π_{n+1} consists of measuring the scattering cross-section for electrons of a specified energy. Obviously, π_1 and π_{n+1} are not equivalent in Σ', since two atoms of different elements may occupy the same volume in a crystal, yet have different scattering cross-sections. As a consequence, π_1 and π_{n+1} belong to two different equivalence classes C_i, C_j with respect to Σ' and, in that range, define two *different* physical quantities. Now, condition *b*) guarantees that C_i, C_j do not both belong to $\{C_k\}$.

There are also many other ways to measure the radius of an atom, each one giving a different result. If we want to insist on speaking of *the* radius of an atom, we must arbitrarily choose one particular equivalence class C_i defined over Σ' and privilege it with respect to the others. In that case, of course, one ought not to forget mentioning C_j, when giving the result of the measurement. Actually, this is what physicists sometimes do, when they say that the radius of such an atom, when measured in such a way, has such a value.

We will say that a physical quantity Q is defined with respect to a physical system σ, when the operational definition of Q includes a class C_k of procedures associated to a class Σ_k of physical systems and $\sigma \in \Sigma_k$. As σ belongs to one Σ_k only, and all procedures are equivalent in C_k, we can define the *result* of Q on σ as being given by $Q(\sigma) = \pi(\sigma)$, where π represents any one of the procedures of C_k.

2·3. *Deterministic and probabilistic quantities.* – Let us now state more precisely what is a *result*. It is convenient to distinguish two kinds of physical quantities:

I) those quantities whose results are commonly thought to be *real numbers*, apart from certain «errors» which depend on the *precisions* (or resolving powers) of the instruments used in the π procedures;

II) those quantities whose results are *probability distributions* of real numbers, apart from a twofold error, determined both by the precisions of the instruments and by the precision with which probability is measured (number of trials).

Obviously, a quantity of the second kind can only be measured on a statistical system, consisting of a number n ($\gg 1$) of similar subsystems.

The quantities of the first kind will be termed *deterministic* (and will be denoted by D_i), while the quantities of the second kind will be termed *probabilistic* (and will be denoted by p_i). It will be emphasized that this is just a matter of nomenclature and *does not imply any theoretical commitment.* Later on, we shall see that a quantity which is deterministic in a given context can become indeterministic in a different context.

Let us first deal with the quantities of the first kind. In this case, the result of the application of a procedure π to a physical system σ is a real number, affected by an error ε (which depends both on σ and on π). As is well known, it is customary to write

$$\pi(\sigma) = r \pm \frac{\varepsilon}{2}, \tag{17}$$

where ε is a positive number.

In other words, $\pi(\sigma)$ is nothing but an *interval* of real numbers of length ε, centred at r.

Of course, when measurements are repeated, with the purpose of increasing the accuracy, one ends up with a statistical (mostly Gaussian) distribution of errors. The real numbers of the interval have different *weights*, and $\varepsilon/2$ represents, say, a convenient multiple of the standard deviation. In this case, the result should be represented by a *fuzzy set* of real numbers rather than by an interval. One might adapt the discussion to this more sophisticated procedure. However, for the sake of simplicity, we will simply refer to intervals of real numbers, as represented by (17). Incidentally, one can always take ε sufficiently large as to be virtually sure that no new measurement will fall outside the interval. This, in turn, means that the probability of a new result falling outside the interval is so small that it can be neglected.

Two real numbers r_1 and r_2 will be said to be *ε-equal* if and only if $|r_1 - r_2| \leq \varepsilon$. In that case we shall write

$$r_1 \underset{\varepsilon}{=} r_2 . \tag{18}$$

Two intervals of real numbers I_1 and I_2 will be said to be ε-equal ($I_1 \underset{\varepsilon}{=} I_2$) if and only if there exists an interval of length ε that contains both of them. Obviously

$$I_1 \underset{\varepsilon}{=} I_2 \quad \text{if and only if } \forall x \in I_1,\ \forall y \in I_2 : x \underset{\varepsilon}{=} y . \tag{19}$$

Further

$$r_1 \underset{\varepsilon}{=} r_2 \quad \text{if and only if } \forall \varepsilon' > \varepsilon : r_1 \underset{\varepsilon'}{=} r_2 . \tag{20}$$

The relation of ε-equality (for real numbers and for intervals alike) is clearly reflexive and symmetric, but it *is not transitive.* Incidentally, this is well known to physicists, as can be illustrated by the problem of the reproduction of standard units: one cannot build, say, 100 standards of length $M_1, \ldots, M_{100}$, by comparing M_1 with the primary standard M_0, then M_2 with M_1, M_3 with $M_2, \ldots, M_{100}$ with M_{99}.

Note that, by virtue of the introduction of the ε-equality, the possible results of the measurement of a D-quantity turn out to represent a *fuzzy space*, a notion which has received some attention in recent times [23, 24]. A fuzzy space is defined as a pair $\langle X, \tau \rangle$, where X is a set with a symmetric reflexive relation $\tau \subseteq X \times X$. As is evident from the above discussion, the space of the measurements of experimental physics represents a particular case of this, X being the set of real numbers (or of intervals of real numbers) and τ being ε-equality. Also a more general notion of *hazy space* has been introduced [24], but we will not make use of it.

Let π and π' represent two measuring procedures belonging to the class $\Pi(\Sigma)$. We will say that π and π' are *ε-equivalent* with respect to Σ if and only if,

A) for every $\sigma \in \Sigma$, the application of π and π' to the same parts of σ yields the same number n of results $\pi_{(1)}(\sigma), \ldots, \pi_{(n)}(\sigma)$ and $\pi'_{(1)}(\sigma), \ldots, \pi'_{(n)}(\sigma)$, respectively;

B') for every pair $\sigma_1, \sigma_2 \in \Sigma$ for which $\pi_{(i)}(\sigma_1) \underset{\varepsilon}{=} \pi_{(i)}(\sigma_2)$ (for all values of i), also $\pi'_{(i)}(\sigma_1) \underset{\varepsilon}{=} \pi'_{(i)}(\sigma_2)$ and *vice versa*.

Evidently, the relation of ε-equivalence is but a particular case of the relation of equivalence, defined with A) and B); one has only to interpret the relation «to be equal results» appearing in B) as the relation «to be ε-equal intervals» appearing in B'). Hence, ε-equivalence is reflexive, symmetric and transitive, and defines for each value of ε a number of equivalence classes $C^{\varepsilon}_a, C^{\varepsilon}_b, \ldots \subset \Pi(\Sigma)$.

A physical quantity Q^{ε} can be defined by the union $C^{\varepsilon} = \bigcup \{C_k\}$ over a set C_k of equivalence classes of measuring procedures, subject to the same conditions a), b), as was $\{C_k\}$. However, since the range of a physical quantity can span several orders of magnitude, it would be convenient to introduce here a small modification in the whole procedure, which does not alter the conceptual frame. We ought to replace the *absolute* precision ε with the *relative* precision $\alpha = \varepsilon/r$ and require that α (not ε) be constant for all the classes and within the classes. But, both for the sake of simplicity and because, in practice, even the relative precision is often allowed to vary in the range, we will not alter the notation.

Obviously, the definition of a physical quantity depends now on the (relative) precision. This is the *largest* value of the precisions of the instruments used in all procedures π of the set that defines the quantity. But, of course, in practical applications, one will be interested in setting this largest value as small as possible.

There is an important consequence of all this. The precisions used in the definition of Q will usually reflect the state of the art reached in a given historical period, that is they will be the best precisions allowed by the instruments. When technology advances, and we pass from ε to $\varepsilon' < \varepsilon$, we may find that one and the same quantity splits into two or more quantities, because a number of procedures, that were ε-equivalent within Σ_k, are not ε'-equivalent.

This situation has occurred several times in macrophysics (*e.g.* in the case of mass and weight measured at different places, or in the case of the specific heats of gases); it *will* occur again if we ever find that inertial and gravitational mass are not identical.

In atomic physics, to say that *the* radius of an atom is on the order of 10^{-8} cm has a correct physical meaning, while it would not make sense to measure it with an accuracy $\varepsilon = 10^{-11}$ cm, for the definition of radius would then be ambiguous. Another striking example is represented by the measurement of the position x at time t of a free mass point m. As long as the precisions are those of macrophysics, a procedure ε-equivalent to the others of C_k^ε can consist in measuring the position x_0 and the momentum p_x at time t_0 and taking $x = x_0 + p_x(t - t_0)/m$. But if we go to microphysics, the situation changes, because, as soon as we measure p_x with sufficient precision, we lose much knowledge of x_0. Therefore, the procedure is no longer ε-equivalent to the more conventional ones.

Let us now turn to the quantities of the second kind, or probabilistic quantities. We shall call ***probabilistic procedure*** of measurement a procedure π which applies to a statistical ensemble $\sigma \in \Sigma$, and whose result is a *probability distribution* $p(x)$ of real numbers, apart from a double error, determined by the resolving power ε^A of the apparatus used and by the accuracy ε^P with which we measure probability. The latter, of course, depends on the number n of trials and is, in general, $O(n^{-\frac{1}{2}})$. Let us make these notions more precise.

If $p(x)$ is a continuous probability distribution, we can plot it against x, in rectangular co-ordinates, and obtain a continuous curve. If from each point of the curve we trace a cross, centred on it, whose horizontal and vertical arms are ε^A and ε^P in overall length, respectively, we obtain a strip $\mathcal{S}$. The result $\pi(\sigma)$ of the application of a probabilistic procedure π to a statistical ensemble σ is represented by the set of all probability distribution $\hat{p}(x)$ whose curves lie entirely within $\mathcal{S}$. Precisely, $\pi(\sigma)$ is the set of all functions $\hat{p}(x) \geqslant 0$, subject to the consistency condition

$$\int_{-\infty}^{+\infty} \hat{p}(x)\,\mathrm{d}x = 1 \tag{21}$$

and such that for every x

$$\text{either} \quad \hat{p}(x) = p[x + \eta(x)\,\varepsilon^A]\,, \quad \text{or} \quad \hat{p}(x) = p(x) + \theta(x)\,\varepsilon^P\,, \tag{22}$$

or both, with $|\eta(x)|$, $|\theta(x)| \leqslant 1$.

Briefly, we will write

$$\pi(\sigma) = p(x) \pm \langle \varepsilon^A, \varepsilon^P \rangle \tag{23}$$

by analogy with (17).

Note that the result of having determined $p(x)$ experimentally by means of an apparatus of resolving power ε^A is not only the indeterminacy of the abscissae by $\pm\,\varepsilon^A/2$, but also the fact that each $p(x)$ represents the result of an integration over an interval ε^A. More precisely, let $e(x)$ denote the response of the apparatus to a δ-function; then $p(x)$ must have the form of a convolution

$$p(x) = \int_{-\infty}^{+\infty} e(y)\, p_0(x - y)\, dy\,. \tag{24}$$

It will be useful to recall that, if $\bar{x}$ and $\bar{x}_0$ denote the mean values of x determined with $p(x)$ and $p_0(x)$, respectively, we have

$$\bar{x} = \bar{x}_0\,. \tag{25}$$

Moreover, if δ, δ_0 and δ_e denote the standard deviations of $p(x)$, $p_0(x)$ and $e(x)$, respectively, we have

$$\delta^2 = \delta_0^2 + \delta_e^2 = \delta_0^2 + O(\varepsilon^{A^2})\,. \tag{26}$$

Finally, we recall that the Fourier spectrum of $p(x)$ is the product of the Fourier spectra of $p_0(x)$ and $e(x)$. Since high frequencies are very little, if at all, represented in the Fourier spectrum of $e(x)$, the same can be said for $p(x)$. As a consequence, $p(x)$ is a fairly well-behaved function; it cannot present very fine spikes or abrupt jumps or very fine ripples.

Of course, in order to ascertain that a probability distribution, found experimentally, belongs to the $\langle \varepsilon^A, \varepsilon^P \rangle$-strip, one must apply induction, as was seen in subsect. 1·3 for the case of the functions $f(q_1, \dots, q_n)$. For the sake of simplicity, we will put ourselves definitely in the case in which the probability of a new point of $p(x)$ not falling within the $\langle \varepsilon^A, \varepsilon^P \rangle$-strip is negligible. It is only a matter of making a sufficient number of measurements. By the same token, we note that the nature of ε^P is essentially different from that of ε^A. As has been mentioned earlier, ε^A reflects the state of the art at a given epoch; there is nothing one can do in order to reduce ε^A, except waiting for the advent of a more refined technology. On the contrary, ε^P can, in many cases, be made as small as is desired; it is only a matter of doing a sufficient number of experiments. For this reason, ε^P can (and will) often be neglected.

One may wonder how strictly the belonging of $\hat{p}(x)$ to the $\langle \varepsilon^A, \varepsilon^P \rangle$-strip determines the relevant parameters of the distribution, such as the mean value of x and the standard deviation, as compared with the corresponding parameters of $p(x)$. Let us assume, without loss of generality, that the mean value of $\bar{x}$ of x for $p(x)$ is zero, and let us call δ and $\hat{\delta}$ the standard deviations of $p(x)$ and $\hat{p}(x)$. For $\hat{p}(x)$ we have

$$(27)\qquad \begin{cases} \hat{\bar{x}} = \displaystyle\int_{-\infty}^{+\infty} x\hat{p}(x)\,\mathrm{d}x\,, \\ \hat{\delta}^2 = \displaystyle\int_{-\infty}^{+\infty} (x-\hat{\bar{x}})^2\hat{p}(x)\,\mathrm{d}x\,. \end{cases}$$

Mathematically, there will probably be some pathological cases, in which these quantities differ abnormally from those of $p(x)$. However, in most cases of physical interest this does not occur. For instance, let us take the very common case in which $p(x)$ is an even-symmetric function, which for $x \to \infty$ vanishes faster than $1/x^2$. This is the case for the Gaussian distribution as well as for the distribution of the particles arriving at a screen at a distance l, after passing through a slit. (Note that in the latter case the usual $\sin^2(\alpha x)/x^2$ distribution must be completed with an additional factor $1/(l^2+x^2)$, due to increasing distance and inclination.) Neglecting ε^P, we have from (22), to the first order in ε^A,

$$(28)\qquad |\hat{\bar{x}}| \leqslant \varepsilon^A\int_{-\infty}^{+\infty}|x|\,|p'(x)|\,|\eta(x)|\,\mathrm{d}x \leqslant \varepsilon^A\int_{-\infty}^{+\infty}|x|\,|p'(x)|\,\mathrm{d}x = -\,\varepsilon^A\int_{-\infty}^{+\infty}xp'(x)\,\mathrm{d}x\,.$$

Upon integration by parts, we obtain

$$(29)\qquad |\hat{\bar{x}}| \leqslant \varepsilon^A$$

and this is exactly what we would have liked to find. A similar calculation carried out for (27) leads to

$$(30)\qquad |\hat{\delta}^2 - \delta^2| \leqslant 3\varepsilon^{A^2} + 2\varepsilon^A\int_{-\infty}^{+\infty}|x|p(x)\,\mathrm{d}x + 2\varepsilon^{A^3}p(0)\,.$$

Note that the integral on the right-hand side is the absolute mean deviation of $p(x)$, and represents an alternative definition of the dispersion of $p(x)$; hence, it will generally be of the same order of magnitude as δ (for instance, for a Gaussian it has the value $(2/\pi)^{\frac{1}{2}}\,\delta$). Hence, one can readily verify that when the order of magnitude of δ is greater than or equal to that of ε^A, and $p(0)$ is on the order of 1, the difference $|\hat{\delta}-\delta|$ is on the order of ε^A. On the other hand, due to (24), $p(0)$ cannot have a very large value (and represent a fine spike), hence pathological cases are excluded.

Two strips $\mathscr{S}_1$, $\mathscr{S}_2$ will be said to be $\langle\varepsilon^A, \varepsilon^P\rangle$-equal when there exists a $\langle\varepsilon^A, \varepsilon^P\rangle$-strip that includes both of them. Similarly, two probability distributions $p_1(x)$, $p_2(x)$ will be $\langle\varepsilon^A, \varepsilon^P\rangle$-equal when there exists an $\langle\varepsilon^A, \varepsilon^P\rangle$-strip that contains both of them. In this case, we shall write

$$p_1(x) \underset{\langle\varepsilon^A,\varepsilon^P\rangle}{=} p_2(x)\,. \tag{31}$$

Let π and π' represent two measuring procedures of the second kind, both belonging to the class $\Pi(\Sigma)$. We will say that π and π' are *$\langle\varepsilon^A, \varepsilon^P\rangle$-equivalent* with respect to Σ if and only if,

A) for every $\sigma \in \Sigma$, the application of π and π' to the same parts of σ yields the same number n of results $\pi_{(1)}(\sigma), \ldots, \pi_{(n)}(\sigma)$ and $\pi'_{(1)}(\sigma), \ldots, \pi'_{(n)}(\sigma)$, respectively;

B'') for every pair $\sigma_1, \sigma_2 \in \Sigma$ for which $\pi_{(i)}(\sigma_1) \underset{\langle\varepsilon^A,\varepsilon^P\rangle}{=} \pi_{(i)}(\sigma_2)$ (for all values of i), also $\pi'_{(i)}(\sigma_1) \underset{\langle\varepsilon^A,\varepsilon^P\rangle}{=} \pi'_{(i)}(\sigma_2)$ and *vice versa*.

One can, therefore, define a probabilistic quantity P in much the same way as was done for a deterministic quantity D. One has only to replace the ε-equivalence with the $\langle\varepsilon^A, \varepsilon^P\rangle$-equivalence.

In conclusion, let σ be a physical system and Q a deterministic (probabilistic) quantity. Then σ belongs to one (and only one) $\Sigma_k \subset \Sigma$, associated with the class C_k of ε-equivalent ($\langle\varepsilon^A, \varepsilon^P\rangle$-equivalent) procedures. Any one of the procedures $\pi \in C_k$ can be applied to σ, giving a result within the accuracy ε ($\langle\varepsilon^A, \varepsilon^P\rangle$). Obviously, due to the ε-equivalence ($\langle\varepsilon^A, \varepsilon^P\rangle$-equivalence), it is immaterial which particular π has been selected out of C_k. We can put $\pi(\sigma) = Q(\sigma)$ and call $\pi(\sigma)$ «a result that can be obtained by measuring the quantity Q in σ».

Clearly, the precision ε ($\langle\varepsilon^A, \varepsilon^P\rangle$) is determined by the specifications of σ and Q. It will be termed the *canonical precision* of Q with respect to σ. A result obtained with this precision is a *proper* result. Of course, in many cases one can use results of less than canonical precision. This can happen, either because the capability of the apparatus has not been fully exploited, or because the results are derived from merely theoretical information. In this case we will talk of *improper* results. A typical example is represented by trivial results (as for instance $D(\sigma) = (-\infty, +\infty)$).

The association of a (proper or improper) result to a physical system σ and a physical quantity Q represents a *measurement*. A *proper measurement* will be the application of a procedure $\pi \in C_k$ to a system $\sigma \in \Sigma_k$, with the canonical precision.

2·4. *Physical states and physical situations.* – Obviously, each procedure π is associated with a certain apparatus A, which, in turn, is a physical system.

In general, there are no logical reasons to discard a situation of strong self-reference, where π is applied to A itself. For instance, a meter stick can obviously measure itself. Later on we will discuss the well-known complications that arise in this connection, in quantum mechanics.

Every measurement requires a certain time, and *time* is also one of the quantities we may want to measure. Again, there does not seem to be any logical difficulty in admitting a form of self-reference. The observer measures both *his time* (during which he does the measurement) and the time of the *outer system* under investigation, by means of a clock A. As the apparatus A is characterized by a certain precision ε_t, time is necessarily measured with a precision not better than ε_t.

We have previously considered a physical system as consisting of all the physical entities that exist in a region of space-time. We will now get nearer the ordinary approach, by talking of *instantaneous* systems. An instantaneous system $\sigma(t_0)$ is, in principle, defined as a constant-time cross-section of a system σ. But, of course, it would be more correct to write $\sigma(t_0 \pm \varepsilon_t)$ and to remember that it is a *slice*, rather than a cross-section, of space-time. Anyway, we will conventionally talk of a system σ at time t_0. As a particular case, σ may consist of an ensemble of identical subsystems, shifted in space and/or time with respect to one another. As an example, one may consider many identical particles shot successively through a hole, or through a two-hole interferometer. In this case, it would be convenient to talk of one and the same time t_0. To this end, the clock will be set at $t = 0$ every time the experiment is repeated.

Suppose we want to measure on $\sigma(t_0)$ a quantity Q defined for σ. The measurement requires a certain time $t_1 - t_0$. If at the end of this time we have obtained a certain result, does the result apply to $\sigma(t_0)$ or $\sigma(t_1)$? Obviously, if $t_1 - t_0 \leqslant \varepsilon_t$, there is no problem. But in a great number of concrete cases we shall have $t_1 - t_0 \gg \varepsilon_t$. In these cases, we will make the convention that the result refers to the final time and will write it in the form $Q[\sigma(t_1)]$ (or, more briefly, $Q(t_1)$).

A most important remark must be made at this point. When we state that the result $Q[\sigma(t_1)]$ of a measurement has been obtained with a procedure π, applied during time $t_1 - t_0$, we must obviously include in π *all* the operations carried out on σ during $t_1 - t_0$. Hence, even before knowing quantum mechanics, one can recognize that, when two or more quantities are measured *simultaneously* (*i.e.* during $t_1 - t_0$), the measurements are not independent of one another. As far as the physical *interactions* with σ are concerned, the procedure π_i selected for each Q_i must be *identical* with the procedures adopted for all the other quantities measured simultaneously; the procedures can differ only in the data reading and processing operations. Moreover, the common part of the π_i's must be applied only *once* for all.

Let us clarify the point with an example. Suppose we want to measure successively (but during $t_1 - t_0$) the momentum p_x and the position x of a particle. The procedure π_p for measuring p_x must consist 1) of all the *physical* operations of a sequence usually applied for measuring p_x, followed by 2) all the *physical* operations usually applied for measuring x, followed by 3) the *data reading and processing* operations needed to obtain p_x. This π_p certainly belongs to one class C_k for the quantity p_x. The procedure π_x for measuring x consists of the same physical operations 1) and 2) in the same order, plus 3′) the reading and processing operations needed to obtain x. In classical physics, all this caution *can be* (but not necessarily *is*) redundant, and the values of p_x and x turn out to be independent of one another. In microphysics, on the contrary, it is of paramount importance, in that the overall procedure affects the canonical precisions. In our example, the precision ε_x for x can be that allowed by the best instruments available, but that of p_x cannot be better than $\hbar/\varepsilon_x$. In other words, a procedure π_p, whose interactional part is the same as that of a π_x, does not fall in the class of procedures applicable to σ with $\varepsilon_p < \hbar/\varepsilon_x$.

An *operational state* ω of a physical system at time t will be defined as a structure of the form

$$\omega \underset{\text{df}}{=} \langle Q_1[\sigma(t)], \ldots, Q_n[\sigma(t)]\rangle \,, \tag{32}$$

where $Q_1[\sigma(t)], \ldots, Q_n[\sigma(t)]$ represents a finite sequence of proper or improper results for the quantities Q_i on $\sigma(t)$, and the order of the measurements is from Q_1 to Q_n.

The operational state ω will be said to be *complete* with respect to the sequence $\langle Q_i\rangle$, when the overall precision of the results is the canonical precision allowed by the sequence $\langle Q_i\rangle$. It is customary to say that a complete state represents the maximum information obtainable about the sequence Q_i. The operational state ω will be *incomplete* with respect to $\langle Q_i\rangle$ when at least one of the quantities is known with less than canonical precision. Of course, incomplete states are not unique.

The concept of complete state of a physical system does not reflect an absolute property (monadic predicate) of the system, because it depends 1) on the apparatus A_i used in each π_i, and 2) on the choice of an order for the Q_i. It seems, therefore, to be rather a *relation* of the system with the whole apparatus + observer [25].

Given a physical system σ, we will denote by $\Omega[\sigma(t)]$ the set of the (complete or incomplete) states that σ can take at time t, and by $\Omega(\sigma)$ the union

$$\Omega(\sigma) = \bigcup_{0 \leqslant t \leqslant T} \Omega[\sigma(t)] \,. \tag{33}$$

Any measurement induces a transformation from $\Omega(\sigma)$ to $\Omega(\sigma)$.

Let π represent a procedure of measurement for the quantity Q_j, and let $Q_j[\sigma(t_1)]$ denote a proper result obtained applying π to σ during time $t_1 - t_0$. If also other quantities have been measured on σ, the system will be in a state of the form

$$\omega[\sigma(t_1)] = \langle Q_1[\sigma(t_1)], \ldots, Q_j[\sigma(t_1)], \ldots, Q_n[\sigma(t_1)]\rangle \, . \tag{34}$$

It is customary to say that π is of the *first kind* if a new measurement of Q_j with a procedure π' (possibly different from π, but with the same canonical precision) applied to $\sigma(t_1)$ during the (« sufficiently brief ») time $t_2 - t_1$ yields the same result. Incidentally, we note that this definition is logically inconsistent, as long as one talks (and thinks) in terms of absolute equality, for it does not allow for the evolution of the state. But it makes sense in terms of ε-equality.

Let π represent a measurement of the first kind. The transformation of $\Omega(\sigma)$ onto $\Omega(\sigma)$ induced by π will transform state ω into state ω' and we will write

$$\omega \underset{\pi}{\Rightarrow} \omega' \, . \tag{35}$$

We will distinguish two cases:

1) ω' differs from ω only with respect to the results concerning a quantity Q_j; in this case we have measured Q_j via π, without « perturbing » the initial state;

2) π' differs from ω not only in regard to Q_j, but also for the results of other quantities Q_i; the application of π to σ has « perturbed » the initial state.

Obviously, two procedures π' and π'' belonging to one and the same equivalence class C_k can bring about different perturbations on the same system $\sigma \in \Sigma_k$. Namely, we can have

$$\omega(\sigma) \underset{\pi'}{\Rightarrow} \omega'(\sigma) \, , \qquad \omega(\sigma) \underset{\pi''}{\Rightarrow} \omega''(\sigma) \, , \qquad \text{with } \omega' \neq \omega'' \, . \tag{36}$$

Hence, in order to be able to speak unequivocally of the transformation induced by a quantity Q_j, we will make (arbitrarily) a canonical selection of a subclass $\bar{C}_k \subset C_k$ of the procedures defining Q_j in σ and will write

$$\Omega(\sigma) \underset{Q_j}{\Rightarrow} \Omega(\sigma) \, . \tag{37}$$

A *physical situation* will be defined as the set of the states taken by a physical system in the time interval $0, T$. The states are determined by the proper measurements made at a number of times $0 < t_1 < \ldots < t_n < T$, and by improper measurements (theory), inbetween.

A physical quantity Q will be said to be defined in a physical situation s when a proper result of Q appears in all the states of s.

3. – Physical truth and physical theories.

3·1. *Truth in physics.* – Physics is widely thought to be the most exact of natural sciences, and to be perfectly reliable in a great number of applications. Nevertheless, many people have rather strange ideas about the validity of its assertions. Ironically, they may be ready to wage their life on a number of facts predicted by physics, while affirming that physics is able to reach at best *approximate truth.* Alternatively, they are likely to say that the laws of physics could be exact, were it not for the fact that they can only be *approximately verified.* This paradoxical situation arises largely because the precisions ε are ordinarily kept *out of the theory.* The theories of physics are then thought to be perfect forms of a Platonic world of ideas, which are inevitably marred in the passage to this mortal world of ours. In other words, the theory is never expected to perfectly agree with the facts!

We are now in a position to take a much more sensible attitude, and to show that physics can arrive at *true* assertions, *i.e.* to assertions corresponding to the facts. This is made possible mainly thanks to our having included the precision ε *in the theory*, from the outset.

It will be necessary first to introduce a formal characterization of the concept of *physical theory.*

As is well known, in standard model theory [26], a *formalized theory* T is usually represented by a pair $\langle \mathrm{FS}, K\rangle$, consisting of a *formal system* FS and the class K of all the *models* of FS. A formal system FS is determined by a triple $\langle L, A, D\rangle$, consisting a formal language L, a particular set A of sentences of L corresponding to the *axioms* of FS, and, finally, a set of (deductive) rules of inference D, which together with *the logical axioms* (contained in A) determine *the logic* of FS. A model of FS is defined as an abstract structure M, in which the axioms of FS are *true.*

In our opinion, this kind of idealization of the concept of « scientific theory », which, so far, has been mainly applied to the logical analysis of mathematical theories, admits of an interesting application also in the case of physical theories. However, in order to obtain a somewhat realistic description of « concrete » physical theories, one has to take into account some specific features of such theories. To this end, it appears necessary to render explicit a number of formal aspects of physical theories, which are ordinarily neglected in standard model theory.

A *formalized physical theory* T can be identified, as in standard model theory, by the usual pair $\langle \mathrm{FS}, K\rangle$, where FS is a *physical formal system* and K the class of the *physical models* of FS. But the notion of physical model requires a specific definition.

Let us first define a *physical structure* M by

$$M = \langle M_0, S, Q_0, \dots, Q_n, \varrho\rangle\,, \tag{38}$$

where

a) M_0 represents the *mathematical part* of M (ordinarily the *standard model* of a mathematical theory);

b) $\langle S, Q_0, ..., Q_n \rangle$ represents the *operational* part of M, S being *a set of physical situations* (see subsect. **2**·4) and each Q_i $(0 \leqslant i \leqslant n)$ representing a *physical quantity*, operationally defined at least in a subset of S (*);

c) ϱ is a function which associates a *mathematical interpretation* in M_0 to the terms of the operational part of M.

A physical structure such as M will be called a *realization* of a formal language L, when

d) M_0 is a realization of the mathematical sublanguage L_0 of L; that is to say that every mathematical symbol of L_0 has an *interpretation* in M_0, according to the standard model-theoretical conventions;

e) L contains, besides the mathematical and logical symbols, a denumerably infinite list of *special physical variables*, corresponding to the quantities $Q_0, ..., Q_n$.

As we want to include nonrelativistic quantum mechanics in the class of the physical theories considered, it is advisable to give *time* Q_0 a special position of its own. For it, the language L will contain the physical variables $t_1, t_2, ...$. For any other deterministic quantity D_i (probabilistic quantity P_i), L contains the physical varaibles $d_{i_1}, d_{i_2} ...$ $(p_{i_1}, p_{i_2} ...)$. The subscript for t and the double subscript for the other quantities originate from the possibility mentioned earlier (subsect. **2**·2) that the same quantity (*e.g.* mass, electric charge) may be measured in different parts of the same physical system. In what follows, we will often retain only one index, for simplicity. In other words, d_i, p_i will stand synthetically each for a finite sequence of d_{i_l}'s and p_{i_l}'s, respectively.

From the semantical point of view, the t_k's are supposed to range over all *ideal values* of the quantity time (*i.e.* real numbers). The d_i's are supposed to range over all functions which associate to any time an *ideal value* for the deterministic quantity D_i (*i.e.* real number). Finally, the p_i's range over all functions which associate to any time an *ideal value* for the probabilistic quantity P_i (*i.e.* a probability distribution of real numbers). We will call the t_k's and the d_i's *physical variables of the firts type* (or else *deterministic variables*), and the p_i's *physical variables of the second type* (or else probabilistic

(*) We stipulate that the deterministic (D_i) and the probabilistic (P_i) versions of the same intuitive physical quantity are labelled by the same index i.

(time dependence, which in this case is inessential, will simply be understood); here λ, w, d will be the variables corresponding to the deterministic quantities Λ, W, D and p will correspond to the probabilistic quantity P, whose ideal value $p(x)\,\mathrm{d}x$ represents the probability that a particle hits the screen between the abscissa x and $x+\mathrm{d}x$ (perpendicular to the slit). Then $\models_s \alpha$, if and only if

a') the deterministic quantities Λ, W, D are defined in s and we can obtain for them (at any time) the intervals $\Lambda(s)$, $W(s)$, $D(s)$; the probabilistic quantity P can be measured, obtaining an $\langle \varepsilon^{\Lambda}, \varepsilon^{P}\rangle$-strip $P(s)$;

b') there exist three ideal values λ, w, d, *i.e.* three real numbers belonging to the intervals $\Lambda(s)$, $W(s)$, $D(s)$, and an ideal value p belonging to the strip $P(s)$, such that

$$p(x) = \frac{\lambda d}{\pi w}\,\frac{\sin^2(\pi(wx/\lambda d))}{x^2}\,. \tag{45}$$

Any formal language which admits of a physical realization, will be called a *physical formal language*. By *physical formal system* we will mean a triple $\langle L, A, D\rangle$, consisting of a physical formal language L, a set of axioms A and a set of rules of inference D.

Any physical formal system FS has a mathematical subsystem FS_0; any element of the set A_0 of the axioms of FS_0 will be either a *logical* or a *mathematical* axiom of FS; any element of $A - A_0$ (which is not a logical axiom) will be called instead a *physical axiom*. The concept of a *theorem* of a physical formal system ($\vdash_{\mathrm{FS}} \alpha$) is defined as usual.

A *model* of a physical formal system FS is a physical realization of the language of FS, such that

I) M_0 is the standard model of the mathematical subsystem of FS,

II) M verifies all the axioms of FS, and the truth in M is preserved by all the rules of inference of FS; in other words, if α is an axiom, $\models_M \alpha$, and if $\alpha_1, \ldots, \alpha_n/\alpha$ is a rule of inference, and $\models_M \alpha_1$, $\models_M \alpha_2$, ..., $\models_M \alpha_n$, then also $\models_M \alpha$.

A *physical theory* T is determined as a pair $\langle \mathrm{FS}, K\rangle$ consisting of a physical formal system FS and the class K of all its models. The theorems and the models of T are assumed to be the theorems and the models of the corresponding FS.

A sentence α of T will be called a *valid* sentence of T ($\models_T \alpha$) if and only if α is true in every model M of T.

By a *partial model* of a theory T we will understand a physical structure M, verifying all the axioms of T that are defined in at least one physical situation s of M, and whose concept of truth is preserved by all the rules of inference of T.

As in standard model theory, one can prove trivially a *soundness theorem*, for any physical theory T (*i.e.* for any sentence α of T, if $\vdash_T \alpha$, then $\models_T \alpha$). But, unlike what happens in standard model theory, a *completeness theorem* (*i.e.*, for any sentence α, if $\models_T \alpha$, then $\vdash_T \alpha$) seems to be unprovable. Indeed, K does not represent, as in the standard case, the class of all « possible worlds » which verify the theory, but only the class of all physical (*i.e.* « real ») worlds which verify the theory. In this situation, a formal completeness property would mean that our theory represents an intuitive « complete » description of the real world; in other words, whatever empirically happens in all parts of the physical world, which corresponds to a model of the theory, should be theoretically predicted by the theory. This may have appealed to some strictly rationalistic philosophers of the past, but today represents, of course, a too strong epistemological requirement. There are also some purely logical reasons speaking against this completeness property. For, we have seen that every physical model M includes the same mathematical part M_0. Now, owing to Gödel's incompleteness theorem, the mathematical subtheory of T will generally (*) contain some undecidable sentences that are true in M_0: as a consequence, some mathematical sentences of T will be true in any physical model of T, yet be not provable in T.

From an intuitive point of view, the notion of truth we have defined represents a sort of *empirical truth*, and it is worth-while to notice that it has some formal properties that are not shared by the classical and standard notion of truth. For instance, for a given sentence α and for a given physical situation s, we may have both $\models_s \alpha$ and $\models_s \neg\alpha$, but *not* $\models_s \alpha\wedge\neg\alpha$. In other words, α and $\neg\alpha$ can be both true, nevertheless the contradiction $\alpha\wedge\neg\alpha$ is not true. To see this, let us take again for α the form (42) which represents the second law of classical mechanics We have $\models_s \alpha$, because there exist three real numbers $r_1 \in F[s(t)]$, $r_2 \in M[s(t)]$, $r_3 \in A[s(t)]$, such that $\models_{M_0} r_1 = r_2 r_3$. But, at the same time, we have $\models_s \neg\alpha$, because there exist three real numbers $r'_1 \in F[s(t)]$, $r'_2 \in M_k[s(t)]$, $r'_3 \in A_p[s(t)]$, such that $\models_{M_0} r'_1 \neq r'_2 r'_3$. Nevertheless, we cannot say $\models_s \alpha\wedge\neg\alpha$, because this would mean that there exist three real numbers $r''_1 \in F[s(t)]$, $r''_2 \in M[s(t)]$, $r''_3 \in A[s(t)]$, such that $\models_{M_0} r''_1 = r''_2 r''_3 \wedge r''_1 \neq r''_2 r''_3$, which is impossible!

As a consequence, in this kind of semantics, the logical connectives turn out to be not *truth-functional*. This means that the truth of a negated sentence is not equivalent to the nontruth of the positive sentence; the truth of a conjunction is not equivalent to the simultaneous truth of both members of the conjunction; and similarly for the other connectives (**).

It is of interest to notice that the failure of the truth-functionality property

---

(*) T must be a « well behaved » theory in the sense that it must satisfy some standard formal requirements. See [27] or a standard textbook in logic, as, for instance, [28].
(**) This situation has been studied in [29].

does not rule out the possibility that the logic of a physical theory T be the classical logic CL. For it is quite possible that all CL laws (expressed in the language of T) are true in any physical model M of T, and consequently valid in T.

3·2. *Theories and subtheories.* – In the framework of our « empirical model theory », we can describe some formal relations between theories, which seem to be characteristic of physical science.

As is well known, a basic concept of standard metamathematics is represented by the relation of theoretical inclusion. A theory T_1 is said to be *theoretically included* in a theory T_2 ($T_1 \subseteq T_2$) if and only if the set of theorems of T_1 is a subset of the set of theorems of T_2.

In standard model theory one can prove that $T_1 \subseteq T_2$ if and only if any model of T_2 (*i.e.* any abstract structure which verifies all the axioms of T_2) is also a model of T_1.

A weaker, but in many cases more interesting relation is that of *relative interpretability* [30]. From an intuitive point of view, T_1 is relatively interpretable in T_2 ($T_1 \leqslant T_2$) if and only if T_2 can prove whatever T_1 proves, except for mere linguistic differences between T_1 and T_2. In other words, there exists a convenient translation of the language of T_1 into the language of T_2, such that the translation of any theorem of T_1 is a theorem of T_2.

The relation $\leqslant$ has proved to be a fundamental concept in the logical analysis of mathematical theories. For instance, many foundational problems, and, to some extent, even the historical progress of mathematical theories, can be described in terms of this theoretical relation.

However, if we turn to the case of physical theories, the same relation seems to cease to have interesting applications. For example, from an informal point of view, one can correctly maintain that « analysis is stronger than arithmetic, and set theory is stronger than analysis ». In much the same way, one would like to be able to say that « both relativistic mechanics and quantum mechanics are stronger than classical (nonrelativistic) mechanics ». But, whereas in the former case the relations hold

$$\text{arithmetic} \leqslant \text{analysis} \leqslant \text{set theory}\,, \tag{46}$$

in the latter case it seems impossible to assert in a reasonable way that

$$\text{classical mechanics} \begin{cases} \leqslant \text{relativistic mechanics}\,, \\ \\ \leqslant \text{quantum mechanics}. \end{cases} \tag{47}$$

Now, in the framework of our empirical model theory, it is possible to define the new theoretical relation *physically weaker than* ($\preccurlyeq$), which seems to be an adequate formal tool for the analysis of the interconnections between concrete physical theories.

From an intuitive point of view, a physical theory T_1 is considered weaker than a theory T_2, when the *domain of validity* of T_1 is included in the domain of validity of T_2. In other words, whenever a class of physical phenomena is explained by T_1, it must be explained also by T_2.

Now, in our empirical model theory, a class of physical phenomena is represented by an operational structure $\langle S, Q_0, \ldots, Q_n \rangle$ within a physical model of the theory under investigation. Of course, the mathematics of a physically stronger theory T_2 may be mathematically stronger than the mathematics of the physically weaker theory T_1. These heuristical considerations suggest, in a natural way, the following formal definition:

T_1 is said to be physically weaker than T_2 ($T_1 \leqslant T_2$) if and only if

a) the mathematical subtheory T_1^0 of T_1 is relatively interpretable in the mathematical subtheory T_2^0 of T_2 (*i.e.* $T_1^0 < T_2^0$),

b) for any physical model $M = \langle M_0, S, Q_0, \ldots, Q_n, \varrho \rangle$ of T_1, there exists a *partial model* $M' = \langle M_0', S', Q_0', \ldots, Q_n', \varrho' \rangle$ of T_2, such that the operational part $\langle S, Q_0, \ldots, Q_n \rangle$ of M coincides with the operational part $\langle S', Q_0', \ldots, Q_n' \rangle$, apart from a possible change of the order of the quantities.

It is worth-while to notice that the definition of $T_1 \leqslant T_2$ inverts the relation between the models of T_1 and T_2, as compared with both $T_1 \subseteq T_2$ and $T_1 < T_2$.

Many physicists, as well as many science historians, construe the progress of science as an increasingly better approximation to the *truth*. We believe that much could be said against this conception, even on purely philosophical grounds. For instance, it is very difficult to see what is really meant by this mythical truth, that we will probably never be able to attain. But even setting aside these objections of principle, we think to be able to describe the evolution of physics in a much more natural and less pessimistic way. Historical progress does not consist in a sequence of theories of increasing approximation, but in a sequence of theories with increasingly larger domains of validity.

It will hardly be necessary to stress that our present aim is to build a suitable *formal scheme*, and not to do the job of the historian. *Real* history can often be at variance with respect to the scheme, especially in the details. But we believe that at least what has been called *internal* history, that is the rational reconstruction of real history, fits fairly well in the scheme.

There is a *historical ladder* of theories

$$T_1 \leqslant T_2 \leqslant \ldots \leqslant T_n \tag{48}$$

which are valid in the corresponding domains

$$\mathscr{D}_1 \subseteq \mathscr{D}_2 \subseteq \ldots \subseteq \mathscr{D}_n\,. \tag{49}$$

The domain $\mathscr{D}_i$ will be defined as the class of all the physical parts (*i.e.* $\langle S, Q_0, ..., Q_n \rangle$) of the models of T_i. By T_n and $\mathscr{D}_n$ we will understand the present state of our knowledge, which, of course, is not the final one. The sequences (48) and (49) are stopped at T_n, $\mathscr{D}_n$, simply because we do not know how to predict the future.

Note that, in this context, a new experiment E whose result is in agreement with T_n does not *confirm* T_n. It only enlarges $\mathscr{D}_n$. By the same token, if the result of E is at variance with the predictions of T_n, we ought not to say that T_n is *falsified*. We can only conclude that E does not fall within $\mathscr{D}_n$. It is then the job of the theoretician to look for a T_{n+1} which can account also for E, *i.e.* whose domain of validity $\mathscr{D}_{n+1} \supset \mathscr{D}_n$ includes E.

3·3. *Deterministic vs. probabilistic theories.* – As an application of our empirical semantics, it will be of interest to deal with a problem, which is much debated in modern physics and presents some puzzling aspects.

As is well known, after the advent of quantum mechanics and of the Copenhagen interpretation, it has been customary for many physicists to think that 1) Nature is indeterministic, 2) however, that part of physics which is termed macroscopic behaves in a deterministic way (though exceptions are known (*)).

There is an apparent contradiction between these two statements. How can a deterministic theory, such as classical mechanics, be in some way a subtheory of an indeterministic theory, such as quantum mechanics? (But note that in the sequel, where we mention an indeterministic theory, we need not necessarily refer to present-day quantum mechanics.) As a matter of fact, if we refer to the relation of theoretical inclusion $\subseteq$, or to the relation of relative interpretability $\leqslant$, this problem can hardly be solved in a satisfactory and rigorous way. But in the framework of our empirical semantics, the difficulty seems to disappear.

To begin with, let us recall, in an informal way, some well-known facts. Any measurement is affected by an *uncertainty*. The uncertainty is due either to the limited resolving power ε^A of the apparatus, or to the fact that its result is a probability distribution (or both). Now, in order to ascertain that we are in the latter case, we need an apparatus of suitable resolving power. For, if the probability distribution is appreciably different from zero only in an interval smaller than ε^A, we cannot tell one case from the other. We must accept this as a result of experience.

As an example, take the case of the beam through a slit, governed by eq. (45). If the particle wavelength λ is much smaller than the slit width w, and w

(*) One has only to think of lasers, superfluids, superconductors which are macroscopic systems governed by quantum mechanics. Alternatively, one can think of the instabilities of macroscopic systems, mentioned by PRIGOGINE [21].

is in turn much smaller than the resolving power ε^A of the instrument with which we scan the screen, we will find that all the particles hit the screen at « the same abscissa » $x = 0 \pm \varepsilon^A$. In other words, we will find the deterministic result of classical mechanics, and the residual uncertainty will be attributed wholly to the limited resolving power of our instrument.

This simple concept, which is part and parcel of the art of the physicist, can offer some difficulty to the formalizer. This is mainly because the value of the resolving power of an instrument is not precisely determined by experience, and moreover depends even on the idiosyncrasies of the observer. For this reason, one must also introduce a certain amount of conventionality. For example, Rayleigh's criterion for the resolving power of an optical instrument is, to some extent, conventional.

Every probability distribution has some degree of concentration around its mean value, which can be characterized in several different ways. In many cases, it is most convenient to use the standard deviation; but there are cases in which a different characterization is better suited. For the sake of simplicity, we will refer here to the standard deviation δ; but the reader is warned that, in some cases, what we denote by δ may have a different definition.

As we have mentioned earlier, experience tells us that, when δ is on the order of ε^A, the probability distribution, in its central part, cannot be distinguished from the normal error distribution (mostly Gaussian) of the instrument. However, this is not always sufficient. One must also make sure that there are not some very rare events in the wings of the distribution, which *can* be distinguished against the normal error background.

These considerations allow us to define a *deterministic* theory in a reasonable way.

A theory T will be said to be deterministic with respect to a model M, if and only if for any probabilistic quantity P of M

1) each result $P[s(t)] = p_0 \pm \langle \varepsilon^A, \varepsilon^P \rangle$ is such that its standard deviation δ_0 is on the order of ε^A (say more precisely $k\delta_0 < \varepsilon^A/2$, where $k > 0$ is determined by convention);

2) in the wings (for $|x - \bar{x}_0| > k\delta_0$) p_0 differs from the normal error distribution of the instrument by less than $\varepsilon^P/2$.

A theory T will be said to be *deterministic*, if and only if it is deterministic with respect to each one of its models.

Intuitively, a theory is deterministic when all the uncertainties of its predictions fall within the resolving powers of the instruments. In this case, each probabilistic quantity can clearly be replaced by a corresponding deterministic quantity. Indeed, let $P[s(t)] = p_0 \pm \langle \varepsilon^A, \varepsilon^P \rangle$ be a probabilistic result which satisfies both conditions 1) and 2). It is natural to say that $P[s(t)]$ determines a $D[s(t)]$, such that

$$D[s(t)] = \bar{x}_0 \pm \frac{\varepsilon^A}{2}, \tag{50}$$

where $\bar{x}_0$ denotes the mean value of x, evaluated by means of the probability distribution $p_0(x)$.

Let T be a deterministic theory; let P denote a probabilistic quantity of M (with corresponding special variables p_i), and D the deterministic quantity of M that corresponds to P (*) (with special variables d_i). It turns out that T must satisfy the following condition:

A) for any physical situation $s(t)$ in which P is defined and for any $\alpha[\mathsf{p}(t)]$, there exists a deterministic sentence $\beta[\mathsf{d}(t)]$, such that

$$\models_{s(t)} \alpha[\mathsf{p}(t)] \leftrightarrow \beta[\mathsf{d}(t)] . \tag{51}$$

Indeed, given $\alpha[\mathsf{p}(t)]$, we must distinguish the case $\models_{s(t)} \alpha[\mathsf{p}(t)]$ from the case $\not\models_{s(t)} \alpha[\mathsf{p}(t)]$. In the first case, let us put

$$\beta[\mathsf{d}(t)] = [\mathsf{d}(t) = \bar{x}_0] . \tag{52}$$

We must show that

$$\models_{s} \alpha[\mathsf{p}(t)] \leftrightarrow \mathsf{d}(t) = \bar{x}_0 \tag{53}$$

or, by the truth definition,

$$\exists p \in P[s(t)]\, \exists r \in D[s(t)] : \models_{M_0} \alpha(p) \leftrightarrow r = \bar{x}_0 . \tag{54}$$

Now, since $\models_{s(t)} \alpha[\mathsf{p}(t)]$, for a certain $p \in P[s(t)]$, there must hold $\models_{M_0} \alpha(p)$. On the other hand, $\bar{x}_0 \in D[s(t)]$, hence (54) is proved. In the case when $\not\models_{s(t)} \alpha[\mathsf{p}(t)]$, we have, by the truth definition,

$$\forall p \in P[s(t)] \not\models_{M_0} \alpha(p) . \tag{55}$$

Let us put in place of (52)

$$\beta[\mathsf{d}(t)] = [\mathsf{d}(t) = r_0] , \tag{56}$$

where $r_0 \notin D[s(t)]$. We must show that

$$\models_{s(t)} \alpha[\mathsf{p}(t)] \leftrightarrow \mathsf{d}(t) = r_0 \tag{57}$$

(*) According to our convention, P and D are labelled by the same index in the model M.

or, by the definition of truth,

(58) $$\exists p \in P[s(t)]\, \exists r \in D[s(t)] : \models_{M_0} \alpha(p) \leftrightarrow r = r_0 \,.$$

Now, let p represent any probability distribution belonging to $P[s(t)]$, and r any real number belonging to $D[s(t)]$; we must have

(59) $$\not\models_{M_0} \alpha(p) \qquad \text{(because } \not\models_{s(t)} \alpha[p(t)]) \,,$$

(60) $$\not\models_{M_0} r = r_0 \qquad \text{(because } r_0 \notin D[s(t)]) \,.$$

Hence

(61) $$\models_{M_0} \alpha(p) \leftrightarrow r = r_0 \,,$$

which proves (58).

It is readily seen that a T, nondeterministic with respect to M, does not, in general, satisfy condition (51). Indeed, for such a T, there is at least a $s(t)$ and at least a P such that the corresponding D is not defined for $s(t)$. In this case, for any choice of $\beta[d(t)]$, the sentence $\alpha[p(t)] \leftrightarrow \beta[d(t)]$ will not be defined in $s(t)$, hence, by the truth definition,

(62) $$\not\models_{s(t)} \alpha[p(t)] \leftrightarrow \beta[d(t)] \,.$$

According to our definition, Newtonian mechanics is deterministic, while statistical and/or quantum mechanics are indeterministic, for they are deterministic in « macroscopic » models and indeterministic in « microscopic » models, the *dividing line* depending on the precisions ε^A. We do not distinguish *in this context* what is unpredictable because of our ignorance of the detailed initial conditions from what is unpredictable because of « intrinsic reasons ».

As an example, consider a physical situation s consisting of a gas of N particles in an (isolated) container, at thermodynamic equilibrium. Let P and D represent respectively the quantity « entropy » in its probabilistic and deterministic versions. Let d_m represent the maximum value for the entropy in s. Statistical mechanics predicts the probability distribution of entropy in $s(t)$. This is asserted by a probabilistic sentence of the form $\alpha[p(t)]$. The result of a measurement $P[s(t)] = p_0 \pm \langle \varepsilon^A, \varepsilon^P \rangle$ will agree with $\models_{s(t)} \alpha[p(t)]$.

Now, if N is on the order of Avogadro's number, both $d_m - \bar{x}_0$ and the standard deviation δ_0 will be much less than any realistic ε^A of the apparatus with which we measure entropy; on the other hand, to reveal any discrepancy in the wing of the distribution would require such a small ε^P, or such a huge number of trials, that the observer should have to live much longer than the Universe; hence we can write the deterministic statement

(63) $$D[s(t)] = d_m \pm \varepsilon^A/2 \,.$$

If $\beta[d(t)]$ represents the deterministic sentence $\neg\, d(t) < d_{\mathrm{m}}$, we will have

$$(64) \qquad \models_{s(t)} \alpha[p(t)] \leftrightarrow \beta[d(t)] \,.$$

In other words, the sentences « entropy never decreases » and « the probability distribution of entropy is p », where p is that predicted by statistical theory, have equivalent meaning with respect to $s(t)$, and are both true.

If, on the contrary, M includes situations with only a few particles, and the initial conditions are taken at random, the theory is not deterministic for M. If $s(t)$ is a gas of, say, $N = 4$ particles, which we know only to be all in the same container, we will have

$$(65) \qquad \text{not} \models_{s(t)} \alpha[p(t)] \leftrightarrow \beta[d(t)] \,.$$

Indeed, $\beta[d(t)]$ will not be defined in $s(t)$. At the same time, we will have

$$(66) \qquad \models_{s(t)} \exists x \{x < d_{\mathrm{m}} - \varepsilon^{A}/2 \wedge p(t)(x) > \varepsilon^{P}/2\} \,.$$

For instance, the probability to find all particles in the same half of the container is 1/16, which only requires $n \gg 16$, in order to be verified by experience.

As we have mentioned earlier, there is a popular tenet that « the macroscopic world is deterministic ». How well is this assertion justified?

Physicists are aware that it cannot represent an absolute truth, for such puzzling cases are known, where a microscopic and unpredictable event can cause a definitely macroscopic event, such as killing a cat or sinking a battleship. After all, our evidence for quantum jumps is macroscopic.

However, it would be awkward to simply say that the above assertion is *false*. It seems to us that there is some analogy with the assertions of statistical mechanics. For example, it is true in classical statistical mechanics that by reversing the velocities of all the molecules in a gas, as suggested by LOSCHMIDT, we have described a system whose entropy is going to decrease. Nevertheless, it is true that, if we choose at random a state of a gas from reality, we are virtually certain that it will not coincide with a Loschmidt state and that it will evolve towards increasing entropy.

Similarly, if we isolate at random a macroscopic part of the real inanimate world and probe it with ordinary instruments, we are virtually certain that it will be possible to describe it as part of a model M_1 of a deterministic theory T_1. At the same time, if we use more refined instruments (*e.g.* capable of counting individual particles), the same part of the world will be part of a model M_2 of an indeterministic theory T_2, such that $T_1 \leqslant T_2$.

But, of course, the box in which Schrödinger's cat is closed is not chosen at random from all possible real boxes. It is built with all the cunning (and cruelty) of which man is capable. No wonder, therefore, that it can behave indeterministically, already at the macroscopic level. It is one of those rare cases, of which « not one in a million » is found in Nature.

4. – Logical problems of quantum mechanics.

4·1. *The logician's dilemma of* QM. – It is well known that quantum mechanics has given rise to some logical difficulties, right from the outset. These difficulties are mainly responsible for the attempt to introduce a new logic, called *quantum logic* (*).

Quantum logic (QL) is different not only from classical logic (CL), but also from the well-known nonclassical logics (such as intuitionistic logic), which so far have found their applications chiefly in the field of the foundations of mathematics.

From a strictly logical point of view, we know today a great deal about QL. We know several different axiomatizations of it and a number of possible semantics; moreover, we are able to interpret formally QL in other better known logics. All these pieces of information have made fairly clear the *logical status* of QL, as well as its precise place in the vast class of the logics weaker than CL [31-34]. As a consequence, the doubts set forth in the past by some scholars, about the legitimacy of considering QL as a « genuine logic », seem today to be completely dispelled.

In spite of all this work of clarification, there is still some argument about the general role that QL should play in QM and the correct way to apply it in the domain of that theory.

In this connection, two main and opposite views of an extreme character have found support:

1) QL is *not* the logic of QM. The logical difficulties of quantum theory (QT) regard the *content* and not the *form* of the theory. As a consequence, there is no hope to overcome those difficulties, by merely adopting a certain logical calculus in place of another.

2) QL represents the « true logic » of QM, in a strong sense. In other words, the entire QT must be formalized in a quantum calculus, and not in the classical one. In this way, most of the apparent logical anomalies of QM automatically disappear.

Between these two extreme views, there is a host of intermediate positions that can be taken. In our opinion, a sensible and realistic attitude, which closely adheres to QT in its present form, is to recognize that *de facto* more than one different logics coexist in the deductive structure of the theory.

Naturally, the coexistence of different logics in the domain of one and the same theory is bound to cause, at least on a first judgement, a lot of perplexity. In this particular case, it gives rise to a number of logical, physical

(*) About quantum logic the reader may see the contributions by MITTELSTAEDT and by BELTRAMETTI and CASSINELLI in this volume.

and epistemological problems. We will try and discuss some of these problems in the framework of the general semantics of physical theories, which we have been describing in the previous sections. It will be shown, in particular, that, in this approach, quite a few logical problems of QM can turn out to be independent of the dilemma QL *vs.* CL.

4'2. *A formal version of nonrelativistic quantum mechanics.* – According to our general semantics, QT can be described as a particular pair, consisting of a formal system FS and a class of physical models K. In this case, K consists of physical structures $M = \langle M_0, S, Q_0, \ldots, Q_n, \varrho \rangle$, which satisfy the following requirements:

1) M_0 is the standard model of functional analysis.

2) For any physical system σ occurring in S, the mathematical interpretation $\varrho(\sigma)$ of σ is a Hilbert space H in M_0; further, the mathematical interpretation $\varrho[\Omega(s)]$ of the set $\Omega(s)$ of all operational states of σ is the set of all *statistical operators* W in H. If W has the form P_ψ, where P_ψ is the projection operator along a (normalized) vector ψ of H, we will say that W represents a *pure state*, otherwise it represents a *mixture*. Owing to the one-to-one correspondence between the set of all P_ψ's and the set of all normalized vectors in H, we may say also that a pure state is represented by a normalized vector ψ of H. We will call $\varrho[\Omega(\sigma)]$ the set of all the ideal states that σ can take.

3) If Q_i (defined in a physical system σ, and different from the quantity time Q_0) represents both the probabilistic version P_i and the deterministic version D_i of a physical quantity, then $\varrho_\sigma(Q_i)$ is a self-adjoint operator in $\varrho(\sigma)$. In the case of the fundamental quantities, such as *position*, *momentum*, *spin*, etc., $\varrho_\sigma(Q_i)$ will be determined according to the standard rules of QT (which need not be recalled here).

4) If W is an ideal state and b is a Borel-set of real numbers, the real number $\mathrm{Tr}\,[WP^{\varrho_\sigma(Q_i)}(b)]$ (where Tr is the trace operator and $P^{\varrho_\sigma(Q_i)}$ is the spectral measure associated with the self-adjoint operator $\varrho_\sigma(Q_i)$) represents the probability that, by performing a measurement for the quantity Q_i, one can find an ideal (first-type) value which is contained in b. Let us denote such probability $\mathrm{Prob}^{Q_i}_W(b)$. In the particular case where W is a pure state of the form $\sum c_j \psi_j$, where the ψ_j's are normalized eigenvectors of $\varrho_\sigma(Q_i)$ with corresponding eigenvalues r_j, and b consists of a single real number r_k, one can prove that $\mathrm{Prob}^{Q_i}_W(b) = |c_k|^2$.

5) There exists at least one ideal translation τ such that, for any system σ, $\tau(\sigma)$ is a function $\hat{W}$ which associates to any time t a statistical operator $W \in \varrho[\Omega(\sigma)]$. W represents an *ideal state* of σ at time t. We will write also

$\tau(\sigma)(t) = \hat{W}(t) = W$ (*). If $\omega[\sigma(t)] = \langle Q_1[\sigma(t)], ..., Q_m[\sigma(t)]\rangle$, an ideal state W associated with $\sigma(t)$ will satisfy the following conditions:

5.1) for any quantity Q_i, which occurs in $\omega[\sigma(t)]$, W determines (according to 4)) a probability distribution p_i^W, which represents a possible ideal (probabilistic) value for Q_i; let Δp_i^W represent the *dispersion* of p_i^W (which is supposed to be defined in some standard way);

5.2) if $Q_i[\sigma(t)]$ is a deterministic result, which has the form $r_i \pm \varepsilon^{A_i}/2$, then

(67) $$\Delta p_i^W \leqslant \frac{\varepsilon^{A_i}}{2} \quad \text{and } \bar{x}_i \in Q_i[\sigma(t)];$$

5.3) if $Q_i[\sigma(t)]$ is a probabilistic result, then

(68) $$p_i^W \in Q_i[\sigma(t)] .$$

6) If σ_1 and σ_2 are two physical systems, and $\sigma_1 + \sigma_2$ represents the compound system, $\varrho(\sigma_1 + \sigma_2) = \varrho(\sigma_1) \otimes \varrho(\sigma_2)$, where $\otimes$ is the tensor product of Hilbert spaces.

If $\tau(\sigma_1)(t) = W_1$ (defined in $\varrho(\sigma_1)$) and $\tau(\sigma_2)(t) = W_2$ (defined in $\varrho(\sigma_2)$), then there exists a statistical operator $W = W_1 \otimes W_2$ of $\varrho(\sigma_1) \otimes \varrho(\sigma_2)$ such that $\tau(\sigma_1 + \sigma_2)(t) = W$. (One can prove that W_1 and W_2 are pure states, if W is the projection along a vector of the form $\psi \otimes \varphi$, where $\psi \in \varrho(\sigma_1)$, $\varphi \in \varrho(\sigma_2)$; otherwise they are both mixtures.)

7) Let $\tau(\sigma)(t_0) = \sum c_j \psi_j$, where the ψ_j's are eigenvectors of $\varrho_\sigma(Q_i)$. Let us suppose that in the time $[t_0, t]$ one carries out a first-kind measurement for the quantity Q_i in $\sigma(t)$, getting the deterministic result $r_k \pm \varepsilon^A/2$, where r_k is an eigenvalue of $\varrho_\sigma(Q_i)$ with corresponding eigenvector ψ_k (for the sake of simplicity we are dealing with a nondegenerate case). Soon after the measurement, at time t, $\sigma(t)$ will be represented by an eigenvector of the operator $P^{\varrho_\sigma(Q_i)}(r_k \pm \varepsilon^A/2)$. In the particular case in which the interval $(r_k \pm \varepsilon^A/2)$ contains only one eigenvalue r_k of $\varrho_\sigma(Q_i)$, we will have $\tau(\sigma)(t) = \psi_k$.

Rule 7) is usually called « von Neumann's projection postulate » or also « reduction (or collapse) of the wave function ».

As to the formal system FS which is characteristic of QT, we will suppose that:

8) The mathematical axioms of FS are the axioms of functional analysis; the set of physical axioms can be described by a unique *metatheoretical schema*

(*) Since all the expressions of functional analysis belong at the same time to the theory and to the metatheory of QT, we will write also in the object-language $\hat{W}(t) = W$.

corresponding to the Schrödinger equation:

$$(69) \qquad \hat{W}(t) = \exp\left[-i\frac{H}{\hbar}t\right]\hat{W}(t_0)\exp\left[i\frac{H}{\hbar}t\right],$$

where H represents the Hamiltonian of the physical system described by $\hat{W}$.

9) The logical axioms and the rules of inference of FS are the axioms and the rules of CL. As a consequence, according to our approach, the *general logic* of QT is CL and not QL. Indeed the general logic of QT must govern at the same time the mathematical and the physical part of FS; but it seems quite improbable that it is convenient to formalize the mathematical subtheory of FS in a quantum-logical calculus. QL will be introduced later and will regard only a particular physical sublanguage of QT. It is worth-while noticing that, in this formal context, the principles which are usually called in the literature « the axioms of QT » have been distinguished in two different classes:

a) specific physical axioms (described by a unique metatheoretical schema),

b) metatheoretical rules concerning the interpretation-function ϱ (which correlates the operational part with the mathematical part of any model of QT).

4'3. *Classical logic and quantum logic in* QT. – We will introduce QL in QT, as a logic which regards a particular physical sublanguage of QT.

Let us assume that $\langle\neg, \wedge, \vee, \rightarrow, \forall, \exists\rangle$ represent the *classical-logical constants* (connectives and quantifiers) of the language L of QT. We can enrich L with a new system of logical connectives and quantifiers $\langle\sim, \&, \curlyvee, \supset, \bigwedge, \bigvee\rangle$ called the *quantum-logical constants*. Let us call L' this extension of L, and let us express QT in L'. One can determine a sublanguage L^* of L' and define, for this sublanguage, the concept of *quantum-logical truth* with respect to the *pure (ideal) physical* states, which belong to the mathematical interpretation $\varrho(\sigma) = H$ of a given physical system σ. We will define L^* as the smallest language which contains, for any quantity Q_i which is defined for σ, all deterministic atomic well-formed formulae of the form $d_i(t) \in b$, $d_i(t) \in x$ (where b is a constant for a Borel-set, and x is a variable ranging over the set B of all Borel-sets of real numbers) and which is closed, according to the standard syntactical conditions, with respect to the quantum-logical constants. Clearly, L^* represents a deterministic sublanguage of L'. Since the time dependence of the d_i's will, in this context, be generally inessential, we will write simply $d_i \in b$ $(d_i \in x)$, or also $P_i b$ $(P_i x)$, because the expression « $d_i \in \ldots$ » can be represented as a monadic predicate of the object-language L^*.

In order to define the concept of quantum-logical truth with respect to the ideal states in H, one has to refer to the *complete orthomodular lattice* L^H

of all closed subspaces of H. Let us denote this algebraic structure by

(70) $$L^H = \langle C, \perp, \sqcap, \sqcup, \bigsqcap, \bigsqcup, 1, 0\rangle ,$$

where

C = the set of all closed subspaces of H,

$X^\perp$ = $[\psi \in H / \forall \varphi \in X(\langle \psi, \varphi\rangle = 0)$ (for any $X \in C$) (in other words, $X^\perp$ contains all the vectors which are orthogonal to any element of X),

$X \sqcap Y$ = $X \cap Y$ (for any $X, Y \in C$),

$X \sqcup Y$ = the smallest subspace $Z \in C$ such that $X \cup Y \subseteq Z$ (for any $X, Y \in C$),

$\bigsqcap\{X_i\}_{i\in I} = \bigcap\{X_i\}_{i\in I}$ (for any class of subpsaces X_i belonging to C),

$\bigsqcup\{X_i\}_{i\in I}$ = the smallest subspace $Z \in C$ such that $\bigcup\{X_i\}_{i\in I} \in Z$ (for any class of subspaces X_i belonging to C),

1 = H (*i.e.* 1 represents the *total space*),

0 = $\{\mathbf{0}\}$, where $\mathbf{0}$ represents the null vector in H.

The algebraic structure L^H can be extended, in a natural way, to an *algebraic realization* A^H of the language L^*, that is to an abstract structure which gives an interpretation to all nonlogical symbols and all sentences of L^*.

This algebraic realization A^H can be determined as a triple $\langle L^H, D, v\rangle$, where D (the *domain of individuals* of A^H) is the set of all Borel-subsets of the real line; v is a function which associates an interpretation to the nonlogical symbols and to the sentences of L^*. In particular, we will have that, for any constant b, $v(b)$ is a Borel-set of real numbers belonging to D (for the sake of simplicity, we will suppose that the language contains a name b for each Borel-set). Further, v will interpret any monadic predicate P_i (corresponding to the expression « $d_i \in \ldots$ ») as a function $v(P_i)$ which associates to any Borel-set b the value $[a\psi \in H/\mathrm{Prob}_\psi^{Q_i}(b) = 1]$; in other words, v interprets the predicate P_i, in the sense that, for any Borel-set b, it determines the set of all physical states ψ for which the probability for the value of the quantity Q_i to be found in b is 1. One can readily prove that $[a\psi \in H/\mathrm{Prob}_\psi^{Q_i}(b) = 1]$ is a closed subspace of H, hence an element of the ortholattice L^H.

Finally, for any sentence α, v associates to α an element of L^H, which represents (intuitively) the *meaning* of α in the realization A^H. The value $v(\alpha)$ is determined according to the following definition (by induction of the length of α):

$$v(P_i b) = v(P_i)(b)\,, \qquad \text{for any atomic sentence } P_i b;$$

$$v(\sim\alpha) = v(\alpha)^\perp;$$

$$v(\alpha \,\&\, \beta) = v(\alpha) \sqcap v(\beta);$$

$$v(\alpha \curlyvee \beta) = v(\alpha) \sqcup v(\beta);$$

$$v(\alpha \supset \beta) = \begin{cases} 1 & \text{if } v(\alpha) \subseteq v(\beta), \\ 0 & \text{otherwise;} \end{cases}$$

$$v(\bigwedge X\alpha) = \bigsqcap \{v(\alpha(b))\}_{b\in D},$$

$$v(\bigvee X\alpha) = \bigsqcup \{v(\alpha(b))\}_{b\in D}.$$

As a consequence, for any atomic sentence $d_i \in b$, we will have $v(d_i \in b) = [a\psi \in H/\mathrm{Prob}_\psi^{Q_i}(b) = 1]$. In other words, the « meaning » of an atomic sentence such as $d_i \in b$ is represented by the set of all physical states for which the value for the quantity Q_i is found in b with probability 1.

Now we are able to define the concept of quantum-logical truth. We will say that a sentence α of L^* is *quantum-logically true* with respect to an ideal state ψ ($\underset{\psi}{\models\!\!\equiv} \alpha$), if and only if $\psi \in v(\alpha)$. Finally, we will say that α is *true in the algebraic realization* A^H ($\underset{H}{\models\!\!\equiv} \alpha$) if and only if $v(\alpha) = 1$. As a consequence, we will have: $\underset{A^H}{\models\!\!\equiv} \alpha$, if and only if, for any state ψ in H, $\underset{\psi}{\models\!\!\equiv} \alpha$.

One can wonder what is the intuitive meaning of the quantum-logical constants, and what is the fundamental difference between the classical and the corresponding quantum-logical constants, which both occur in L'. From an intuitive point of view, the sentences of L^* represent particular examples of *deterministic* sentences, which do not involve *probabilistic* variables. Roughly speaking, we are here isolating, in an indeterministic framework (quantum theory), a class of deterministic sentences; and we are trying to give a meaning to the assertion that « a deterministic sentence is *a priori certainly true* with respect to an ideal physical state ».

In order to see this, in an intuitive way, let us consider a simple example. Let Q_i and Q_j be two quantities with corresponding operators $\varrho(Q_i)$ and $\varrho(Q_j)$, and let ψ be an ideal physical state such that $\psi = \Sigma_{c_r} \varphi_r$ and $\psi = \Sigma_{c_s} \chi_s$, where the φ_r's and the χ_s's are respectively eigenvectors of $\varrho(Q_i)$ and $\varrho(Q_j)$ (for the sake of simplicity we are considering a nondegenerate case). Let us consider the following sentences of L^*: $d_i = r_i$, $\sim d_i = r_i$, $d_i = r_i \;\&\; d_j = r_j$, $d_i = r_i \curlyvee d_j = r_j$ ($d_i = r_i$ is of course equivalent to the standard form $d_i \in \{r_i\}$, where $\{r_i\}$ represents a particular example of a Borel set).

According to the definition of quantum-logical truth and to the axiomatization of QT, we will obtain

$$\underset{\psi}{\models\!\!\equiv} d_i = r_i \tag{71}$$

if and only if $\psi \in v(d_i = r_i)$,

if and only if $\psi \in [a\psi/\mathrm{Prob}_\psi^{Q_i}(\{r_i\}) = 1]$, which means:

with *certainty* the value for the quantity Q_i in the state ψ is r_i;

(72) $$\models_{\psi} \sim d_i = r_i$$

if and only if $\psi \in v(d_i = r_i)^{\perp}$,

if and only if $\psi \in [a\psi/\text{Prob}_{\psi}^{Q_i}(\{r_i\}) = 1]^{\perp}$,

if and only if ψ is orthogonal to the (normalized) eigenvector of $\varrho(Q_i)$ with eigenvalue r_i,

which means:

it is *impossible* that the value for the quantity Q_i in the state ψ is r_i;

(73) $$\models_{\psi} d_i = r_i \,\&\, d_j = r_j$$

if and only if $\psi \in v(d_i = r_i) \sqcap v(d_j = r_j)$,

if and only if $\psi \in v(d_i = r_i) \cap v(d_i = r_j)$,

which means:

with *certainty* the value for the quantity Q_i in the state ψ is r_i and with *certainty* the value for Q_j is r_j;

(74) $$\models_{\psi} d_i = r_i \curlyvee d_j = r_j$$

if and only if $\psi \in v(d_i = r_i) \sqcup v(d_j = r_j)$,

if and only if ψ belongs to the smallest subspace which includes $v(d_i = r_i) \cup v(d_j = r_j)$,

if and only if ψ belongs to the subspace which is spanned by the (normalized) eigenvector of $\varrho(Q_i)$ with eigenvalue r_i and by the (normalized) eigenvector of $\varrho(Q_j)$ with eigenvalue r_j,

which means:

with *certainty* the value for Q_i in ψ *could* be r_i, or the value for Q_j in ψ could be r_j.

In other words, the quantum-logical connectives seem to behave, in this context, as a kind of « modalization » of the classical-logical connectives.

Whereas the quantum conjunction seems to correspond perfectly to classical conjunction, the quantum negation can be interpreted as the assertion of an *impossibility* (that means the *necessity* of a classical negation); moreover, the quantum disjunction seems to assert that necessarily the classical disjunction between two *possibilities* must hold.

So far, such a modal interpretation of the quantum-logical connectives seems to represent only an intuitive description of what happens in a particular case. Later on, we will see that the whole framework of quantum logic admits of a modal interpretation, which represents a formal generalization of the particular example sketched above.

Let us now point out some immediate consequences of the definition of quantum-logical truth.

The « strong *tertium non datur principle* », *i.e.* the *metatheoretical* assertion according to which « for any sentence α, either α or its negation is true », is here generally violated. Indeed we may have, for a certain α and a certain ψ,

$$\text{(75)} \qquad \text{not} \models_{\psi} \alpha \quad \text{and not} \models_{\psi} \sim \alpha .$$

This will hold when $\psi \notin v(\alpha)$ and $\psi \notin v(\alpha)^{\perp}$ (recall that $X \cup X^{\perp}$ is properly included in H, if both X and $X^{\perp}$ are different from H). Nevertheless, the « weak *tertium non datur* principle », *i.e.* the theoretical assertion « α or not α », must be true for any sentence α. In other words,

$$\text{(76)} \qquad \models_{\psi} \alpha \curlyvee \sim \alpha .$$

Indeed, $v(\alpha) \sqcup v(\alpha)^{\perp} = H$. Consequently, ψ must belong to $v(\alpha \curlyvee \sim \alpha)$. As a consequence of (75), we will have that the « non truth of a sentence » does not coincide with the « truth of the negation of the sentence ». In other words, there are three possible *states of truth* for a given sentence α with respect to a physical state ψ: *true* (when $\models_{\psi} \alpha$), *false* (when $\models_{\psi} \sim \alpha$), *indetermined* (when neither $\models_{\psi} \alpha$ nor $\models_{\psi} \sim \alpha$).

The conjunction of conditions (75) and (76) gives rise to an apparently curious logical situation. However, such a situation is historically not new. As is well known, already ARISTOTLE recognized that, when we are interested in a *contingent* sentence α concerning the *future*, the application of the metatheoretical *tertium non datur* becomes critical. According to a modern interpretation of Aristotle's *De Interpretatione*, one may suppose that ARISTOTLE himself did, in a sense, realize the possibility of a logical situation, founded on the conjunction of (75) and (76). Indeed he seemed to suggest that neither the sentence « tomorrow there will be a sea-battle » nor the sentence « tomorrow there will not be a sea-battle » is *today* true. Nevertheless, the sentence « either tomorrow there will be a sea-battle or tomorrow there will not be a sea-battle » must be *today* true.

Conditions (75) and (76) show at the same time that the truth of a disjunction does not generally imply the truth of at least one of the members of the disjunction. In other words,

(77) $$\underset{\psi}{\models}\alpha \curlyvee \beta \not\Rightarrow \underset{\psi}{\models}\alpha \quad \text{or} \quad \underset{\psi}{\models}\beta .$$

On the other hand, there holds

(78) $$\underset{\psi}{\models}\alpha \quad \text{or} \quad \underset{\psi}{\models}\beta \Rightarrow \underset{\psi}{\models}\alpha \curlyvee \beta ,$$

whereas for the conjunction we have

(79) $$\underset{\psi}{\models}\alpha \,\&\, \beta \quad \text{if and only if } \underset{\psi}{\models}\alpha \text{ and } \underset{\psi}{\models}\beta .$$

This asymmetrical behaviour of conjunction and disjunction determines the failure of the *distributivity* laws. Indeed, we will have

(80) $$\underset{\psi}{\models}(\alpha \,\&\, \beta) \curlyvee (\alpha \,\&\, \gamma) \supset \alpha \,\&\, (\beta \curlyvee \gamma) .$$

But generally, not the other way around!

As a counterexample, consider the case where

$$v(\alpha) \sqcap \big(v(\beta) \sqcup v(\gamma)\big) \not\subseteq \big(v(\alpha) \sqcap v(\beta)\big) \sqcup \big(v(\alpha) \sqcap v(\gamma)\big)$$

(which is possible, owing to the nondistributivity of the orthomodular lattice L^H). Hence, by the truth definition: $v(\alpha \,\&\, (\beta \curlyvee \gamma) \supset (\alpha \,\&\, \beta) \curlyvee (\alpha \,\&\, \gamma)) = \{\mathbf{0}\}$ and not $\underset{\psi}{\models}\alpha \,\&\, (\beta \curlyvee \gamma) \supset (\alpha \,\&\, \beta) \curlyvee (\alpha \,\&\, \gamma)$ (because ψ, being a normalized vector, is different from the null vector $\mathbf{0}$).

The dual form of (80) will be

(81) $$\underset{\psi}{\models}\alpha \curlyvee (\beta \,\&\, \gamma) \supset (\alpha \curlyvee \beta) \,\&\, (\alpha \curlyvee \gamma) .$$

But generally, not the other way around!

One can easily find a number of physical examples, which show that (75)-(81) are quite reasonable principles in the framework of QT.

For instance, let ψ represent a spin-$\frac{1}{2}$ particle with a well-determined value for the quantity « x-component of spin ». In other words we will have

(82) ψ is an eigenvector of the operator $\varrho(Q_i)$ corresponding to the quantity the « x-component of spin » with eigenvalue $r_1 = +\frac{1}{2}$.

At the same time, we can write

(83) $\psi = c_1\psi_1 + c_2\psi_2$, where ψ_1 and ψ_2 are eigenvectors of the operator $\varrho(Q_j)$ (corresponding to the quantity « z-component of spin ») with eigenvalues $r_1 = +\frac{1}{2}$ and $r_2 = -\frac{1}{2}$.

Let us consider the following atomic sentences of L^*:

$$\mathsf{d}_i = r_1 \quad \text{(the value for the } x\text{-component of spin is } +\tfrac{1}{2}\text{)},$$

$$\mathsf{d}_j = r_1 \quad \text{(the value for the } z\text{-component of spin is } +\tfrac{1}{2}\text{)},$$

$$\mathsf{d}_j = r_2 \quad \text{(the value for the } z\text{-component of spin is } -\tfrac{1}{2}\text{)}.$$

We will have

$$\text{(84)} \qquad \text{not } \models_{\psi} \mathsf{d}_j = r_1 \quad \text{and} \quad \text{not } \models_{\psi} \sim \mathsf{d}_j = r_1$$

for neither « ψ is an eigenvector of $\varrho(Q_j)$ with eigenvalue r_1 » nor « ψ is orthogonal to ψ_1 »;

$$\text{(85)} \qquad \models_{\psi} \mathsf{d}_j = r_1 \vee \sim \mathsf{d}_j = r_1,$$

because $\psi \in X_{\psi_1} \sqcup X_{\psi_1}^{\perp}$ (X_{ψ_1} represents the unidimensional subspace which contains ψ_1);

$$\text{(86)} \qquad \models_{\psi} \mathsf{d}_j = r_1 \vee \mathsf{d}_j = r_2,$$

because $\psi \in X_{\psi_1} \sqcup X_{\psi_2}$; hence

$$\psi \in v(\mathsf{d}_j = r_1) \sqcup v(\mathsf{d}_j = r_2).$$

In other words, necessarily « the state ψ can either have spin up or have spin down ».

Further

$$\text{(87)} \qquad \text{not } \models_{\psi} \mathsf{d}_j = r_1 \quad \text{and} \quad \text{not } \models_{\psi} \mathsf{d}_j = r_2,$$

because ψ is not an eigenvector of $\varrho(Q_j)$;

$$\text{(88)} \qquad \models_{\psi} \mathsf{d}_i = r_1 \,\&\, (\mathsf{d}_j = r_1 \vee \mathsf{d}_j = r_2),$$

because

$$\psi \in v(\mathsf{d}_i = r_1) \quad \text{and} \quad \psi \in v(\mathsf{d}_j = r_1 \vee \mathsf{d}_j = r_2).$$

In other words, it is true that the state ψ has value r_1 for the x-component of spin, whereas for the z-component it can have either spin up or spin down.

On the other hand, we will have

$$\text{(89)} \qquad \text{not } \models_{\psi} (\mathsf{d}_i = r_1 \,\&\, \mathsf{d}_j = r_1) \vee (\mathsf{d}_i = r_1 \,\&\, \mathsf{d}_j = r_2).$$

Indeed,

$$v(\mathsf{d}_i = r_1 \,\&\, \mathsf{d}_j = r_1) = v(\mathsf{d}_i = r_1) \cap v(\mathsf{d}_j = r_1) = X_{\psi} \cap X_{\psi_1} = \{\mathbf{0}\}.$$

Similarly,

$$v(d_i = r_1 \,\&\, d_j = r_2) = \{\mathbf{0}\} .$$

Hence,

$$v(d_i = r_1 \,\&\, d_j = r_1) \sqcup v(d_i = r_1 \,\&\, d_j = r_2) = \{\mathbf{0}\} ,$$

which proves (89). In other words, the sentence which asserts that the state ψ either can have value r_1 for the x-component and at the same time spin up for the z-component, or it can have value r_1 for the x-component and at the same time spin down for the z-component, is not true. From an intuitive point of view, the non truth of our sentence sounds very reasonable, owing to the incompatibility of the two quantities Q_i and Q_j.

So far, we have defined the concept of quantum-logical truth, only with respect to the ideal states. But one can extend this definition, in a natural way, also to the case of operational states. If $\omega[\sigma(t)]$ is an operational state and α is a sentence of L^*, we will say that α is *quantum-logically true* in $\omega[s(t)]$ and in the physical situation $s(t)$ which contains only $\omega[\sigma(t)]$ ($\underset{\omega[\sigma(t)]}{\models\!\!\equiv} \alpha$ and $\underset{s(t)}{\models\!\!\equiv} \alpha$), if and only if 1) α is defined in $\sigma(t)$, 2) there exists an ideal state ψ of $\omega[\sigma(t)]$, such that $\underset{\psi}{\models\!\!\equiv} \alpha$.

One may wonder what is the relation between the concept of quantum-logical truth and that of empirical truth (defined in subsect. **3·1**). Of course, this comparison makes sense only for the case of atomic deterministic sentences, which belong both to L and to L^*. One can readily recognize that the quantum-logical truth represents a stronger concept than the empirical truth. Namely there holds

$$\text{(90)} \qquad \underset{s(t)}{\models\!\!\equiv} d_i \in b \;\Rightarrow\; \underset{s(t)}{\models} d_i \in b ,$$

but, generally, not the other way around. Indeed, let $\underset{s(t)}{\models\!\!\equiv} d_i \in b$. By definition of quantum-logical truth, there exists an ideal state ψ of $\omega[\sigma(t)]$ such that $\underset{\psi}{\models\!\!\equiv} d_i \in b$ (that means $\mathrm{Prob}_{\psi}^{Q_i}(b) = 1$); further D_i is defined in $\sigma(t)$, and gives rise to a deterministic result $D_i[\sigma(t)] = r \pm \varepsilon^{A_i}/2$. Let p_i^{ψ} be the probability distribution (corresponding to the quantity Q_i) which is determined by ψ. There holds $\bar{x}_i \in b$. Since ψ is an ideal state of $\omega[\sigma(t)]$, we must have

$$\Delta p_i^{\psi} \leqslant \varepsilon^{A_i}/2 \qquad \text{and} \qquad \bar{x}_i \in D_i[\sigma(t)] .$$

As a consequence, by definition of empirical truth we will obtain $\underset{s(t)}{\models} d_i \in b$. A counterexample, which shows that the quantum-logical truth is stronger than the empirical truth, can be obtained as follows: let us consider two incompatible quantities Q_i and Q_j such that $\varepsilon^{A_i} \cdot \varepsilon^{A_j} < \hbar$, where ε^{A_i} and ε^{A_j} represent the precisions of the two instruments by which we can measure *sepa-*

rately Q_i and Q_j. Let us consider an operational state (with respect to the two quantities Q_i and Q_j) $\omega[\sigma(t)] = \langle Q_j[\sigma(t)], Q_i[\sigma(t)]\rangle$. We suppose $\models_{s(t)} d_i = r$ (where $s(t)$ is the physical situation which contains only $\omega[\sigma(t)]$). As a consequence, by definition of empirical truth, $Q_i[\sigma(t)]$ is a deterministic result $(r_i \pm \varepsilon^{A_i}/2)$ and $r \in Q_i[\sigma(t)]$. In this case, $Q_j[\sigma(t)]$ must be a probabilistic result; otherwise the uncertainty principle would be violated. Let us suppose $Q_j[\sigma(t)] = = p_0 \pm \langle \varepsilon^{A_j}, \varepsilon^{P}\rangle$, where the dispersion Δp_0 of p_0 is supposed to be the optimal one (*i.e.* $\Delta p_0 = \hbar/\varepsilon^{A_i}$). As a consequence, no eigenvector of $\varrho(Q_i)$ can be an ideal state of $\omega[\sigma(t)]$. Indeed, for any eigenvector ψ of $\varrho(Q_i)$ we will have $\Delta p_j^{\psi} \gg \hbar/\varepsilon^{A_i}$. Therefore, by (30), p_j^{ψ} cannot belong to the strip $Q_j[\sigma(t)]$. This shows that $\not\models_{s(t)} d_i = r$.

4·4. *A modal interpretation of* QL. – In the previous subsection, we have seen that our concept of quantum-logical truth can violate some laws of CL (for instance, the distributivity laws). One can ask: what is the logic which governs this particular subtheory of QT, which we have expressed in the language L^*? The answer is: quantum logic, which turns out to be a weaker logic than CL. From a purely logical point of view, QL admits of a number of different syntactical and semantical descriptions (which turn out to be all equivalent). In this framework, the most intuitive approach to QL can be obtained as a natural generalization of our physical starting point.

As we have seen, in any Hilbert space H we can define a complete orthomodular lattice L^H, and this lattice can be extended in a natural way to an algebraic realization for the language L^*. Now, any L^H represents nothing but a particular case in the class of all abstract structures, which are called *complete orthomodular lattices.* We will say that an abstract structure of the kind

$$A = \langle B, \perp, \sqcap, \sqcup, \bigsqcap, \bigsqcup, 1, 0\rangle \tag{91}$$

is a complete orthomodular lattice when

a) B is a nonempty set;

b) the substructure $\langle B, \sqcap, \sqcup\rangle$ is a *lattice* (*i.e.* $\sqcap$ and $\sqcup$ are binary operations in B which satisfy both the associative and the commutative property; further there holds $a \sqcap (a \sqcup b) = a$ and $a \sqcup (a \sqcap b) = a$. We will call $a \sqcap b$ $(a \sqcup b)$ also the *infimum* (the *supremum*) of a and b. One can define a *lattice relation* $a \leqslant b$ as $a \sqcap b = a$ ($\leqslant$ turns out to be a partial ordering in B);

c) 1 and 0 represent respectively the *maximum* and the *minimum* of A (*i.e.* $\forall x \in B (x \leqslant 1$ and $0 \leqslant x)$);

d) the 1-ary operation $\perp$ is an orthocomplement (*i.e.* $a \sqcup a^{\perp} = 1$, $a \sqcap a^{\perp} = 0$, $a = a^{\perp\perp}$, $a \leqslant b \rightarrow b^{\perp} \leqslant a^{\perp}$);

e) the *completeness* property means that any set X of elements of B has in B an *infimum* $\sqcap X$ (*i.e.* the greatest element, with respect to $\leqslant$, which is smaller than or equal to any element of X) and a *supremum* $\sqcup X$ (*i.e.* smallest element which is greater than or equal to any element of X);

f) the *orthomodularity* property means that for any $a, b \in B : a \leqslant b \Rightarrow b \leqslant a \sqcup (a^\perp \sqcap b)$.

One can easily recognize that our *concrete* orthomodular lattice L^H does satisfy the properties *a*)-*f*).

Let us now refer to a generic first-order language L with n-adic predicates P^n_m and with the quantum-logical constants. We can define the concept of *algebraic orthorealization* for L as follows: an algebraic orthorealization for L is a structure

$$O = \langle A, D, v \rangle \tag{92}$$

where

a) A is a complete orthomodular lattice $\langle B, {}^\perp, \sqcap, \sqcup, \bigsqcap, \bigsqcup, 1, 0 \rangle$;

b) D is a nonempty set (the set of *individuals* of O); let $L^\#$ be an extension of L containing an individual name d for any $d \in D$;

c) v interprets the nonlogical constants of $L^\#$ and associates to each sentence α of $L^\#$ a value $v(\alpha)$ in B. For any d, $v(\mathsf{d}) = d$. For any predicate P^n_m, $v(P^n_m)$ is an *n-ary attribute* on O, that is a function which associates to any n-ple $\langle d_1, \ldots, d_n \rangle$ of elements of D an element of B (in other words, $v(P^n_m)$ is an element of B^{D^n}). Further, for any sentence α of $L^\#$, $v(\alpha)$ is determined as follows:

$$\begin{aligned}
v(P^n_m \mathsf{d}_1 \ldots \mathsf{d}_n) &= v(P^n_m)\big(v(\mathsf{d}_1), \ldots, v(\mathsf{d}_n)\big)\,, \\
v(\sim \beta) &= v(\beta)^\perp\,, \\
v(\beta \mathbin{\&} \gamma) &= v(\beta) \sqcap v(\gamma)\,, \\
v(\beta \curlyvee \gamma) &= v(\beta) \sqcup v(\gamma)\,, \\
v(\beta \supset \gamma) &= \begin{cases} 1 & \text{if } v(\beta) \leqslant v(\gamma)\,, \\ 0 & \text{otherwise}\,, \end{cases} \\
v(\bigwedge x \beta) &= \bigsqcap \{v(\beta(x/\mathsf{d}))\}_{d \in D}\,, \\
v(\bigvee x \beta) &= \bigsqcup \{v(\beta(x/\mathsf{d}))\}_{d \in D}\,.
\end{aligned}$$

A sentence α will be said to be *true* in an orthorealization O ($\models_O \alpha$) if and only if $v(\alpha) = 1$; α will be said to be *quantum-logically valid* or also a *quantum-logical law* ($\models_{\mathrm{QL}} \alpha$) if and only if, for any algebraic orthorealization O, $\models_O \alpha$.

In this way we have completely characterized QL, by means of a semantical approach. QL turns out to be a sublogic of CL, because all quantum laws are classical laws, but not *vice versa*.

One could try and describe the « meaning » of the quantum-logical constants, with the « eyes » of a classical logician, that is within a theoretical framework, where we are using CL and not QL. Formally, this can be done, by using a particular superlogic of CL, which is represented by a form of modal logic [32, 35]. As is well known, in recent times, modal logic has found a number of different applications to the logical analysis of physical theories. This might appear, at first sight, somewhat surprising, if one recalls that modal logic, so far, has hardly been applied to foundational problems of mathematics. An intuitive argument that could, to some extent, explain the reason of such a successful application, has been pointed out by VAN FRAASSEN [36]. The notion of *possible world* (which is a basic concept of the modal semantics), in spite of its metaphysical appearance, allows a number of « empirical applications ». For instance it may be interpreted on some occasions as « physical state » or « physical situation », etc.

Let L_M be a *classical modal language*, whose nonlogical constants are the same as in L. The logical constants of L_M are the following:

the classical logical connectives and quantifiers: $\neg, \wedge, \vee, \rightarrow, \forall, \exists$;

the modal operators: L (*necessarily*), M (*possibly*);

an entailement-connective: $\rightrightarrows$ ($\alpha \rightrightarrows \beta$ means intuitively α *entails* β).

One can define a translation of the quantum language L (with quantum-logical constants) into the classical modal language L_M. If α is a formula of L, we will denote by α^* the translation of α in L_M:

$$
\begin{aligned}
p^* &= LMp\,, \quad \text{for any atomic formula } p \text{ of } L,\\
(\sim \alpha)^* &= L \neg \alpha^*\,,\\
(\alpha \mathbin{\&} \beta)^* &= \alpha^* \wedge \beta^*\,,\\
(\alpha \curlyvee \beta)^* &= LM(\alpha^* \vee \beta^*)\,,\\
(\alpha \supset \beta)^* &= \alpha^* \rightrightarrows \beta^*\,,\\
(\bigwedge x\alpha)^* &= \forall x \alpha^*\,,\\
(\bigvee x\alpha)^* &= LM\,\exists x \alpha^*\,.
\end{aligned}
$$

In other words, our translation interprets any quantum negation as the *necessity* of a classical negation, any quantum conjunction as a classical conjunction, any quantum disjunction as the *necessity of the possibility* of a classical disjunction, any quantum implication as a classical *entailment*, any quantum universal quantifier as a classical universal quantifier, any quantum existential quantifier as the *necessity of the possibility* of a classical existential quantifier. Such a modal interpretation of the quantum-logical language represents nothing but a generalization of the actual meaning of the quantum-logical constants, in the case of those particular sentences of QT which we have studied in the previous subsection.

So far, we have only defined a linguistic translation of a quantum language into a classical language. In order to obtain a correct interpretation of QL in a form of modal logic, one has to determine a modal logic ML such that, for any sentence α of the quantum language L, its modal translation α^* is a valid sentence of ML, if and only if α is quantum-logically valid. In other words,

$$(93) \qquad \models_{\mathrm{QL}} \alpha \quad \text{if and only if} \models_{\mathrm{ML}} \alpha^* .$$

Only on this ground one can say that the concept of « quantum-logical law » has been completely reduced to a classical concept. A modal logic ML, which turns out to be apt to this aim, can be found. This is an extension of a well-known modal system, which in the modal literature is usually called the « Brouwer's system » or also simply the « system B » [37]. Semantically, this ML can be characterized by defining its class of realizations K_{ML}: we will say that a sentence α of ML is valid in ML ($\models_{\mathrm{ML}} \alpha$) if and only if, for any realization M belonging to K_{ML}, α is true in M ($\models_M \alpha$). The class K_{ML} represents a particular class of a kind of models, which are usually called *Kripkian models*, after KRIPKE, who proposed this type of models for the semantical investigation of modal and intuitionistic logic [38, 39]. The characteristic feature of the Kripkian realizations is to contain a certain class of *possible worlds*, which are correlated by a well-determined relation, called *the accessibility relation*, between the worlds.

Omitting all logical details (which are, in this framework, inessential), we will recall only an important theorem. One can prove that any algebraic orthorealization O of L can be transformed into a modal realization M^O of L_M, such that

$$(94) \qquad \models_O \alpha \quad \text{if and only if} \models_{M^O} \alpha^*$$

for any sentence α of L.

Vice versa, any modal realization M of L_M can be transformed into an algebraic realization O^M of L, such that

$$(95) \qquad \models_M \alpha^* \quad \text{if and only if} \models_{O^M} \alpha$$

for any sentence α of L.

On this ground, one can readily prove (93). Indeed, let $\models_{\mathrm{QL}} \alpha$ and let us suppose $\not\models_{\mathrm{ML}} \alpha^*$. Consequently, there exists a modal realization M, such that $\not\models_M \alpha^*$ and $\not\models_{O^M} \alpha$, against the hypothesis $\models_{\mathrm{QL}} \alpha$. The reverse is proved similarly.

These results give a complete answer to our question concerning the classical meaning of our quantum-logical constants.

Let us now go back to QT, and ask what is the physical meaning of the modal realization M^{A^H}, which we have associated, in this way, to any algebraic realization A^H of the sublanguage L^* of QT.

As a consequence of the proof of (94), we obtain that the *possible worlds* of the modal realization M^{A^H} are represented by all nonnull closed subspaces of H; in other words, any nonnull subspace represents, in the Kripkian realization M^{A^H}, a « possible physical world ». Among the physical worlds it is interesting to consider the *unidimensional* physical worlds X_ψ, which contain only one normalized vector ψ. As a consequence of our results, we will have: for any sentence α of L^*, α is quantum-logically true in the state ψ, if and only if the modal translation α^* is *classically true* in the physical world X_ψ.

As a conclusion of this subsection, let us ask whether these model-theoretical results may have some bearing on the general philosophy of quantum mechanics. It seems to us that all this can lead, in a natural way, to the following claim: from a logical point of view quantum logic is not really *essential* to the logical development of quantum mechanics, since the role played by this logic can be equivalently replaced by a form of classical modal logic. Perhaps, this conclusion can give some psychological help to those people who think that a theoretical situation essentially founded on a *mixture of logics* is very disagreeable. And without any doubt, if quantum logic were essential to the logical development of quantum mechanics, quantum mechanics, as a whole, would be founded on a mixture of logics; for, so far, it is quite improbable that we will formalize the whole of mathematics needed for quantum mechanics, in a quantum-logical calculus such as QL. In spite of this, we must say that we do not feel any particular allergy to a situation of mixture of logics. On the contrary, it seems to us that, generally, a form of plurality of logics cannot be avoided in modern science. In any case, even if, philosophically, we do not trust a unique privileged logic, from a logical point of view any reduction of a logic to another logic represents a relevant result. However, in this particular case, concerning quantum logic, can we really assert that we have completely reduced quantum logic to classical logic? As we have seen, we have only proved the following metatheorem: $\models_{\mathrm{QL}} \alpha$ if and only if $\models_{\mathrm{ML}} \alpha^*$. But ML is not *simply* classical logic. Indeed it represents a very particular extension of classical logic. One could ask: why just the modal system ML fits with quantum mechanics? Thus, the general question concerning the disagreeable mixture of logics in quantum mechanics seems to remain, in a sense, still open.

4·5. *Logical self-reference, set-theoretical paradoxes and the measurement problem in* QT. – Not all logical problems of QT have directly to do with what we have called the logician's dilemma « QL or CL ». For instance, the most debated question concerning the measurement problem in QT appears to be, from a logical point of view, a typical question of *semantical closure* of a theory [40]. To what extent a consistent theory (in this case QT) can be *closed* with respect to the *objects* and the *concepts* which are described and expressed in its metatheory? As is well known, the *limitative* theorems of logic and the paradoxes of set theory teach us that there are some definite limits to the semantical closure of any consistent theory (which satisfies some standard formal requirements). In particular, the theory can never express and prove *all* that is expressed and proved in its metatheory; further, it cannot generally describe (up to certain limitations) its universe as its *own object*. From an intuitive point of view, one could recall a number of reasons why a well-behaved scientific theory cannot be logically *self-sufficient*. In particular, a quite disagreeable situation is met when a physical theory (QT), owing to purely *logical* reasons, turns out to be submitted to some limitations concerning its capability of describing and expressing certain specific *physical* objects and concepts. This is the curious situation arising from the measurement problem of QT.

We will discuss von Neumann's « measurement paradox » [5], by repeating a well-known argument in the framework of our formal semantics for QT. Let $M = \langle M_0, S, Q_0, \ldots, Q_n, \varrho \rangle$ be a model of QT, and let σ be a (microphysical) system in S. Suppose we carry out a first-kind and first-type measurement concerning the quantity Q_i, during the (« sufficiently short ») time interval $[t_0, t]$. The measurement is performed by using an apparatus A, which occurs in the metatheoretical definition of Q_i. Since A itself represents a physical system (consisting of elementary particles), we may suppose that also A belongs to S. On other words, A can be, at the same time, an object of QT and of its metatheory (MT). As a consequence, the compound system $\sigma + A$ will be in S. Let $\varrho(\sigma) = H_1$, $\varrho(A) = H_2$, and let us suppose that $\tau(\sigma)(t_0)$ and $\tau(A)(t_0)$ are both pure states; for instance, $\tau(\sigma)(t_0) = \hat{\psi}(t_0) \in H_1$ and $\tau(A)(t_0) = \hat{\varphi}(t_0) \in H_2$. By 4.2.6) we will have

$$\tau(\sigma + A)(t_0) = \hat{\psi}(t_0) \otimes \hat{\varphi}(t_0) \in H_1 \otimes H_2 \,. \tag{96}$$

It is fairly natural to assume the following necessary condition in order for $\tau(A)$ to represent (in $H_1 \otimes H_2$) a measurement apparatus for the quantity $\varrho(Q_i)$ (in H_1) (for the sake of simplicity we are supposing that the spectrum of $\varrho(Q_i)$ is discrete and nondegenerate):

1) There exists in H_2 a self-adjoint operator E (whose spectrum is discrete and nondegenerate) with eigenvectors φ_j and corresponding eigenvalues a_j such that

1.1) the spectrum of E less a_0 is in a one-to-one correspondence with the spectrum of $\varrho(Q_i)$,

1.2) $\tau(A)(t_0) = \varphi_0$,

1.3) if $\tau(\sigma)(t_0) = \psi_k$ (ψ_k eigenvector of $\varrho(Q_i)$ with corresponding eigenvalue r_k), then $\tau(\sigma)(t) = \psi_k$; if $\tau(\sigma)(t) = \psi_k$, then $\tau(A)(t) = \varphi_k$.

From an intuitive point of view, condition 1) asserts that $\tau(A)$ may assume different *states* (eigenstates of E) that correspond to different « recording states » of $\varrho(Q_i)$-values. In state φ_k, A « records » the value r_k; φ_0 represents the initial state, when the apparatus A has not yet read any value.

Let us now imagine a concrete case where

$$(97) \qquad \tau(\sigma)(t_0) = \hat{\psi}(t_0) = \sum c_j \psi_j \quad (\psi_j \text{ eigenvectors of } \varrho(Q_i)) \, .$$

By 1.2) we will have $\tau(A)(t_0) = \varphi_0$ and thus by 4.2.6)

$$(98) \qquad \tau(\sigma + A)(t_0) = \hat{\psi}(t_0) \otimes \hat{\varphi}(t_0) = \sum c_j \psi_j \otimes \varphi_0 = \hat{\Psi}(t_0) \, .$$

Further, let us suppose that, by performing a measurement in $\sigma(t_0)$ (using the apparatus A), we get the deterministic result $r_k \pm \varepsilon^A/2$. For the sake of simplicity, we suppose that r_k is the only eigenvector of $\varrho(Q_i)$ contained in the interval $r_k \pm \varepsilon^A/2$. As a consequence, by the projection postulate 4.2.7) and by 1.3) we will have $\tau(\sigma)(t) = \psi_k$ and $\tau(A)(t) = \varphi_k$. Thus, by 4.2.6)

$$(99) \qquad \tau(\sigma + A)(t) = \hat{\psi}(t) \otimes \hat{\varphi}(t) = \psi_k \otimes \varphi_k = \hat{\Psi}(t) \, .$$

At the same time, $\hat{\Psi}(t_0)$ is submitted to axiom (69) (Schrödinger equation). In other words there holds

$$(100) \qquad \hat{\Psi}(t) = U(t)\hat{\Psi}(t_0) = U(t) \sum c_j \psi_j \otimes \varphi_0 \, ,$$

where $U(t)$ represents the linear operator $\exp[-i(H/\hbar)t]$, and H is the Hamiltonian of the compound system $\sigma + A$. Consequently, one obtains (by the linearity of $U(t)$)

$$(101) \qquad \hat{\Psi}(t) = \sum c_j U(t) \psi_j \otimes \varphi_0 \, .$$

Hence, by 1.3) and 4.2.6)

$$(102) \qquad \hat{\Psi}(t) = \sum c_j \psi_j \otimes \varphi_j \, .$$

But $\sum c_j \psi_j \otimes \varphi_j \neq \psi_k \otimes \varphi_k$. Therefore,

$$(103) \qquad \hat{\Psi}(t) \neq \hat{\Psi}(t) \; !$$

Thus we have proved (in the metatheory of QT) a contradiction, which shows the inconsistency of QT in this formulation.

One can distinguish the different hypotheses involved in our proof of the contradiction, in two different groups: I) « axioms » of QT, II) special hypotheses, in particular the following ones: *a*) the apparatus A is a physical system in S; that means it is a physical system described by the theory, and, as such, it can be translated as a pure state, which satisfies condition 1); *b*) the evolution of the compound system $\sigma + A$ (during the measurement process) is a *physical transformation*, which is completely described by the equation of motion of the theory (Schrödinger's axiom). In other words, the axioms of the theory describe, in the same way, « normal » physical systems such as σ and « strange » systems such as A and $\sigma + A$.

As it usually happens, the different ways out of the paradox are founded on different choices of the « guilty » hypothesis, to be rejected (or at least weakened). Let us first discuss those views which accept the whole standard axiomatization of QT and try and weaken only our special hypothesis. Formally, these hypotheses seem to be somewhat similar to two corresponding hypotheses which are used in set theory (ST) in order to derive Russell's antinomy:

a) certain « strange » collections which are objects of the metatheory of ST (such as the collection of all sets described by the theory) are also *objects* of the theory (that means, they are *sets*, and thus belong to the *universe* of ST),

b) all axioms of ST hold, in the same way, for *all* the objects of the theory (for « normal » sets, as well as for « strange » sets).

From an abstract point of view, in both cases (QM and ST) there are only three possible ways out:

I) rejecting *a*) and accepting *b*),

II) accepting *a*) and rejecting *b*),

III) rejecting both *a*) and *b*).

As is well known, in the case of ST, I) corresponds to Zermelo's solution (the « strange » collections are not objects of the theory), II) to the von Neumann-Bernays-Gödel solution (the usual axioms of ST do not hold, generally, for the strange collections, but only for normal sets), finally III) corresponds, in some way, to Russell's solution (which limits the universe of sets and, at the same time, relativizes the axioms to specific classes of sets). (See, *e.g.*, [41].)

In QT, solution I) has been maintained, for instance, by BOHR. In our language we could describe Bohr's view as follows: « strange » objects, such as the measurement apparatus A, are *essentially only metatheoretical*, since they represent *classical* and not *quantum* systems. Solution II) has been maintained by VON NEUMANN: the « strange » system A is a legitimate object of QT (since it consists of elementary particles, and as a particular microsystem must be described by QT). Thus $A \in S$, and A can be represented as a pure state satisfying condition 1). At the same time, the axioms of the theory do not generally hold for such strange objects: in particular, the evolution of a compound system such as $\sigma + A$ (corresponding to a measurement process) is not governed by Schrödinger's axiom (this is sufficient to prevent the derivation of (100)).

One can easily recognize that, in a sense, VON NEUMANN has proposed exactly the same « logical operation » for ST as for QT. As is well known, the intuitive reason why the evolution of the compound system from $\hat{\psi}(t_0) \otimes \hat{\varphi}(t_0)$ to $\psi_k \otimes \varphi_k$ fails to be governed by Schrödinger's axiom can be explained by means of an « informational interpretation ». One claims that this evolution does not represent a « physical transformation » of the system, but rather a transformation of our knowledge about the system. This interpretation is apparently sufficient in order to overcome the logical difficulty; nevertheless, it gives rise to some physical and epistemological problems. The most serious one seems to be the following: the compound system $\sigma + A$ has undergone some physical change during the time $[t_0, t]$ (since A has recorded a value, and A might have perturbed σ). If the transformation $\hat{\psi}(t_0) \otimes \hat{\varphi}(t_0) \Rightarrow \psi_k \otimes \varphi_k$ does not obey the Schrödinger equation, what kind of physical law does it obey? A purely subjectivistic answer (according to which the reduction of the wave function represents a matter of « consciousness » and not a physical process) is, of course, quite unsatisfactory for many epistemological reasons. One may observe that this way of thinking, if maintained also in the case of ST, could lead to very strange conclusions. Think for instance that an assertion such as « $V \in V$ » (the total class belongs to itself) is not a theorem of ST and cannot be a set-theoretical truth (in any model of ST); nevertheless, we can grasp in our consciousness the claim that the class of all sets belongs to itself!

However, in spite of these epistemological difficulties, one can give a *purely logical interpretation* to von Neumann's thesis, which is completely free of any subjectivistic and spiritualistic connotation. Let us first note that a measurement apparatus A (object of the metatheory) can always be identified with the *observer* himself. In the case in which $A \notin S$, $\tau(\sigma)(t_0) = \sum c_j \psi_j$, and by performing a Q_i-measurement we get the result $r_k \pm \varepsilon^A/2$ (and consequently, by the projection postulate, $\tau(\sigma)(t) = \psi_k$), we can say that the « apparatus-observer A has realized the reduction of the wave function ». Now, von Neumann's solution can be simply interpreted as follows: if the apparatus-observer A is an object of the theory (*i.e.* $A \in S$), then A *cannot realize* the reduction of the wave function. This is feasible only by another A' which is « external » with respect to the universe of the theory. In other words, any apparatus, as a particular physical system, can be an object of the theory (that means, can belong to S). Nevertheless, *any apparatus which realizes the reduction of the wave function is necessarily only a metatheoretical object.*

Such a conclusion does not sound as a subjectivistic claim; on the contrary, it seems to be very close to some similar limitative results that we have accepted in logic. For instance, the following ones:

1) to realize a proof of the consistency of a well-behaved scientific theory, one must be « external » with respect to the theory, in the sense that one cannot use *only* the proof-theoretical tools allowed by the theory (Gödel's theorem);

2) to « grasp » the concept of truth for a well-behaved theory one cannot speak *only* the language of the theory (Tarski's theorem).

von Neumann's solution (in this reformulation) turns out to be a « purely logical solution »: it is sufficient to avoid the paradox; nevertheless, it does not give any explanation of what happens *physically* in the transformation from $\hat{\varphi}(t_0)\otimes\hat{\varphi}(t_0)$ to $\psi_k\otimes\varphi_k$.

Some solutions of type I) or essentially of type III) arise from the attempt to give such an explanation. Let us try and describe them in a somewhat abstract way. The intuitive starting point of a number of solutions of type III) can be simplified as follows: the apparatus A is to be represented as a quantum system, whereas the transformation $\hat{\varphi}(t_0)\otimes\hat{\varphi}(t_0) \Rightarrow \psi_k\otimes\varphi_k$ is to be described as a physical relation. Nevertheless, both A and the wave function reduction relation cannot be described directly by the theory, since they require a more sophisticated physical theory, which includes a theory of the irreversible processes. From a logical point of view, such a way of thinking admits of at least two different formal descriptions:

1) One may propose the following formal relations between different physical theories:

$$\begin{array}{ccc} \mathrm{QT}+X & \subseteq & \mathrm{MT}\,, \\ & \preccurlyeq\ \mathrm{MP}\ \succcurlyeq & \end{array}$$

where

QT represents quantum theory (in the standard axiomatization developed in our formalization),

MT is the metatheory (in a formal version),

MP represents the macrophysics (including thermodynamics) which is needed in order to develop an adequate theory of the measurement processes,

$\subseteq$ is the standard logical relation of *theoretical inclusion*,

$\preccurlyeq$ represents the relation of *physically weaker than* (defined in subsect. **3**·2),

X represents a set of special hypotheses (for instance, ergodiclike hypotheses and/or hypotheses concerning the actual state of the Universe).

2) A much simpler formal relation that can be proposed is the following

$$\mathrm{QT}\subseteq\mathrm{QMD}\subseteq\mathrm{MT}\,,$$

where QMD represents *quantum macrodynamics*, which includes an adequate theory of the measurements processes and is included in the metatheory of QT. Our first possibility seems to be a good schematization, for instance, of Jordan's and Ludwig's approaches [42], whereas the second possibility is rather close to the Daneri-Loinger-Prosperi solution [43].

As we have seen, von Neumann's solution gives rise to a kind of *regressus ad infinitum* concerning the measurement apparatus: the infinite chain of the metatheories of QT cannot be interrupted, not only in their mathematical parts (owing to the limitative theorems of logic), but also in the physical parts (in other words, the physical part of the metatheories of this infinite chain cannot be constant). Is this kind of « physical regressus ad infinitum » unavoidable also for solutions of type III)? In other words, let us assume, for instance, our second formal possibility (QT $\subseteq$ QMD $\subseteq$ MT). Is QMD necessarily submitted to the same limitations as QT, in the sense that QMD cannot describe completely its own measurement processes? This question seems to be, so far, open. Any answer depends, of course, essentially on the kind of formal version which one associates with QMD. At any rate, the theory seems still to be very far from the formal axiomatic structure developed for QT.

Finally, we want to deal with those ways out of the paradox which try and weaken the « axioms » of QT. Of course, the critical axiom is represented by the projection postulate. A weak form of this postulate, which has been proposed in the literature, sounds as follows:

4.2.7*) Let $\hat{\varphi}(t_0) = \sum c_j \psi_j$ (ψ_j being eigenvectors of $\varrho(Q_i)$) and let us suppose that one carries out, in the state $\hat{\varphi}(t_0)$, a first-kind measurement for the quantity Q_i. Then, *the probability that soon after the measurement (at time t) the state $\psi(t_0)$ transits to the state ψ_k is given by* $|c_k|^2$.

Actually, 4.2.7*) represents a very ambiguous formulation; and it is not easy to read it univocally (in our metatheoretical language). As a limiting case one could assert that 4.2.7*) represents only a different formulation of a particular case of 4.2.4). As a consequence, according to this interpretation, the projection postulate would be simply dropped. However, one could propose also a different interpretation of 4.2.7*), namely the following: if in the time t_0 we translate σ as $\sum c_j \psi_j$, and if we carry out a measurement for Q_i in σ, then the probability that soon after the measurement we will translate σ as ψ_k is given by $|c_k|^2$. This formulation seems to suggest that the concept of probability can be applied to our translation function τ: in other words, any τ (as a function from $\Omega(\sigma)$ into $\varrho(\Omega(\sigma))$) should have a definite probability value. However, it is really very hard to see how such a probability could be defined in a reasonable way!

A weaker formulation of the projection postulate (which is formally more satisfactory) is the following:

4.2.7′) Let $\tau(\sigma)(t_0) = \sum c_j \psi_j$ (ψ_j eigenvectors of $\varrho(Q_i)$). Let us carry out a first-kind measurement for the quantity Q_i in $\sigma(t_0)$, using an apparatus A. Then, if $\tau(A)(t_0) = \varphi_0$ (and consequently, by 4.2.6) (69), and condition 1) $\tau(\sigma + A)(t) = \sum c_j \psi_j \otimes \varphi_j$, where $[t_0, t]$ represents the time of the measurement process), $\tau(\sigma)(t)$ and $\tau(A)(t)$ correspond respectively to the mixtures $\sum |c_j|^2 P\psi_j$ and $\sum |c_j|^2 P\varphi_i$.

Apparently 4.2.7′) does not give rise to any logical difficulty. Nevertheless, it cannot be reasonably termed a « reformulation of the projection postulate » since it turns out to be a theorem derivable from the standard axiomatization of QT without projection (+ condition 1)).

Dropping the projection postulate represents, of course, a quite satisfactory solution from a logical point of view, as far as it prevents the derivation of any *known* paradox. Nevertheless, as is well known, it gives rise again to some epistemological difficulties. The most important one seems to be the following: apparently, after a measurement process of the sort described, we *do* find the apparatus A (and the system σ) in a definite eigenstate, since we « read » a definite result. How does this situation *de facto* agree with the prediction of the theory, according to which the state of the apparatus (and of the system) after the measurement is necessarily a mixture?

One can give different possible answers to this question. It seems to us that (within the framework of this kind of approach) the most satisfactory answer is represented by the so-called « modal interpretation » proposed by VAN FRAASSEN (*). In our metalanguage, one can reformulate Van Fraassen's main thesis as follows: as we already know, it may happen

$$\models_{s(t)} \mathsf{d}_i = r_k \qquad \text{and} \qquad \models_{A(t)} \mathsf{e} = a_k$$

(where e is a special variable corresponding to the observable E concerning our apparatus A); nevertheless,

$$\text{not} \models\!\!\!\equiv_{s(t)} \mathsf{d}_i = r_k\,, \qquad \text{not} \models\!\!\!\equiv_{A(t)} \mathsf{e} = a_k\,, \qquad \text{not} \models\!\!\!\equiv_{s(t_0)} \mathsf{d}_i = r_k\,.$$

In other words, a system may have (empirically) a certain value for a given quantity (empirical truth at time t), and, in spite of this, one cannot assert the *quantum-logical truth* of the corresponding sentence, neither at time t_0

(*) See [44, 45]. Notice that the « modal interpretation of QM » proposed by VAN FRAASSEN has to be distinguished from the « modal interpretation of QL » (see subsect. **4**·4).

nor at time t. That means: *events* may happen, without being *a priori certain*.

One could notice that the intuitive consequence of this kind of solution (founded on a rejection of the projection postulate) seems to be not very far from our « logical interpretation » of von Neumann's solution (which instead accepts the projection). Indeed, in both cases, one is led to the conclusion that any physical system which realizes an unambiguous reading on the measurement apparatus must be « external » with respect to the theory. And this « external condition » is asserted in a purely « logical way » (*i.e.* without any attempt to describe the physical change that takes place, during the measurement process, in the compound system « system + apparatus »).

As a final question, let us ask whether or not one can maintain a « quantum-logical solution » to the measurement paradox. One can recognize that mere logic cannot be the only « cause » of the paradox. Indeed, one can easily transform our derivation in a *formal proof*, and it turns out that one has used a very weak logic (which is a sublogic of classical logic, quantum logic and intuitionistic logic at the same time). Thus (unlike what happens, for instance, in the case of the « two-hole paradox ») *changing logic* is not sufficient to avoid, in a simple way, the proof of our contradiction. Nevertheless, quantum logic (if assumed in the metatheory of QT) can help us in a particular interpretation of the measurement problem, namely the so-called « ignorance interpretation ». As is well known, the « ignorance interpretation » arises, in this context, as an attempt to give a satisfactory intuitive meaning to the mixture $\sum|c_j|^2 P\psi_j$ and $\sum|c_j|^2 P\varphi_j$ (into which, according to 4.2.7′), the two systems σ and A have to be translated soon after the end of the measurement). According to this interpretation, the system $\sigma(t)$ $(A(t))$ *is* in a definite eigenstate ψ_k (φ_k), although *we do not know* in which state it is. In other words, mixtures (unlike superpositions) reflect only our *ignorance* about the system, and not a real indetermination of the system itself. As is well known, the ignorance interpretation gives rise to a deep logical difficulty. Indeed, generally, a mixture W is not determined, in a unique way, by a set of eigenvectors. In other words, we could have, for two incompatible quantities $\varrho(Q)$ and $\varrho(Q')$ (with eigenvectors respectively ψ_j and ψ'_h),

$$W = \sum |c_j|^2 P\psi_j = \sum |c'_h|^2 P\psi'_h .$$

As a consequence, if the ignorance interpretation were right, the system under consideration should be in two incompatible eigenstates at the same time! However, if one is thinking (in the metatheory) *quantum logically* (instead of *classically*), such paradoxical consequence seems to disappear. Indeed, quantum logic does not permit the following classical inference

$$\frac{\bigvee x\,\alpha(x),\ \bigvee y\,\beta(y)}{\bigvee x \bigvee y\,(\alpha(x)\ \&\ \beta(y))}\,.$$

In other words, we might have:

1) the system is in an eigenstate of $\varrho(Q)$

$$\left(\bigvee \psi_j(\tau(\sigma(t))) = \psi_j\right),$$

2) the system is in an eigenstate of $\varrho(Q')$

$$\left(\bigvee \psi'_h(\tau(\sigma(t))) = \psi'_h\right),$$

3) nevertheless the system is not at the same time in an eigenstate of $\varrho(Q)$ and in an eigenstate of $\varrho(Q')$

$$\left(\text{not } \bigvee \psi_j \bigvee \psi'_h(\tau(\sigma(t))) = \psi_j \,\&\, \tau(\sigma(t)) = \psi'_h\right).$$

From a formal point of view, such an argument is apparently correct. Nevertheless, from an intuitive point of view, quantum logic can be hardly described as a good solution of the « ignorance difficulties ». Indeed as we have seen, the quantum-logical constants do not have the same meanings as the classical ones. What does really mean an expression like $\bigvee \psi_j(\tau(\sigma)(t) = \psi_j)$ (where $\bigvee$ is the quantum-logical existential quantifier)? According to our modal interpretation (described in the previous section), our critical expressions would simply mean « *necessarily* the system σ at time t *could* be in an eigenstate ψ_j ». And, of course, such a reading does not give any help to the ignorance interpretation.

REFERENCES

[1] D. Hilbert: *Arch. Math. Phys.*, **1**, 44 (1901).
[2] G. Hamel: *Die Axiome der Mechanik*, in *Handbuch der Physik*, Vol. **5** (Berlin, 1955), p. 1.
[3] J. McKinsey, A. Sugar and P. Suppes: *Journ. Rat. Mech. Anal.*, **2**, 253 (1953).
[4] C. Carathéodory: *Math. Ann.*, **67**, 355 (1909).
[5] J. von Neumann: *Mathematische Grundlagen der Quantenmechanik* (Berlin, 1932).
[6] A. N. Kolmogorov: *Grundbegriffe der Wahrscheinlichkeitsrechnung* (Berlin, 1933).
[7] P. C. Suppes: *Introduction to Logic* (New York, N. Y., 1957).
[8] J. D. Sneed: *The Logical Structure of Mathematical Physics* (Dordrecht, 1971).
[9] M. Przelecki: *The Logic of Empirical Theories* (London, 1969).
[10] M. L. Dalla Chiara and G. Toraldo di Francia: *Riv. Nuovo Cimento*, **3**, 1 (1973).
[11] R. Wójcicki: *Journ. Phil. Logic*, **3**, 337 (1974).
[12] G. Toraldo di Francia: *Riv. Nuovo Cimento*, **4**, 144 (1974).

[13] G. Toraldo di Francia: *The concept of progress in physics*, in *Italian Studies in the Philosophy of Science*, Boston Studies in the Philosophy of Science (to be published).
[14] G. Toraldo di Francia: *What have physicists learned from experience about inductive inference?*, in *Formal Methods in the Methodology of Empirical Sciences* (Dordrecht, 1976), p. 330.
[15] W. V. Quine: *From a Logical Point of View* (Cambridge, Mass., 1953), p. 20.
[16] W. V. Quine: *Necessary truth*, in *The Ways of Paradox* (New York, N. Y., 1966), p. 50.
[17] J. Nicod: *Foundations of Geometry and Induction* (New York, N. Y., 1930), p. 219.
[18] C. G. Hempel: *Recent problems of induction*, in *Mind and Cosmos* (Pittsburg, Penn., 1966), p. 120.
[19] C. G. Hempel: *Journ. Symb. Logic*, **8**, 122 (1943).
[20] I. Prigogine: this volume, p. 308.
[21] N. Goodman: *Fact, Fiction and Forecast* (Cambridge, 1955).
[22] C. G. Hempel: *Recent problems of induction*, in *Mind and Cosmos* (Pittsburg, Penn., 1966).
[23] T. Poston: *Fuzzy geometry*, thesis (University of Warwick, 1971).
[24] C. T. J. Dodson: *Bull. London Math. Soc.*, **6**, 191 (1974).
[25] G. Toraldo di Francia: *L'indagine del mondo fisico* (Torino, 1976), p. 408.
[26] C. C. Chang and H. J. Keisler: *Model Theory* (Amsterdam, 1973).
[27] K. Gödel: *Monatsh. Math. Phys.*, **39**, 173 (1931).
[28] E. Mendelson: *Introduction to Mathematical Logic* (Princeton, N. J., 1964).
[29] M. L. Dalla Chiara: *A multiple sentential logic for empirical theories*, in *Formal Methods in the Methodology of Empirical Sciences* (Dordrecht, 1976), p. 43.
[30] A. Tarski, A. Mostowski and R. Robinson: *Undecidable Theories* (Amsterdam, 1953).
[31] H. Dishkant: *Stud. Logica*, **30**, 23 (1972).
[32] R. I. Goldblatt: *Journ. Phil. Logic*, **3**, 19 (1974).
[33] G. Hardegree: *The conditional in quantum logic*, in *Logic and Probability in Quantum Mechanics* (Dordrecht, 1976), p. 55.
[34] M. L. Dalla Chiara: *Stud. Logica*, **35**, 23 (1976).
[35] M. L. Dalla Chiara: *Journ. Phil. Logic*, **6**, 391 (1977).
[36] B. C. Van Fraassen: *Semantic analysis of quantum logic*, in *Contemporary Research in the Foundations and Philosophy of Quantum Theory* (Dordrecht, 1973), p. 80.
[37] G. E. Hughes and M. J. Cresswell: *An Introduction to Modal Logic* (London, 1968).
[38] S. Kripke: *Acta Phil. Fenn.*, **16**, 83 (1963).
[39] S. Kripke: *Semantic analysis of intuitionistic logic*, I, in *Formal Systems and Recursive Functions* (Amsterdam, 1965), p. 92.
[40] M. L. Dalla Chiara: *Journ. Phil. Logic*, **6**, 331 (1977).
[41] A. Fraenkel and Y. Bar-Hillel: *Foundations of Set Theory* (Amsterdam, 1958).
[42] G. Ludwig: *Phys. Bl.*, **11**, 489 (1955).
[43] A. Daneri, A. Loinger and G. M. Prosperi: *Nucl. Phys.*, **33**, 297 (1962).
[44] B. C. Van Fraassen: *A formal approach to philosophy of science*, in *Paradigms and Paradoxes: The Philosophical Challenge of the Quantum Domain* (Pittsburg, Penn., 1972), p. 303.
[45] B. C. Van Fraassen: *The Einstein-Podolski-Rosen paradox*, in *Logic and Probability in Quantum Mechanics* (Dordrecht, 1976), p. 283.

Some Foundational Problems in the Special Theory of Relativity.

M. JAMMER

Bar-Ilan University - Ramat-Gan, Israel

1. – Introduction.

Just as the discovery of non-Euclidean geometry in the early nineteenth century gave rise to the development of fundational research in mathematics and thus led eventually to the establishment of metamathematics, a highly respected autonomous discipline and now part of the regular curriculum of modern mathematical instruction, so did the advent of special relativity, more than any other theory in physics, contribute to the study of the foundations of physics. True, important work on the foundations of physics had been carried out already at the end of the last century by theoreticians such as HELMHOLTZ, MACH, KIRCHHOFF and HERTZ. But the deeper import of their work could be fully understood only after the rise of the special theory of relativity at the beginning of the present century. Moreover, it is an historical fact that the work of the great originators of quantum mechanics, like HEISENBERG, BOHR, JORDAN or DIRAC, has been greatly influenced, as admitted by themselves, by the conceptual impact of Einstein's special relativity. If in the not too distant future—it is hopefully only a question of time—the systematic study of the conceptual foundations of physics becomes institutionalized into an autonomous discipline, the logical, epistemological and methodological problems raised by the theory of relativity will play a prominent role in the subject-matter of this study.

Although, admittedly, questions related to the foundations of quantum mechanics are presently at the forefront of general interest in foundational research and constitute the object of a unprecedented dissension [1] among the experts in this field, it would be rash to conclude that the foundations of special relativity, despite its status of being an exceptionally well established theory, are a matter of universal consensus and unaminity. Although about three quarters of a century have passed since Albert EINSTEIN first proposed the theory in 1905 [2], many textbooks and even monographs of the theory contain mispresentations, if not misconceptions, of some of its more delicate issues.

It is the purpose of this paper to draw attention to some of these issues and to contribute possibly to their clarification.

The advent of a novel theory, although constructed usually to resolve previously not encountered internal inconsistencies exhibited by the older theory or to remove an incompatibility of the latter with experience, leads not only to revisions of former results, but often also to the recognition that what previously had been regarded as an unproblematic issue is in fact a matter of profound intricacies. A new physical insight does not only open up new vistas for the future, it also acts retroactively in our understanding of ideas of the past.

This applies to special relativity with respect to one of the most basic discoveries in physics, Olaf Roemer's determination of the finite velocity of light (1676). Eddington's statement, « Time, as we now understand it, was discovered by Roemer », a statement which, if exhaustively interpreted, is one of the most profound statements ever made in physics, indicates the intimate connection between Roemer's discovery [3] and special relativity which according to Adrian Daniel FOKKER [4] should have been called « chronogeometry », that is a theory of the measurements of time and space. Even though EDDINGTON intended, first of all, to emphasize the distrinction between the subjectively perceived « Here-Now » and the (time-retarded) objectively conceived « Worldwide-Now », his statement, by referring implicitly to the discovery of the *finiteness* of the fastest signal, also hinted to the definitional (conventionalistic) nature of distant simultaneity; and thereby it alluded to the radical modification of our conception of time as entailed by the special theory of relativity.

To fully understand how special relativity affects the theory of Roemer's discovery, it should be recalled that one of the most fundamental ideas underlying the conceptual edifice of relativity, as repeatedly stressed by Hans REICHENBACH [5] or Adolf GRÜNBAUM [6], is the conventionality ingredience of intrasystemic distant simultaneity. In fact, this notion, even more than the notion of the intersystemic relativity of simultaneity, played, as we shall see, an important role in Einstein's pioneering paper [2].

It should be recalled that all the well-known methods of measuring the velocity of light (in vacuo or air) such as Fizeau's by use of a toothed wheel (1849), Focault's by use of rotating mirrors (1850), Newcomb's by use of a revolving prism (1885) or Michelson's various precision measurements (1882, 1885), yielded only the averaged round-trip velocity of light and not its one-way velocity [7]. To measure the latter, two distant synchronized clocks are required. But to synchronize such clocks, as is well known, one has to know the velocity of light.

Aware of this vicious circle and its unavoidability, EINSTEIN realized that the determination of the one-way velocity of light requires a definitional or conventional stipulation. He thus posited, *i.e.* declared by convention, that the time required by light to travel from point A to point B equals the time required by light to travel from point B to point A, implying that no experimental procedure exists which makes it possible to determine unambiguously and without any convention the one-way velocity of light.

More precisely, EINSTEIN pointed out [2] that a clock stationary at A establishes, in the usual way, an « A-time », and a clock stationary at B similarly a « B-time », the assumption being made that identical clocks, if stationary in the same reference system, have the same « rate » but may differ in their « setting » (zero). Intrasystemic synchronization consists therefore only in relating the « zeros » of the clocks, the operation being called « setting the clocks ».

EINSTEIN now emphasized that an « A-time » and a « B-time » does not yet constitute a common « time ». This latter, he declared, can now (« nun » in the German original) be established by stipulating by definition that the « time » which the light requires to travel from A to B is equal to the « time » it requires to travel from B to A. It has already been pointed out by Charles SCRIBNER [8] that the widely used Perrett-Jeffery translation [9] of Einstein's 1905 paper has « somewhat altered » the « exact meaning » of the original, for it reads that the common time cannot be established « unless we establish by definition that the 'time' required by light ... ». What happened seems to be that these translators misread the German word « nun » as « nur », which they correctly translated by « unless ». But through this error the translation turned out to be precisely contradictory to the original intention which implied the admission of alternative choices, later called « nonstandard » signal synchronizations.

For notational convenience let us apply the Reichenbach notation in reviewing the issue under discussion. If clock A (*i.e.* a clock stationary at point A in an inertial system S) emits a light pulse (signal) at time t_A as measured by A and if this signal is received by clock B at time t_B as measured by B, and immediately reflected to A, where it is received at time t'_A as measured by A, then the two clocks are *defined* as being synchronous if the following condition is satisfied:

$$t'_A - t_B = t_B - t_A \,, \tag{1.1}$$

which, in the Reichenbach notation, can also be written

$$t_B = t_A + \varepsilon(t'_A - t_A) \tag{1.2}$$

with $\varepsilon = \frac{1}{2}$ (see fig. 1).

ε will be called the Reichenbach synchronization coefficient. The « thesis of the conventionality of intrasystemic distant simultaneity » or briefly « conventionality thesis » consists in the statement that the numerical value of ε need not necessarily be $\frac{1}{2}$, but may be any number in the open interval between 0 and 1, *i.e.* $0 < \varepsilon < 1$, without ever leading to any conflict with experience. The restriction imposed on ε is a consequence of causality require-

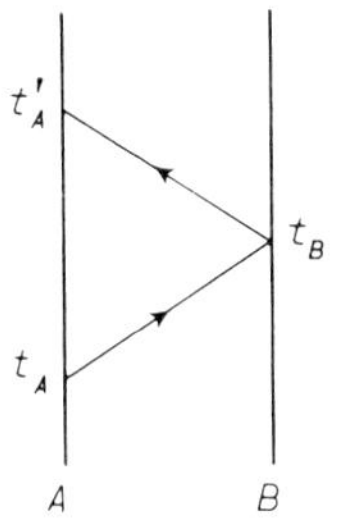

Fig. 1.

ments. The finiteness of the empirically measured round-trip velocity implies that $t'_A > t_A$. Hence the assumption $\varepsilon \geqslant 1$, or equivalently $t_B - t_A \geqslant t'_A - t_A$, would entail that the one-way time (from A to B) is less than the round-trip time (from A to B and back to A). On the other hand, the assumption $\varepsilon \leqslant 0$, or equivalently $t_B \leqslant t_A$, would imply that the one-way time from A to B is either zero or negative, which means that the signal arrives at the same time when it was emitted or even earlier than that. Neither assumption is compatible with causality.

To conclude our introductory considerations, let us point out that the contention that the one-way velocity of light can be measured by experiment flagrantly contradicts the conventionality thesis. For, on the assumption of this contention, let v_{AB} denote the empirically established velocity of light in its passage from A to B. Then by definition of « velocity »

$$v_{AB} = \frac{AB}{t_B - t_A}, \tag{1.3}$$

where AB denotes the distance between A and B. Since this distance is also an empirically measurable quantity, an unambiguous (unique) value of $t_B - t_A$ would be obtained. Since, moreover, $t'_A - t_A$ in (1.2) is likewise an empirically measurable quantity, for it can be measured by the use of one single clock, the value of ε in (1.2) would be uniquely determined, contrary to the conventionality thesis.

2. – Roemer's determination of the velocity of light.

Let us now review, in the context of the preceding considerations, Roemer's discovery, based as it was on the astronomical observations of the eclipses of the satellites of the planet Jupiter. Since his method is well known, we shall sketch it only briefly, but emphasize some points of interest in our present context (see fig. 2).

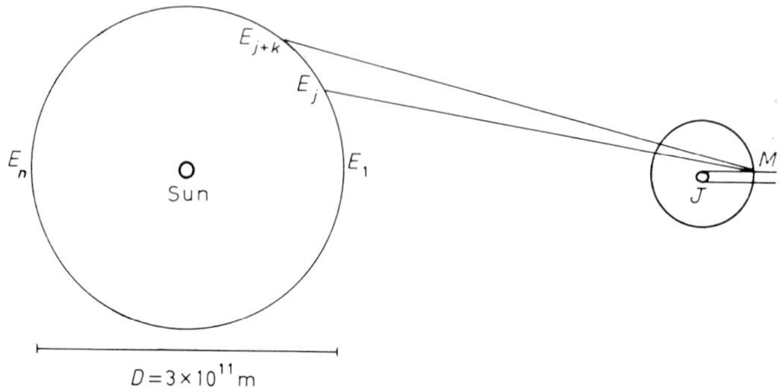

Fig. 2.

Let T_0 denote the period of the satellite M (Io) of Jupiter J whose position in the solar system is, for simplicity, assumed to be stationary; and let t_k denote the time interval between k consecutive emersions of M from the shadow behind J. Finally, let $\Delta_k L$ denote the change in distance of the Earth E from M (at its point of emersion) during t_k. Clearly then

$$t_k = kT_0 + \frac{\Delta_k L}{c}, \tag{2.1}$$

where c denotes the velocity of light.

This equation contains two unknowns, T_0 and c. T_0 can be determined as follows. Assume that during m emersions E returns to its initial position, say E_1, so that $\Delta_m L = 0$. Then eq. (2.1) yields

$$t_m = mT_0, \tag{2.2}$$

which shows that measurements of m and of t_m determine T_0. Now, c can be determined as follows. Assume that during n emersions E moves from conjugation E_1 to opposition E_n so that $\Delta_n L = D$, the diameter of the orbit of E. Then (2.1) yields

$$c = \frac{D}{t_n - nT_0}, \tag{2.3}$$

which may be called « Roemer's equation ».

To obtain a deeper insight into the physical significance of Roemer's equation let us put $t_n = nT = D/v$, so that T denotes the average period of occultation (during t_n) and v denotes the average velocity of a fictitious earth moving along D (*i.e.* from E_1 to E_n is a straight line). Finally, let $\nu = T^{-1}$ and $\nu_0 = T_0^{-1}$. Then clearly (2.3) reduces to

$$\nu = \nu_0(1 - \beta)\,, \tag{2.4}$$

where, as usual, β denotes v/c. But (2.4) is the well-known (nonrelativistic) Doppler equation for a source at rest and an observer in motion with velocity v. Roemer's method turns out to be an application of the Doppler effect with c as the only unknown.

Since in this method only *one* clock is used and no reflections are involved, it seems plausible to contend that Roemer's method is a means to measure the one-way velocity of light. In fact, such a claim has been made by prominent physicists such as Eugene FEENBERG [10], Resa MANSOURI and Roman U. SEXL [11]. Also Percy William BRIDGMAN [12] once expressed some sympathy with this point of view, adding however that « the situation is not as straightforward as it might seem ». Most eloquent, perhaps, on this point was G. Burniston BROWN [13]. To support his claim he suggested the following terrestrial analogous thought-experiment (fig. 3).

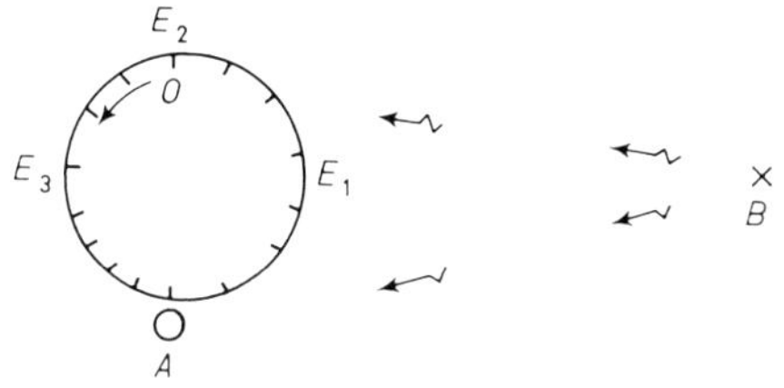

Fig. 3.

The periodic eclipses are replaced by a beacon B, flashing at intervals whose equality is controlled by a clock at A. The observer O is carried around on the edge of a circular rotating table and makes a mark on the stationary surrounding rim every time he observes a flash. These marks get further apart between E_1 and E_2, corresponding to the increase in eclipse time periods in Roemer's discovery. The clock A, at rest with respect to B and the centre of the rotating table, makes marks on the table edge, the distance between which can be used as a test for the uniformity of the rotation as well as for the purpose of converting the distances between the marks on the stationary rim into time intervals. The distance E_1E_3 is measured with a meter bar. From these data, BROWN contended, the one-way velocity of light can be ascertained in an empirical way. « Roemer's method ... nullifies Einstein's contention, repeated by Eddington and others, that we only know the out-and-return velocity, not the one-way velocity, so that the time of arrival of a signal at a distant point is never known from observation but can only be a convention », he concluded.

Roemer's method posed a problem, of course, also to the « conventionalists ». Already in 1925 REICHENBACH discussed this problem from his point of view [14]. More recently, L. KARLOV of the University of Sydney studied the problem with the technique of modern nonstandard synchronization methods and arrived at the conclusion that the velocity which Roemer's method yields is « expressly not the one-way speed of light, but the average speed over a closed path » [15].

We are, of course, aware of the fact that the literature in relativity abounds with pseudoscientific papers written by quacks and charlatans. But we wish to stress that all the names mentioned in the present context are those of competent and serious physicists. We must, therefore, conclude that the conceptual status of Roemer's determinations of the velocity of light, the very beginning of the prehistory of special relativity, is a matter of a serious dispute. This to show was the objective of the present section of this paper.

3. – The rise of special relativity.

It is fortunate for the student of physics that a systematic presentation of a physical theory need not reflect its historical development. For the latter is usually a very complicated subject of study. This does not mean, however, that some knowledge about the vicissitudes preceding the advent of a new physical theory and some study of its historical evolution, particularly during the early phases of its conceptual consolidation, does not contribute to a more profound understanding of the theory itself. In general, such an historical approach consists in the study of the contradictions and inconsistencies, either internal or with experience, of the older theory and of the various solutions proposed to overcome these difficulties, as well as of the reasons that led to the acceptance of the new theory in favour of possible alternatives.

Due to the fact that the special theory of relativity had its origin in a conflict between two disparate theories, classical (Newtonian) mechanics, on the one hand, and optics or, more generally, Maxwell's theory of the electromagnetic field, on the other, the prehistory of special relativity is a particularly intricate story. Much has been written on this subject, and even more on the notorious priority question: EINSTEIN *vs.* POINCARÉ. We do not intend to indulge in this subject. We shall make, however, some remarks on a very small number of historical issues which seem to be little known despite their importance precisely for the study of the foundations of the theory.

If one were asked what single experiment had been most influential, directly or indirectly, on the development of the special theory of relativity, he would undoubtedly answer: the Michelson-Morley experiment, even though it may have played only a negligible role in Einstein's reasoning in 1905. In fact, many textbooks take this experiment as their point of departure for their exposition of the theory. If one were asked what theoretical feature had been most instrumental for the development of this theory, he would undoubtedly answer: the lack of Galilean invariance of the Maxwell equations.

Now, it is an undeniable historical fact that both these issues, the idea of detecting experimentally the ether drift as attempted by the Michelson-Morley experiment and the absence of the Galilean invariance of the equations of the electromagnetic field, have their origin and source in the work of one man: James Clerk MAXWELL who died in the year EINSTEIN was born.

On March 19, 1879, a few months before he died, MAXWELL sent a letter to David Peck TODD, the director of the Nautical Almanac Office in Washington, in which he proposed a method to detect the Earth's motion through the ether by an extension of Roemer's method. MAXWELL explained that an accurate timing of the eclipses of Jupiter's satellite at different epochs of the Earth's orbital motion should result in different values of the thus-measured velocity of light, depending on whether the Earth moves in the direction of the propagation of light or in the direction opposite to it. « If it were possible, he wrote to TODD, to measure the one-way velocity of light between two terrestrial stations in each of these two cases, the difference between the two transit times would depend linearly on the ratio of the Earth's velocity v and the velocity of light c with respect to the ether. This would be a first-order effect .... But in the terrestrial methods of determining the velocity of light, the light comes back along the same path again, so that the velocity of the Earth with respect to the ether would alter the time of the double passage by a quantity depending on the square of the ratio of the Earth's velocity to that of light, and this is quite too small to be observed » [16].

As it happened, MICHELSON, who had been transferred to Todd's office only a short time before Maxwell's letter arrived, read it and was so much impressed by its contents that he took Maxwell's final remarks as a challenge which incited him to design the experiment that carries his name.

As is well known, before 1861 MAXWELL would have written, had he used modern notation, the equations of the electromagnetic field as follows (free field):

$$\text{div}\,\boldsymbol{E} = 0\,, \qquad \text{div}\,\boldsymbol{B} = 0\,, \tag{3.1, 2}$$

$$\text{curl}\,\boldsymbol{E} = \frac{\partial B}{\partial t}\,, \qquad \text{curl}\,\boldsymbol{B} = 0\,. \tag{3.3, 4}$$

Now, these equations, as can be easily checked, are covariant under the Galilean transformation

$$\boldsymbol{r}' = \boldsymbol{r} - \boldsymbol{v}t\,, \qquad t' = t\,, \tag{3.5, 6}$$

(*i.e.* $\boldsymbol{\nabla}' = \boldsymbol{\nabla}$, $\partial_{t'} = \partial_t + \boldsymbol{v}\cdot\boldsymbol{\nabla}$), provided the field transforms according to the equations [17]

$$\boldsymbol{E}' = \boldsymbol{E} + \nabla \times \boldsymbol{B}\,, \qquad \boldsymbol{B}' = \boldsymbol{B}\,. \tag{3.7, 8}$$

Even if sources (*i.e.* charge and current densities) are taken into consideration, so that (3.1) and (3.4) are replaced by

$$\operatorname{div} \boldsymbol{E} = \varrho/\varepsilon_0 \quad \text{and} \quad \operatorname{curl} \boldsymbol{B} = \mu_0 \boldsymbol{j}\,,$$

the field equations remain Galilean invariant, provided the charge and current densities transform according to

$$\varrho' = \varrho + \varepsilon_0\mu_0 \nabla \cdot \boldsymbol{j}\,, \qquad \boldsymbol{j}' = \boldsymbol{j}\,.$$

The last equations are obviously incompatible with the conception of a current as being a transport of charges. It is, therefore, not surprising that the condition of local conservation of charges, *i.e.* the conservation equation

$$\partial\varrho/\partial t + \operatorname{div} \boldsymbol{j} = 0\,,$$

suggested the introduction of the displacement current and thus led MAXWELL to the well-known form of his equations. But it was precisely the introduction of the displacement current (density) in eq. (3.4) that destroyed the Galilean invariance of the equations. Thus, Maxwell's introduction of the displacement current in 1861 [18] prepared the ground for the theoretical conflict mentioned above. It makes sense to speak of Maxwell as the « spiritual father » of special relativity.

But the story has not yet ended. It is an open question whether MAXWELL was, partly at least, motivated to postulate the displacement current for reasons of symmetry. Norman CAMPBELL, in fact, has argued in favour of such a suggestion when he wrote [19]:

« Suppose you found a page with the following marks on it—never mind if they mean anything, » on the left the eqs. (3.1)-(3.2) and on the right the equations

$$\operatorname{div} \boldsymbol{E} = 0\,, \qquad \operatorname{div} \boldsymbol{B} = 0\,,$$

$$\operatorname{curl} \boldsymbol{E} = \partial\boldsymbol{B}/\partial t\,, \qquad \operatorname{curl} \boldsymbol{B} = -\,\partial\boldsymbol{E}/\partial t\,.$$

« I think you would see that the set of symbols on the right side are 'prettier' in some sense than those on the left; they are more symmetrical. Well, the great physicist, James Clerk MAXWELL, about 1870, thought so too; and by substituting the symbols on the right for those on the left, he founded modern physics ... ».

It has been repeatedly pointed out that a similar quest for symmetry was one of the motivations that prompted EINSTEIN to write his historic 1905 paper [2]. Even though Maxwell's equations, wrote EINSTEIN in the introductory section of this paper, exhibit a high symmetry, their application within the theory « as usually understood at the present time » leads to « asymmetries which do not appear to be inherent in the phenomena ». This he exemplified by a reference to the interaction between a magnet and a conductor, where the observable phenomenon depends only on the relative motion of the two, whereas the theoretical explanation of the induction current produced for the case when the magnet is moving and the conductor at rest differs essentially from the explanation for the case when the conductor is moving and the magnet is at rest. Considerations of symmetry, one may thus conclude, may have been operative both for posing and for resolving the problem which lies at the heart of special relativity, which, as every theory of invariants, can itself be regarded as a theory of symmetries.

Einstein's 1905 relativity paper [2], undoubtedly one of the most important papers ever written in physics, can serve as a useful point of departure for a systematic study of the foundational problems of special relativity. True, as a trailblazing and pioneering paper it is somewhat deficient in elegance and thought economy. But just because of this fact it contains virtually all the raw material for such a study. An analysis of the postulates explicitly stated in it and of the assumptions tacitly made in it, and of the logical relations among these, leads directly to the principal problems of such a foundational study. A logical analysis of this paper has been carried out by a number of authors (A. I. MILLER, J. MERLEAU-PONTY, H. M. SCHWARTZ, M. JAMMER). Limitations of space do not allow us to present such an analysis. We shall confine ourselves, therefore, to merely pointing out on what premises its conclusions have been obtained.

In the introductory section the « conjecture » that « the same laws of electrodynamics and optics will be valid for all frames of reference for which the equations of mechanics hold good » is raised to the status of a postulate, called the « principle of relativity » (PR) (assumption A_1). The frames of reference referred to are, of course, inertial systems (in which the law of inertia is valid), a term not used in the paper. This relativity postulate is followed by a second postulate « which is only apparently irreconcilable with the former, namely, that light is always propagated in empty space with a definite velocity c which is independent of the state of motion of the emitting body ». We shall call this the « light postulate » (LP_1) (assumption A_2). In the first section of the *Kinematical Part* (intrasystemic) distant simultaneity is defined, as stated above (assumption A_3), and it is assumed (« wir nehmen an ») that the synchronism thus defined is symmetric (A_4) and transitive (A_5). It is further assumed (« wir setzen noch der Erfahrung gemäss fest ») that (in the notation of (1.2))

$$\frac{2AB}{t'_A - t_A}$$

is a « universal constant » c, the velocity of light in empty space (A_6), this being the two-way light principle. In sect. **2** the PR is restated in a slightly more general formulation and the second postulate is reformulated as follows: « Any ray of light moves in the 'stationary' system of co-ordinates with the determined velocity c, whether the ray be emitted by a stationary or by a moving body » (A_7 or LP_2).

In sect. **3** the « transformation of co-ordinates and times from a stationary system to another system in uniform motion of translation relatively to the former » (the term « Lorentz transformation », coined by POINCARÉ, is not used) are derived on the assumption (A_8) of the homogeneity of space and time, required to account for the linearity of these transformations. The derivation makes use also of assumptions A_2 and A_3. Furthermore, to derive $\varphi(v) = 1$, an essentially group-theoretical argument is applied combined with the tacit rejection of time-reversal (A_9). By this step a recourse to applying the principle of « reciprocity », to be explained below, is avoided. In the rest of the *Kinematical Part* no further assumptions are made; all relativistic effects, such as length contraction, clock retardation and the composition of velocities, are derived as logical consequences of the transformation equations. The same applies to the *Electrodynamical Part.* In the final section of the paper, dealing with the dynamics of a charged particle in an electromagnetic field, the only link with experience is established by showing how the velocity of such a particle can be calculated from the data of such a field.

To complete our list of assumptions made in this paper it should be pointed out that Einstein's definition of intrasystemic distant simultaneity rested on two further tacit assumptions which at that time had no empirical warrant at all: the rate of moving clocks depends on the path and/or velocity of their transport (A_{10}); and, finally, light is the fastest signal (A_{11}).

It is no exaggeration to say that almost the whole foundational research in special relativity deals with the analysis of the eleven assumptions A_1 to A_{11} and their logical interrelation such as consistency, redundancy, eliminability and completeness.

In fact, Hans Reichenbach's *Axiomatization of the Theory of Relativity* [20], the earliest and, notwithstanding Hermann Weyl's depreciative critique [21], one of the best investigations of this kind, does just this.

Let us examine, for example, Einstein's assumption A_4 concerning the symmetry of synchronization. Let $A \,\mathrm{s}\, B$ denote the fact that clock B has been synchronized with clock A as described above so that (fig. 4) $t_B = \frac{1}{2}(t_A + t'_A)$ and, for the same reason, $t'_B = \frac{1}{2}(t'_A + t''_A)$. If we posit that $t'_A - t_A = t''_A - t'_A$, which amounts to the assumption that the result of a measurement does not depend on the epoch at which it is being performed, then the elimination of t_A and t''_A from the preceding equations yields $t'_A = \frac{1}{2}(t_B + t'_B)$, which means $B \,\mathrm{s}\, A$ and proves that s is a symmetric relation. By a similar, though slightly more complicated, analysis we can derive the transitivity of this relation by postulating that the round-trip time of the propagation of light among three different points is independent of the direction. However, if it is postulated from the very beginning that light has « always » (EINSTEIN) the constant velocity c, A_4 and A_5 would follow directly from this postulate. But such a postulate presupposes, through its use of the concept of velocity, that the concept of « time » (EINSTEIN) has been defined already. This example shows how careful one has to be in the logical analysis of the basic assumptions.

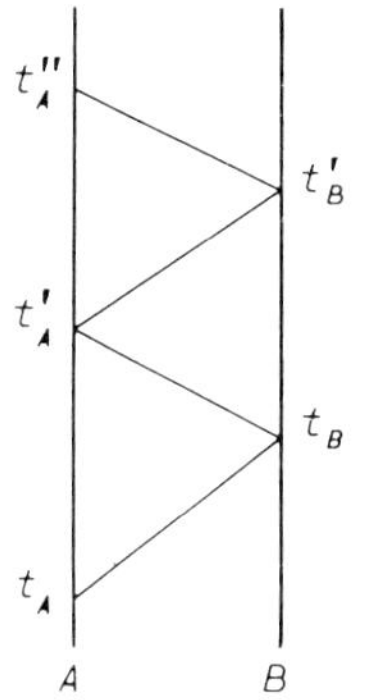

Fig. 4.

Another example which to some extent is a matter of dispute still today concerns the logical relation between A_2 (LP_1) and A_7 (LP_2). Strictly speaking, LP_1 merely excludes the ballistic conception of light, such as proposed in 1908 by Walter RITZ, whereas LP_2, the so-called « second postulate », is usually interpreted as stating that the velocity of light is a universal constant for all inertial systems. This distinction between LP_1 and LP_2 was emphasized already in 1910 by Richard C. TOLMAN [22], who pointed out that LP_2 is a logical consequence of PR and LP_1. LP_1 in his view agrees with the ether theory of light, whereas PR, by denying any physical meaning to the absolute velocity of the source, agrees with the point of view of the emptiness of free space. It should, therefore, not be surprising, he concluded, « that the combination of principles of such different character should have led to a modification in our ideas as to the nature of time and space » [23]. The cogency of Tolman's interpretation has recently been questioned as shown by the discussion led by K. F. SCHAFFNER [24], S. WYKSTRA [25] and R. B. WILLIAMSON [26].

4. – The group-theoretical approach.

In contrast to axiomatizations of formal uninterpreted systems, such as encountered in pure mathematics, in axiomatizations of the special theory of relativity (A. A. ROBB (1921, 1936), H. REICHENBACH (1924), E. A. MILNE (1948), P. SUPPES (1959), J. W. SCHUTZ (1970)) questions concerning the consistency or independence of the postulates are usually given much less attention, if any at all, than questions concerning their completeness. Take, *e.g.*, Reichenbach's *Axiomatization* and you will see that his main concern was to exhaust the completeness problem; with the requirements of consistency and independence (and redundancy) he dealt in only a few lines (sect. **29** of his book).

There are various reasons for this state of affairs. Roughly speaking, the physical theory itself, in its operational sense, may be regarded as an empirically well-confirmed model (interpretation, realization) of the formal axiomatic system and hence as a warrant for the consistency of its postulates. Strictly speaking, independence is not a requirement which, if not satisfied, discards the system completely. On the other hand, one of the main purposes of such axiomatizations is to find out whether or not certain assumptions have been surreptitiously introduced which perhaps are not fully empirically warranted. In the case of a full-fledged theory, such as special relativity, this explains of course the emphasis on the completeness problem.

During the formative stages of the theory the situation was quite the opposite. In fact, the proof that the two major postulates of the theory, the principle of relativity and the light principle (PR and LP_1 or PR and LP_2), are « only apparently irreconcilable » [2] played a dominant role in the early sections of Einstein's 1905 paper and was achieved through his demonstration of the relativity of intersystemic simultaneity. Clearly, an inconsistency between these two postulates, as recently (erroneously) claimed by J. M. Kingsley [27], according to whom the thruth of PR implies the falsity of LP_2 and the truth of LP_2 implies the falsity of PR, would have been fatal to Einstein's theory. Like Euclid's axiomatization of geometry, Einstein's postulates as explicitly formulated in his 1905 paper did not satisfy the completeness condition, which of course is the general fate of all pioneering works.

The problem of the independence of Einstein's postulates may be said to have been raised, even though only in an indirect way as we shall presently see, for the first time four or five years after the advent of the theory. For, contrary to such a problem in the case of uninterpreted axiomatic systems, where it is solved by providing an interpretation (model) in which the postulate whose relative independence is under test is not satisfied but all the other postulates are, it consisted in the study of the question whether the theory could be constructed *without* the use of LP_2. Had the answer been affirmative, it would have shown that LP_2 is redundant and hence either logically not related to the body of the theory or dependent on the other postulate (postulates).

However, the study of this problem did not originate within the context of a logical investigation but as a result of physical considerations.

It will be recalled that in 1909 Constantin Carathéodory published his classical paper *Grundlagen der Thermodynamik* [28] in which he succeeded in deriving the second law of thermodynamics from a rather general abstract postulate (of the inaccessibility of certain thermodynamic states), disencumbering thereby the logical foundations of thermodynamics from the specific properties of a particular process, the Carnot cycle.

Meanwhile, it had been realized that the theory of relativity, although officially not yet recognized as an autonomous discipline (it was usually classified as part of electrodynamics), has a much broader scope and field of applications than originally thought. In particular, it was felt that the propagation of light, after all only a specific, particular process in physics, plays a role of excessive importance in the logical construction of a theory of such far-reaching consequences as, *e.g.*, the mass-energy relation. Carathéodory's banishment of the steam engine from the foundations of thermodynamics suggested a similar feat for the foundations of relativity by relieving them from the individual properties of light.

Precisely such an attempt was made for the first time in December 1909 by the Tiflis-born mathematician Woldemar von Ignatowsky in an address delivered in Moscow. In the first publication of theses ideas [29], which he wrote in Berlin, he studied the problem: « What transformations can be derived from the principle of relativity alone? » In his view the principle of relativity implies some kind of « reciprocity » (we shall have to say more on this point later on) which he defined with reference to two inertial systems S and S' (in the usual standard configuration) as follows:

1) If in S a physical quantity E is a function of certain parameters a_j, *i.e.* if $E = \varphi(a_1, a_2, \dots)$, then in S' the corresponding quantity E' is given by $E' = \varphi(a'_1, a'_2, \dots)$, where a'_j is the transformed a_j.

2) If $E' = f(a_1, a_2, \dots)$, then $E = f(a'_1, a'_2, \dots)$.

Although marred by a number of obscurities which Philipp Frank tried to clarify in an (unpublished) exchange of letters with him, Ignatowsky's paper culminated in the surprising discovery that without any recourse to the light principle the following « Lorentz-type » equations for the transformations $x, t \to x', t'$ can be derived:

$$x' = \frac{x - vt}{\sqrt{1 - kv^2}}, \qquad t' = \frac{t - kvx}{\sqrt{1 - kv^2}}, \tag{4.1,2}$$

where v is the velocity of S' relative to S and k is a constant.

Frank, in collaboration with his student-fellow from Vienna, Hermann Rothe [30], soon realized that the approach initiated by von Ignatowsky is based on the theory of continuous groups which had been developed by Sophus Lie and began to gain wide-spread attention, primarily through the publication of the textbook by Lie and Scheffers [31]. The idea of deriving the Lorentz-type transformation equations without recourse to the properties of light attracted much attention and numerous papers on this subject were published in the course of the next few decades, competing with each other in the claim of having further reduced the number of presuppositions to be made. A closer analysis of many of these papers shows, however, that their authors were not aware of the tacit assumptions contained in their derivations. Thus, for example, the Cambridge physicist and sometime President of Jesus College, L. A. PARS [32], declared solemnly in the beginning of his paper that « the existence of an invariant velocity follows from the mere hypothesis of relativity »; his proof, however, was based on the implicitly made assumption that space and time are isotropic, that the equations are linear, etc. Often authors published derivations without realizing that the same derivation, in merely another notation, had been published before. Sometimes, as for instance in the case of Ernst Benjamin ESCLANGON, Edouard LE ROY and Charles LALAN [33], authors squabbled with each other about the priority of their proof.

An important issue in these group-theoretical derivations of the Lorentz-type transformation equations was the question whether the linearity of these equations could be derived from more general assumptions. The frequently made argument that, since a uniform motion (inertial motion) in S must be mapped into a uniform motion in S', the transformation must be linear seemed to be reducible to a more rigorous reasoning. Lalan's just-mentioned paper may be credited with having provided the earliest proof that this linearity is a consequence of the homogeneity of space and time, an argument which more recently has been thoroughly studied by V. BERZI and V. GORINI [34].

The proof of this contention can be sketched as follows. Let

$$\begin{pmatrix} x' \\ t' \end{pmatrix} = F(v)\begin{pmatrix} x \\ t \end{pmatrix}, \quad \text{or, briefly,} \quad X' = F(X), \tag{4.3}$$

represent the transformation. By assumption, the space-time translation operators T_a and $T_{a'}$ in S and S', respectively, leave (4.3) invariant. Hence

$$T_{a'}X' = F(T_a X), \tag{4.4}$$

or

$$F(T_a X) = T_{a'}F(X). \tag{4.5}$$

Thus

$$F(X + a) = F(X) + a', \tag{4.6}$$

where

$$a = (a_x, a_t) \quad \text{and} \quad a' = (a_{x'}, a_{t'}) \tag{4.7}$$

are independent of X. For $X = 0$ we obtain

$$F(a) = F(0) + a' . \tag{4.8}$$

Hence

$$F(X + a) = F(X) + F(a) - F(0) . \tag{4.9}$$

With $G(X)$ defined by

$$G(X) = F(X) - F(0) \tag{4.10}$$

subtraction of $F(0)$ on both sides of (4.9) yields

$$G(X + a) = G(X) + G(a) . \tag{4.11}$$

But this leads, by the usual method, to the functional equation

$$G(\varrho X) = \varrho G(X) , \tag{4.12}$$

which shows that $G(X)$ and with it $F(X)$ are linear [35].

If one takes into consideration that Hans FREUDENTHAL has shown in 1964 [36] that (under the topological assumptions always made in our context) the homogeneity of space (*not* space-time!) is a consequence of the isotropy of space, he may further reduce the number of assumptions required for this proof.

It must be admitted that all derivations of the Lorentz-type transformation published so far show only the *sufficiency* of the postulates made. A systematic study of their *necessity* seems not yet to have been carried out. This applies also to the more recent derivations by M. STRAUSS (1946), Y. P. TERLETSKII (1968), G. SÜSSMANN (1969) and A. R. LEE and T. M. KALOTAS (1975).

After this lengthy discourse on generalities of the group-theoretical derivations of the Lorentz-type transformation, it seems not inappropriate to present a fully worked-out example.

Let there be given two inertial systems S and S' and let v denote the velocity of S' relative to S (*i.e.* as measured in S). As usual, we confine ourselves for the sake of mathematical convenience to the case of only a two-dimensional space-time. Our derivation will be based on four assumptions. Assuming homogeneity of space and time, and hence linearity of the transformation equations, (P1), we can write

$$\begin{pmatrix} x' \\ t' \end{pmatrix} = \begin{pmatrix} f_{11}(v) & f_{12}(v) \\ f_{21}(v) & f_{22}(v) \end{pmatrix} \begin{pmatrix} x \\ t \end{pmatrix}, \quad \text{or, briefly,} \quad \begin{pmatrix} x' \\ t' \end{pmatrix} = F(v) \begin{pmatrix} x \\ t \end{pmatrix} . \tag{4.13}$$

We further postulate that the transformations $F(v)$ form a group (P2), that reciprocity holds (P3), *i.e.* $F^{-1}(v) = F(-v)$, and, finally, that space is isotropic (P4), *i.e.*

$$\begin{pmatrix} -1 & 0 \\ 0 & 1 \end{pmatrix} F(v) = \bar{F}(v) \begin{pmatrix} -1 & 0 \\ 0 & 1 \end{pmatrix}. \tag{4.14}$$

For notational convenience we write

$$f_{jk}(v) = f_{jk}, \quad f_{jk}(-v) = \bar{f}_{jk}, \quad F(v) = F_v = F \quad \text{and} \quad F(-v) = \bar{F}.$$

That (4.14) expresses indeed spatial isotropy can be see as follows. The transformation $x \to -x$, $x' \to -x'$, $v \to -v$ implies

$$\begin{pmatrix} -x' \\ t' \end{pmatrix} = \begin{pmatrix} \bar{f}_{11} & \bar{f}_{12} \\ \bar{f}_{21} & \bar{f}_{22} \end{pmatrix} \begin{pmatrix} x \\ t \end{pmatrix}, \tag{4.15}$$

which with (4.13) yields $f_{11} = \bar{f}_{11}$, $f_{22} = \bar{f}_{22}$, $f_{12} = -\bar{f}_{12}$ and $f_{21} = -\bar{f}_{21}$, which is identical with (4.14).

P3 implies for the determinant of F that $\|F\| = \|\bar{F}\|^{-1}$. P4 implies $\|F\| = \|\bar{F}\|$. Hence

$$\|F\| = 1. \tag{4.16}$$

P3 combined with (4.16) yields

$$F^{-1} = \begin{pmatrix} f_{22} & -f_{12} \\ -f_{21} & f_{11} \end{pmatrix} = \begin{pmatrix} \bar{f}_{11} & \bar{f}_{12} \\ \bar{f}_{21} & \bar{f}_{22} \end{pmatrix}.$$

Hence

$$f_{11} = \bar{f}_{22}. \tag{4.17}$$

On the other hand, P4 implies that

$$\begin{pmatrix} -f_{11} & -f_{22} \\ f_{21} & f_{22} \end{pmatrix} = \begin{pmatrix} -\bar{f}_{11} & \bar{f}_{12} \\ -\bar{f}_{21} & \bar{f}_{22} \end{pmatrix}.$$

Hence

$$f_{22} = \bar{f}_{22}. \tag{4.18}$$

From (4.17) and (4.18) we see that $f_{11} = f_{22}$, so that

$$F = \begin{pmatrix} f_{11} & f_{12} \\ f_{21} & f_{11} \end{pmatrix}. \tag{4.19}$$

Now, from $x' = f_{11}x + f_{12}t$ and $v = (x/t)\big|_{x'=0} = -f_{12}/f_{11}$ or $f_{12} = -vf_{11}$ it follows that

$$F = \begin{pmatrix} f_{11} & -vf_{11} \\ f_{21} & f_{11} \end{pmatrix}. \tag{4.20}$$

Let w_v^2 be defined by $w_v^2 = -vf_{11}/f_{21}$ or

$$f_{21} = -vf_{11}/w_v^2. \tag{4.21}$$

Then

$$F = f_{11}\begin{pmatrix} 1 & -v \\ -v/w_v^2 & 1 \end{pmatrix}. \tag{4.22}$$

But from (4.16)

$$1 = f_{11}^2(1 - v^2/w_v^2)$$

or

$$f_{11} = (1 - v^2/w_v^2)^{-\frac{1}{2}} \underset{\mathrm{Df}}{=} \Gamma_v. \tag{4.23}$$

Hence

$$F = \Gamma_v\begin{pmatrix} 1 & -v \\ -v/w_v^2 & 1 \end{pmatrix}. \tag{4.24}$$

Applying the group postulate P2, we can write

$$F_vF_{v'} = \Gamma_v\Gamma_{v'}\begin{pmatrix} 1 + vv'/w_{v'}^2 & -(v + v') \\ -(v/w_v^2 + v'/w_{v'}^2) & 1 + vv'/w_v^2 \end{pmatrix} = \tag{4.25}$$

$$= F_{v''} = \Gamma_{v''}\begin{pmatrix} 1 & -v'' \\ -v''/w_{v''}^2 & 1 \end{pmatrix}.$$

Comparison of the diagonal elements yields

$$w_v^2 = w_{v'}^2 \underset{\mathrm{Df}}{=} w^2,$$

so that

$$w^2 = \text{const} \tag{4.26}$$

and hence

$$F = \Gamma_v\begin{pmatrix} 1 & -v \\ -v/w^2 & 1 \end{pmatrix}, \tag{4.27}$$

or in detail

$$x' = \Gamma_v(x - vt)\,, \qquad t' = \Gamma_v\left(t - \frac{v}{w^2}\,x\right), \tag{4.28}$$

which are the Lorentz-type transformation equations, where

$$\Gamma_v = (1 - v^2/w^2)^{-\frac{1}{2}}\,. \tag{4.29}$$

We shall now show that causality implies that w^2 is positive. To this end assume that $w^2 = -u^2$, where u is a positive (real) number. It is then obvious that in this case (4.28) can be represented by the equations

$$\begin{cases} x' = x \cos a - ut \sin a\,, \\ t' = x \sin a + ut \cos a\,, \end{cases} \tag{4.30}$$

where $\operatorname{tg} a = v/u$. This set of equations can be interpreted as a rotation about the angle a with $-\frac{1}{2}\pi < a < \frac{1}{2}\pi$. For any fixed v, by the group postulate P2, a succession of n transformations with the same relative v will also be given by (4.30) with a replaced by na. For $v \neq 0$ and $u \neq \infty$ it always possible to find a n such that $\frac{1}{2}\pi < na < \frac{3}{2}\pi$. But then $\cos a$ is negative and with it the coefficient of t in the second equation of (4.30). It follows that the temporal order of events taking place, *e.g.*, at the spatial origin of S ($x = 0$) will be reversed in S', a result which entails causal anomalies. Only if w^2 is nonnegative can these be avoided.

If in (4.28) we set $w = \infty$, we obtain the Galilean transformations

$$x' = x - vt\,, \qquad t' = t\,, \tag{4.31}$$

and time becomes absolute (*i.e.* system independent). Conversely, if we assume the absoluteness of time (*i.e.* $t = t'$), then clearly $\Gamma_v = 1$, which implies that $w = \infty$ and the transformation must be Galilean.

If in (4.28) we set $w = c$, we obtain the ordinary Lorentz transformation

$$x' = \gamma(x - vt)\,, \qquad t' = \gamma\left(t - \frac{v}{c^2}\,x\right), \tag{4.32}$$

where $\gamma = \gamma_v = \Gamma_w\big|_{w=c}$.

Furthermore, if we regard v'' as a composition of v and v', *i.e.*

$$v'' = v \overset{*}{+} v'\,,$$

comparison of the matrix elements in (4.25) shows that

$$\Gamma_{v''} = \Gamma_v \Gamma_{v'}\left(1 + \frac{vv'}{w^2}\right) \tag{4.33}$$

and also that

$$\Gamma_{v''} v'' = \Gamma_v \Gamma_{v'} (v + v') , \tag{4.34}$$

from which it follows by division that

$$v'' = \frac{v + v'}{1 + vv'/w^2} , \tag{4.35}$$

which is the well-known composition law of velocities in special relativity.

Finally, since by (4.35)

$$w \overset{*}{+} v = \frac{w + v}{1 + wv/w^2} = w , \tag{4.36}$$

w has to be identified as a « limiting velocity ».

The preceding group-theoretical derivation clarifies the role which the « second postulate » (LP_2) plays in special relativity. It eliminates the possibility of Galilean transformations and it determines the value of the « limiting velocity » w ($w = c$). It will now also be understood, however, that any relativistic theorem which differs from its nonrelativistic analogue or has no such analogue, as, *e.g.*, the velocity dependence formula of inertial mass or the mass-energy relation $E = mc^2$, can replace LP_2 in the logical construction of the special theory of relativity.

Let us conclude this section with a few remarks (as promised above) about « reciprocity » of which use has been made in postulate P3. Contrary to widespread opinion, the reciprocity principle is *not* a logical consequence of the principle of relativity (PR). The use of the reciprocity principle made above amounts to the statement that, if $v_{SS'}$ is the velocity of S' relative to S (*i.e.* as measured in S), and $v_{S'S}$ is the velocity of S relative to S' (*i.e.* as measured in S'), then

$$v_{SS'} = - v_{S'S} . \tag{4.37}$$

This relation, which may be called the « reciprocity of relative velocities », is only one example of the reciprocity principle. Another example is the « reciprocity of relative lengths »: of two identical rods (*i.e.* of the same proper-length or rest-length) the length of the one, if at rest in S', as measured in S is equal to the length of the other, if at rest in S, as measured in S'. Other examples are the « reciprocity of relative time intervals », the temporal analogue of the preceding, or the « reciprocity of relative passage times » which deals with the passage times of identical rods past fixed points in the inertial systems. None of these is a logical consequence of the principle of relativity. Space does not allow us to present a general proof which must be based on the relation-theoretical analysis of the concepts involved. Suffice it to point out that reciprocity of relative velocities, as defined above, holds between two inertial systems only if both are standard signal synchronized (*i.e.* $\varepsilon = \varepsilon' = \frac{1}{2}$), a condition not implied by the principle of relativity (which means, of course, that the principle of relativity is logically compatible with the nonfulfilment of the synchronization requirement). For details the reader is referred to [37] (especially p. 124) and [38].

5. – The light-geometric approach.

In contrast to the group-theoretical approach which tries to completely ignore the properties of light, the light-geometric approach aims at establishing the special theory of relativity exclusively on the basis of the properties of light signals. To counter the above-mentioned objection that the properties of light are those of only a particular physical phenomenon, the light-geometric approach may defend its position on the grounds of the following counter-arguments. Light, or electromagnetic radiation in general, belongs to that branch of physics which is best understood and experimentally best confirmed; the theory of the electromagnetic field, as far as we presently know, is the most fundamental theory—from the conceptual point of view—for it is the only one whose laws of motion (Maxwell's equations) are formulizable, as J. A. SCHOUTEN and J. HAANTJES and D. VAN DANTZIG [39] and others have shown, in a way independent of metrical geometry, that is without recourse to any metric or parallelism. Furthermore, all information we obtain about the external world is received *via* the electromagnetic field; this refers not only to vision but also to haptic (touch, smell) and acoustic sensations as well as to bubble chambers just as to X-ray astronomy. In other words, the role which light plays in physics is of a far greater importance than that of any other physical phenomenon.

It would lead us too far astray from our subject to discuss here the question to what extent, if at all, these contentions are justified. There is, however, one further reason why the light-geometric approach gained wide popularity, especially among the authors of textbooks on special relativity: it provides a simple derivation of the Lorentz transformation.

As an example, let us recall the well-known derivation, based on the light postulate LP_2, which may be sketched as follows. Defining $x_4 = ict$, we derive from the light postulate ($x \equiv x_1$)

$$x_1^2 + x_4^2 = x_1'^2 + x_4'^2 , \tag{5.1}$$

which represents in the (x_1, x_4)-plane a rotation

$$\begin{cases} x_1' = x_1 \cos a + x_4 \sin a , \\ x_4' = - x_1 \sin a + x_4 \cos a . \end{cases} \tag{5.2, 3}$$

For $x_1' = 0$,

$$\operatorname{tg} a = - x_1/x_4 = i\beta \qquad (\beta = v/c)\,.$$

Hence

$$\cos a = (1 + \operatorname{tg}^2 a)^{-\frac{1}{2}} = (1 - \beta^2)^{-\frac{1}{2}} = \gamma$$

and $\sin a = i\beta\gamma$. (5.2)-(5.3) can thus be written

$$x' = \gamma(x - vt)\,, \qquad t' = \gamma\left(t - \frac{v}{c^2}x\right), \tag{5.4, 5}$$

which are the Lorentz transformation equations.

Logically more rigorous constructions of the special theory of relativity, on the basis of the properties of light alone, have been proposed by numerous investigators amongst whom we mention only H. REICHENBACH [20], C. CARATHÉODORY [40], E. A. MILNE [41] and L. PAGE [42].

Most interesting in this context is the derivation of the Lorentz transformation on the basis of the so-called « k-calculus », which has been developed, originally in connection with a discussion of the notorious « clock paradox », by Hermann BONDI [43], under the influence of MILNE.

Two observers A and B, whose world-lines are depicted in fig. 5, move with relative velocity v and exchange light signals as shown in the diagram.

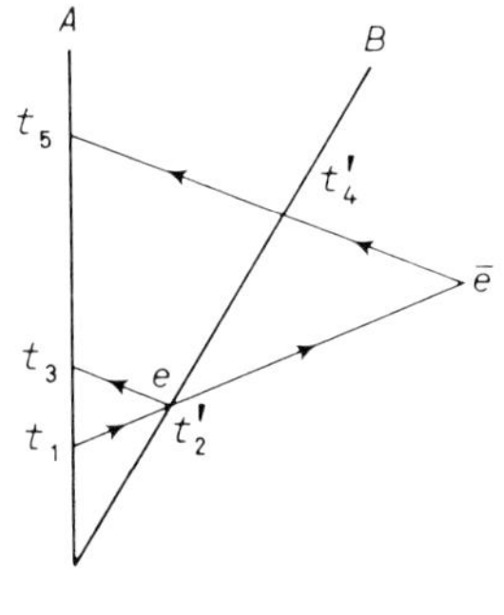

Fig. 5.

From the equivalence of the observers it follows that, if $t_2' = kt_1$, where $k = k(v)$ is a function of v, then also $t_3 = kt_2'$, so that $t_3 = k^2 t_1$. Observer A assigns to event e the time $t_e = \frac{1}{2}(t_3 + t_1)$ and the distance $x_e = \frac{1}{2}c(t_3 - t_1)$, since in the light-geometric approach any distance is given by the product of c with the time it takes for the signal to traverse the distance. Hence

$$v = \frac{x_e}{t_e} = c\,\frac{t_3 - t_1}{t_3 + t_1} = c\,\frac{k^2 - 1}{k^2 + 1}$$

or

$$k = \sqrt{\frac{c+v}{c-v}}\,. \tag{5.6}$$

A simple calculation shows that

$$\tfrac{1}{2}(k+k^{-1}) = (1-v^2/c^2)^{-\frac{1}{2}} = \gamma\,, \qquad \tfrac{1}{2}(k-k^{-1}) = \gamma\beta\,. \tag{5.7}$$

Similarly, A assigns to event $\bar{e}$ the time $t = \frac{1}{2}(t_5 + t_1)$ and the distance $x = \frac{1}{2}c(t_5 - t_1)$, so that

$$t_5 = t + (x/c)\,, \qquad t_4 = t - (x/c)\,. \tag{5.8}$$

B assigns to event $\bar{e}$ the time $t' = \frac{1}{2}(t_4' + t_2')$ and the distance $x' = \frac{1}{2}c(t_4' - t_2')$, so that

$$t_4' = t' + (x'/c)\,, \qquad t_2' = t' - (x'/c)\,. \tag{5.9}$$

But $t_5 = kt_4'$ and $t_1 = k^{-1}t_2'$, so that

$$t + (x/c) = k\,(t' + x'/c)\,,$$
$$t - (x/c) = k^{-1}(t' - x'/c)\,.$$

By addition and subtraction of the last two equations and by taking account of (5.7) we obtain

$$x = \gamma(x' + vt')\,, \qquad t = \gamma\left(t' + \frac{v}{c^2}x'\right), \tag{5.10}$$

which are the Lorentz transformations for $x', t' \to x, t$.

Mathematical deductions, such as the preceding, carried out in accordance with the light-geometric approach, are often very ingenious and even impressive. But if critically analysed, in an axiomatic fashion, it turns out that one cannot dispense with rigid rods [20, sect. **18**].

The reason for this conceptual deficience of the light-geometric approach lies in the fact, discovered 1910 by Ebenezer CUNNINGHAM [44] and Harry BATEMAN [45], that the Maxwell equations are covariant not only under the Lorentz group, but also under the more general group of conformal transformations in four-dimensional space-time. While under the Lorentz group a fixed configuration is mapped into a configuration every point of which has the same velocity of translation, the simplest (nontrivial) operation of the conformal group transforms such a configuration into a system which as a whole is expanding or contracting. It had been known for quite some time (LIOUVILLE, DARBOUX) that every conformal transformation is the product of translations, rotations, magnifications, reflections and inversions. To prove the covariance of the Maxwell equations under the conformal group it sufficed, therefore, to prove their covariance only with respect to a single inversion, the centre of which is the origin. This precisely has been done by BATEMAN and CUNNINGHAM.

Prior to their discovery it had been taken for granted that the Lorentz transformations alone leave the Maxwell equations covariant. It is interesting to see that this erroneous view could have been corrected, in principle at least, much earlier—in fact, even before the advent of the special theory of relativity. Since the Lorentz transformations form a (linear) subgroup of the conformal group, this conclusion has interesting conceptual-historical implications.

To understand this point, let us recall that the electromagnetic-field tensor can be derived as the four-dimensional curl of the electromagnetic four-potential A_μ ($\boldsymbol{A}$, $A_4 = i\varphi$). In virtue of this fact the Maxwell equations (in the Lorentz gauge) are equivalent to the four-dimensional d'Alembertian equation

$$\Box A_\mu = 0\,, \qquad \mu = 1, 2, 3, 4\,, \tag{5.11}$$

where the d'Alembertian differential operator $\Box$ stands for

$$\frac{\partial^2}{\partial x_\nu \partial x_\nu} \equiv \nabla^2 - c^{-2}\frac{\partial^2}{\partial t^2}\,. \tag{5.12}$$

(5.11) is, therefore, a four-dimensional generalization of the Laplace equation which played a prominent role in potential theory. Now, in the three-dimensional case the so-called « Kelvin transformation » of one harmonic function (*i.e.* a function satisfying the Laplacian equation) into another such function was well known ever since Lord KELVIN (William THOMSON) had published this result in a letter to LIOUVILLE as early as 1847 [46]. In fact, KELVIN had shown that, if $f(x, y, z)$ is a harmonic function of x, y, z in a domain D, then also

$$g(x', y', z') = \frac{a}{r'} f\left(\frac{a^2 x'}{r'^2}, \frac{a^2 y'}{r'^2}, \frac{a^2 z'}{r'^2}\right) \tag{5.13}$$

is a harmonic function of x', y', z' in the domain D' into which D is carried over by the inversion

$$x = \frac{a^2}{r'^2}x'\,, \qquad y = \frac{a^2}{r'^2}y'\,, \qquad z = \frac{a^2}{r'^2}z'\,, \tag{5.14}$$

where

$$r' = (x'^2 + y'^2 + z'^2)^{\frac{1}{2}}\,, \qquad r = (x^2 + y^2 + z^2)^{\frac{1}{2}} \quad \text{and} \quad rr' = a^2\,.$$

Furthermore, elaborating on results obtained by LIOUVILLE and DARBOUX, LIE and Felix KLEIN [47] demonstrated already in 1871 that in a space of n dimensions conformal transformations which map the expression

$$dx_1^2 + dx_2^2 + \dots + dx_n^2$$

into a multiple of itself form a group of $(n+1)(n+2)/2$ parameters.

Reviewing these results one may conclude that in principle the Bateman-Cunningham discovery could have been made already in the early seventies of the last century, in fact shortly after the discovery of the Maxwell equations themselves. If it would have been possible to present the Maxwell equations in terms of co-ordinates of a four-dimensional space [48]—a step which only Hermann MINKOWSKI carried out in 1908—then (for $n = 4$) one could have proved that there exists a 15-parameter group G_{15} (the conformal group) under which the Maxwell equations are covariant.

Surprisingly, the Bateman-Cunningham discovery remained almost completely unnoticed at its time. The only exception was Philipp FRANK, who proved [49], by using the by now obsolete Sommerfeld formulation of tensor analysis in terms of six-vectors etc., that among all *linear* transformations only the Lorentz transformations leave the Maxwell equations covariant. A modern version of Frank's demonstration may be based on a paper by K. STIEGLER [50], who derived the Lorentz transformation from the Maxwell equations, the principle of relativity and the assumption of the linearity of space-time transformations between inertial reference systems (in addition to the tacit presupposition of the isotropy of space).

When Felix KLEIN drew Einstein's attention to the Bateman-Cunningham discovery, EINSTEIN replied: « Die Transformation durch reziproke Radien wahrt zwar die Form der Maxwellschen Gleichungen, nicht aber den Zusammenhang zwischen Koordinaten und Massergebnissen von Masstäben » [51]. When Lancelot Law WHYTE discussed this issue in 1935 with MILNE, who (as mentioned above) was one of the foremost proponents of the light-geometric approach, MILNE made the confession « that he had not yet reached a firm opinion on this point » [52].

More recent results such as those obtained by D. C. CASHMORE [53] fully corroborate the conclusion that the light-geometric approach can serve as a basis for the construction of relativistic physics (including relativistic mechanics) *only* if it is « contaminated » with an element alien to the spirit of light-geometry (rigid rods).

This does not mean, however, that this approach is conceptually inferior to the group-theoretical approach as described in the preceding section. For, as we have seen, the group postulate together with all the other general assumptions concerning the homogeneity and/or isotropy of space or time lead only to the Lorentz-*type* transformation equations. To obtain *physically* significant transformations, that is to interpret the formal Lorentz-type transformation as a physically significant transformation (which can be done only by determining the specific numerical value of the « limiting velocity » w), some concrete physical phenomenon has to be referred to. One may, therefore, equally well contend that the group-theoretical approach can serve as a basis for the construction of relativistic *physics only* if it is « contaminated » with an element alien to the spirit of group theory.

6. – The nature of length contraction.

So far we have been mostly concerned with problems connected with the derivation of the Lorentz transformation equations. Let us now discuss a problem connected with the consequences of these transformation equations. First of all, it is clear that these conclusions themselves must also be consistent among each other, for otherwise the set of postulates would be afflicted with internal inconsistencies. This point is of importance in connection with the « clock paradox », which according to some authors is the most serious problem pertaining to the consequences of the Lorentz transformation, namely the well-known effect of « time retardation ». In fact, the self-contradiction of the alleged « conclusion » that of two twins, who once have departed from each other and later have come together again, each must be younger than the other, has been interpreted by certain authors as an indication that the basic postulates of the theory lack logical consistency. But much, indeed too much, has been written about the clock paradox [54]. Since, moreover, problems touching upon the issues of consistency and independence have already been dealt with to some extent, let us discuss the spatial analogue of time dilatation, the relativistic length contraction, which, although also a logical consequence of the Lorentz transformation, gives us the opportunity to deal with problems of a different character,—especially since very little has been written about the problem we shall now discuss.

It concerns the nature, and in particular the ontological status, of the relativistic length contraction.

It should not be ignored, however, that this problem and the clock paradox have much in common as far as their history is concerned. Both can be traced back to Einstein's 1905 relativity paper. Both became the subject of general attention at the same time, six years after the publication of Einstein's paper. The first detailed discussion of the clock paradox (then called the problem of « Langevin's travellers ») was published in 1911 by Paul LANGEVIN in his classic paper *L'evolution de l'espace et du temps* [55].

The first published discussion of the problem of length contraction likewise dates back to the year 1911. It was a controversy between EINSTEIN and the mathematician Vladimir VARIĆAK of the University of Zagreb concerning the « reality » of this effect and was brought into the open by papers published in 1911 in the *Physikalische Zeitschrift.*

The problem had its origin in the question, much discussed at that time, how to define in relativity a « rigid » body without contradicting the relativistic disallowance of infinitely fast causal propagations. For, obviously, a strictly rigid rod (in the classical sense of the word), if pushed at one end, would, due to the fact that it does not undergo any deformation (contraction or expansion), constitute a means to transmit a signal with infinite velocity. The issue of relativistic rigidity is a highly complicated subject and cannot be discussed in our paper. Suffice it to point out that it led to what became known as the « Ehrenfest paradox » [56]. EHRENFEST considered a disk of radius R (when at rest) and assumed the disk to be gradually set in rotational motion around its axis. Calling R' the radius of the disk (when in motion) as measured by an observer at rest (in the inertial system), EHRENFEST pointed out that R' has to satisfy two contradictory conditions: 1) since every element of the periphery moves in its own direction with a momentary velocity $R'\omega$ (ω being the angular velocity), the periphery should experience a Lorentz contraction in consequence of which $2\pi R' < 2\pi R$; 2) since a radial element moves in a direction normal to its extension, it does not contract at all in consequence of which $R' = R$.

To resolve this contradiction by showing that Ehrenfest's is only a pseudo-problem, VARIĆAK claimed that the relativistic or « Einsteinian » contraction is « merely an apparent, subjective phenomenon, resulting from the kind of our clock synchronization and length measurement » [57]. In other words, VARIĆAK contended that the relativistic contraction is merely what REICHENBACH later called a « metrogenic » phenomenon and not a real physical process. He distinguished it from the pre-relativistic FitzGerald-Lorentz contraction, which had been proposed to account for the null result of the Michelson-Morley experiment and which was subsequently regarded as explainable as a result of the motion through the ether, affecting the electromagnetic forces that exist among the molecules of the moving body.

In reaction to Varićak's paper EINSTEIN published in 1911 [58] a short essay in which he criticized these ideas, « lest they may cause confusion ». To ask whether the contraction is « real » (« wirklich ») or not, EINSTEIN declared, is misleading. It is not real in so far as it does not exist for a co-moving observer; it is « real » in the sense of being verifiable in principle by physical means (« durch physikalische Mittel »). In this context it is worth-while to recall that Einstein's gradual apostasy from Mach's positivistic epistemology began only about 1917. In fact, the words just quoted have still an outspoken positivistic ring.

The dispute between VARIĆAK and EINSTEIN may be regarded as the beginning of the controversy concerning the ontological status of relativistic length contraction.

To prove his point, EINSTEIN suggested the following thought-experiment. Two rods, $A'B'$ and $A''B''$, of equal proper-length are supposed to move along the x-axis of an inertial system with constant equal but opposite velocities in S. In the course of their motion the two extremities A' and A'' will meet at a point A and the other extremities at a point B of the axis (fig. 6). According to the theory, EINSTEIN argued, the distance AB is smaller than the proper-length of either rod and this fact can be verified by applying one of the rods in the state of rest to the line AB.

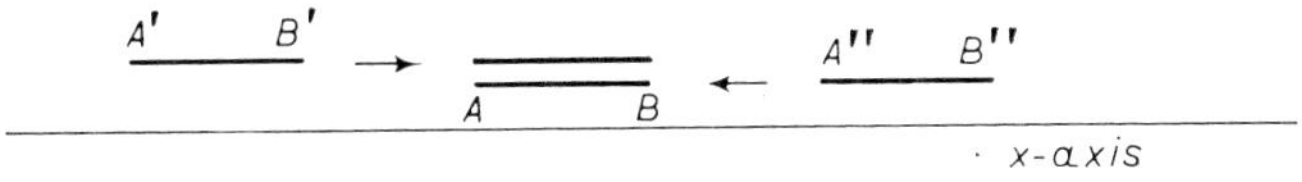

Fig. 6.

It is clear that what EINSTEIN wanted to show by this experiment was that the length contraction is operationally verifiable without any recourse to clock synchronization. But did EINSTEIN really prove his point? To answer this question let us assume that rod $A'B'$ moves with velocity v' to the right and rod $A''B''$ with velocity v'' to the left, their common proper-length being L_0. Let e_A denote the event marked by the coincidence of A' with A'' and e_B the event marked by the coincidence of B' with B''. Finally, let

$$\alpha' = (1 - v'^2/c^2)^{\frac{1}{2}} \quad \text{and} \quad \alpha'' = (1 - v''^2/c^2)^{\frac{1}{2}}.$$

A simple calculation shows that the time lapse Δt_{AB} between e_A and e_B, relative to S, is given by

$$\Delta t_{AB} = L_0 \frac{\alpha'' - \alpha'}{v' + v''} \tag{6.1}$$

and the spatial distance between these events, relative to S, is given by

$$\Delta x_{AB} = L_0 \frac{\alpha' v'' + \alpha'' v'}{v' + v''}. \tag{6.2}$$

It follows from (6.1) that e_A and e_B are simultaneous in S if and only if $v' = v''$ $(= v)$ in which case, as shown by (6.2),

$$\Delta x_{AB} = \alpha L_0, \tag{6.3}$$

where

$$\alpha = (1 - v^2/c^2)^{\frac{1}{2}}. \tag{6.4}$$

We thus arrive at the following result. If Einstein's thought-experiment was designed to verify the quantitative result that the length of each rod decreases in the ratio α, then the velocities of the two rods had to be equal (in absolute value), a requirement which necessitates the use of clocks. To take refuge to a dynamical principle such as the principle of action and reaction (two equal masses at the ends of a compressed spring which is being released) would involve a vicious circle (to verify that the masses are equal we have to measure, perhaps by the same experimental arrangement, their relative velocities, cf. Mach's determination of mass [59]; but to measure a velocity requires a clock, in fact, two synchronized clocks). If, however, Einstein's experiment was designed to verify only the qualitative result that $\Delta x_{AB} < L_0$, then indeed no clock would have been needed; for this result follows from (6.2) whatever the values of v' and v''.

Max von Laue, who discussed this problem in the early editions of his well-known book on relativity, the first textbook of this subject [60]—the first edition appeared in 1911—, disagreed both with Varićak and with Einstein. He distinguished between the case of a rod set in motion with respect to a rod at rest and the case of a rod at constant motion with respect to a rod at rest, all rods being of equal proper-length. In the former case, he contended, the contraction is a physically real process, pertaining to the mechanics of elastic continua, in the latter case it is « a consequence of the different application of the space and time scheme by differently moving observers ».

The problem was also studied by Reichenbach and others and became in the early forties the subject of a heated discussion between Herbert Dingle and Paul S. Epstein, part of which has been published in *The American Journal of Physics* [61]. Since the Dingle-Epstein controversy came to an imbroglio with no solution in sight (just as the Dingle-McCrea controversy on the clock paradox published in a flurry of papers in *Nature* a few years later), the editor of *The American Journal of Physics* asked Leopold Infeld to serve as a dispassionate umpire between the two disputants. Infeld, after having given much thought to this problem, arrived at the conclusion that to decide between Epstein and Dingle would be to decide between realism and idealism; and, he added, he did not believe that « this decision can be made on the plane of logical arguments » [62].

When in 1951, on the occasion of the centennial of George Francis Fitz-Gerald's birth, the Royal Society of Dublin (where FitzGerald was born and died) published a commemorative issue of its *Proceedings*, a number of essays in it were devoted to various aspects of the length contraction and its ontological status in physical theory (Synge, Ives, Whitrow, McCrea). We cannot go into details concerning the various views presented on this topic. It may be mentioned, however, that McCrea [63] also discussed in this connection the visual appearance (for one single observer) of this contraction and anticipated results later achieved in the classical papers by J. Terrel (1959), R. Penrose (1959) and others.

A correct assessment of the « nature » of relativistic length contraction, I believe, can be obtained only by asking the « right » question. This question, it seems to me, is this: What are the conventional, and what are the factual, ingredients in the relativistic phenomenon of length contraction? Thanks to the recent analyses of the conventional content of modern physical theories, and of the special theory of relativity in particular, such as carried out by Adolf GRÜNBAUM, Wesley C. SALMON, John A. WINNIE and others, we are today in a position to answer this question.

Our analysis will be based on defining the length of a moving rod with respect to a stationary inertial system S as the product of the velocity v of the rod relative to S and the passage time Δt, that is the time, as measured by a clock U stationary in S, it takes for the rod to pass past the clock U:

$$L = v\Delta t \,. \tag{6.5}$$

For the sake of completeness let us recall the proof of the preceding formula. Let L_A^B denote the length of a rod as measured in the inertial system B if the rod is at rest in the inertial system A.

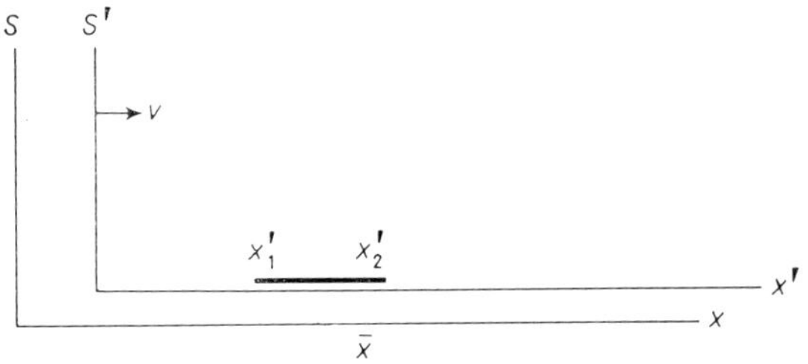

Fig. 7.

In accordance with the Lorentz transformation $x' = \gamma(x - vt)$ (fig. 7)

$$L_{S'}^{S'} = x_2' - x_1' = \gamma(x_2 - v\bar{t}) - \gamma(x_1 - v\bar{t}) = \gamma(x_2 - x_1)\,.$$

Hence

$$L_{S'}^{S'} = \gamma L_{S'}^{S}\,, \tag{6.6}$$

where, according to the usual definition of the length of a moving rod, the positions of x_1' and x_2' have been measured in S at the same instant of time $\bar{t}$ with respect to S.

Alternatively, if Δt denotes the passage time of the rod past a fixed position $\bar{x}$ in S, as measured by a clock U fixed at $\bar{x}$, then

$$L_{S'}^{S'} = x_2' - x_1' = \gamma(\bar{x} - vt_2) - \gamma(\bar{x} - vt_1) = \gamma v(t_1 - t_2)\,.$$

Hence

(6.7) $$L_{S'}^{S'} = \gamma v \Delta t\,.$$

Comparison of (6.6) with (6.7) shows that $L_{S'}^{S} = v\Delta t$ as asserted in (6.5). We also shall make use of the thesis of conventionality as described in sect. **1**.

To make our exposition self-contained we have to refer to the situation described in fig. 1. Let $x_A < x_B$ be the positions of A and B on the x-axis of the system S. According to the two-way light principle

(6.8) $$c = \frac{2AB}{t'_A - t_A}\,.$$

If c_+ denotes the velocity of light travelling in the positive direction along the x-axis (from A to B), it follows from (1.2) that

(6.9) $$c_+ = \frac{AB}{t_B - t_A} = \frac{AB}{\varepsilon(t'_A - t_A)} = \frac{c}{2\varepsilon}\,;$$

similarly for c_-, the velocity of light in the opposite direction,

(6.10) $$c_- = \frac{c}{2(1-\varepsilon)}\,.$$

Consider now a particle P moving with constant velocity v (in standard synchrony: $\varepsilon = \frac{1}{2}$) from A to B, leaving A at A-time $t_A = 0$ and arriving at B at B-time $\bar{t}_B$, the clock at B being standard-synchronized with that at A. Assume that at A-time $t_A = 0$ a light signal has also been sent to B, where it arrives at B-time t_B (fig. 8).

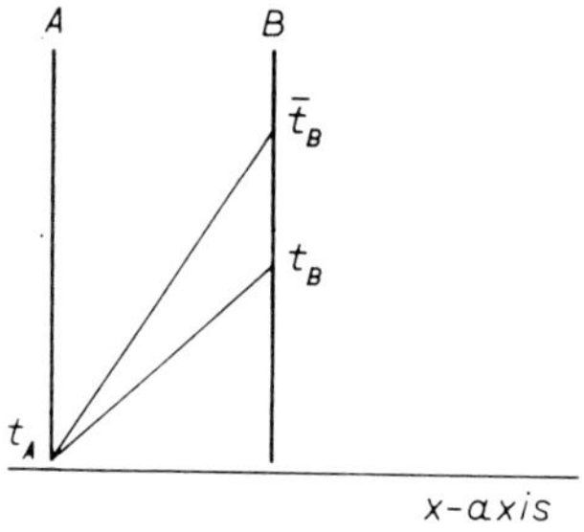

Fig. 8.

In the case of standard synchrony

(6.11) $$\bar{t}_B = \frac{AB}{v} \quad \text{and} \quad t_B = \frac{AB}{c}\,,$$

whereas in the case of ε-synchrony

$$\bar{t}_B = \frac{AB}{v_+} \quad \text{and} \quad t_B = \frac{AB}{c_+}, \tag{6.12}$$

where v_+ denotes the velocity of P as determined by a particular choice of ε. However, $\bar{t}_B - t_B$ (just like Δt above) denotes the difference between two readings of one and the same clock and must therefore be synchrony-independent. It hence follows from the last two equations that

$$\frac{1}{v} - \frac{1}{c} = \frac{1}{v_+} - \frac{1}{c_+}. \tag{6.13}$$

It thus also follows from (6.9) that

$$v_+ = \frac{cv}{c + (2\varepsilon - 1)v}. \tag{6.14}$$

An analogous calculation for the velocity of a particle moving from B to A, that is in the negative direction along the x-axis, yields

$$v_- = \frac{cv}{c - (2\varepsilon - 1)v}. \tag{6.15}$$

We now turn to the calculation of the length contraction of a rod of proper-length L_0, moving with constant velocity v (in standard synchrony) in S along the positive direction of the x-axis. In the case of standard synchrony, its length in S is given by

$$L = v\,\Delta t \tag{6.5}$$

with

$$L = L_0(1 - v^2/c^2)^{\frac{1}{2}}. \tag{6.16}$$

But in ε-synchrony it is given by

$$L_+ = v_+\,\Delta t \tag{6.17}$$

with v_+ as in (6.14).

From eqs. (6.5), (6.14), (6.16) and (6.17) we obtain

$$L_+ = \frac{L_0(1 - \beta^2)^{\frac{1}{2}}}{1 + \beta(2\varepsilon - 1)} = \alpha_+ L_0, \tag{6.18}$$

or expressed in terms of v_+

$$L_+ = L_0 \sqrt{\frac{(c - v_+[2\varepsilon - 1])^2 - v_+^2}{c^2}}. \tag{6.19}$$

Analogously, for a rod moving in the negative direction of the x-axis

$$L_- = L_0 \sqrt{\frac{(c + v_-[2\varepsilon - 1])^2 - v_-^2}{c^2}}. \tag{6.20}$$

For standard synchrony ($\varepsilon = \frac{1}{2}$) formulae (6.19) and (6.20) reduce to (6.16) as required by the Lorentz transformation. As we see, the length contraction factor depends on ε as well as on the direction of motion. In fact, it can be shown that for a suitable choice of ε the factor can be made larger than 1, leading to a *length dilatation* rather than to a *length contraction*.

Furthermore, as seen from (6.18) and its analogue

$$L_- = \frac{L_0(1-\beta^2)^{\frac{1}{2}}}{1-\beta(2\varepsilon - 1)} = \alpha_- L_0 , \tag{6.21}$$

the factor equals unity, for motions in the positive direction of the x-axis if and only if

$$\varepsilon = \bar{\varepsilon}_+ \equiv \frac{\beta - 1 + (1-\beta^2)^{\frac{1}{2}}}{2\beta}, \tag{6.22}$$

and for motions in the negative direction of the x-axis if and only if

$$\varepsilon = \bar{\varepsilon}_- = \frac{\beta + 1 - (1-\beta^2)^{\frac{1}{2}}}{2\beta}, \tag{6.23}$$

where

$$0 < \bar{\varepsilon}_+, \bar{\varepsilon}_- < 1 \text{ as long as } 0 < v < c .$$

It thus proves it possible to eliminate length contraction for each particular motion (*i.e.* for each v) in the *positive* direction of the x-axis by choosing $\varepsilon = \bar{\varepsilon}_+$. And the same applies for each particular motion in the *negative* direction of the x-axis by choosing $\varepsilon = \bar{\varepsilon}_-$. But, since

$$\bar{\varepsilon}_+ + \bar{\varepsilon}_- = 1 , \tag{6.24}$$

it is impossible to find a value of ε such that the relativistic length contraction could be eliminated in *both* directions at a time. Whatever the value of ε chosen, there are always motions for which the measured length of a rod differs from its proper-length.

The uneliminability of length contraction shows that, even though in its quantitative amount it is synchrony-dependent and hence convention-laden, it is an effect the very existence of which is convention-free. It is in this sense —but also only in this sense—that EINSTEIN was right when he rejected Varićak's interpretation as merely « resulting from the kind of our clock synchronization ».

REFERENCES

[1] M. JAMMER: *The Philosophy of Quantum Mechanics* (New York, N. Y., 1974).
[2] A. EINSTEIN: *Ann. der Phys.*, **17**, 891 (1905).
[3] O. ROEMER: *Journ. des Sçavans*, **10**, 233 (1676).
[4] A. D. FOKKER: *Synthese*, **9**, 442 (1953).
[5] H. REICHENBACH: *The Philosophy of Space and Time* (New York, N. Y., 1958).
[6] A. GRÜNBAUM: *Philosophical Problems of Space and Time* (Dordrecht, 1973).
[7] E. BERGSTRAND: *Encyclopedia of Physics*, edited by S. FLÜGGE, Vol. **54** (Berlin, 1956).
[8] C. SCRIBNER jr.: *Amer. Journ. Phys.*, **31**, 398 (1963).
[9] A. EINSTEIN, H. A. LORENTZ, H. MINKOWSKI and H. WEYL: *The Principle of Relativity*, translated by W. PERRETT and G. B. JEFFERY (New York, N. Y., 1952).
[10] E. FEENBERG: *Found. Phys.*, **4**, 121 (1974).
[11] R. MANSOURI and R. U. SEXL: *Vienna reports in gravitation and cosmology*, preprint (1975).
[12] P. W. BRIDGMAN: *A Sophisticate's Primer of the Theory of Relativity* (Middletown, Conn., 1962), p. 43.
[13] G. B. BROWN: *Bull. Phys.*, **18**, 71 (1967).
[14] H. REICHENBACH: *Zeits. Phys.*, **33**, 628 (1925).
[15] L. KARLOV: *Austral. Journ. Phys.*, **23**, 243 (1970).
[16] D. P. TODD: *Nature*, **21**, 315 (1880).
[17] M. LE BELLAC and J.-M. LÉVY-LEBLOND: *Nuovo Cimento*, **14** B, 217 (1973).
[18] J. C. MAXWELL: *Phil. Mag.*, **21**, 161, 291, 338 (1862).
[19] N. R. CAMPBELL: *What is Science?* (London, 1921), p. 155.
[20] H. REICHENBACH: *Axiomatik der relativistischen Raum-Zeit-Lehre* (Braunschweig, 1924); *Axiomatization of the Theory of Relativity* (Berkeley, Cal., 1969).
[21] H. WEYL: *Deutsche Literaturzeitschrift 1924*, p. 2122.
[22] R. C. TOLMAN: *Phys. Rev.*, **31**, 26 (1910).
[23] R. C. TOLMAN: *Relativity, Thermodynamics and Cosmology* (Oxford, 1934), p. 15.
[24] K. F. SCHAFFNER: *Brit. Journ. Phil. Sci.*, **25**, 53 (1974).
[25] S. WYKSTRA: *Brit. Journ. Phil. Sci.*, **27**, 259 (1976).
[26] R. B. WILLIAMSON: *Studies Hist. Phil. Sci.*, **8**, 49 (1977).
[27] J. M. KINGSLEY: *Found. Phys.*, **5**, 295 (1975); see also **6**, 677, 681 (1976).
[28] C. CARATHÉODORY: *Math. Ann.*, **67**, 355 (1909).
[29] W. VON IGNATOWSKY: *Phys. Zeits.*, **11**, 972 (1910).
[30] P. FRANK and H. ROTHE: *Ann. der Phys.*, **34**, 825 (1911); *Phys. Zeits.*, **13**, 750 (1912).
[31] S. LIE and G. SCHEFFERS: *Vorlesungen über kontinuierliche Gruppen* (Leipzig, 1893).
[32] L. A. PARS: *Phil. Mag.*, **42**, 249 (1921).

[33] E. B. Esclangon: *Compt. Rend.*, **202**, 708 (1936); E. Le Roy: *Compt. Rend.*, **202**, 794 (1936); Ch. Lalan: *Compt. Rend.*, **203**, 1491 (1936).
[34] V. Berzi and V. Gorini: *Journ. Math. Phys.*, **10**, 1518 (1969).
[35] Cf. also M. Podlaha and E. Novratil: *Acta Phys. Austriaca*, **24**, 99 (1966).
[36] H. Freudenthal: *Adv. Math.*, **1**, 145 (1964).
[37] B. Ellis and P. Bowman: *Phil. Sci.*, **34**, 116 (1967).
[38] P. Mittelstaedt: *Der Zeitbegriff in der Physik* (Mannheim, 1976).
[39] J. A. Schouten and J. Haantjes: *Physica*, **1**, 869 (1934); D. van Dantzig: *Proc. Camb. Phil. Soc.*, **30**, 421 (1934).
[40] C. Carathéodory: *Sitz. Preuss. Akad. Wiss.*, **12**, 12 (1924).
[41] E. A. Milne: *Relativity, Gravitation and World Structure* (Oxford, 1935).
[42] L. Page: *Phys. Rev.*, **49**, 254 (1936).
[43] H. Bondi: *Discovery*, **18**, 505 (1957); *Lectures on General Relativity* (Brandeis Institute) (Englewood Cliffs, N. J., 1965), p. 375.
[44] E. Cunningham: *Proc. London Math. Soc.*, **8**, 77 (1910).
[45] H. Bateman: *Proc. London Math. Soc.*, **8**, 223 (1910).
[46] W. Thomson: *Journ. de Math.*, **12**, 256 (1847).
[47] S. Lie: *Math. Ann.*, **5**, 145 (1871); F. Klein: *Math. Ann.*, **5**, 157 (1871).
[48] For four-dimensional (n-dimensional) generalizations of mechanics see, *e.g.*, R. Lipschitz: *Liouville Journ.*, **74**, 116 (1872) and papers by R. S. Heath, W. Killing and N. I. Stringham.
[49] P. Frank: *Ann. der Phys.*, **35**, 599 (1911).
[50] K. Stiegler: *Proc. Phys. Soc.*, **71**, 512 (1958).
[51] F. Klein: *Gesammelte Mathematische Abhandlungen*, Vol. **1** (Berlin, 1921).
[52] L. L. Whyte: *Brit. Journ. Phil. Sci.*, **4**, 160 (1953).
[53] D. C. Cashmore: *Proc. Phys. Soc.*, **81**, 181 (1963).
[54] L. Marder: *Time and the Space-Traveller* (London, 1971).
[55] P. Langevin: *Scientia*, **10**, 31 (1911); republished, *Scientia*, **108**, 221 (1973); English translation: *Scientia*, **108**, 289 (1973).
[56] P. Ehrenfest: *Phys. Zeits.*, **10**, 918 (1909).
[57] V. Varićak: *Phys. Zeits.*, **12**, 169 (1911).
[58] A. Einstein: *Phys. Zeits.*, **12**, 169 (1911).
[59] M. Jammer: *Concepts of Mass in Classical and Modern Physics* (Cambridge, Mass., 1961; New York, N. Y., 1964).
[60] M. von Laue: *Das Relativitätsprinzip* (Braunschweig, 1911).
[61] See *Amer. Journ. Phys.*, **10**, 1, 203, 205 (1942).
[62] L. Infeld: *Amer. Journ. Phys.*, **11**, 219 (1943).
[63] W. H. McCrea: *Proc. Roy. Dublin Soc.*, **26**, 27 (1952).

The Importance of Being (a) Constant (*).

J.-M. LÉVY-LEBLOND (**)

Laboratoire de Physique Théorique, Université Paris VII - Paris, France

1. – The changing constants of physics.

Physical constants are among the commonest of the various quantities used by theoretical physics. There is no calculation whatsoever which does not rely on one or more such constants (even though some of them are well hidden, as I will show later on). Nevertheless, it is suprising how little thought has been devoted to these constants as such. Most physical notions, such as energy, or length, or frequency, etc., have been subjected to historical and epistemological inquiries. The constants which appear in the physical laws, that is in the relationships connecting the basic physical notions, deserve as well to be investigated from a general point of view.

Many questions indeed may be asked immediately: are all the constants appearing in a standard table to be put on the same footing? For example, is the mass of the proton of the same nature as the elementary electric charge or Planck constant? How comes that there are fundamental physical constants, such as the velocity of light or Planck constant, in modern physics, in contradistinction with classical mechanics for instance? Is there a limit to the number of possible fundamental constants? Why is it that the list of the physical constants considered as fundamental ones changes with time, as shown by a comparison of standard tables at various epochs? Have the thermodynamical constants k (Boltzmann) and J, appearing mostly today as mere unit conversion factors, anything in common with the modern constants $\hbar$ and c? What is the meaning of taking as unity the numerical value of some constant, or, in a more puzzling way, to let it «go to zero (or infinity)» in the special circumstances of approximate validity for some limit theory?

(*) After the title of Oscar Wilde's play, *The Importance of Being Earnest*, as translated in French by *L'importance d'être Constant*.

(**) Postal address: Laboratoire de Physique Théorique des Hautes Energies, Université Paris VII, Tour 33-43, 1er étage, Place Jussieu, 75221 Paris Cedex 05, France.

These, and others, are the questions I intend to explore here. Let me stress right at the start that it is my intention to show how the epistemological status of the physicals constants bears witness to the development of physical science in general. In other words, I am not looking for an abstract, ideal and definitive scheme. Rather, I would like to propose these reflexions on the physical constants in order to emphasize the changing nature of physical ideas. As I hope to show, the important conceptual role played by some of the physical constants permits, through their consideration, an original probing of the various historical and conceptual strata which build up physical theory. Indeed, the main theme of my lectures will be that physical constants change, and very much so, in their status, if they do not vary in their values.

A classification. I will first propose a classification of the various physical constants into three types. In order of increasing generality, I distinguish:

A) *Properties of particular physical objects* considered as fundamental constituents of matter; for instance, the masses of « elementary particles », their magnetic moments, etc.

B) *Characteristics of classes of physical phenomena*: Today, these are essentially the coupling constants of the various fundamental interactions (nuclear, strong and weak, electromagnetic and gravitational), which, to our present knowledge, provide a neat partition of all physical phenomena into disjoint classes.

C) *Universal constants*, that is constants entering universal physical laws, characterizing the most general theoretical frameworks, applicable in principle to any physical phenomenon; Planck constant $\hbar$ is a typical example, to be discussed later on.

This classification is not supposed to assign a definite and definitive place to each constant, once and for all. Quite on the contrary, its purpose is to permit a discussion of the changing status of these constants. The simplest example of such changes is that of constants dropping out altogether from the table of physical constants. This happens mainly to constants of type *A*, when the objects they are attached to cease to be considered as « elementary » or « fundamental ». For instance, while the atomic weights of chemical elements some decades ago were considered as fundamental constants, it is known today that they are to be expressed in terms of the nucleonic and electronic masses and the strong and electromagnetic coupling constants. It is not necessary (and fortunately so!) that a complete calculation be carried out: an explanation of principle is sufficient. In a similar way, all macroscopic properties of matter in bulk (viscosity, elasticity, density, etc.) have disappeared from tables of fundamental constants, since they are supposed to depend essentially on the atomic theory of matter. If a theory of hadrons ever comes to maturity, the masses and electromagnetic moments of the nucleons will be eliminated in favour of coupling constants and quark masses, elementary lengths or whatever else.

A similar situation would prevail, although for different reasons, if a « constant » was found by experiment to show an actually variable numerical value, exhibiting, for instance, a cosmological dependence (as it has been proposed by DIRAC for G, by GAMOW for e, etc., although these hypotheses presently seem to be ruled out, in their simplest form at least). The law expressing the time variation of the ex-constant would then contain one or more numerical constants which would replace the dishonoured one in the table of fundamental constants. These new constants might indeed be novel ones, previously unknown to physics, or combinations of standard ones; as an example of the last case consider Dirac's hypothesis, according to which $G \simeq (\hbar e^2/m^3 c^2) t^{-1}$, where t is the (variable) age of the Universe and m a typical mass for elementary particles

More interesting is the case of fundamental constants experiencing a change in their type—according to the above classification. I will discuss at length this phenomenon in the following lectures on the example of two important constants, c and G. But a first, and simple, example is furnished by the elementary charge e. When it first occurred in physics, through the discovery of the electron and the measurement of its charge, it clearly was a type-A constant, a property of that particular object, the electron. After it came to be realized that any electric charge, on any kind of particle, was an integer multiple of e, this constant appeared as characterizing the whole class of electromagnetic phenomena. In its present role as the elementary coupling constant of the electromagnetic field to a charged object, e clearly is to be thought of as a type-B constant. This situation, of course, might well change in the future. If electromagnetic interactions are unified in a consistent theory with other ones (for instance, weak interactions, as present-day gauge theories seem to offer the prospect), e could be eliminated in favour of a deeper constant. Another possibility might be (and has been, in the past) conceived, even if very unplausible: could all interactions be explained in terms of electromagnetism, e would characterize all possible physical phenomena and should become a type-C universal constant.

I will concentrate, from now on, on the universal constants. Considering Planck constant first, I will, on this particular example, discuss the conceptual role of universal constants in general, as well as some aspects of quantum mechanics which appear in a new light from the proposed point of view. In a third section, the existence will be shown of hidden universal constants, forgotten ones in the realm of classical physics, as well as overlooked ones in modern physics. The *so-called* velocity of light will be studied in the fourth section, again both as an example of general considerations on universal constants, and as a way to approach some epistemological problems of « special relativity ». The final section will be devoted to Newton gravitational constant in connection with the interpretation of « general relativity ».

* * *

These lectures are based on a reorganization of J.-M. LÉVY-LEBLOND: *Riv. Nuovo Cimento*, **7**, 187 (1977). Some material has been added, notably the end of sect. **3** and all of sect. **5**, while some topics have been left out, such as the discussion of the physical meaning of the formal limit processes on constants (whereby they go to zero, or infinity).

Very little has been written about the conceptual nature of physical constants. The only recent reference which was brought to my attention is a paper by C. BERNARDINI: *La Fisica nella Scuola*, **4**, 8 (1975).

2. – Universal constants and conceptual synthesis; the example of $\hbar$ and quantum mechanics.

In order to understand the role of the physical constants, let us consider the particular case of the Planck constant. It may be interesting to note that it was not considered by PLANCK, at least when he first introduced it, as a *fundamental* constant. He rather understood it as an expression of the complex interaction between radiation and matter, which, according to him, gave rise to the quantization of energetic exchanges. In this view, the value of h was to be explained in terms of other constants (one may suppose that such a theory would be expected to explain the numerical coefficient $2\pi/\alpha \simeq 861$ in the expression $h = (2\pi/\alpha)(e^2/c)$ ...). A similar standpoint is taken today by those who, interpreting quantum mechanics as an incomplete and phenomenological theory, try to replace it by a supposedly deeper theory; such is the case of the investigations on « stochastic electrodynamics », for instance.

I will stick here to the conventional view that quantum theory, in its most general form, is presently a universal theoretical framework, which applies to any physical phenomenon whatsoever, so that $\hbar$ indeed is a universal constant (*). It was first introduced into physics through the Planck-Einstein relationship, $E = \hbar\omega$. This relationship is customarily interpreted as associating an energy E with the pulsation of an undulatory physical phenomenon. The connection thus established between a concept of particle mechanics, the energy

(*) It should be clear today that the « good » Planck constant is not h, but $\hbar = h/2\pi$. Indeed, in all theoretical expressions, such as Schrödinger equation, commutation relations or Heisenberg inequalities, as well as in dimensional analyses, numerical coefficients are of order unity only if use is made of $\hbar$. Instead, h introduces spurious powers of 2π (consider, for instance, the ionization energy of hydrogen: $E = \frac{1}{2}me^4/\hbar^2 \simeq \simeq 20me^4/h^2$). Without further ado, we thus take $\hbar$ as *the* Planck constant.

of a discrete entity, and one of wave theory, the pulsation, leads to the idea of wave-particle duality in quantum physics and, further on, to the philosophy of complementarity. Such an interpretation was quite natural in the early days of quantum theory, when one had to approach this new unknown theory from the old classical ones. Duality and complementarity served the very useful purpose of letting physicists use the classical concepts in the quantum domain as far as possible, while taking into account their limits of validity as imposed by these general principles. In such a way, many quantum results were obtained, or at least qualitatively understood, without using a yet-to-be-developed full quantum theory. Most of Bohr's theoretical work is a magnificent example of such a line of thought. It is to be realized today, however, that quantum theory does exist and that its concepts, after fifty years of collective practice, are deeply rooted in the present common sense of working physicists. These concepts need no longer be approached from classical ones but may, and should, be taken at their face value [1, 2]. Such an intrinsically quantum understanding leads one to recognize that the objects of quantum physics are not either waves or particles, as duality would want us to believe; they are *neither waves, nor particles*, even though they do exhibit, under very particular circumstances, two types of limit behaviour as (classical) waves, or (classical) particles. It has been proposed to stress this ontological point by calling them quantons, or « partiqles ». But it may be too late to change the terminology, and while we will probably stick to the customary one, let it be recognized explicitly for one of those necessary abuses of language which are so common in the discourse of physics (as for any science). Coming back to Planck constant, the relationship $E = \hbar\omega$, according to this point of view, is not to be interpreted as linking two classical concepts, but rather as transcending them through their synthesis, to establish a new single concept, with a broader scope. The quantum energy indeed is a new concept, since it associates to any physical state a whole spectrum of numerical values and is to be represented by a Hermitian operator, as opposed to the numerical function which represents energy in classical mechanics. Here again, a new name should have been given to stress the emergence of this concept, as an intrinsic one. Energy and pulsation then appear as two particular facets of a more general notion, each of which being the only visible one, from either one of two quite specific points of view. The role played here by Planck constant in bringing together these two facets is characteristic of universal constants. Any universal constant may be so described as a « concept synthesizer », expressing the unification of two previously unconnected physical concepts into a single one of extended validity.

Of course, any one universal constant usually brings about several such synthetical concepts; Planck constant also unifies momentum and wave number through the de Broglie relationship $\boldsymbol{p} = \hbar\boldsymbol{k}$. This is easily understood, since any physical concept by essence belongs to a theoretical framework which relates it to other concepts. The synthesis of two concepts thus is a local aspect of a more global unification of two pre-existing consistent theoretical structures. Bringing them into contact at one point usually requires the fitting together of other parts as well. This is what happens when the spatio-temporal consistency of particle mechanics, on the one hand, and wave theory, on the other, requires $\hbar$ to play the same role with respect to momentum and wave number (space aspect) as it does with respect to energy and pulsation (time aspect). Following this point of view may lead to a better understanding of the new concepts. It is to be stressed, for instance, how the de Broglie relationship, $\boldsymbol{p} = \hbar\boldsymbol{k}$, exhibits the intrinsically nonclassical nature of the quantum concept it establishes. Indeed, the classical wavelength is independent of the reference frame, or Galilean-invariant, while the momentum of the classical particle with mass m changes according to $\boldsymbol{p}' = \boldsymbol{p} + m\boldsymbol{v}$, under a Galilean transformation with velocity $\boldsymbol{v}$. The de Broglie relationship $\boldsymbol{p} = \hbar\boldsymbol{k}$ thus seems inconsistent with Galilean invariance, that is with the structure of space-time in « nonrelativistic » physics [3]. This pseudoparadox, far from dismissing the de Broglie relationship as LANDÉ maintained, points to a conceptual difference between classical wave number which indeed is Galilean-invariant, and quantum « wave number » which is not [4]. The quantum « wavelength » transforms according to $\boldsymbol{k}' = \boldsymbol{k} + m\boldsymbol{v}/\hbar$, which does not reduce to the classical limit $\boldsymbol{k}' = \boldsymbol{k}$ when $\hbar$ « goes to zero ». This fact is related to the quantum « waves » being represented by complex numbers, in contradistinction with the real amplitudes of classical waves (whether they be acoustical or hydrodynamical).

Indeed let

$$f(x, t) = a \sin(\omega t - kx) \tag{2.1}$$

be the amplitude of a classical harmonic travelling wave, in a certain inertial frame R. In another inertial frame R', related to the first one by a Galilean transformation with velocity v, the wave will be described by an amplitude f' such that

$$f'(x', t') = f(x, t)\,, \tag{2.2}$$

where (x', t') are the co-ordinates in R' of the point with co-ordinates (x, t) in R:

$$\begin{cases} x' = x - vt\,, \\ t' = t\,. \end{cases} \tag{2.3}$$

In other words,

$$f(x', t') = a \sin[\omega t' - k(x' + vt')] = a \sin[(\omega - kv)t' - kx']\,. \tag{2.4}$$

This again is a harmonic travelling wave, with a different frequency however,

$$\omega' = \omega - kv\,, \tag{2.5}$$

which is but the Doppler effect, and the same wave number:

$$k' = k\,. \tag{2.6}$$

Empirically, it is quite clear that the distance between the crests of successive waves here on the Como lake is the same when measured on a photograph taken from a mountain or from a flying plane. Now, let $\psi(x, t)$ be the wave function for a quantum particle in an eigenstate of energy and momentum, that is of pulsation $\omega = E/\hbar$ and wave number $k = p/\hbar$:

$$\psi(x, t) = A \exp\left[i(\omega t - kx)\right]. \tag{2.7}$$

Can the preceding derivation be repeated? No! Indeed, all we can require from Galilean invariance is the equality of the probabilities in the two reference frames, and not the amplitude themselves:

$$|\psi'(x', t')|^2 = |\psi(x, t)|^2\,. \tag{2.8}$$

This allows for a phase factor, possibly depending upon the transformation considered:

$$\psi'(x', t') = \exp\left[if_v(x, t)\right]\cdot\psi(x, t)\,. \tag{2.9}$$

This relationship, in the quantum case, replaces (2.2). The phase factor f_v is severely constrained by group-theoretical arguments and can be computed as

$$f_v(x, t) = \frac{m}{\hbar}\left(vx + \frac{1}{2}v^2 t\right), \tag{2.10}$$

where m is the mass of the particle. This can also be proven by requiring the invariance under (2.9) of the Schrödinger equation for the free particle,

$$i\hbar\frac{\partial\psi}{\partial t} = -\frac{\hbar^2}{2m}\frac{\partial^2\psi}{\partial x^2}, \tag{2.11}$$

which is *not* invariant under the mere substitution of co-ordinates (2.3). The necessity of the phase factor (2.10) is due to a deep property of the Galilei group which has nontrivial projective (*i.e.* « up to a factor ») representations as its physical representations [5]. It becomes clear, thus, that a quantum « wave function » and the associated « wave number » are specific concepts, with properties differing from those of a classical wave function and its wave number. The constant $\hbar$ in effect operates a synthesis which leads to novel concepts, in a truly dialectical *aufhebung*.

As another example of the role played by $\hbar$ as a conceptual tool for building quantum theory, let us sketch a heuristic theory of angular momentum [6]. Full Galilean invariance implies, besides time and space translations, space rotations. This, in classical mechanics, leads to the conservation of angular momentum, in addition to energy and momentum. In the same way as quantum theory, through $\hbar$, relates states with proper values of energy or momentum to harmonic « waves » in space or time, it must relate a state with a proper value L of the angular momentum along some axis to a harmonic « wave » in the angular co-ordinate around that axis. We then write

$$L = \hbar m \,, \tag{2.12}$$

where

$$m = \frac{2\pi}{\alpha} \tag{2.13}$$

with α the period of the « angular wave » (see table I).

TABLE I.

	Classical wave concepts		Classical particle concept	Quantum synthesis
	Period	Pulsation		
Time displacements	τ	$\omega = 2\pi/\tau$	Energy E	$E = \hbar\omega$ (Planck-Einstein)
Space displacements (lengths)	λ	$k = 2\pi/\lambda$	Momentum p	$p = \hbar k$ (de Broglie)
Space rotations (angles)	α	$m = 2\pi/\alpha$ (integer)	Angular momentum L	$L = \hbar m$

It is quite clear that α has to be a submultiple of 2π, so that *m is an integer*. We thus recover in an elementary way the quantization of angular momentum with $\hbar$ as its unit. It is clear also that this heuristic argument is in close parallel with the formal treatment, since quantization of angular momentum in fact comes from the compactness of the rotation group or, equivalently, of the angular co-ordinate space. This approach can be pursued to obtain, at the same heuristic level, Heisenberg-like inequalities expressing the incompatibility of two angular-momentum components, and the quantization of the angular-momentum modulus. It is seen how $\hbar$, by linking the classical quantities L and m, transforms them in a unique and novel concept, the quantum angular momentum.

More generally, we see that $\hbar$ deserves to be called *the* quantum constant, not so much because it is itself the unit of quantized « action » (as in early quantum theory), but because its very existence leads to the specificity of the quantum concepts. Let us conclude, then, that universal constants express the synthetical transcending not only of pairs of concepts, but of whole conceptual arrays. In this sense, a universal constant is a « theory synthesizer », more than a mere « concept synthesizer ».

3. – Hidden universal constants; from classical to modern physics.

3·1. *The fate of universal constants.* – I will now investigate other universal constants and show that they can be understood in much the same way as $\hbar$, namely as underlying the foundations of new physical concepts. Now, it is an elementary but crucial remark that « novelty » is not invariant under time translations! It fades away as time goes on, so that we should not be surprised to see a universal constant experience a change in status after the concept it synthetizes has been firmly established. Indeed, when a sufficient familiarity has been acquired through years of experimenting, theorizing and teaching, one no longer needs to reach the « new » concept through the ancient ones it is based upon, as related by the constant considered. One simply uses the concept as such. The constant then appears as a mere numerical conversion factor, enabling one to express a given physical quantity in terms of different units, connected to one or the other previous aspect of this quantity. No deep conceptual role is any more attributed to the constant, since the synthesis it symbolizes is, so to speak, achieved from the start. In other words, the theoretical status of a universal constant decreases as its practical importance increases.

A good example of this situation is afforded by the classical thermodynamical constants J and k. The first one served to unify heat with work through the relationship $W = JQ$, while the second one showed that temperature was but a statistical aspect of kinetic energy, as expressed by $E = kT$ (up to some numerical factor depending on the number of degrees of freedom). Of course, as emphasized earlier, J and k not only introduced new concepts, but whole new theories: thermodynamics for the first one, statistical mechanics for the second. We are so accustomed today to these ideas that they are incorporated into the implicit background of physical theory. Theoreticians almost automatically choose a convenient system of units such that $k = 1$, since they know that energy and temperature, or work and heat in fact (now) *are* but a single concept (note, however, that in the International System of Units, the fundamental units, beyond the metre, the second and the kilogram, comprise the degree Kelvin—and the candela). In such a way, these constants gradually fade out of sight in quite a literal sense: less and less are they written in formal expressions.

From this point of view, it is seen that J and k indeed are universal constants in the very same fundamental way as $\hbar$. Only does our long collective practice of the concepts they express enable us to forget about their nature and to consider them as mere conversion factors. It must be said that such a process today is well under way concerning $\hbar$. While all textbooks and articles of the twenties kept a detailed record of all $\hbar$'s in their formulae, it is the common use today to take them as unity, which only means adopting a more adapted system of units. This convention has become almost tacit in the recent years, so that, except perhaps at the educational level, it will soon be obvious that there is no difference of nature between $\hbar$, on the one hand, and k, J, on the other. The same remarks, of course, hold for c.

This, then, is the ordinary fate of universal constants: to see their nature as concept synthesizers be progressively incorporated into the implicit common background of physical ideas, then to play a role of mere unit conversion factors and often to be finally forgotten altogether by a suitable redefinition of physicals units. Once this is realized, one may well ask how much of these forgotten universal constants are lying around. It is appropriate to recall here the *Parable of the Surveyors* due to Taylor and Wheeler [7]:

Once upon a time there was a Daytime surveyor who measured off the king's lands. He took his directions of north and east from a magnetic compass needle. Eastward directions from the center of the town square he measured in meters (x in meters). Northward directions were sacred and were measured in a different unit, in miles (y in miles). His records were complete and accurate and were often consulted by the Daytimers.

Nighttimers used the services of another surveyor. His north and east directions were based on the North Star. He too measured distances eastward from the center of the town square in meters (x' in meters) and sacred distances north in miles (y' in miles).

His records were complete and accurate. Every corner of a plot appeared in his book with its two co-ordinates, x' and y'. One fall a student of surveying turned up with novel openmindedness.

Contrary to all previous tradition he attended both of the rival schools operated by the two leaders of surveying. At the day school he learned from one expert his method of recording the location of the gates of the town and the corners of plots of land. At night school he learned the other method. As the days and nights passed the student puzzled more and more in an attempt to find some harmonious relationship between the rival ways of recording location. He carefully compared the records of the two surveyors on the locations of the town gates relative to the center of the town square.

In defiance of tradition, the student took the daring and heretical step to convert northward measurements, previously expressed always in miles, into meters by multiplication with a constant conversion factor, K. He then discovered that the quantity $[(x_A)^2 + (Ky_A)^2]^{\frac{1}{2}}$ based on Daytime measurements of the position of gate A had exactly the same numerical value as the quantity $[(x'_A)^2 + (Ky'_A)^2]^{\frac{1}{2}}$ computed from the readings of the Nighttime surveyor for gate A.

He tried the same comparison on the readings computed from the recorded positions of gate B, and found agreement here too. The student's excitement grew as he checked his scheme of comparison for all the other town gates and found everywhere agreement. He decided to give his discovery a name. He called the quantity

$[x^2 + (Ky)^2]^{\frac{1}{2}}$ the *distance* of the point (x, y) from the center of town. He said that he had discovered the *principle of the invariance of distance*; that one gets exactly the same distances from the Daytime co-ordinates as from the Nighttime co-ordinates, despite the fact that the two sets of surveyors' numbers are quite different.

We may now realize that there is at least a most important universal constant in the simplest of all physical theories, namely plane Euclidean geometry. This constant expresses the theoretical assertion of space isotropy; this physical property is what enables us to synthesize the concepts of northward and eastward distances into the single general concept of distance, independent of the orientation. Of course, the numerical value of the constant K is without physical significance and due to mere historical contingencies. Its very existence, however, is a fundamental aspect of geometry. In the above parable, both distances, eastward and northward, in the town of the parable, in due time, came to be measured with the same unit, so that the constant K disappeared and the discovery of the bright student faded into oblivion: was it not an « obvious » result? This parable was intended to stress, and rightly so, the analogous role played today by the constant c with respect to space-time (see below). But the parable of the surveyors also compels us, conversely, to unearth many of these forgotten universal constants, incorporated as they are into what now seems to be immediate truth, but once was the object of a lengthy and difficult working out. Another purely geometrical example is given by the evaluation of areas. Indeed let us choose, as a unit of area, the area of some arbitrary plane figure, for instance an average human hand palm. It is then a « law of physics » (or a theorem of geometry) that the area A of a square with a side of length L, measured with some length unit, for instance the foot, is given by $A = \alpha L^2$, where α is some universal constant. With our units its value is roughly $\alpha = 0.11$ hand palm per squared foot. One *then* redefines the unit of area as the area of the square with unit side so that α vanishes from sight. It should not be forgotten, however, that this constant expressed the now « obvious » synthesis of areas with squares of lengths. The same could be said of volumes, of course, and is not without factual relevance today. After all, the Anglo-Saxon traditional units of volumes, gallons or pints, are *not* defined by cubing the lengths units, foot or inch; the universal constant β entering the relationship $V = \beta L^3$ has not been taken as unity. We will see below that, even within the scientific metric system, the constant β cannot be forgotten altogether. It should now be clear that many such hidden universal constants lie at the core of the main statements of classical physics, all the more so in the oldest theories such as geometry or Newtonian mechanics, where a long practice has led to the complete incorporation of their significance at an all-implicit level. The absence of universal constants in this part of physics is but an apparent privilege of old age. One might thus classify the universal constants (type C above) into three subclasses according to their historical status:

i) the *modern* ones, such as $\hbar$ (and c, as will be seen), the conceptual role of which is still dominant;

ii) the *classical* ones, such as k or J, which today appear essentially as unit conversion factors, their conceptual role having become almost implicit, and

iii) the *archaic* ones, which have been so well assimilated and digested as to become totally invisible.

3·2. *The point of view of practice.* – The story, however, is not that simple. It is only from the theorist's point of view that the life of a universal constant reaches the happy end of such a drift into the Nirvana of unity and oblivion. The experimentalists working in the lab, when making measurements, must use concrete definitions of their units and cannot at will identify two operationally independent standards as do the theorists on the paper. It is a fact that, whatever fundamental system of units is adopted, based on the theoretical knowledge of the time, the use of units belonging to various other systems adapted to such and such domain of physics cannot be eliminated altogether. There are two reasons for this state of affair. The first one is historical social inertia, which, for instance, forces the experimental physicists on the other side of the Atlantic to plan and order the nuts, bolts, plates, rods, etc. of their apparatus by stating their dimensions in foots and inches rather than metres and centimetres. The universal constant χ entering the relationship $l_{\text{US}} = \chi l_{\text{EU}}$ between the length of some object in the United States and the length of the same one in Europe (so that the subscript EU refers to us, while US refers to some of you) thus can be taken as unity in principle—but in principle *only*. This constant, once more, does express a fundamental law of physics, namely the homogeneity of space, enabling one to define the concept of the length of an object independently of its location. But there is a second reason for the persisting of nonorthodox units, which is due to the nature of experimental physics proper, of which metrology is a fundamental aspect. Once a system of units is chosen (such as the modern International System of Units), every unit of a physical magnitude not belonging to the fundamental ones may be derived from the fundamental units. However, these derived units are often defined in such a way as to be of a very awkward use, or even as to lack of the required precision, in a given domain of physics. Easier and better measurements may be done with the use of an independently defined local system of units. It then remains as a task for experimental metrology to relate these local units to the fundamental system through the experimental determination of conversion factors, which, as shown before, are nothing but genuine universal constants. A simple example may be given here. The relationship between volumetric measurements with the litre as a primary unit and linear measurements in metres (for instance) requires the experimental determination of the universal constant in the expression $V = \beta L^3$ between volumes and lengths; the constant is in fact $\beta = (1.000\,028 \pm 0.000\,004)$ litre·dm^{-3}, which, of course, can be taken as unity—by the theorist. Whe whole field of photometry also furnishes a strong case for the necessity of keeping apparently redundant magnitudes and units.

After all, the present point cannot be expressed better than by the craftsmen themselves. Here is a lengthy quote from a paper by Cohen and DuMond [8]:

> The tendency of the physical sciences to split up into smaller and smaller domains of specialization and the demand for ever increasing precision in many of these domains has forced upon us the adoption of local unit systems, each chosen for use in a given domain, so as to afford the highest degree of relative precision which metrological techniques permit for comparing magnitudes in that domain. But since the physical sciences constitute an indissoluble unit, over-all consistency including all domains is also a mandatory requirement (in fact, it is the only way of verifying the success of our efforts both in theory and in experiment) so that we are obliged with whatever precision can be obtained, to link the entire system to a set of fundamental units which we usually refer to as « absolute units » and we do this by determining « conversion factors » ... .
>
> We have to live with almost all the presently existing local units because 1) their abolition would lead to concealment of the metrological-operational truth as to how measured values were obtained, 2) their abolition would frequently entail a sacrifice of real observational precision of great value in the localized domains. Note that these are in both cases requirements imposed by practical reasons having little or nothing to do with physical theory or conceptual simplicity... (*note*: It is for this reason that theoreticians are particularly unaware of the need for local units & conversion factors. For them the difference between lengths & volumes is merely the writing of a superscript 3) ... .
>
> It is worthwhile to call attention to a few specific cases of presently indispensable local units and conversion factors so as to illustrate this necessity:
>
> 1) Conversion Factor from Linear to Volumetric Measure,
>
> 2) Relative & Absolute Nuclidic Mass Scales,
>
> 3) The X-Unit-to-Milliangström Conversion Factor,
>
> 4) The Acceleration of Terrestrial Gravity,
>
> 5) « As Maintained » & Unqualified Absolute Electrical Units.
>
> It is strange how often we encounter proposals on the part of physicists & chemists to solve such situations by framing verbal conventions or by defining the numerical value of conversion coefficient by decree! A difficult problem of experimental precision metrology cannot be solved by verbal definition or decree any more than a state legislature was able to decree the legal value of the ratio of the circumference of a circle to its diameter. (*note*: Action on House Bill 246, Indiana State Legislature (1897), was postponed indefinitely on 11 February 1897, after preliminary acceptance by a unanimous vote of 67-0 nine days earlier. This bill would have defined π exactly, and granted to the citizens of the state the royalty-free use of this value.)

The structure of physics, as far as we know it, admits of only a small numbers of arbitrarily defined units, and once these are fixed, the system becomes overdetermined and loses its consistency if others are arbitrarily injected without the introduction of conversion factors, factors which must be *experimentally determined.*

Let me only add that these « conversion factors », in order to be « experimentally determined », must have been theoretically defined, as universal constants.

It should be added that, exactly as the practical imperatives of experimental physics prevent a naive dismissing of « classical » constants by a change in unit conventions, more general social conditions can impose the persistence of « unnatural » archaic constants. Two simple historical examples of such a situation come to mind immediately. The first one is, once more, the question of volume measurements. Its scientific, metrological, aspect, discussed earlier, corresponds to a much more general fact, valid since the highest antiquity; namely volumes usually are *not* determined by geometrical means from length measurements. Indeed, most volume determinations in practical life concern flowing materials, liquid or granular, such as beverages or grains (solids are mainly considered according to their weight). This is why independent volume units, defined by the capacity of some standard containers, have been the rule until the advent of the metric system (and even then—see above). For instance, in the Anglo-Saxon system, it has been said, pints are not related to inches. As a consequence, even though, as far as orders of magnitude are concerned, the volume unit usually has been comparable to the cube of the length unit, there was no real need to redefine these units in such a way that the appropriate universal constant be unity. A similar situation prevails for surface measurements, for instance with special units for land areas. In medieval Europe, common units for the areas of farmlands were based on human labour time; England had a «daieswork» (250 m^2), Germany a «tagwerk» —occasionally still used in Bavaria recently—, and the French vineyards were measured in « ouvrée hommée » (17 m^2) [9]. Other area units were based on how much land a man could plow in a given time.

The second example is furnished by the constant χ expressing the homogeneity of space, that is the possibility of using the same units of length at each and every point of space. The case of the Anglo-Saxon system apart, this constant now is almost universally taken as unity due to the international adoption of the metric system. This is a recent occurrence on the historical scale however, and beyond a doubt came about very much later than the theoretical understanding of such a possibility. The point is that such a unification was not necessary, since space indeed was heterogeneous, socially, if not geometrically. Localized and rather autonomous social entities, from tribes to cities, were the rule in human society until less than a few centuries ago. The progressive unification of the social space is related to the rise of merchant and industrial capitalism. It is the needs of this new social order which brought about the redefinition of local units of lengths so that the corresponding universal constant took on the status of an archaic one by apparently vanishing out of sight.

3'3. *The hidden constants of particle physics.* – One should not believe, however, that all hidden universal constants are archaic ones. Quite on the contrary, the most recently discovered universal constants of physics (« hypermodern » ones, to speak like chess players) have been hidden right from the start, due to the special circumstance of their occurrence within a quantum context. Indeed, consider some of the novel physical quantities introduced in elementary-particle physics: baryon number B, hypercharge Y, isospin I, strangeness S. These are *a priori* heterogeneous quantities, independently defined, each of which needs a special unit—and a special « dimension » from the point of view of dimensional analysis. It so happens that all of these quantities are discretely quantized, and may be expressed as integer (or half-integer) multiples of some basic value. The corresponding choice of unit then is tacitly made and no change of unit is ever contemplated, so that one is led to think of any such quantity as having a dimensionless value, *i.e.* as a pure (integer) number. The relationships between these quantities, such as the Gell-Mann–Nishijima formula, $2Y = S + B$, then appear as mere numerical expressions. Were it not for the quantized nature of these quantities, however, one would have been forced to choose explicit units in a more or less arbitrary way. This is particularly clear for the isospin, by comparison with angular-momentum spin. If the latter takes on integral or half-integral values in the natural quantum unit of angular momentum $\hbar$ (so that, as I have stressed, $\hbar$ eventually hides away), there is no such pre-existing unit of isospin, a new quantity, with no connection with space-time. Just try to think of the way isospin would, if it could, manifest itself at a macroscopic level, in a continuous classical approximation.... Isospin thus deserves a new dimension and a special unit; one may propose to define the Heisenberg as the isospin of the nucleon, in which case isospin will take on integral values only. The customary choice rather is to choose as a unit the isospin of the pion, let us call it the Wigner (1 Wigner = 2 Heisenberg). In a similar way, one should define a unit of baryon number, for instance that of the nucleon, call it the Pauli, and a unit of strangeness, for instance that of the kaon, to be called, of course, the Gell-Mann. Hypercharge is to be measured as ordinary charge, in unit of the elementary charge e. Several universal constants now appear, in relationships such as the defining one for hypercharge, $I_3 = \varrho(Q - Y)$, or $2Y = \sigma S + \tau B$; they have numerical values $\varrho = 1$ Wigner/e (or 2 Heisenberg/e), $\sigma = 1$ e/Gell-Mann and $\tau = 1$ e/Pauli. Besides their conceptual value, these remarks might be a little more than purely formal in some cases where change of units might be considered. For instance, a natural enough (although conventional, and not fundamental) unit of baryon number, adapted to macroscopic physics, could be defined as 1 Loschmidt $= 6 \cdot 10^{23}$ Pauli. Or, in a different context, the existence of quarks with so-called « fractional » charges should lead to the introduction of a new unit of charge, 1 Joyce $= e/3$. The hadronic universal constants would then take on new numerical value: $\varrho = \frac{1}{3}$ Wigner/Joyce, $\sigma = 3$ Joyce/Gell-Mann, $\tau = 3$ Joyce/Pauli.

These considerations answer the concern, often expressed, that we are running short of universal constants or, equivalently, of independent physical dimensions. Indeed, it is argued sometimes, by using $\hbar$ and c as units there remains but one free dimension (mass or length, for instance). A new constant would then completely fix a natural system of units and, it is feared, freeze any further research and mean the end of physics. It is true that a specific domain of physics may be closed in such a way as far as the dimensionality questions are considered. This has been the case long ago for geometry, where, as shown above, all universal constants have been eliminated and all units, whether of angles, areas, volumes, expressed in terms of length only. It might also occur in some dynamical theories: the « geometrodynamical » view of space-time and mechanics proposes as natural units the Planck's system based on $\hbar$, c and G as standards, yielding derived units of mass, length and time. Nevertheless, as shown by the previous example in hadronic physics, there is no reason to expect or fear a shortage of new physical concepts, requiring new dimensions, new units and yielding, through their relationships, new universal constants. There are no grounds either to believe that all forthcoming new concepts will necessarily be caracterized by an integral quantization, which, unfortunately, hides the whole question under a veil of triviality.

4. – The case of c; velocity of light (or is it?) and special relativity.

The constant c, as proved by its very name—the « velocity of light »—has entered physics as a type-A constant, namely a specific property of a particular physical object, light—and observe that is *visible* light only which is meant here. After it was measured for the first time by RÖMER, for a long period its status was no different from that of the speed of sound, for example. This is all the more true since the velocity of light—as that of sound—depended on the transmitting medium, as became rapidly known, and c was but its velocity in a particular medium, the ether. One could hope to explain it and even to compute its value, from the underlying properties of the ether, exactly as the velocity of sound in some material was rapidly understood in terms of its density and compressibility. The prospect then (at least a retrospective prospect, from our present viewpoint) was to see c dropping altogether from the list of the fundamental constants. Quite the opposite evolution took place and c has experienced an impressive gain in status.

A first indication that c could be related to other phenomenona than visible light came from KIRCHHOFF, around 1855. At that time, the laws of electrostatics and magnetostatics had been established—as belonging to two separate theories, of course. One knew the two constants, which in modern notation are written ε_0 and μ_0, entering the laws of electricity and magnetism, respectively. From the present point of view, these would be constants of type B, since they are essentially coupling constants, characterizing the strength of electric and magnetic forces. Now, KIRCHHOFF made the simple but deep remark that the combination $(\varepsilon_0\mu_0)^{-\frac{1}{2}}$ had the dimensions length/time, that is of a velocity. This constant was measured from electrostatic and magnetostatic experiments by WEBER and KOHLRAUSCH (1856). The resulting value was $3.1\cdot10^8\ \mathrm{ms}^{-1}$, that is equal to the velocity of light within the experimental errors. KIRCHHOFF was impressed by the coincidence and MAXWELL used this idea as an argument to infer « that light consists in transverse undulations of the same medium which is the cause of electric and magnetic phenomena » [10].

MAXWELL was right of course, as his theory eventually showed. Electricity and magnetism were unified, and c appeared as the velocity of electromagnetic waves in general—no longer as that of visible light only. It now characterized the whole theory of electromagnetism. It must be emphasized here that, in this theoretical framework, c takes on a more general meaning than a mere velocity of propagation, even though its dimensions are those of a velocity. Indeed, it shows up in many expressions such as, for example, the one for the energy density of an electromagnetic field,

$$w = \frac{\varepsilon_0}{2}\boldsymbol{E}^2 + \frac{1}{2\mu_0}\boldsymbol{B}^2 = \frac{\varepsilon_0}{2}(\boldsymbol{E}^2 + c^2\boldsymbol{B}^2)\,, \tag{4.1}$$

and may thus enter calculations dealing with purely static phenomena, where no actual propagating electromagnetic wave is present. From that point of view, it appears that the conventional terminology, that is calling c the velocity of light, is somewhat too restrictive, already in classical electrodynamics. After all, measurements of c can be performed by purely electrostatic and magnetostatic means, instead of the more usual direct measurements of c as a velocity. Such indirect measurements at some times have given the most precise experimental values. Anyway, it is clear that c now has to be considered at least as a type-B constant, characterizing electromagnetic phenomena in general.

The ensuing theoretical developments are well known, which, from the invariance of c, lead to special relativity. Now, although this was the actual historical path, there is no reason why it should be the best one from a logical point of view—all the more so since there is probably no way to define « the » most logical, or conceptually most satisfactory way to build a theory. Rather, we might and should try to re-build the foundations of our theories as they develop, incorporating along modern discoveries, at least when they are well established. In the present case, the relationship between special relativity and fundamental particle physics is so close, both experimentally and theoretically, that the latter necessarily must have some conceptual influence on the former. A consistent contemporary view of relativity theory, I contend, can no longer be based on the invariance of c as a property of electromagnetic waves. Let me give but two arguments for a change of perspective. The first one is of a very general character. Nature, to our present knowledge, seems to present us with four completely different classes of phenomena, associated with four fundamental interactions. Electromagnetic interactions are but one of these four; we also know of nuclear interactions, strong and weak, and gravitational ones. Now it is a fact that the structure of space-time, as expressed by special relativity, up to now seems to be universal, that is independent of the physical processes themselves. Strong and weak interactions are consistently considered to be invariant under the same Lorentz transformations as electromagnetic ones. It may even be said that relativistic invariance is one of the main tools of contemporary theoretical physics, expressing, as WIGNER aptly puts it, a « super-law » to which all physical laws have to obey. Gravitational interactions, at least within a particular interpretation of « general relativity », can be understood in a similar way also (as will be shown in sect. **5**). It thus appears as if space-time was an all-embracing abstract arena, possessing an intrinsic structure (the Lorentz-invariant structure of Minkovski space), quite independent of the concrete and various phenomena taking place in this common arena: the scenery of the stage does not depend on the play, nor on the actors. Such a situation may be philosophically unsatisfactory—and contrary views have been repeatedly advanced (for instance by MACH)—, it may well be scientifically a provisory situation, but it is the present one. Now, if one insists upon founding special relativity on electromagnetism, its universality becomes difficult to understand and to accept. There is no logical inconsistency here, but a real conceptual difficulty. It shows off very often in the teaching of relativity theory, where students taught in the conventional way come to associate special relativity with electromagnetism to the extent of being quite puzzled when they learn its relevance for nuclear and particle physics as well. One should, therefore, establish relativity theory on a more general, interaction-independent basis, than some property of electromagnetic radiation.

A second and more factual reason to play down the role played by the speed of light in special relativity is that in fact the constant c is *not* the velocity of light, at least in the sense that

1) c is probably not only the velocity of light,

2) c is perhaps not even the velocity of light.

Indeed, from our present point of view, that is using our knowledge of relativistic dynamics, we understand the velocity of electromagnetic radiation to be the invariant velocity c as a consequence of the fact that the photon has a vanishing mass. But any zero-mass particle, according to the standard theory, has to travel with the same invariant velocity c. The neutrinos, for instance, if their masses do indeed vanish, also propagate at the velocity c, although they have nothing to do with electromagnetism. Furthermore, the photon mass has a zero mass only within our experimental precision. It will *never* be known to have an exactly zero mass, while, of course, it could well be found to have, in fact, an nonzero one [11]. The constancy of the velocity of light signals, used as the basic hypothesis for the building of special relativity, then either will never be ascertained, or will be disproved. However, the breakdown of this hypothesis does not necessarily imply the ruin of its consequences—if these can be derived from other, more general hypotheses.

This indeed is the case, as special relativity may be built upon axioms dealing only with the abstract structure of space-time, and, in particular, relying on group-theoretical considerations [12]. One then reverses the historical order of deduction: from the space-time structure, one derives a general theory of particle dynamics. This theory allows for zero-mass particles, which are then shown to travel with the invariant velocity c. A (quantum) field theory of such particles if they are of a vector nature (spin 1) then leads back to the classical Maxwell theory of electromagnetism. The situation is summarized by fig. 1.

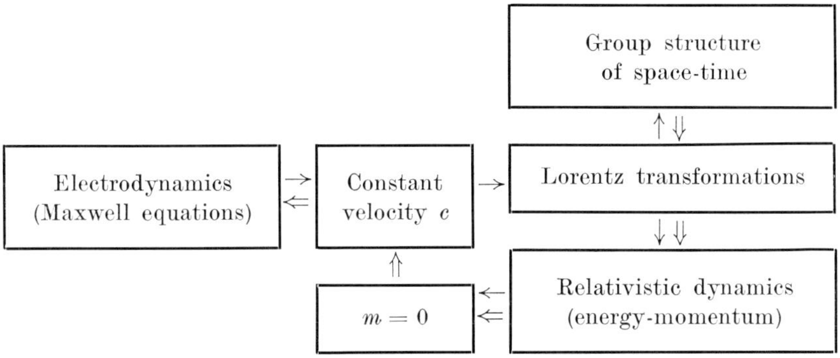

Fig. 1. – Conceptual relationships between electrodynamics and special relativity: $\rightarrow$ the historical path, $\Rightarrow$ the modern path.

I may now come back to my leitmotiv (or should I write here light-motiv?). The constant c, in the general group-theoretical approach, appears as an abstract constant: it has the dimensions of a velocity, but is not expected necessarily to be the velocity of real physical objects. Rather, it is a universal constant; as $\hbar$, or the other universal constants (k, J and the hidden ones), it is a conceptual synthetizer. This most clearly appears in the expression for the invariant space-time interval:

$$(\Delta s)^2 = (\Delta t)^2 - c^2(\Delta x)^2 , \tag{4.2}$$

where it is seen how c permits the synthesis of previously unconnected concepts, namely time intervals and space intervals, in a single, more general one. The situation is perfectly similar to that of ordinary plane geometry, as emphasized by the Parable of the Surveyors already mentioned (see sect. **3**). In other words, while it so happens that photons travel with the velocity c, c is no more the speed of light than $\hbar$ is the spin of light, even though photons indeed have a spin $\hbar$. The constant c definitely is a type-A constant, connected with the universal structure of space-time. As all such constants, it synthetizes various concepts; besides spatio-temporal intervals (4.2), its most important role perhaps is to generate the mass-energy concept through the best known physical formula of all times:

$$E = mc^2 . \tag{4.3}$$

This aspect may be stressed in a most interesting way by sketching still another approach to special relativity. The idea is to build directly relativistic dynamics, independently of space-time considerations. The advantage of such an approach is to keep close contact with the actual practice of contemporary physics where particle dynamics certainly is the largest consumer of relativistic concepts as well as the best supplier of empirical evidence for their validity. Following DAVIDON [13], let us then look for the functions $E(v)$ and $p(v)$ giving the energy and momentum of a particle as functions of its velocity. We first *define* the « inertia » $I(v)$ of the particle by the expression

$$p = I(v)\cdot v . \tag{4.4}$$

We now assume that the inertia depends on the energy, in the simplest possible way, namely that any increase $\mathrm{d}E$ in the energy produces a proportional increase $\mathrm{d}I$ in the inertia:

$$\mathrm{d}I = a\cdot \mathrm{d}E \tag{4.5}$$

with a a positive constant. This idea, as stressed by DAVIDON, originated at the end of the last century in the studies of the classical electron theory (LORENTZ, ABRAHAM, etc.). It certainly belongs to one of the several paths leading to the special relativity, although an ignored one. Finally, we have the standard relationship between velocity, energy and momentum (Hamilton's equation):

$$v = \frac{\mathrm{d}E}{\mathrm{d}p} . \tag{4.6}$$

From (4.5) and (4.6), we derive

$$\mathrm{d}I = av\,\mathrm{d}p , \tag{4.7}$$

so that, using (4.4), we get

$$I\,\mathrm{d}I - ap\,\mathrm{d}p = 0 \tag{4.8}$$

and

$$I^2 - ap^2 = I_0^2 \tag{4.9}$$

(I_0^2 being a constant of integration)

We finally obtain, through elementary calculations,

$$I = \frac{I_0}{\sqrt{1-av^2}}, \qquad E = \frac{I_0/a}{\sqrt{1-av^2}}, \qquad p = \frac{I_0 v}{\sqrt{1-av^2}}, \tag{4.10}$$

in the case where $I_0 \neq 0$.

It is clear that I_0 may be identified with the (invariant) mass of the particle, for instance by looking at the low-velocity approximation of (4.10) and comparing it with the Galilean formulae. What is more interesting for us is the fundamental meaning of the universal constant a, expressing through (4.5) the synthesis of energy and inertia. We may, of course, notice that a has the dimension of a velocity and write

$$a = c^{-2}, \tag{4.11}$$

thus recovering standard expressions. And now, there is the possibility, of course, that $I_0 = 0$, so that (4.9) becomes

$$I = c^{-1}p, \tag{4.12}$$

leading, from (4.5), to

$$E = cp. \tag{4.13}$$

According to (4.4), we obtain

$$v = c, \tag{4.14}$$

so that particles with zero mass indeed have an invariant velocity. It is interesting to derive the Lorentz transformations from the expressions (4.10), that is to obtain kinematics from dynamics. This leads to a new circuit in fig. 1, which might be of great interest.

Let me finally use the example of the constant c to enrich my characterization of universal constants. The point is that their role in the synthesis and unification of previously unrelated concepts or sets thereof, if it is the prime one in historical order of appearance, has for a corollary the fact of their leading to split and separate previously fused, if not confused, concepts. Two simple examples in relativity theory may be given here. The first one deals with the impossibility in Einsteinian relativity of a concept with the following two properties of the velocity in Galilean relativity: i) to be an additive quantity, that is the canonical parameter of inertial transformations, obeying the simple composition law $v_{12} = v_1 + v_2$, and ii) to give the time rate of spatial change, namely $v = \mathrm{d}x/\mathrm{d}t$, for uniform motions. In Einsteinian relativity, if the second property is used as a definition of what we will keep calling « velocity » v, the first one will hold true for another quantity, the so-called « rapidity » φ. The two quantities are related by $v = \mathrm{tgh}\, \varphi$, or, with dimensional notations, $v = c\, \mathrm{cotgh}\, (v/c)$, which makes apparent their fusion in the limit $c \to \infty$. The introduction of the concept of rapidity is of a major help for educational purposes. It not only yields a more compact and more significant expression for Lorentz transformations via hyperbolic functions, but it explains away the pseudoparadoxes associated to the idea of a limiting velocity or nonadditivity of velocities, as simply due to a bad choice of parameter, such as would occur if rotations where labelled through the tangent of the angle instead of the angle itself. In recent years, the concept of rapidity also has been fruitfully used in high-energy phenomenology. A similar clarification may be achieved in relativistic dynamics, by introducing, with the concepts of energy and mass, the one of inertia, defined as above, that is the coefficient of the velocity in the expression for the momentum. It is seen, then, that inertia is to be identified with energy in Einsteinian relativity, but with mass in Galilean relativity [14]. The occurrence of the universal constant c splits inertia from mass at the same time as it fuses it with energy. In the description of space-time, the splitting of the two categories of pairs of simultaneous events from that of invariant intervals (now the lightlike ones), may be interpreted in quite the same way. Other examples can be found at will. To use a material metaphor, it may be said that the fitting of two conceptual structures, while bringing into contact previously separated pieces, also generates stresses requiring various splits within the new body.

5. – Newton constant G; gravitation and/or general relativity.

Gravity certainly is the first natural phenomenon to have been harnessed by physical theory. It is not surprising, therefore, that its specific constant has a complex historical and epistemological status.

The story begins with GALILEO establishing the laws of free-fall motion. The « acceleration of gravity » g there appears as a universal constant, since it is the same for all falling bodies—that is for all bodies! Free fall has to be regarded then as an intrinsic and general property and g as a type-C constant. Since it relates velocity and time of fall

$$v = gt\,, \tag{5.1}$$

it certainly obeys our preceding description.

However, NEWTON comes and dismisses g altogether from its role as a fundamental constant, by explaining it as a purely local number, characterizing only earthly physics, and expressing it in terms of the Earth mass M and radius R, and of a new fundamental constant G:

$$g = G\frac{M^2}{R}. \tag{5.2}$$

What place now is to be attributed to this G in our classification of physical constants? When it appears, in Newton's theory G certainly is to be thought of as *universal* constant, of type C. Indeed, no force other than the gravitational ones at the time were known to physics. Newton's achievements naturally were taken as offering the prospect of a complete understanding of the physical world. The law of « *universal* attraction », it was felt, could explain the cohesion of matter at the microscopic level as well as the behaviour of the solar system. There was but one force, and G was its symbol. The theory of gravitation was the whole theory of physics, which indeed gave to G the status of a universal constant. To interpret it within our characterization of universal constants as conceptual synthesizers, it suffices to distinguish the two concepts of *inertial* mass and *gravitational* mass. The inertial mass m of an object, its « coefficient of inertia », determines, in Newtonian mechanics, its acceleration γ in response to any given force F:

$$\gamma = \frac{1}{m}F. \tag{5.3}$$

Its gravitational mass $\mathscr{M}$, on the other hand, fixes the gravitational forces it exerts or undergoes (*). The most natural choice of units for the gravitational mass consists in writing the Newton force (5.2) in the form

$$F_N = \frac{\mathscr{M}_1\mathscr{M}_2}{d^2}. \tag{5.4}$$

The constant G now appears through the fundamental statement that gravitational mass is identical to inertial mass. This conceptual identity is translated into numerical equality by taking into account the difference in the units independently chosen for m and $\mathscr{M}$. The existence thus is required of a constant relating the two quantities:

$$\mathscr{M} = \Gamma m. \tag{5.5}$$

(*) A more careful discussion could even distinguish the active and passive gravitational masses.

Clearly, this is Newton's constant in disguise, and one has

$$G = \Gamma^2 . \tag{5.6}$$

By the end of the XVIII century, electricity and magnetism were in turn brought within the realm of physical theory. They previously appeared as natural forces of a rather vitalist (« biological », as we would say today) character. It was a major breakthrough when they were shown, through Coulomb's law, to « yield to the power of calculus », according to the admirative words of DELAMBRE in 1810. But, from then on, gravity lost its role as the unique ruler of the physical world, and was confined to the domain of heavenly bodies and their motions. Down on earth, the XIX century saw the rising and triumph of electromagnetic theory with its own constants. During that period, G appears as a constant of type B, characterizing a special class of physical phenomena, namely gravitational interactions.

The third episode unfolds when EINSTEIN, trying to generalize Newton's theory of gravity in accordance with his own theory of relativity (instead of Galileo's), hits upon « general relativity », as a purely geometrical theory. Insofar as gravitation now was to be described in spatio-temporal terms, G anew became a type-C universal constant. Indeed, according to the conventional view of general relativity, the curvature of space-time expresses the presence of mass (rather than being simply caused by it). Einstein's equations

$$R_{\mu\nu} - \tfrac{1}{2} g_{\mu\nu} R = - 8\pi G T_{\mu\nu} \tag{5.7}$$

can be viewed as identifying two previously independent concepts, namely length (or, more generally, space-time distance) and mass. This is particularly clear, in the following expression, directly derived from (5.7) upon contraction:

$$R = 8\pi G T , \tag{5.8}$$

where R is the curvature scalar and T the mass-energy density (invariant trace of the energy-momentum tensor: $T = T^{\mu}{}_{\mu}$). Dimensionally, R is an inverse squared length and T a mass density:

$$[R] = \mathscr{L}^{-2} , \qquad [T] = \mathscr{M}\mathscr{L}^{-3} , \tag{5.9}$$

so that

$$[G] = \mathscr{L}\mathscr{M}^{-1} \tag{5.10}$$

directly connecting mass and length (we have here assumed $c = 1$ so that $\mathscr{L} = \mathscr{T}$). It is worth emphasizing that a consistent application of this point of view leads to think of mass (and energy, etc.) as purely geometrical properties. All of physics ultimately is contained in, and explained by, the structural properties of space-time. Geometrodynamics, as advocated by WHEELER, is the systematic expression of this line of thought.

To repeat: gravity now is but a way to describe the spatio-temporal arena. As such, it holds a prominent position and submits to its rule all specific phenomena taking place in the common arena. Even though these phenomena at the moment seem to belong to various classes, not (not yet?) unified, namely electromagnetic, strong and weak, all of them have to abide by the universal laws of gravitation, as expressed by general relativity.

This truly imperialistic point of view has been repeatedly challenged. Especially since particle physics has come to occupy the first place at the hit-parade of theoretical physics, it has tried to free itself from the grip of gravity —even though it was an purely abstract domination, since most recent only are the issues where gravity and other interactions have to be simultaneously taken into account, such as the physics of black holes. The idea was to bring back some uniformity in the status of the known physical interactions, namely to put gravitation on the same footing as the other ones.

Such a program can be carried out [15]. A theory of gravitation can be constructed as a standard classical field theory through the following steps: 1) start with the usual Minkowski space-time, endowed with its flat metric $\eta_{ij} = (1, -1, -1, -1)$; 2) consider gravitation as mediated through a classical field with mass as its source, and propagating in space-time; 3) look for the tensorial properties of the field, eliminate scalar (spin 0), for it fails to explain the advance of planetary perihelions, and vector (spin 1), for it cannot give a purely attractive force (see the electromagnetic field); 4) decide for the next simplest possibility, namely a symmetrical tensor (spin 2) $g_{\mu\nu}(x)$; 5) write the Lagrangian for the free field by invariance arguments; 6) introduce in the Lagrangian interaction with matter by coupling the field to the energy-momentum tensor of matter. Now comes the crucial point: because of the (universal) identity of inertial and gravitational mass, the energy-momentum tensor which is the source of the gravitational field must include the energy-momentum density of the gravitational field itself! In other words, the gravitational field is self-generating (a similar situation would prevail for electromagnetism if the electromagnetic field carried an electric charge—which of course yields an inconsistent theory). This fundamental property gives rise to an intrinsic nonlinearity of the field equations which turn out—lo and behold!—to be absolutely identical with Einstein equations (5.7). The final touch consists in identifying the « field » $g_{\mu\nu}(x)$ of this theory with the « metric » $g_{\mu\nu}(x)$ of the usual one; the argument is that, due to the universal coupling of matter with gravity, the field $g_{\mu\nu}(x)$ acts on all physical systems in such a way that any length or time measurement is re-scaled, and that the effective metric, as measured, is not the underlying and unobservable flat η, but is given by g itself.

A further step can be taken, establishing gravitation theory not as a classical field theory, but as a quantum one. The specific properties of a spin-2 massless particle, the graviton, and, especially, the associated gauge invariance, are used to build a theory which, once more, is formally identical with Einstein's one [16].

It should be added that this approach to gravitation theory is not the only heterodox one to give the orthodox answer. In order to keep as close as possible to conventional field theory (essentially electromagnetic field theory), it relies heavily on the energy-momentum conservation law as expressed through the energy-momentum tensor. The very concepts of energy and momentum for the gravitational fields, however, have often been criticized. Interpretations of Einstein's theory even have been proposed in which the gravitational field carries no energy—so that, as a consequence, there are no gravitational waves (in the sense of « free » propagating fields) [17]. In such interpretations, however, gravitation is « de-geometrized » as well and reduced to the status of a particular interaction, with G, then, as a type-B constant.

From the precedingly described point of view, it is clear that general relativity *a*) is not a *relativity*, in that its geometrical interpretation and its invariance under co-ordinate transformation just « turn out », without any deep physical meaning; *b*) is not *general*, in that it is just the theory of gravitation, on the same level as the theory of electromagnetism or the theory (?) of strong interactions. In other terms, the constant G here reverts to its type-B status.

These considerations might appear as a purely formal game, and are indeed often considered with some contempt, as mere « philosophy », by some physicists. There are, however, at least two good reasons for keeping in mind the existence of alternative interpretations of one and the same theory.

For educational purposes first, it is not a matter of indifference which viewpoint is adopted in teaching the theory. The access may be eased or hindered, depending on the agreement, or otherwise, of the student general conception of physics with that adopted in this domain. To people educated in particle physics, for instance, the *a priori* geometrical interpretation usually is very uncongenial, while, to mathematically minded lovers of differential geometry, the field-theoretical interpretation looks rather awkward. The second motivation to favour pluralism in epistemological matters concerns the future of physics. As long as the theory holds good, indeed, the various viewpoints differ little, if at all, in the concrete applications of the theory. However, when the time comes from a change, it may prove much easier to modify the theory starting from one interpretation than from the other (*). A pluralistic approach is an absolute necessity for science to keep alive:

(*) Just think of how difficult it is to adapt directly the action-at-a-distance Newtonian formalism of mechanics ($F = m\gamma$) to Einsteinian relativity, while the same problem is easily solved in the Lagrangian formalism ($\delta A = 0$).

« The essential values in the pursuit of knowledge are heuristic values; fecundity, creative conceptualization, transitive power, extent and rightness of the novelty. These values have a direct link with plurality: plurality of intellectual informing is the very condition for cognitive inventing; and, conversely, knowledge is pluralistic in its ways and contents—not monolithic and exclusive. » [18]

REFERENCES

[1] M. BUNGE: *Philosophy of Physics* (Dordrecht, 1973).
[2] J.-M. LÉVY-LEBLOND: *Dialectica*, **30**, 162 (1976); and in *Quantum Mechanics Half a Century Later*, edited by J. LEITE LOPES and M. PATY (Dordrecht, 1977), p. 171.
[3] A. LANDÉ: *Amer. Journ. Phys.*, **43**, 701 (1975).
[4] J.-M. LÉVY-LEBLOND: *Amer. Journ. Phys.*, **44**, 1131 (1976).
[5] See, for a review, with the original references there J.-M. LÉVY-LEBLOND: in *Group Theory and Its Applications*, Vol. **2**, edited by E. LOEBL (New York, N. Y., 1971).
[6] J.-M. LÉVY-LEBLOND: *Amer. Journ. Phys.*, **44**, 719 (1976).
[7] E. F. TAYLOR and J. A. WHEELER: *Spacetime Physics* (San Francisco, Cal., 1966).
[8] E. R. COHEN and J. W. M. DUMOND: *Rev. Mod. Phys.*, **37**, 537 (1965).
[9] See H. A. KLEIN: *The World of Measurements* (New York, N. Y., 1974), p. 75.
[10] E. WHITTAKER: *A History of the Theories of Aether and Electricity* (London, 1951).
[11] A. S. GOLDHABER and M. M. NIETO: *Rev. Mod. Phys.*, **43**, 277 (1971).
[12] See, for example, at a tutorial level, A. R. LEE and T. M. KALOTAS: *Amer. Journ. Phys.*, **43**, 434 (1975); J.-M. LÉVY-LEBLOND: *Amer. Journ. Phys.*, **44**, 271 (1976). See also M. Jammer's lecture: this volume, p. 202.
[13] A. DAVIDON: *Found. Phys.*, **5**, 525 (1976).
[14] J.-M. LÉVY-LEBLOND: in *Group Theoretical Methods in Physics*, edited by A. JANNER, T. JANSSEN and M. BOON, Lecture Notes in Physics, Vol. **50** (Berlin, 1976).
[15] S. N. GUPTA: *Phys. Rev.*, **96**, 1683 (1954); *Rev. Mod. Phys.*, **29**, 334 (1957); W. THIRRING: *Ann. of Phys.*, **16**, 96 (1961); R. U. SEXL: *Fortschr. Phys.*, **15**, 269 (1967); S. DESER: *Gen. Rel. Grav.*, **1**, 9 (1970); C. W. MISNER, K. S. THORNE and J. A. WHEELER: *Gravitation* (San Francisco, Cal., 1973), p. 424; Y. AVISHAI and H. EKSTEIN: *Comm. Math. Phys.*, **37**, 193 (1974).
[16] S. WEINBERG: *Phys. Rev.*, **138**, 3988 (1965); D. G. BOULWARE and S. DESER: *Ann. of Phys.*, **89**, 193 (1974).
[17] A. QUALE: *Fortschr. Phys.*, **21**, 265 (1973); *Gen. Rel. Grav.*, **6**, 63 (1975). See also N. ROSEN: *Lett. Nuovo Cimento*, **19**, 249 (1977); J. RAYSKI: to be published.
[18] J. SCHLANGER: *Le comique des idées* (Paris, 1977).

Quantum Logic.

P. MITTELSTAEDT

Institut für Theoretische Physik der Universität - Köln, Germany

Introduction.

The subspaces of Hilbert space constitute an orthocomplemented quasi-modular lattice L_q if one considers between two subspaces A, B the relation $A \leqslant B$ and the operations $A \wedge B$, $A \vee B$, $\neg A$. Furthermore, since subspaces can be interpreted as quantum-mechanical propositions and since the operations $\wedge$, $\vee$, $\neg$ have some similarity with the logical operations « and », « or » and « not », respectively, the question arises whether the lattice L_q can be interpreted as the logical calculus of quantum-mechanical propositions, called quantum logic.

From a formal point of view there are some similarities between the lattice L_q and lattices which are known to have a logical interpretation, *e.g.* the Boolean lattice L_B of classical logic and the implicative lattice L_i of intuitionistic logic. Many syntactic requirements which can be formulated on the basis of L_B and L_i are also fulfilled in L_q. More important, however, is the question whether a logical semantic can be found for the lattice L_q.

A Boolean lattice L_B can be interpreted by a two-valued truth function. It is well known that neither a two-valued truth function nor a conveniently generalized truth function exist on the lattice L_q. On the other hand, also the lattice L_i of intuitionistic logic cannot be interpreted by truth values. However, an implicative lattice can be considered as a logical calculus if one takes into account a more general semantic—the operational interpretation which makes use of the dialogic technique.

Although this interpretation cannot directly be applied to the lattice L_q, it turns out that a generalization of the dialogic method can be used as an interpretation of a lattice L_{qi}, which may be considered as the intuitionistic part of L_q and which differs from L_q by the relation $\vee = A \vee \neg A$. In order to demonstrate this possibility, we first introduce some obvious modifications of the dialogic method which come from the possible incommensurability of propositions about quantum-mechanical systems. With the aid of this generalized dialogic method we derive a propositional calculus Q_{eff} which is similar to the calculus of effective (intuitionistic) logic, but contains a few restrictions which are based on the incommensurability of quantum-mechanical propositions. Therefore, this calculus will be called the calculus of effective quantum logic.

The calculus Q_{eff} is a model of the lattice L_{qi}, which differs from the lattice L_{q} by the axiom $\mathsf{V} = A \vee \neg A$, the logical interpretation of which is called the *principle of excluded middle*. This principle is based on the value-definiteness of the propositions considered and cannot be justified by means of dialogs alone. However, it can be shown within the framework of the calculus Q_{eff} that the value-definiteness of the elementary propositions which are proved by quantum-mechanical propositions is inherited by all finite compound propositions. In this way one arrives at the calculus Q of full quantum logic which incorporates the principle of excluded middle for all propositions and which is a model for the lattice L_{q}.

1. – The lattice L_{q} of subspaces of Hilbert space.

1'1. *The Hilbert space.* – The abstract Hilbert space $\mathcal{H}$ is a set of elements called *vectors* $f, g, \ldots$, which satisfy the following axioms [1, 2]:

i) *$\mathcal{H}$ is a linear vector space with complex coefficients*, *i.e.* for any two elements $f, g \in \mathcal{H}$, there exists an element $(f+g) \in \mathcal{H}$ and for every element $f \in \mathcal{H}$ and every complex number λ there exists an element $\lambda f \in \mathcal{H}$. Furthermore, there exists a zero element $\underline{0}$ such that for all $f \in \mathcal{H}$ we have $\underline{0} + f = f$ and $0 \cdot f = \underline{0}$.

ii) *There exists a positive-definite scalar product in $\mathcal{H}$.* For any two elements $f, g \in \mathcal{H}$ a complex-valued function (f, g) is defined which satisfies the following relations: $(f, g) = (g, f)^*$, $(f, g + h) = (f, g) + (f, h)$, $\lambda(f, g) = (f, \lambda g)$, $(f, f) \equiv \|f\|^2 > 0$ unless $f = \underline{0}$.

iii) *$\mathcal{H}$ is separable*, *i.e.* there exists a sequence $\{f_n\}$ of elements $f_n \in \mathcal{H}$ $(n = 1, 2, \ldots)$ with the property that, for any $f \in \mathcal{H}$ and $\varepsilon > 0$, there exists at least one element $f_n \in \{f_i\}$ such that $\|f - f_n\| < \varepsilon$.

iv) *$\mathcal{H}$ is complete*, *i.e.* any sequence $\{f_n\}$ of elements $f_i \in \mathcal{H}$ with the property $\lim\limits_{n,m\to\infty} \|f_n - f_m\| = 0$ has a uniquely defined limit $f \in \mathcal{H}$ such that $\lim\limits_{n\to\infty} \|f - f_n\| = 0$.

We shall now introduce the concept of a *subspace*. A subset $\mathcal{M}$ of $\mathcal{H}$ is called a *linear manifold*, if

$$f \in \mathcal{M} \curvearrowright \lambda f \in \mathcal{M}\,,$$

$$f, g \in \mathcal{M} \curvearrowright (f + g) \in \mathcal{M}\,.$$

A linear manifold $\mathcal{M}$ is called a *closed linear manifold*, if for any sequence $\{f_n\}$

with $f_i \in \mathcal{M}$, which has a limit vector $f \in \mathcal{H}$, also this limit vector belongs to $\mathcal{M}$, *i.e.* $f \in \mathcal{H}$. Closed linear manifolds will also be called ***subspaces*** and will be denoted by $A, B, C, \ldots$.

Let A be a subspace. The set $\mathcal{M}_{\neg}(A)$ of elements which are orthogonal to all vectors of A, *i.e.*

$$\mathcal{M}_{\neg}(A) := \{f : (f, g) = 0 \text{ for all } g \in A\},$$

is also a subspace, which will be denoted by $\neg A$. The subspace $\neg A$ is also said to be *completely orthogonal to* A. From the definition it is easily seen that

$$A = \neg\neg A .$$

Furthermore, if A and B are subspaces such that $A \leqslant B$ (we use the sign $\leqslant$ for the set-theoretical inclusion), it follows that

$$A \leqslant B \curvearrowright \neg B \leqslant \neg A .$$

If A and B are subspaces, the set $\mathcal{M}_{\wedge}(A, B)$ of elements

$$\mathcal{M}_{\wedge}(A, B) := \{f : f \in A \text{ and } f \in B\}$$

which corresponds to the intersection of the sets A and B is also a subspace, and will be denoted by $A \wedge B$. Obviously we have

$$A \wedge B \leqslant A , \qquad A \wedge B \leqslant B .$$

Furthermore, it is easily seen that the intersection $A \wedge B$ is the largest subspace contained in A and B (infimum).

If A and B are subspaces, the set $\mathcal{M}_{\vee}(A, B)$ consists of all elements $h = \alpha f + \beta g$ with $f \in A$, $g \in B$ and the limit vectors. Therefore, $\mathcal{M}_{\vee}(A, B)$ is a closed linear manifold, which will be denoted here by $A \vee B$. The subspace $A \vee B$ is said to be spanned by the elements $f \in A \cup B$, where $A \cup B$ is the set-theoretical union of A and B. Obviously we have

$$A \leqslant A \vee B , \qquad B \leqslant A \vee B .$$

Furthermore, one finds that $A \vee B$ is the smallest subset which contains A A and B (supremum).

1·2. *The lattice of subspaces.* – We will now consider the set $\mathcal{S}$ of subspaces A, B, C. According to the above-mentioned results, on the set $\mathcal{S}$ we have

1) 1-place operation

$$\theta_\neg : \mathscr{S} \to \mathscr{S} \,,$$

2) 2-place operations

$$\theta_\wedge : \mathscr{S} \times \mathscr{S} \to \mathscr{S} \,, \qquad \theta_\vee : \mathscr{S} \times \mathscr{S} \to \mathscr{S} \,,$$

and a 2-place relation $R \subseteq \mathscr{S} \times \mathscr{S}$, which is given by the set-theoretical inclusion $\leqslant$.

From the definition of the relation R it follows

(1.1) $$A \leqslant A \,,$$

(1.2) $$A \leqslant B \,,\ B \leqslant C \rightharpoonup A \leqslant C \,,$$

i.e. $\mathscr{S}$ is a partially ordered set. Furthermore, from the definitions of the operations $\theta_\wedge$ and $\theta_\vee$ we obtain

(2.1) $$A \wedge B \leqslant A \,,$$

(2.2) $$A \wedge B \leqslant B \,,$$

(2.3) $$C \leqslant A \,,\ C \leqslant B \rightharpoonup C \leqslant A \wedge B \,,$$

(3.1) $$A \leqslant A \vee B \,,$$

(3.2) $$B \leqslant A \vee B \,,$$

(3.3) $$A \leqslant C \,,\ B \leqslant C \rightharpoonup A \vee B \leqslant C \,.$$

These relations state that for any two elements $A, B \in \mathscr{S}$ in respect to the partial-ordering relation $\leqslant$ there exists a greatest lower bound $A \wedge B$ (infimum) and a least upper bound $A \vee B$ (supremum). A partially ordered set $\mathscr{S}$ which has this property is called a *lattice* $L = L(\mathscr{S})$ [3]. In addition, the lattice of subspaces considered here has a zero element $\bigwedge = \underline{0}$ and a unit element $\bigvee = \mathscr{H}$, which satisfy the relations

(4.0) $$\bigwedge \leqslant A \,, \qquad A \leqslant \bigvee \,.$$

The operation $\theta_\neg$, which maps a subspace $A \in \mathscr{S}$ to the completely orthogonal subspace $\neg A \in \mathscr{S}$, is an automorphism which satisfies the relations

(4.1) $$A \wedge \neg A \leqslant \bigwedge \,,$$

(4.2) $$\bigvee \leqslant A \vee \neg A \,,$$

(4.3) $$A = \neg\neg A \,,$$

(4.4) $$A \leqslant B \rightharpoonup \neg B \leqslant \neg A \,.$$

If, in a lattice L with elements $\wedge$ and $\vee$, an automorphism $A \to \neg A$ is defined which fulfills the conditions (4.1)-(4.4) this lattice is said to be *orthocomplemented.* The element $\neg A$ is called the *orthocomplement* of A [3].

Furthermore, it can be shown for subspaces of Hilbert space that, in addition to the axioms of an orthocomplemented lattice, the law

$$(5) \qquad B \leqslant A\,,\;\; C \leqslant \neg A \curvearrowright A \wedge (B \vee C) \leqslant B$$

is valid. An orthocomplemented lattice which has this property (5) is called *quasi-modular.* Therefore, the subspaces of Hilbert space constitute an *orthocomplemented quasi-modular* lattice, which will be denoted here by $L_{\mathfrak{q}}$ [2, 4]. There are, of course, other interesting properties (σ-completeness, atomicity, covering law) of the lattice of subspaces, which however will not be considered here [2, 5].

For every subspace $A \in L_{\mathfrak{q}}$ there exists precisely one bounded linear operator

$$P_A : \mathscr{H} \to A$$

with the domain $D(P_A) = \mathscr{H}$ and the range $\Delta(P_A) = A$, which fulfills the relation

$$P_A = P_A^+\,, \qquad P_A^2 = P_A\,.$$

This operator P_A is called projection operator, since an arbitrary vector is « projected » by P_A into the subspace $\Delta(P_A)$, *i.e.* $P_A f = f_A$, $f_A \in A$. Moreover, the operator P_A is self-adjoint with eigenvalues 0 and 1. Therefore, it may be considered as an *observable* with measuring values 0 and 1. If a quantum-mechanical system is described by a state vector $\varphi \in \mathscr{H}$, the equation $P_A\varphi = \varphi$ (eigenvalue 1) means that $\varphi \in A$, and the equation $P_A\varphi = 0$ (eigenvalue 0) that $\varphi \in \neg A$, which means that the vector φ has no component in the subspace A.

For this reason subspaces $A, B, C \in \mathscr{S}$ are often considered as *properties* of a physical system S with state vector φ [1, 2]. According to this interpretation the system $S(\varphi)$ is said to possess the property E_A if and only if $\varphi \in A$, *i.e.* if $P_A\varphi = \varphi$, which means that the observable P_A has the eigenvalue 1. If $P_A\varphi = 0$, *i.e.* $\varphi \in \neg A$, the system will be said to have the property $E_{\neg A}$, which obviously excludes the property E_A and *vice versa.* The statement that « the system has the property E_A » will be called the *proposition* $A(S, \varphi)$. The concept of proposition will become very important in the next sections.

1·3. *The relation of commensurability.* – The correspondence between subspaces $A \leqslant \mathscr{H}$ and physical properties E_A or propositions $A(S, \varphi)$ becomes particularly apparent if the concept of *commensurability* is introduced into the lattice $L_{\mathfrak{q}}$ of subspaces. This can be done in the following way. We start from an orthocomplemented lattice L_o which fulfills the axioms (4.0)-(4.4). In this lattice we define a 2-place relation $K \subseteq L_o \times L_o$ (commensurability) by

$$(A, B) \in K \leftrightarrow A = (A \wedge B) \vee (A \wedge \neg B)$$

and denote it by $A \sim B$. It is easily seen that the partial-ordering relation R is below K, *i.e.*

$$R \subseteq K \subseteq L_o \times L_o .$$

In fact, it follows from $A \leqslant B$ by means of (2.3) that $A \leqslant A \wedge B$, which implies $A \leqslant (A \wedge B) \vee (A \wedge \neg B)$. The inverse, *i.e.* $(A \wedge B) \vee (A \wedge \neg B) \leqslant A$, is always true in L_o.

In L_o the relation K is not symmetric. The symmetry of the commensurability relation is in L_o rather a condition, which is equivalent to the quasi-modularity [6, 7]. In fact, we have the

Theorem. In an orthocomplemented lattice L_o the relation K is symmetric, *i.e.* $A \sim B \rightarrow B \sim A$, if and only if the lattice is quasi-modular, *i.e.* $L_q(5)$ holds.

The proof of this theorem is straightforward but lengthy and, therefore, will not presented here [6, 7].

This theorem illustrates the importance of the quasi-modular law $L_q(5)$. Since the commensurability as a physical relation between two properties of a system should always be symmetric, the quasi-modularity is in fact a very important condition which should be taken into account in every axiomatization of the lattice of subspaces.

The lattice L_q is not distributive, *i.e.* the distributive law $A \wedge (B \vee C) = (A \wedge B) \vee (A \wedge C)$ does not hold generally in the quasi-modular lattice L_q. However, if the elements B and C are commensurable with A, the distributivity can be demonstrated, *i.e.* we have the

Theorem. In the lattice L_q

$$B \sim A \ , \ C \sim A \rightarrow A \wedge (B \vee C) = (A \wedge B) \vee (A \wedge C) .$$

Proof. From the premises

$$B = (B \wedge A) \vee (B \wedge \neg A) \ , \ C = (C \wedge A) \vee (C \wedge \neg A)$$

it follows

$$A \wedge (B \vee C) = A \wedge \{(B \wedge A) \vee (B \wedge \neg A) \vee (C \wedge A) \vee (C \wedge \neg A)\} .$$

Applying the quasi-modular law to the right-hand side, we get

$$A \wedge (B \vee C) = (A \wedge B) \vee (A \wedge C)$$

completing the proof. In order to further explain the meaning of this theorem, it should be mentioned here that the relation of commensurability is closed under the operations $\wedge$, $\vee$ and $\neg$, *i.e.*

$$A \sim B \curvearrowright A \sim \neg B\,,$$

$$A \sim B\,, A \sim C \curvearrowright A \sim (B \wedge C) \text{ and } A \sim (B \vee C)\,.$$

From this closure property together with the above-mentioned theorem it follows that three elements $A, B, C \in L_q$, which are pairwise commensurable, will generate a sublattice $L(A, B, C) \subseteq L_q$ which is orthocomplemented and distributive, *i.e.* a Boolean sublattice $L_B(A, B, C) \subseteq L_q$. Conversely, given a Boolean sublattice $L_B \subseteq L_q$, any two elements $A, B \subseteq L_B$ will be commensurable, *i.e.* $A \sim B$.

Summarizing the results of this section, we arrive at the following statements: In the lattice L_q, the relation $K \subseteq L_q \times L_q$ fulfills four conditions

$K1)$ K is symmetric;

$K2)$ $R \subseteq K$;

$K3)$ if $S \subseteq L_q$ is a subset of elements with $S \times S \subseteq K$, then S generates a Boolean sublattice $L_B(S) \subseteq L_q$;

$K4)$ if $L_B \subseteq L_q$ is a Boolean sublattice, the elements of any subset $S \subseteq L_B$ are commensurable, *i.e.* $S \times S \subseteq K$.

Conversely, it can easily be shown [6] that due to these conditions the relation K is uniquely defined in L_q. Finally, it follows that an orthocomplemented lattice L_0 in which there exists a relation K satisfying $K1)$-$K4)$ is quasi-modular.

1·4. *The material quasi-implication.* – In a Boolean lattice the relation R can also be defined as an operation in the following sense. For any two elements $A, B \in L_B$ of a Boolean lattice L_B there exists an element $A \to B$ which satisfies the conditions

$$\text{I)}\quad A \wedge (A \to B) \leqslant B\,,$$

$$\text{II)}\quad A \wedge C \leqslant B \curvearrowright C \leqslant A \to B\,.$$

In a Boolean lattice the element $A \to B$ is uniquely determined by these conditions and is given by $A \to B = \neg A \vee B$. This element is related to the relation R by the condition

$$\text{III)}\quad A \leqslant B \leftrightarrow \bigvee \leqslant A \to B\,.$$

In the logical interpretation of the lattice L_B, the relation $\leqslant$ is called « implication » and correspondingly the operation $A \to B$ « material implication ».

In this interpretation, condition III) states that $A \to B$ is « true » if $A \leqslant B$ holds, and condition I) corresponds to the « modus ponens » law which is essential for any logical inference [8].

The existence of a material implication is very nearly connected with the distributive law. In fact, the following theorem holds: If in an orthocomplemented lattice L_0 for any two elements A, B the existence of an element $A \to B$ is required which satisfies I) and II), the distributive law is valid and, hence, the lattice is Boolean. Therefore, we find that the existence of an element $A \to B$, which is very important for the logical interpretation, is necessary and sufficient for the distributivity of the lattice considered.

A similar situation can be found in the lattice L_q. For a logical interpretation one must require the existence of an element $A \to B$ which satisfies at least the conditions I) and III). It is obvious that condition II), which is of minor importance for the logical interpretation, cannot be fulfilled, since in that case the lattice would be distributive. However, it turns out that a somewhat weaker condition II′) can be formulated which fulfills all the necessary requirements [9]. In fact, we have the

Theorem. In an orthocomplemented, quasi-modular lattice L_q for any two elements A, B there exists an element $A \to B$ which satisfies the conditions

$$\text{I)}\quad A \wedge (A \to B) \leqslant B\,,$$

$$\text{II}')\quad A \wedge C \leqslant B \curvearrowright \neg A \vee (A \wedge C) \leqslant A \to B\,,$$

and which is given by $A \to B = \neg A \vee (A \wedge B)$.

Proof.

I) $A \wedge (A \to B) = A \wedge (\neg A \vee (A \wedge B)) = A \wedge B \leqslant B$ using the quasi-modular law.

II′) From $A \wedge C \leqslant B$ it follows $A \wedge C \leqslant A \wedge B$ and hence $\neg A \vee (A \wedge C) \leqslant \leqslant \neg A \vee (A \wedge B) = A \to B$.

Furthermore, it follows from the conditions I) and II′) that $A \to B = = \neg A \vee (A \wedge B)$ satisfies also the condition

$$\text{III)}\quad \bigvee \leqslant A \to B \leftrightarrow A \leqslant B$$

in the lattice L_q. Since the element $A \to B = \neg A \vee (A \wedge B)$ fulfills conditions I) and III) and at least the weaker condition II′), it has almost the properties of the material implication in L_B, and will, therefore, be called *material quasi-implication.*

In addition to these statements, it can be shown that in the lattice L_q the material quasi-implication is even *uniquely* determined by the conditions I) and II′). The proof of this important fact is tedious and will not be presented here. It can be found in the literature [9].

In an orthocomplemented quasi-modular lattice, the conditions I) and II′) are not only fulfilled by the material quasi-implication, they are even the strongest condition which can be postulated for $A \to B$. In fact, we have the following

Theorem. An orthocomplemented lattice L_0 with the property that for any two elements $A, B \in L_0$ there exists an element $A \to B$ which satisfies conditions I) and II′) is quasi-modular, *i.e.* for arbitrary elements A, B, C we have

$$A \leqslant C\,,\;\; B \leqslant \neg C \curvearrowright C \wedge (A \vee B) = A\,.$$

Proof. From the premises we get $C \wedge A \leqslant A$ and $C \wedge B \leqslant A$ and with II′)

$$\neg\, C \vee (C \wedge A) \leqslant C \to A\,, \qquad \neg\, C \vee (C \wedge B) \leqslant C \to A\,.$$

On account of

$$A \leqslant A \wedge C \leqslant \neg\, C \vee (C \wedge A) \leqslant C \to A\,, \qquad B \leqslant \neg\, C \leqslant \neg\, C \vee (C \wedge A) \leqslant C \to A\,,$$

it follows

$$A \vee B \leqslant C \to A \quad \text{and} \quad C \wedge (A \vee B) \leqslant C \wedge (C \to A) \leqslant A\,,$$

completing the proof.

This theorem gives rise to an alternative axiomatization of L_q. Instead of the quasi-modular law $L_q(5)$ one could require for all pairs $A, B \in L_0$ the existence of an element $A \to B$ which satisfies the axioms $L_q(5.1) = \text{I})$ and $L_q(5.2) = \text{II}')$. This axiomatization is of particular interest for the logical interpretation of the lattice L_q.

2. – The logical interpretation of the lattice L_q.

2·1. *The relation between lattice theory and logic.* – In the framework of quantum theory there exists a model for the lattice L_q. The elements $A, B, \ldots \in L_q$ are interpreted as propositions $A(S) \rightleftharpoons P_A\varphi(S) = \varphi(S)$ about a quantum-mechanical system S in the state $\varphi(S)$. Therefore, the lattice L_q can be considered as a propositional lattice. Furthermore, there are some formal similarities between L_q and lattices which have logical calculi as models. As examples we mention here the Boolean lattice L_B of classical logic and the implicative lattice L_i of intuitionistic logic. Therefore, it could be conjectured that also a logical calculus (quantum logic) can be found which is a model for L_q.

It is obvious that not every lattice can be interpreted as a logical calculus. First of all, there are some formal (syntactic) requirements which must be fulfilled by the lattice. Since in logical systems the implication is often used in iterated form, for instance in the *modus ponens law*, it must be required that the partial-ordering relation $A \leqslant B$ of the lattice can also be defined as a 2-place operation, denoted by $A \to B$ [9]. This condition is fulfilled in a Boolean lattice L_B, where the *material implication* is given by $\neg A \vee B$ and, as we have seen in the last section, in the lattice L_q, where the *material quasi-implication* is equal to $\neg A \vee (A \wedge B)$. It is also fulfilled in the implicative lattice L_i of the intuitionistic or effective logic. A lattice is called *implicative* [8]—or *relatively pseudocomplemented* [3]—if for any two elements A, B there exists an element $A \to B$, the material implication, which satisfies the conditions

I) $A \wedge (A \to B) \leqslant B$,

II) $A \wedge C \leqslant B \curvearrowright C \leqslant A \to B$.

If there is a zero element $\bigwedge$ in L_i, an element $\neg A = A \to \bigwedge$—the pseudocomplement—can be defined which satisfies the relation $A \bigwedge \neg A \leqslant \bigwedge$ (law of contradiction) but not the relation $\bigvee \leqslant A \vee \neg A$ (*tertium non datur*). In this case, the element $A \to B$ cannot be expressed by the other operations $\wedge$, $\vee$, $\neg$, which is, however, not essential for the possibility of a logical interpretation.

More important is the question of the semantic interpretation of the lattice L_q. A Boolean lattice L_B can be interpreted by a two-valued truth function. It has been shown by GLEASON [10] and KAMBER [6] that neither a two-valued truth function nor a conveniently generalized truth function does exist on the lattice L_q. On the other hand, it is well known that also the lattice L_i of the intuitionistic logic cannot be interpreted by truth values [11]. In this case, it is the missing principle of excluded middle which makes a valuation of L_i by a two-valued truth function impossible.

However, an implicative lattice can be considered as a logical calculus if one takes into account a more general method—the operational method which makes use of the dialogic technique [12-14]. Although this interpretation cannot directly be applied to the lattice L_q, it turns out that a generalization of the dialogic method can be used as an interpretation of a lattice L_{qi}, which can be considered as the intuitionistic part of L_q, *i.e.* if one adds to the axioms of L_{qi} the « tertium non datur » $\bigvee \leqslant A \vee \neg A$, the resulting lattice is isomorphic to L_q [15-19]. Finally, it can be shown that the dialogic interpretation can be extended also to the lattice L_q [20].

In order to demonstrate the possibility of a logical interpretation of L_{qi} and of L_q by means of dialogs, we proceed in the following way: We start with time-dependent (material) propositions about a quantum-mechanical system which can be tested by a measuring process. In the next step we define time-independent connected propositions by the possibilities of proving these propositions in a dialog. Already in this stage of the discussion, the mutual incommensurability of quantum-mechanical propositions becomes important, and must be incorporated into the detailed rules of a dialog-game. The rules of this *material* (*quantum*) *dialog-game* consist essentially of a concise formulation of the conditions of the possibilities of proving connected material propositions in a dialog. Starting from these rules, one can establish a *formal* (*quantum*) *dialog-game* which allows for the construction of all connected propositions which can be proved irrespective of the material propositions contained in it. These *logical propositions* can be summarized in the calculus Q_{eff} of *effective quantum logic*, which turns out to be a model for the lattice L_{qi} mentioned above.

2·2. *Elements of a language of quantum physics.* – We start from a quantum-mechanical system S (atom, nucleus, elementary particle) the properties of which can be tested by experiments. A proposition $a = a(S, t)$ which asserts that the system S at the time t has a certain measurable property will be called an *elementary proposition*. Elementary propositions will be denoted by $a, b, c \ldots$. We will assume that an experimental procedure M is known which can be considered as a *proof* of the proposition a, and another procedure $\overline{M}$ which can be considered as a *disproof*. Propositions which have these properties are said to be *proof-definite* and *disproof-definite*, respectively. We use the following notation:

$\vdash a$ proposition a has been proved by M,

$\vdash\!\!\!\top a$ proposition a has been disproved by $\overline{M}$.

The concepts of proof and disproof must be defined such that there is no system S for which the proposition considered can be proved and disproved at the same time.

For a proposition a which is as well proof-definite as disproof-definite a *counter-proposition* $\bar{a}$ can be defined such that

$$\vdash a \leftrightarrow \vdash\!\!\!\top \bar{a}\,, \qquad \vdash \bar{a} \leftrightarrow \vdash\!\!\!\top a\,.$$

As an example we consider the proposition a which corresponds to the projection operator P_a with eigenvalues 0 and 1. If φ is the state of the system, we have the relations

$$\vdash a \leftrightarrow P_a \varphi = \varphi \leftrightarrow (1 - P_a)\varphi = 0 \leftrightarrow \vdash\!\!\!\top \bar{a}\,,$$

$$\vdash \bar{a} \leftrightarrow P_{\bar{a}} \varphi = \varphi \leftrightarrow (1 - P_{\bar{a}})\varphi = 0 \leftrightarrow \vdash\!\!\!\top a\,,$$

which show that the counter-proposition $\bar{a}$ corresponds to the projection operator $P_{\bar{a}} = 1 - P_a$.

In this example the propositions a and $\bar{a}$ have the additional property that for every system an experimental test gives a well-defined result, that is $\vdash a$ or $\vdash \bar{a}$. Propositions which have this property will be called *value-definite*. In the following we will assume that elementary propositions a, b, c are always value-definite.

In the next step we define compound propositions by means of dialogs [15-18]. Since dialogs serve as proof procedures for compound propositions, these propositions are said to be *dialog-definite*. A dialog is a formalized kind of discussion between two participants, the proponent P, who asserts a certain proposition, and the opponent O, who attempts to refute this assertion. Without specifying how arguments are to be handled in the course of the dialog, the *concept of dialog* can be defined operationally by some rules which explain how a dialog should be carried out and which will be called *frame-rules* [14, 18]:

*F*1) At the beginning of the dialog the proponent P asserts the initial argument. In this way the initial position of the dialog is established.

*F*2) The opponent O attempts to refute this assertion. The dialog then consists of a series of arguments which are stated in turn by the two participants P and O, and which obey certain rules.

*F*3) Arguments are either attacks on or defences of previous arguments, but not both.

*F*4) *a*) The participants have the right to invoke an attack at any position of a dialog.

b) Having been attacked, the participants are obliged to defend in the reverse succession of the respective attacks, at the latest when there is no opportunity of attack left, *i.e.* the latest obligation has to be performed first.

*F*5) If one of the participants has no argument to continue, he loses the dialog. In this case, the other one wins and the final position of the dialog is established.

In the framework of the dialog defined by *F*1)-*F*5), arguments are either attacks on or defences of certain propositions. The manner in which a proposition can be defended or attacked depends on the possibilities of proving or disproving this proposition. These possibilities will be formulated for various types of propositions by means of *argument-rules* which specify and complete the frame-rules of the dialog. For elementary propositions a, b, c the possibilities of proving or disproving have already been mentioned and can be summarized in the following table:

Elementary proposition	Possibility of attack	Possibility of defence
a	$a?$	$a!$
$\bar{a}$	$\bar{a}?$	$\bar{a}!$

By $a?$ we denote the challenge to prove a, by $a!$ the successful proof of a. The proof of $\bar{a}$ is given by a disproof of a.

Compound propositions are not proof-definite but dialog-definite. This means that compound propositions are *defined* by the possibilities of attack and defence, *i.e.* by the respective argument-rule. Here we consider those compound propositions which correspond to the logical connectives:

Connective			Attack	Defence
conjunction	$a \wedge b$	(a and b)	1? 2?	$a!$ $b!$
disjunction	$a \vee b$	(a or b)	?	$a!$ $b!$
material implication	$a \rightarrow b$	(a then b)	$a?$	$b!$
negation	$\neg a$	(not a)	a	

The colloquial denotations « a and b », « a or b », « a then b » and « not a » are added here only for mnemonic support. They are without any influence on the use of the connectives in a dialog, which is governed exclusively by the argument-rule formulated in the above table.

In order to illustrate the meaning of this argument-rule we consider, for instance, the material implication $a \rightarrow b$. If the proponent asserts the proposition $a \rightarrow b$, then he assumes the obligation, in the case that a can be proved by the opponent, of justifying b. Hence a dialog in which P successfully defends the proposition $a \rightarrow b$ reads:

	O	P
0	[ ]	$a \rightarrow b$
1	$a(0)$	[ ]
2	$a!\langle 1\rangle$	$a?(1)$
3	[ ]	$b\ \langle 0\rangle$
4	$b?(3)$	$b!\langle 3\rangle$

Here we have used a schematic representation of the dialog by means of two columns. The left column consists of all arguments of O, the right column of all arguments of P. The succession of arguments is enumerated by rows which begin at zero. The argument of P in row zero is the initial argument, whereas the argument of O in row zero is the empty argument [ ]. If an argument is an attack against a statement which has been asserted in row x, we write (x) on the right-hand side of the argument. If an argument is a defence of an assertion which has been stated (in row y) and attacked later in the dialog, we write $\langle y \rangle$ on the right-hand side of the defending argument. If a participant does not defend, the empty argument is placed in the respective row, and he continues in the next row.

2·3. *Commensurability and incommensurability.* – In the above dialog about $a \to b$ we have assumed that the proponent P can actually perform the proof of b in $P4$, and thus is successful in the defence of the initial argument $a \to b$ up to row 4. There is, however, no limitation on the length of the dialog and the opponent could attack again assuring proposition a, like in row $O1$. One could dispense with a repetition of the experimental proof for b only if there would be any guarantee that the result $b!$ of the b-measurement in $P4$ is still valid after the new a-attack of the opponent and therefore still *available* for the proponent. However, the *availability* of a proposition once proved in the course of the dialog is no longer guaranteed if other propositions have been proved by experiments at a later stage of the dialog. We know from quantum mechanics that, due to the incommensurability of measurable properties, the result of the measurement of a certain observable 0_a can be completely destroyed by the measurement of another observable 0_b which is not commensurable with 0_a.

However, at the present stage of our discussion we do not like to refer to any empirical knowledge. Therefore, as a next step we introduce two *availability propositions* $k(a, b)$ and $\bar{k}(a, b)$ which state whether a given proposition a is still available after an experimental test of proposition b, or not. The availability propositions must be proved or disproved outside of the dialog by procedures which follow from the precise definition of the propositions $k(a, b)$ and $\bar{k}(a, b)$ [20].

The propositions $k(a, b)$ and $\bar{k}(a, b)$ will be called *commensurability* and *incommensurability*, respectively, and will be defined in the following way:

$k(a, b)$: Two propositions a and b are said to be *commensurable*—$k(a, b)$ has been proved—if in a given system S the propositions a and b can be measured in arbitrary sequence without thereby influencing the result of the measurement.

$\bar{k}(a, b)$: Two propositions a and b are said to be *incommensurable*—$\bar{k}(a, b)$ has been proved—if in a given system S the result of a measurement of the one proposition can be changed by a measurement of the other proposition.

It is a consequence of these definitions that $k(a, b)$ and $\bar{k}(a, b)$ are counter-propositions, *i.e.* $\vdash k(a, b) \leftrightarrows\, \mid\!\!\!\top \bar{k}(a, b)$ and $\vdash \bar{k}(a, b) \leftrightarrows\, \mid\!\!\!\top k(a, b)$. Similar as for elementary propositions, the possibilities of proving or disproving propositions $k(a, b)$ and $\bar{k}(a, b)$ in a dialog can be summarized in the following table:

Available propositions	Attack	Defence
commensurability $k(a, b)$	$k(a, b)?$	$k(a, b)!$
incommensurability $\bar{k}(a, b)$	$\bar{k}(a, b)?$	$\bar{k}(a, b)!$

Since the proofs for elementary propositions and for commensurabilities have to be performed outside the dialog, these two kinds of propositions will be called *material propositions*. The possibilities for attacking and defending material propositions, *i.e.* $a?$, $k(a,b)?$, $\bar{k}(a,b)?$ and $a!$, $k(a,b)!$, $\bar{k}(a,b)!$, respectively, are not arguments in the strict sense of the frame-rules; the following weakening of F 4*b*) will be formulated here as an argument-rule for material propositions [18, 20]:

If a participant cannot defend against an attack of a material proposition (a, $k(a,b)$ or $\bar{k}(a,b)$), he may assume a previous obligation of defence.

The reason for this liberalization of the dialog rules is based on the fact that neither the proponent nor the opponent has any influence on the proof-result of a material proposition. Therefore, one should allow the participants to check the truth of several propositions by the respective proof-attempts.

If once the popositions $k(a,b)$ and $\bar{k}(a,b)$ have been defined, the infinite dialogs about $a \wedge b$, $a \vee b$ and $a \rightarrow b$ can then be replaced by finite dialogs. This can be achieved if commensurabilities are introduced into the *argument-rule* as additional possibilites of attack and defence in the following way [20]:

Connective	Attack	Defence
$a \wedge b$	$1?$, $2?$, $k(a,b)?$	$a!$, $b!$, $k(a,b)!$
$a \vee b$	$?$	$a!$, $b!$, $\bar{k}(a,b)!$
$a \rightarrow b$	a, $k(a,b)?$	$b!$, $k(a,b)!$
$\neg a$	a	—

As an illustration of this reformulation of the argument-rule, we consider again a successful dialog about $a \rightarrow b$:

	O	P
0	[ ]	$a \rightarrow b$
1	$a(0)$	[ ]
2	$a!\langle 1\rangle$	$a?\,(1)$
3	[ ]	$b\langle 0\rangle$
4	$k(a,b)?\,(0)$	$k(a,b)!\langle 0\rangle$
5	$b?\,(3)$	$b!\langle 3\rangle$

Before challenging b the opponent asks for $k(a,b)$, which is proved by the proponent in $P4$. After the defence $b!$ in $P5$ the dialogic proof of $a \rightarrow b$ can be considered as complete, since according to the proof of $k(a,b)$ the result $b!$ in row $P5$ will be available in the further course of the dialog, even if the opponent attacks again proving proposition a.

From the redefinition of the connectives $a \wedge b$, $a \vee b$ and $a \to b$ by means of *finite* dialogs which contain the availability propositions $k(a, b)$ and $\bar{k}(a, b)$, the following result concerning the time dependence of compound propositions can be obtained. Since elementary propositions are proved in a dialog in a certain temporal sequence, one might argue that a compound proposition, *e.g.* $a \to b$, is related to two time values t_a and t_b, respectively, the order of which might be important. However, since in the course of a successful dialog the commensurability $k(a, b)$ must be justified, the temporal order of the proofs of propositions a and b is not longer essential. Therefore, the time difference $\delta t = t_b - t_a$ between the proofs could—in principle—be made to vanish. For this reason the logical connectives will be considered here as compound propositions, the elementary subpropositions of which are related to the same time-value—just as it is the case in ordinary logic. A detailed exposition of this problem can be found in ref. [21].

2·4. *The material dialog-game.* – Elementary propositions $a, b, c, \ldots$, availability propositions $k(a, b)$ and $\bar{k}(a, b)$ and the logical connectives $a \wedge b$, $a \vee b$, $a \to b$ and $\neg a$ are the basic elements of a scientific language of quantum physics. We now extend the set S_e of elementary propositions first by incorporating arbitrary iterations of the logical connectives. The resulting set S of propositions is then given by the following inductive scheme:

$$\left.\begin{array}{l} \text{I) Elementary propositions } a \in S_e \text{ are propositions.} \\ \text{II) If } A \text{ and } B \text{ are propositions, } i.e.\ A \in S,\ B \in S, \text{ then } A \wedge B,\ A \vee B, \\ \quad A \to B,\ \neg A \text{ are propositions.} \end{array}\right\} S$$

From an algebraic point of view, the logical connectives are 1- and 2-place operations on the set S. In a second step, we extend the set S by incorporating also the commensurabilities and its iterations. The extended set S^* is then given by the inductive scheme

$$\left.\begin{array}{l} \text{I) Elementary propositions } a \in S_e \text{ are propositions.} \\ \text{II) If } A \text{ and } B \text{ are propositions, } i.e.\ A \in S^*,\ B \in S^*, \text{ then } A \wedge B, \\ \quad A \vee B,\ A \to B,\ \neg A,\ k(A, B), \bar{k}(A, B) \text{ are propositions.} \end{array}\right\} S^*$$

Apart from the logical connectives on the set S^*, also the commensurabilities $k(A, B)$ and $\bar{k}(A, B)$ are 2-place operations on the set S^*. The reason why we have carried out the extension of S_e in two steps will become clear in the course of the following considerations.

We next introduce the concept of a material dialog-game. The *material dialog-game* D_m or D^*_m is a proof procedure for propositions $A \in S$ or $A \in S^*$, respectively. Since compound propositions $A \in S$ contain material propositions, *i.e.* elementary propositions and commensurabilities k and $\bar{k}$, which have to be proved outside of the dialog, this game will be called *material* dialog-game or more precisely material quantum dialog-game. Within the frame-rules $F1$)-$F5$) of the dialog, material dialogs are defined by the possibilities of proving or disproving propositions $A \in S$ or $A \in S^*$, which will be formulated in the *argument-rules of the material (quantum) dialog-game.* These rules will be formulated here in respect to the extended set S^*. However, for the formulation of a formal quantum logic, it will turn out to be sufficient to consider propositions $A \in S$ and the dialog-game D_{m}. Therefore, in the following investigations we will restrict ourselves essentially to S and D_{m}, respectively. The extended material dialog-game D^*_{m} will be needed only for the derivation of a special result in sect. **4**.

We are now going to formulate the argument-rules A_{m} of the material dialog-game. These rules summarize the results of the preceding section in a systematic form: In $A_{\mathrm{m}}(1)$, $A_{\mathrm{m}}(2)$ and $A_{\mathrm{m}}(3)$ the possibilities of proving material propositions are formulated. In $A_{\mathrm{m}}(1)$ the counter-proposition $\bar{a}$ which belongs to an elementary proposition is no longer mentioned. The reason is that within the scope of the material dialog-game and for a value-definite elementary proposition a the respective counter-proposition $\bar{a}$ can be completely replaced by the negation $\neg a$. The proof of this *dialog-equivalence* is easy and will not be given here (cf. ref. [19]). The rules $A_{\mathrm{m}}(4)$ a)-d) are concerned with the possibilities of proving the logical connectives by means of the availability propositions k and $\bar{k}$. The reason for this generalization of the usual definition of the connectives is the potential incommensurability of quantum-mechanical propositions, which must be taken into account already in the definitions of the logical connectives.

The essential difference between the argument-rules for quantum-mechanical propositions and the respective rules for propositions of classical physics is the *restricted availability* of propositions expressed in the rule $A_{\mathrm{m}}(5)$. According to this rule, the availability of a proposition A which has been asserted at an earlier stage of the dialog must always be checked by testing the commensurabilities $k(A, B_i)$ ($i = 1, \ldots, n$) of this proposition A and all propositions B_i which have been asserted later in the dialog. The reason for this restriction is obvious: In case A and B_k are incommensurable, the result of a successful proof of A, which consists of a series of measurements, could be completely destroyed by measurements which belong to the proof of B_k.

In the argument-rules of the material (quantum) dialog-game no special empirical assumptions have been made concerning the commensurability of propositions. Instead, we have rather incorporated into the dialog-rules an additional testing procedure which decides in every case about the availability of a proposition. For that reason the material dialog-game for quantum-mechanical propositions presented here is not restricted to propositions about quantum-mechanical systems, but can be applied as well to propositions of classical physics.

The argument rules of the material dialog game D^*_m.

$A_m(1)$	Elementary propositions	Attack	Defence
	a	$a?$	$a!$
$A_m(2)$	Availability propositions	Attack	Defence
$a)$	$k(A, B)$	$k(A, B)?$	$k(A, B)!$
$b)$	$\bar{k}(A, B)$	$\bar{k}(A, B)?$	$\bar{k}(A, B)!$
$A_m(3)$	If a participant cannot defend against an attack of a material proposition (*i.e.* a, $k(A, B)$, $\bar{k}(A, B)$), he may assume a previous obligation of defence.		
$A_m(4)$	Connectives	Attack	Defence
$a)$	$A \wedge B$	$1?$, $2?$, $k(A, B)?$	A, B, $k(A, B)!$
$b)$	$A \vee B$	$?$	A, B, $\bar{k}(A, B)!$
$c)$	$A \rightarrow B$	A, $k(A, B)?$	A, $k(A, B)!$
$d)$	$\neg A$	A	—
$A_m(5)$	If a participant attacks a proposition A, the other participant may check the availability of A by using the arguments $k(A, B)?$. B is any proposition asserted in the dialog after A, and the attack $k(A, B)?$ may be used once for every B.		

By means of the argument rules $A_m(1)$-$A_m(5)$ material dialogs can be performed in order to prove or disprove compound propositions. The question whether a certain proposition A can successfully be defended in a material dialog will —in general—depend on the truth or falsity of the elementary propositions contained in A. There are, however, a few compound propositions which can be defended in a material dialog, irrespective of the material propositions contained in it. In order to illustrate this important fact, let us consider as an example the proposition $A \equiv (a \wedge b) \rightarrow (a \rightarrow b)$ and its dialog in D_m.

	O	P
0	[ ]	$(a \wedge b) \rightarrow (a \rightarrow b)$
1	$a \wedge b$ (0)	[ ]
2	$a! \langle 1 \rangle$	$1?$ (1)
3	$b! \langle 1 \rangle$	$2?$ (1)
4	$k(a, b)! \langle 1 \rangle$	$k(a, b)?$ (1)
5	[ ]	$a \rightarrow b \langle 0 \rangle$
6	$k(a, b)?$ (5)	$k(a, b)! \langle 5 \rangle$
7	$a(5)$	$b \langle 5 \rangle$
8	$b?$ (7)	$b! \langle 7 \rangle$

Here we have assumed that the opponent can prove the propositions a, b and $k(a, b)$ in $O2$, $O3$ and $O4$, respectively. This is the situation most disadvantageous for P. Nevertheless, P wins the dialog. He can prove the propositions $k(a, b)$ and b, which are necessary for a successful defence, in $P6$ and $P8$, since just these propositions have been shown to be valid previously by the opponent. Furthermore, for the proof of b in $P8$, availability problems do not occur. The proposition a which is asserted in $O7$ does not influence the availability of proposition b which has been proved in $O3$—since also the availability proposition $k(a, b)$ has been proved in $O4$. The proponent can either refer to the result of row $O3$ or he can repeat the proof of b, for which he gets the same result. Hence, P wins the dialog. If, on the other hand, O cannot prove any one of the propositions a, b, or $k(a, b)$, he loses the dialog already in row 2, 3 or 4, respectively.

This example shows that there are in fact compound propositions which can be defended in the material dialog-game D_{m} irrespective of the elementary propositions contained in it. In the next section, formal procedures will be established which allow for the construction of all propositions of this kind.

3. – The effective quantum logic.

3·1. *Formally true propositions.* – In the last section we have seen that, within the scope of the material quantum dialog-game D_{m}, there are compound propositions which can be successfully defended in a material dialog irrespective of the material propositions contained in it. Propositions of this kind will be said to be *formally true*. In principle one could investigate the formal truth of a given proposition by considering all possibilities of a dialog in respect to the proof-results of the elementary propositions. However, in general it would be difficult to examine all such possibilities and to formalize this procedure in an appropriate manner. Therefore, we will chose a different way: We formulate the general possibilities of proving *formally true* propositions in a dialog, using considerations which have been applied already in the example mentioned above. These possibilities will then be summarized in *formal argument-rules* which constitute the so-called *formal quantum dialogs*.

The general validity of a formally true proposition means that P wins the dialog independently of the evidence for the elementary propositions. Therefore, a proposition is formally true if P has a strategy of success even in the situation which is most unfavourable for him. This is the case if elementary propositions asserted by the opponent can always successfully be proved by him. Consequently, in the formal dialog game we will assume that O can in fact prove all proposed elementary propositions. On the other hand, a strategy of success can only exist if the proponent can never be committed by the opponent to prove a proposition explicitly—since the proposition in question could be false and in that case the proponent would lose the dialog. This is the case if P asserts only such propositions which he can take over from the opponent who can then not question them. Hence we will assume that P may only assert elementary propositions that have already been maintained by O in the dialog.

The proponent should, however, be allowed to refer to an elementary proposition which has been asserted previously by the opponent *only* if the respective proposition is *available*. As mentioned above, this means that the proof-result obtained previously is still valid in the respective position of the dialog. The availability of a proposition a is given if between the proof and its citing no proposition incommensurable with a has been proved.

These aspects can now be used in order to formulate *argument-rules* of the *formal quantum dialog-game* D_f. Again these rules characterize the possibilities of proving some kind of propositions, that is the *formally true propositions*. The argument rules read:

$A_f(1)$ *a*) Elementary propositions are not attackable.

b) O is allowed to state elementary propositions in every position of a dialog; P is allowed to state elementary propositions only if they have been asserted by O previously and if they are still available in the respective position of the dialog.

Similarly, in the formal quantum dialog-game one has to take into account restrictions in the proponent's possibilities to attack a proposition, which again come from the incommensurability of quantum-mechanical propositions. After having stated a proposition in a dialog, a participant is committed to defend this proposition against an attack only as long as it is still available. Therefore, we must restrict P's possibilities to attack previous propositions of O by the following argument rule:

$A_f(2)$ P is allowed to attack propositions of O only as long as they are available in the respective position of the dialog.

At the present stage of our discussion, the availability of a proposition A which has been asserted by the opponent can only be demonstrated by proving all the material propositions $k(A, B_i)$ $(i = 1, \ldots, n)$, where B_i are the propositions which have been stated between the proof of A and its citing. However, in the next subsection we will eliminate also the proofs of the availability propositions $k(A, B)$, thus eliminating the proofs of all material propositions in the formal dialog-game D_f.

For the logical connectives we use the same dialogic definitions in D_f as in the material quantum dialog-game D_m. However, we will not consider here the availability propositions as possibilities of attacks and defences, as has been done in $A_m(4)$, since the material propositions $k(A, B)$ and $\bar{k}(A, B)$ will be eliminated in the formal dialog-game. Instead we go back to the original definition of the logical connectives, the dialogs of which are of course infinite. Therefore, we get the argument-rule:

$A_f(3)$	Connectives	Attacks	Defences
a)	$A \wedge B$	1? 2?	A, B
b)	$A \vee B$	?	A, B
c)	$A \to B$	A	B
d)	$\neg A$	A	—

The argument rule $A_f(2)$ which restricts the possibilities of attacks to a proposition in a dialog is somewhat too restrictive and must be weakened if one really wants to obtain *all* propositions which are formally true. The reason is simply that the validity of a proposition A, which has been proved at a certain stage of the dialog, is not *completely* destroyed by the proof of a proposition B, which is not commensurable with A, but in general only *partially*. This means that, usually, there is still some information left about A which can be used in the further course of the dialog. However, the rules of the formal dialog-game which are concerned with these *partial commensurabilities* are rather complicated and will not be needed in our further discussion. Therefore, we will not formulate these rules here, but refer instead to the literature [18].

3·2. *The formal dialog-game* D_f. – In the material quantum dialog-game D_m there are two kinds of material propositions, the elementary propositions and the commensurabilities k and $\bar{k}$. In the formal quantum dialog-game D_f the proof of material propositions, which have to be performed outside of the dialog, must be eliminated. The proofs of elementary propositions have already been eliminated in the last subsection by establishing the argument rules $A_f(1)$ and $A_f(2)$. Therefore, dialogs which use these rules instead of the respective rules of the material dialog-game are independent of elementary propositions. However, there are commensurability propositions still contained in the dialogs, the proof of which has to be carried out in proof procedures outside of the dialog. Consequently, in the next step, commensurability propositions $k(A, B)$ which depend on the elementary propositions of which A and B consist must be eliminated in the formal quantum dialog-game.

There are, however, certain commensurabilities $k(A, B)$ which can be demonstrated irrespective of the elementary propositions contained in A and B. It is obvious that these *formal commensurabilities* must not be eliminated but rather incorporated into the rules of the formal dialog-game. On the other hand, *formal incommensurabilities* do not exist. In order to demonstrate a formal commensurability $k(A, B)$, one has to recall the definition of $k(A, B)$ mentioned above: Material dialogs about A and B are carried out in turn; if it is certain that all material dialogs about A and B have always the same result, the formal commensurability $k(A, B)$ is proven. Using this definition one finds that the commensurabilities $k(A, A)$, $k(A, k(A, B))$ and $k(A, k(B, A))$ are formally true. Furthermore, it can be shown that $k(A, A \to B)$ is true. In addition to these formal commensurabilites, some relations between commensurabilities can be shown to be true, irrespective of the proof-results of the elementary propositions, *e.g.* $k(A, B) \curvearrowright k(\neg A, B)$. These relations will be used here as rules of a calculus K, the beginnings of which are given by the formal commensurabilities mentioned above. In this calculus the rules will be denoted by the double arrow $\Rightarrow$; several premises will be connected by a double comma ,,. The calculus K of formal commensurabilities then reads:

$K(1.1)$	$k(A, A)$,	
$K(1.2)$	$k(A, k(A, B))$,	
$K(1.3)$	$k(A, k(B, A))$,	
$K(1.4)$	$k(A, A \to B)$,	
$K(2)$	$k(A, B) \Rightarrow k(B, A)$,	
$K(3)$	$\vdash_{D_f} A \to B \Rightarrow k(A, B)$,	
$K(4)$	$k(A_1, B),, k(A_2, B) \Rightarrow k(A_1 * A_2, B)$	with $* \in \{\wedge, \vee, \to\}$,
$K(5)$	$k(A, B) \Rightarrow k(\neg A, B)$.	

This calculus may be considered as a formal tool which, starting from the beginnings $K(1.1)$-$K(1.4)$, allows for the construction of new formal commensurabilities. If a formal commensurability $k(A, B)$ can be derived in K, we write $\vdash_K k(A, B)$. It is not stated here that this calculus K is complete in any sense. However, for the following considerations the calculus K will turn out to be sufficiently large.

Formal commensurabilities have to be proved by a derivation in the calculus K, *i.e.* by a proof procedure outside of the dialog. Therefore, within the framework of the formal dialog-game, formal commensurabilities must still be considered as material propositions. However, it can be shown that formal commensurabilities are in some sense equivalent to dialog-definite compound propositions. This means that the proof of a material commensurability proposition can be eliminated in any dialog. In fact we have the following

Theorem.

a) If the formal commensurability $k(A, B)$ can be proved, *i.e.* $\vdash_K k(A, B)$, there then exists a strategy of success for the dialog about the proposition $A \to (B \to A)$.

b) If there exists a strategy of success for the dialog about the proposition $A \to (B \to A)$, the obligation to prove $k(A, B)$ in a dialog can be circumvented.

The proof of this theorem is complicated and will not be presented here (cf. ref. [18]). The theorem can be generalized to serveral commensurabilities $k(A_1, B)$, $k(A_2, B)$, ... and to propositions $A_1 \wedge A_2 \ldots \to (B \to A_1 \wedge A_2 \ldots)$, respectively. We will not go into the details of this problem and refer to the literature.

Due to this theorem the material proofs of formal commensurabilities can now be completely eliminated by the dialogic proofs of certain compound propositions. In order to incorporate all formal commensurabilities which can be proved by means of the calculus K into the rules of the formal quantum dialog-game, we formulate the additional argument-rule:

$A_f(7)$ *a*) O is not allowed to attack the initial argument $A \to \to ((A \to B) \to A)$.

b) P is allowed to replace the initial argument $A \to (B \to A)$ by the new initial argument $B \to (A \to B)$.

c) P is allowed to replace the initial argument $(A_1 * A_2) \to (B \to \to (A_1 * A_2))$ by the new initial arguments $A_1 \to (B \to A_1)$ or $A_2 \to (B \to A_2)$ with $* \in \{\wedge, \vee, \to\}$, where O has to choose between the two possibilities.

d) P is allowed to replace the initial argument $\neg A \to (B \to \neg A)$ by the new initial argument $A \to (B \to A)$.

By means of this argument-rule all formal commensurabilities which can be derived in K are now incorporated in the formal quantum dialog-game D_f. Due to the special form $A \to (B \to A)$ of the proposition which is equivalent to $k(A, B)$, the commensurabilities $K(1.2)$, $K(1.3)$ and the rule $K(3)$ can be proved in a formal dialog and, therefore, are not mentioned in $A_f(7)$. The numbering of the argument rules is according to ref. [18]. As mentioned above, the argument-rules which are concerned with details of the availability problem, *i.e.* $A_f(4)$, $A_f(5)$ and $A_f(6)$, are not presented here.

3·3. *The calculus Q_{eff} of effective quantum logic.* – After having formulated the argument-rules of the formal quantum dialog-game, we are now going to establish a propositional calculus with the aid of which all those propositions can be derived, which can be successfully defended in the formal quantum dialog-game D_f. This calculus will be called the *calculus of effective quantum logic Q_{eff}*. The formulae of Q_{eff} consist first of *beginnings*, *i.e.* assertions of the kind that $\vdash_{D_f} A$ is valid, and secondly of *rules*, *i.e.* implications of the kind $\vdash_{D_f} A \curvearrowright \vdash_{D_f} B$, which will be written in the calculus with the double arrow $\Rightarrow$. Beginnings $\vdash_{D_f} A$ must be proved by a successful defence of A in a formal dialog. The rule $\vdash_{D_f} A \Rightarrow \vdash_{D_f} B$ states that, if proposition A can be defended in a formal dialog, then proposition B can also be justified dialogically. For the proof of the rule $\underset{D_f}{\vdash} A \Rightarrow \underset{D_f}{\vdash} B$ the proposition A will, therefore, be presupposed by the opponent as a *hypothesis* before the dialog. The proponent has then to defend the proposition B in a formal dialog whereby he may refer to the hypothesis A, which has been accepted by the opponent. It is obvious that the referability of the hypothesis A is not restricted in any way.

For a formulation of the calculus Q_{eff} which is most convenient for a comparison with a lattice, it is useful to introduce two special propositions $\bigvee$ (truth) and $\bigwedge$ (falsity). The use of both of these propositions within the framework of the dialogic method shall be established in such a way that $\bigvee$ cannot be questioned by either participant of the dialog and that whoever maintains $\bigwedge$ shall have lost the dialog.

From this definition it follows that the propositions $A \to \bigvee$ and $\bigwedge \to A$ can be defended in a dialog for all propositions A. In fact, in the dialogs

	O	P
0	[ ]	$\bigwedge \to A$
1	$\bigwedge$	

	O	P
0	[ ]	$A \to \bigvee$
1	A	$\bigvee$

P has a strategy of success. The first dialog is lost by O since he attacks by asserting $\bigwedge$. The second dialog is lost by O since he cannot attack $\bigvee$.

For the formulation of the calculus of effective quantum logic, it is useful, in addition to the operation $\wedge$, $\vee$, $\to$, $\neg$ on the set S of propositions, to define a 2-place relation $R \leqslant S \times S$ by

$$A \leqslant B \rightleftharpoons \underset{D_f}{\vdash} A \to B\ .$$

This relation will be called « implication » and must be distinguished from the operation $A \to B$, denoted here as « material implication ». According to the definition, the relation $A \leqslant B$ between the propositions A and B holds if and only if the proposition $A \to B$ can be defended in a formal dialog. If for two propositions $A \leqslant B$ and $B \leqslant A$ are valid, we write $A = B$.

By using the relation $\leqslant$, it follows from the two dialogs mentioned above that we have the general validity of the implication

$$A \leqslant \bigvee\ , \qquad \bigwedge \leqslant A\ .$$

In particular it follows from these relations together with the commensurability rule $K(3)$ that $\bigvee$ and $\bigwedge$ are commensurable with all propositions. Furthermore, according to the dialogic definition of $\bigvee$ we find that for an arbitrary proposition A the statement $\underset{D_f}{\vdash} A$ is equivalent to $\underset{D_f}{\vdash} \bigvee \to A$. On the other hand, $\underset{D_f}{\vdash} \bigvee \to A$ is equivalent to the relation $\bigvee \leqslant A$. Therefore, a proposition A can be defended in a dialog if and only if $\bigvee \leqslant A$ holds, *i.e.*

$$\vdash_{Df} A \leftrightarrows \bigvee \leqslant A \ .$$

From the dialogic definition of $\bigwedge$ we find that for an arbitrary proposition A the statement $\vdash_{Df} \neg A$ is equivalent to $\vdash_{Df} A \to \bigwedge$. Since $\vdash_{Df} A \to \bigwedge$ can be replaced by $A \leqslant \bigwedge$, we obtain the result

$$\vdash_{Df} \neg A \leftrightarrows A \leqslant \bigwedge \ .$$

By means of these formal tools, the calculus of effective quantum logic can now be formulated in a most convenient manner. For the formula of the calculus we use combinations of the symbols $\bigvee, \bigwedge, A, B, C, \ldots$ with $\wedge, \vee, \to, \neg$ and $\leqslant$ and the bracket symbol. With these formulae we establish the rules for the derivation of implications $A \leqslant B$. For the designation of the rules we use the double arrow $\Rightarrow$ and the double comma ,, .

In order to illustrate the dialogic proofs for beginnings and rules of Q_{eff}, we will demonstrate here by means of formal quantum dialogs two examples, the first of which is the beginning (modus ponens law)

$$A \wedge (A \to B) \leqslant B \ .$$

The dialogic proof uses the formal commensurability

$$k(A, A \to B)$$

in $P4$, when P takes over proposition A from $O2$.

	O	P
0	[ ]	$(A \wedge (A \to B)) \to B$
1	$A \wedge (A \to B)$ (0)	[ ]
2	$A \langle 1 \rangle$	1? (1)
3	$A \to B \langle 1 \rangle$	2? (1)
4	$B \langle 3 \rangle$	A (3)
5	[ ]	$B \langle 0 \rangle$

The second example which we will discuss here is the rule

$$A \wedge C \leqslant B \Rightarrow A \to C \leqslant A \to B \ .$$

In the dialogic proof the opponent has to presuppose the hypothesis $A \wedge C \to B$ in row -1. In $P3$ the modus ponens law has been used, which, however, could also have been proved in a subdialog. It is essential for this proof that A and C need not be commensurable.

	O	P
-1	$A \wedge C \rightarrow B$	
0	[ ]	$(A \rightarrow C) \rightarrow (A \rightarrow B)$
1	$A \rightarrow C$ (0)	$A \rightarrow B \langle 0 \rangle$
2	A(1)	[ ]
3	[ ]	$A \wedge C$ (-1)
4	2? (3)	[ ]
5	$C \langle 1 \rangle$	A(1)
6	$B \langle -1 \rangle$	$C \langle 3 \rangle$
7		$B \langle 1 \rangle$

In case that A and C are commensurable, we have $C \leqslant A \rightarrow C$ and the rule can be strengthened to

$$A \wedge C \leqslant B \Rightarrow C \leqslant A \rightarrow B .$$

However, this rule, which is well known from ordinary effective logic, cannot be demonstrated in the formal quantum dialog-game. It is, in fact, the decisive difference between ordinary logic and quantum logic. This difference will be investigated from an algebraic point of view in the next section.

We will now present without further proofs the calculus Q_{eff} of effective quantum logic. The calculus reads:

$Q_{\text{eff}}(1.1)$ $A \leqslant A$,

$Q_{\text{eff}}(1.2)$ $A \leqslant B,, B \leqslant C \Rightarrow A \leqslant C$,

$Q_{\text{eff}}(2.1)$ $A \wedge B \leqslant A$,

$Q_{\text{eff}}(2.2)$ $A \wedge B \leqslant B$,

$Q_{\text{eff}}(2.3)$ $C \leqslant A,, C \leqslant B \Rightarrow C \leqslant A \wedge B$,

$Q_{\text{eff}}(3.1)$ $A \leqslant A \vee B$,

$Q_{\text{eff}}(3.2)$ $B \leqslant A \vee B$,

$Q_{\text{eff}}(3.3)$ $A \leqslant C,, B \leqslant C \Rightarrow A \vee B \leqslant C$,

$Q_{\text{eff}}(4.1)$ $A \wedge (A \rightarrow B) \leqslant B$,

$Q_{\text{eff}}(4.2)$ $A \wedge C \leqslant B \Rightarrow A \rightarrow C \leqslant A \rightarrow B$,

$Q_{\text{eff}}(4.3)$ $A \leqslant B \to A \Rightarrow B \leqslant A \to B$,

$Q_{\text{eff}}(4.4)$ $B \leqslant A \to B,, C \leqslant A \to C \Rightarrow B * C \leqslant A \to (B * C)$ with $* \in \{\wedge, \vee, \to\}$,

$Q_{\text{eff}}(5.0)$ $\bigwedge \leqslant A$,

$Q_{\text{eff}}(5.1)$ $A \wedge \neg A \leqslant \bigwedge$,

$Q_{\text{eff}}(5.2)$ $A \wedge B \leqslant \bigwedge \Rightarrow A \to B \leqslant \neg A$,

$Q_{\text{eff}}(5.3)$ $A \leqslant B \to A \Rightarrow \neg A \leqslant B \to \neg A$.

In this calculus $Q_{\text{eff}}(1.1)$, (2.1), (2.2), (3.1), (3.2), (4.1), (5.0), (5.1) are beginnings and $Q_{\text{eff}}(1.2)$, (2.3), (3.3), (4.2), (4.3), (4.4), (5.2), (5.3) are rules. These rules, which are used here for the formulation of the calculus, will be called *constitutive rules*. It is obvious that there are further rules which can be deduced from the constitutive rules by purely logical inferences. These rules are said to be *deducible*. However, for the following discussions it is not required to consider the totality of these rules, since the constitutive rules are sufficient in a certain sense.

It follows, from the dialogic proof procedure of the beginnings and the constitutive rules of Q_{eff}, that every implication derivable from Q_{eff} can be proved in a formal quantum dialog, *i.e.*

$$\underset{Q_{\text{eff}}}{\vdash} A \leqslant B \curvearrowright \underset{D_f}{\vdash} A \to B \, .$$

For this reason the calculus Q_{eff} will be called *consistent* with regard to the class of quantum dialogically provable implications. Furthermore, the calculus Q_{eff} is also *complete* with regard to this class, *i.e.* every quantum dialogically provable implication can be derived from Q_{eff}. Therefore, we also have the implication

$$\underset{D_f}{\vdash} A \to B \curvearrowright \underset{Q_{\text{eff}}}{\vdash} A \leqslant B \, .$$

The proof of this completeness property of Q_{eff} is rather complicated and cannot be presented here. It has been performed by STACHOW and can be found in ref. [18].

4. – The full quantum logic.

4·1. *The quasi-implicative lattice.* – The calculus Q_{eff} of effective quantum logic given by $Q_{\text{eff}}(1.1)$-(5.3) consists of beginnings and of constitutive rules. Deducible rules of the form $\alpha \Rightarrow \beta$ are then to be proved in the following way: After having added the implication α to the beginnings of Q_{eff}, the implication β must be derived within the calculus Q_{eff}. Therefore, for a deducible rule we use the notation $\alpha \underset{Q_{\text{eff}}}{\vdash} \beta$. Apart from deducible rules, there exists another kind of rules which in Q_{eff} must be considered as valid in a certain sense and which will be called *admissible rules*. A rule $\alpha \Rightarrow \beta$ is said to be admissible if, after the addition of $\alpha \Rightarrow \beta$ to the constitutive rules of Q_{eff}, no implication γ can be deduced which is not also deducible in Q_{eff} itself. For an admissible rule we use the notation $\underset{Q_{\text{eff}}}{\vdash} \alpha \Rightarrow \beta$.

The calculus of effective quantum logic Q_{eff}(1.1)-(5.3) has the special and important property that a rule which is admissible in respect to the constitutive rules can also be deduced in Q_{eff}, *i.e.* we have the

Theorem. In the calculus Q_{eff} of effective quantum logic

$$\underset{Q_{\text{eff}}}{\vdash} \alpha \Rightarrow \beta \text{ implies } \alpha \underset{Q_{\text{eff}}}{\vdash} \beta \, .$$

This property will be called *syntactic completeness*. It means that the beginnings and the constitutive rules of Q_{eff} are complete in the sense that by starting from these rules all admissible rules can be deduced. Therefore, it is no longer necessary to consider special proof procedures for the admissibility. The proof of this completeness theorem can be found in ref. [19].

The property of syntactic completeness has the consequence that the calculus Q_{eff} can be completely replaced by a *lattice*. In fact, if we consider the implication $A \leq B$ as a partial-ordering relation on the set S of propositions, the connectives $A \wedge B$ and $A \vee B$ as operations on S (infimum and supremum) and the double arrow $\Rightarrow$ as logical inference, then the rules Q_{eff}(1.1)-(3.3) are axioms which constitute a lattice. The rules Q_{eff}(4.1)-(5.3) further specialize this lattice by postulating a zero element $\wedge$ and by defining the operations $A \rightarrow B$ and $\neg A$. The lattice which is obtained in this way will be called *quasi-implicative lattice* and will be denoted by L_{qi}. (The reason for this terminology will become clear in the following subsection.) It is obvious that the calculus Q_{eff} can be *completely* replaced by the lattice L_{qi} only if the double arrow $\Rightarrow$ can actually be considered as a logical inference, *i.e.* if the calculus Q_{eff} is *syntactically complete*. Otherwise the admissible rules of Q_{eff} would get lost in replacing the calculus by a lattice, and the replacement would not be complete.

The *quasi-implicative* lattice L_{qi} is given by axioms which are formally equivalent to the rules Q_{eff}(1.1)-(5.3) and which will be denoted here by L_{qi}(1.1)-(5.3), respectively. The axioms of L_{qi} read:

The lattice L_{qi} is a set $S = \{A, B, \ldots\}$ of elements with a partial-ordering relation $\leq$ such that

L_{qi}(1.1) $\qquad A \leq A$,

L_{qi}(1.2) $\qquad A \leq B,\ B \leq C \curvearrowright A \leq C$.

Furthermore, for any two elements $A, B \in L_{\text{qi}}$ and in respect to the relation $\leq$ there exists an element $A \wedge B$ (infimum) and an element $A \vee B$ (supremum)

such that

$L_{qi}(2.1)$ $\quad A \wedge B \leqslant A$;

$L_{qi}(2.2)$ $\quad A \wedge B \leqslant B$;

$L_{qi}(2.3)$ $\quad C \leqslant A,\ C \leqslant B \frown C \leqslant A \wedge B$;

$L_{qi}(3.1)$ $\quad A \leqslant A \vee B$;

$L_{qi}(3.2)$ $\quad B \leqslant A \vee B$;

$L_{qi}(3.3)$ $\quad A \leqslant C,\ B \leqslant C \frown A \vee B \leqslant C$.

In addition to these axioms which constitute a lattice, for any two elements $A, B \in L_{qi}$ there exists an element $A \to B$ called *quasi-implication* which satisfies the axioms

$L_{qi}(4.1)$ $\quad A \wedge (A \to B) \leqslant B$;

$L_{qi}(4.2)$ $\quad A \wedge C \leqslant B \frown A \to C \leqslant A \to B$;

$L_{qi}(4.3)$ $\quad A \leqslant B \to A \frown B \leqslant A \to B$;

$L_{qi}(4.4)$ $\quad B \leqslant A \to B,\ C \leqslant A \to C \frown$

$$\frown B * C \leqslant A \to (B * C) \qquad \text{with } * \in \{\wedge, \vee, \neg\}.$$

Moreover, the lattice L_{qi} has a zero element $\bigwedge$ and for any element $A \in L_{qi}$ there exists an element $\neg A$ which will be called *quasi-pseudocomplement* and which satisfies the axioms

$L_{qi}(5.0)$ $\quad \bigwedge \leqslant A$,

$L_{qi}(5.1)$ $\quad A \wedge \neg A \leqslant \bigwedge$,

$L_{qi}(5.2)$ $\quad A \wedge C \leqslant \bigwedge \frown A \to C \leqslant \neg A$,

$L_{qi}(5.3)$ $\quad A \leqslant B \to A \frown \neg A \leqslant B \to \neg A$.

Furthermore, it can be shown that the *quasi-implication* $A \to B$ is uniquely defined by the axioms $L_{qi}(4.1)$-(4.4) and that the *quasi-pseudocomplement* $\neg A$ is also uniquely defined by $L_{qi}(5.0)$-(5.3). The proofs of these uniqueness theorems are tedious and will not be presented here. They can be found in ref. [16, 22].

4·2. *The relation between L_{qi} and the lattices L_i and L_q.*

4·2.1. The lattices L_{qi} and L_i. In this subsection we consider the relation between L_{qi} and the *implicative lattice* L_i, which was already briefly mentioned in subsect. **2·1**. If we replace the axioms $L_{qi}(4.1)$-(4.4) and $L_{qi}(5.0)$-(5.3) of the quasi-implicative lattice L_{qi} by the axioms

$L_i(4.1)$ $\quad A\wedge(A\to B)\leqslant B\ ,$

$L_i(4.2)$ $\quad A\wedge C\leqslant B \curvearrowright C\leqslant A\to B\ ,$

$L_i(5.0)$ $\quad \bigwedge \leqslant A\ ,$

$L_i(5.1)$ $\quad A\wedge\neg A\leqslant \bigwedge\ ,$

$L_i(5.2)$ $\quad A\wedge C\leqslant \bigwedge \curvearrowright C\leqslant \neg A\ ,$

we obtain an *implicative lattice* L_i with zero element. Here the element $A\to B$ is called *implication* and the element $\neg A$ the *pseudocomplement of* A. The implication $A\to B$ as well as the pseudocomplement $\neg A$ are uniquely defined by the the axioms $L_i(4.1)$-(4.2) and $L_i(5.0)$-(5.2), respectively.

That property of an implicative lattice which is most important for the following considerations is that in an implicative lattice the distributive law $A\wedge(B\vee C)\leqslant(A\wedge B)\vee(A\wedge C)$ is valid for all elements. Conversely, a finite distributive lattice or a completely distributive lattice (in which the distributive law holds even for an infinite series of elements) is implicative, *i.e.* the axioms $L_i(4.0)$-(5.2) can be deduced.

The proofs of these results can be found in ref. [3, 8].

In the lattice L_{qi} the commensurability relation $K\subseteq L_{qi}\times L_{qi}$ is given by

$$A\sim B \rightleftharpoons A\leqslant B\to A\ .$$

It follows from the axioms $L_{qi}(4.3)$, (4.4), (5.3) that this commensurability relation is symmetric and closed under the operations $\wedge$, $\vee$, $\to$ and $\neg$. Consequently, for three elements $A, B, C\in L_{qi}$ which are pairwise commensurable, *i.e.* for which the relations $A\sim B$, $A\sim C$, $B\sim C$ hold, it follows that also the relations

$$A\sim\neg B\ ,\quad A\sim\neg C\ ,\quad A\sim B*C\ ,\quad B\sim\neg A\ ,\quad B\sim\neg C\ ,\quad B\sim A*C\ ,$$
$$C\sim\neg A\ ,\quad C\sim\neg B\ ,\quad C\sim A*B\quad \text{with}\quad *\in\{\wedge,\vee,\to\}$$

are satisfied. This means that all elements of L_{qi} which are formed by A, B, C and arbitrary iterations of the operations $\wedge$, $\vee$, $\to$, $\neg$ are mutually commensurable.

Three elements $A, B, C\in L_{qi}$ which are pairwise commensurable thus generate a sublattice $L(A, B, C)\subseteq L_{qi}$ of elements which are mutually commensurable. Since for commensurable elements from $L_{qi}(4.2)$ and $L_{qi}(5.2)$ the stronger axioms $L_i(4.2)$ and $L_i(5.2)$, respectively, can be deduced, it follows that this sublattice is implicative. Therefore, for all elements of $L(A, B, C)$ the distributive law holds. On the other hand, since from $L_i(4.2)$ the relation $B \leqslant A \to B$ follows, the elements $A, B \in L_i$ of an implicative lattice are always commensurable. Therefore, summarizing these results, we obtain the following

Theorem.

a) If three elements $A, B, C \in L_{qi}$ are pairwise commensurable, then the lattice which is generated by these elements is an *implicative sublattice* $L_i(A, B, C) \subseteq L_{qi}$ of the lattice L_{qi}.

b) If L_i is an *implicative sublattice* of L_{qi}, *i.e.* $L_i \subseteq L_{qi}$, and if A, B, C ... are elements of L_i, then these elements are pairwise commensurable.

If all elements of a lattice L_{qi} are pairwise commensurable, *i.e.* if we can add the relation $A \leqslant B \to A$ to the axioms, the lattice in question is implicative. Furthermore, it can easily be shown that all axioms of L_{qi} are satisfied in an implicative lattice L_i [22]. Therefore, the *quasi-implicative* lattice L_{qi} is in fact a weakening of the *implicative* lattice L_i.

This circumstance also justifies the terminology used here.

4·2.2. The lattices L_{qi} and L_q. In this subsection we consider the relation between L_{qi} and the orthocomplemented quasi-modular lattice L_q, which was already discussed in detail in sect. **1**. Similarly as in the preceding subsection we will show here that L_{qi} is a weakening of the lattice L_q.

The lattice L_q is *orthocomplemented*, *i.e.* in addition to the axioms of L_{qi} the axioms

$L_q(4.2)$ $$\bigvee \leqslant A \vee \neg A\,,$$

$L_q(4.3)$ $$A = \neg\neg A\,,$$

$L_q(4.4)$ $$A \leqslant B \curvearrowright \neg B \leqslant \neg A$$

are valid in L_q. However, in the lattice L_{qi} it can easily be shown [16] that the relations

$$\bigvee \leqslant A \vee \neg A \curvearrowright A = \neg\neg A\,,$$

$$A \leqslant B \curvearrowright \neg B \leqslant \neg A$$

are satisfied. Therefore, if one adds nothing but the relation $\bigvee \leqslant A \vee \neg A$ (*tertium non datur*) to the axioms of L_{qi}, one arrives at an orthocomplemented lattice. The quasi-pseudocomplement $\neg A$ defined within the framework of L_{qi} then goes over into an orthocomplement.

Furthermore, the lattice L_q is quasi-modular, *i.e.* in addition to the axioms of L_{qi} also the quasi-modular law

$L_q(5)$ $$B \leqslant A,\ C \leqslant \neg A \curvearrowright A \wedge (B \vee C) = B$$

is valid in L_q. However, already in the framework of the lattice L_{qi} the *weak quasi-modularity*

$$B \leqslant A,\ C \leqslant \neg A \curvearrowright A \wedge (B \vee C) = B$$

can be shown to be valid [16, 22], whereby $\neg A$ is the quasi-pseudocomplement. Consequently, since, after the addition of the relation $\bigvee \leqslant A \vee \neg A$ to the axioms of L_{qi}, the quasi-pseudocomplement goes over into an orthocomplement, the quasi-modular law $L_q(5)$ is already valid in the lattice L_{qi} with the additional axiom $\bigvee \leqslant A \vee \neg A$. Therefore, if to the axioms of L_{qi} the *tertium non datur* $\bigvee \leqslant A \vee \neg A$ is added, the resulting lattice is orthocomplemented and quasi-modular, *i.e.* all axioms of L_q are fulfilled. (The quasi-implication $A \to B$ defined within the lattice L_{qi} goes then over into the element $\neg A \vee (A \wedge B)$.)

On the other hand, it can easily be shown that, in an orthocomplemented quasi-modular lattice L_q, the axioms $L_{qi}(1.1)$-(5.3) of L_{qi} are satisfied. Thus we arrive at the conclusion that the lattice L_{qi} is in fact a weakening of the lattice L_q from which is differs only by the *tertium non datur* law $\bigvee \leqslant A \vee \neg A$.

4·3. *The principle of excluded middle*. – The lattice L_{qi} differs from L_q by the axiom $\bigvee \leqslant A \vee \neg A$. Correspondingly in the calculus Q_{eff}, which is a model of L_{qi}, the principle of excluded middle, *i.e.* the implication $\bigvee \leqslant A \vee \neg A$, is not included. Therefore, if one would add this implication as a new beginning to the rules of Q_{eff}, one would obtain a calculus Q, which could serve as a logical interpretation of the lattice L_q. However, this calculus Q, which will be called the *calculus of full quantum logic*, cannot be justified on the basis of formal quantum dialogs alone. Within the framework of the formal dialog-game D_f there is no strategy of success for the principle of excluded middle, *i.e.* $\vdash_{D_f} A \vee \neg A$ cannot be proved.

However, as already mentioned in subsect. **2·2**, elementary propositions $a, \bar{a}$ are value-definite, *i.e.* there exists either a proof for a or a disproof for a which is equivalent to a proof for $\bar{a}$. If we use this property of value-definiteness, in the framework of the material dialog-game D_m there is a strategy of success for the proposition $a \vee \bar{a}$. In fact the dialog reads

	O	P
0	[ ]	$a \vee \bar{a}$
1	? (0)	$a \langle 0 \rangle$
2	a? (1)	[ ]
3	[ ]	$\bar{a} \langle 0 \rangle$
4	$\bar{a}$? (3)	$\bar{a}! \langle 3 \rangle$

Since by assumption there exists either a proof or a disproof of a, P has either a strategy of success for a in row 2, or a strategy of success for $\bar{a}$ in row 4. Therefore, P has a strategy of success for $a \vee \bar{a}$.

Furthermore, since for value-definite propositions $a, \bar{a}$ within the material dialog-game D_m the two propositions $\bar{a}$ and $\neg a$ are dialog-equivalent, it follows that there is also a strategy of success for $a \vee \neg a$, *i.e.* we have $\vdash_{D_m} a \vee \neg a$. Conversely, if there is a strategy of success for $a \vee \neg a$ within the material dialog-game D_m, the proposition a is value-definite. Therefore, the material property of value-definiteness of a is equivalent to the material truth of the proposition $a \vee \neg a$, which is also called the *principle of excluded middle.*

In the framework of the formal dialog-game, the value-definiteness of the elementary propositions a means that the opponent O possesses a material proof for $a \vee \neg a$ and the proponent P is always allowed to refer to $a \vee \neg a$ as a proposition of O. Therefore, in the formal dialog-game for all value-definite propositions $a, b, c \ldots$ the propositions $a \vee \neg a$, $b \vee \neg b$, $c \vee \neg c \ldots$, respectively, have to be presupposed as hypothesis in the dialog by the opponent. However, it can be seen by a detailed inspection of the dialoging method that, starting from the hypotheses $a \vee \neg a$ and $b \vee \neg b$, there is no strategy of success within D_f for the principle of excluded middle of the connectives $a \wedge b$ $a \vee b$, $a \to b$, *i.e.* the propositions $(a * b) \vee \neg (a * b)$, $* \in \{\wedge, \vee, \to\}$, cannot be defended successfully by P. In the framework of the calculus Q_{eff} this means that the rule

$$\vee \leqslant A \vee \neg A,, \vee \leqslant B \vee \neg B \Rightarrow \vee \leqslant (A * B) \vee \neg (A * B)$$

cannot be deduced from Q_{eff}(1.1)-(5.3).

The reason why the value-definiteness of the elementary propositions is not inherited by the connectives can be illustrated by the following argument. Let a and b be value-definite propositions with the possible truth values (0, 1). The truth value of a compound proposition $a * b$ will then be defined to be 1 or 0, if P wins or loses the material dialog about $a * b$, respectively. From the material dialog about $a \wedge b$ and the argument rule $A_m(4a)$ one thus derives the following « truth table »:

$a \wedge b$	(?)	0	0	0
a	1	1	0	0
b	1	0	1	0

If a and b have both the truth values 1, the truth value of $a \wedge b$ is not determined since it depends on the truth of the material proposition $k(a, b)$. It would be determined if, in addition to the elementary propositions, also the commensurability proposition $k(a, b)$ would be value-definite. This is, however, not the case. Since the material proposition $k(a, b)$ is defined by an infinite proof procedure, it will not have definite truth values even if the elementary propositions a and b are value-definite.

Remark. The argument that the lack of the value-definiteness of $k(a, b)$ is the reason for the missing principle of excluded middle in Q_{eff} is for the present merely heuristic. A conclusive justification of this argument will be given in the next subsection.

4·4. *The calculus of full quantum logic.* – Let us assume as a hypothesis that the material propositions $k(A, B)$ and $\bar{k}(A, B)$ are in fact value-definite. In order to investigate the consequences of this hypothetical assumption within the calculus Q_{eff}, we first extend the material dialog-game D_{m} and the formal dialog-game D_{f} to the dialog-games D^*_{m} and D^*_{f}, respectively, by incorporating the material propositions $k(A, B)$ and $\bar{k}(A, B)$. We thus arrive at the extended calculus Q^*_{eff} of effective logic which, in addition to the rules $Q_{\text{eff}}(1.1)$-(5.3) of Q_{eff}, contains also the following beginnings and rules [20]:

$Q^*_{\text{eff}}(6.1)$ $\quad A \wedge B \leqslant k(A, B)\,,$

$Q^*_{\text{eff}}(6.2)$ $\quad \bar{k}(A, B) \leqslant A \vee B\,,$

$Q^*_{\text{eff}}(6.3)$ $\quad \neg\,\bar{k}(A, B) = \neg\,\neg\,k(A, B)\,,$

$Q^*_{\text{eff}}(6.4)$ $\quad \vee \leqslant A \vee \neg A,, \ \vee \leqslant B \vee \neg B \Rightarrow k(A, B) =$

$$= (A \wedge B) \vee (A \wedge \neg B) \vee (\neg A \wedge B) \vee (\neg A \wedge \neg B)\,.$$

This extended calculus Q^*_{eff} can again be shown to be consistent and complete with respect to the extended formal dialog-game D^*_{f} [20].

If one assumes that the propositions $k(A, B)$ and $\bar{k}(A, B)$ are value-definite, it follows by means of Q^*_{eff} that the proposition $\bar{k}(A, B)$ can be eliminated by $\neg\,k(A, B)$. In fact, within the extended calculus Q^*_{eff} the rule

$$\vee \leqslant k(A, B) \vee \bar{k}(A, B) \Rightarrow \neg\,k(A, B) = \bar{k}(A, B)$$

can be deduced. Consequently, it follows from the assumed value-definiteness of $k(A, B)$ and $\bar{k}(A, B)$ that the principle of excluded middle

$$\vee \leqslant k(A, B) \vee \neg\,k(A, B)$$

is valid for $k(A, B)$.

If the principle of excluded middle for the commensurability proposition $k(A, B)$ is presupposed, it can be shown within the framework of the calculus Q^*_{eff} that the value-definiteness of two propositions A and B is inherited by the connectives $A \wedge B$, $A \vee B$ and $A \rightarrow B$. In fact, we have the following

Theorem. In the calculus Q^*_{eff} the rule

$$\bigvee \leqslant A \vee \neg A,, \bigvee \leqslant B \vee \neg B,, \bigvee \leqslant k(A, B) \vee \neg k(A, B) \Rightarrow$$

$$\Rightarrow \bigvee \leqslant (A * B) \vee \neg (A * B) \qquad \text{with } * \in \{\wedge, \vee, \rightarrow\}$$

can be deduced [20].

Starting from value-definite elementary propositions, it follows by means of this theorem that also all finite compound propositions are value-definite. Therefore, the relation $\bigvee \leqslant A \vee \neg A$ would then be valid for all propositions and the calculus Q_{eff} could be extended to the calculus Q of full quantum logic.

In order to justify the principle of excluded middle for all propositions, it is still necessary to legitimate the assumption that $k(A, B)$ and $\bar{k}(A, B)$ are value-definite. This can be done by the following reasoning. The proof procedures of $k(A, B)$ and $\bar{k}(A, B)$ which are performed outside of the dialog require an infinite number of steps which cannot be performed in practice. Therefore, in order to establish the commensurability of two propositions, the participants (O and P) may introduce the following strategy: They postulate that $k(A, B)$ and $\bar{k}(A, B)$—as material propositions—can be *confirmed* by a sufficiently large but finite number of dialogs. Furthermore, if A and B are value-definite, any finite series of proofs will itself be value-definite. Therefore, if A and B are value-definite, it follows that $k(A, B)$ is also value-definite, *i.e.* a proof procedure decides between $k(A, B)$ and $\bar{k}(A, B)$. Consequently, if the participants are using the ***strategy of confirmation***, Q^*_{eff} can be extended by the additional rule

$$Q^*_{\text{eff}}(C) \qquad \bigvee \leqslant A \vee \neg A,, \bigvee \leqslant B \vee \neg B \Rightarrow \bigvee \leqslant k(A, B) \vee \bar{k}(A, B)\,,$$

from which we immediately obtain

$$\bigvee \leqslant A \vee \neg A,, \bigvee \leqslant B \vee \neg B \Rightarrow \bigvee \leqslant k(A, B) \vee \neg k(A, B)\,.$$

By means of this result and the above-mentioned theorem, it follows that in the framework of Q^*_{eff} with the extension $Q^*_{\text{eff}}(C)$ the value-definiteness of two propositions A and B is in fact inherited by all finite compound propositions, *i.e.* we have the

Theorem. In the calculus Q^*_{eff} with the extension $Q^*_{\text{eff}}(C)$ the rule

$$\bigvee \leqslant A \vee \neg A,, \bigvee \leqslant B \vee \neg B \Rightarrow (A * B) \vee \neg (A * B)$$

can be deduced.

Elementary propositions $a, b, \ldots$ are value-definite. This value-definiteness can be expressed by the relations $\bigvee \leqslant a \vee \neg a$, $\bigvee \leqslant b \vee \neg b$, etc. If the calculus

Q^*_{eff}, $Q_{\text{eff}}(C)$ is now extended to the calculus Q^{**}_{eff} by incorporating the relations $\bigvee \leqslant a \vee \neg a$ for all elementary propositions, it follows by means of the theorem last mentioned that in Q^{**}_{eff} the principle of excluded middle $\bigvee \leqslant A \vee \neg A$ is valid for all finite compound propositions. Therefore, Q^{**}_{eff} is equivalent to the calculus Q of full quantum logic, which thereby can be justified. On the other hand, the calculus Q of full quantum logic is a model of the lattice L_q, the logical interpretation of which is thus performed.

REFERENCES

[1] J. von Neumann: *Mathematical Foundations of Quantum Mechanics* (Princeton, N. J., 1955).
[2] J. M. Jauch: *Foundations of Quantum Mechanics* (Reading, Mass., 1968).
[3] G. Birkhoff: *Lattice Theory*, 3rd edition, Amer. Math. Soc. Publ., Vol. **25** (Providence, R. I., 1973).
[4] G. Birkhoff and J. von Neumann: *Ann. of Math.*, **37**, 823 (1936).
[5] J. M. Jauch and C. Piron: *Helv. Phys. Acta*, **36**, 827 (1963).
[6] F. Kamber: *Math. Ann.*, **158**, 158 (1965).
[7] M. Nakamura: *Kodai Math. Series Rep.*, **9**, 158 (1957).
[8] H. B. Curry: *Foundations of Mathematical Logic* (New York, N. Y., 1963).
[9] P. Mittelstaedt: *Zeits. Naturforsch.*, **27** a, 1358 (1972).
[10] A. M. Gleason: *Journ. Math. Mech.*, **6**, 885 (1957).
[11] P. Lorenzen: *Formal Logic* (Dordrecht, 1965).
[12] P. Lorenzen: *Metamathematik* (Mannheim, 1962).
[13] W. Kamlah and P. Lorenzen: *Logische Propädeutik* (Mannheim, 1967).
[14] K. Lorenz: *Arch. Math. Logik und Grundlagenforsch.*, **10**, **11** (1968).
[15] P. Mittelstaedt: *Philosophical Problems of Modern Physics* (Dordrecht, 1976).
[16] P. Mittelstaedt and E. W. Stachow: *Found. Phys.*, **4**, 355 (1974).
[17] P. Mittelstaedt: *Quantum Logic*, in: PSA 1974, edited by R. S. Cohen *et al.* (Dordrecht, 1976), p. 501.
[18] E. W. Stachow: *Journ. Philos. Log.*, **5** (1976).
[19] E. W. Stachow: *Journ. Philos. Log.*, **7**, 347 (1978).
[20] P. Mittelstaedt and E. W. Stachow: *Journ. Philos. Log.*, **7**, 181 (1978).
[21] P. Mittelstaedt: in *Proceedings of the Symposium on Quantum Logic, Bad Homburg, Germany, 1976, Journ. Philos. Log.*, **6**, 463 (1977).
[22] P. Mittelstaedt: *Quantum Logic* (Dordrecht, 1978).

Galilean and Lorentz Particles: A New Approach of Quantization.

C. Piron

Department of Theoretical Physics, University of Geneva - 1211 Geneva 4, Switzerland

This is the plan of the seminar:

1) an introduction to the formalism of the quantum physics,

2) the notion of imprimitivity system,

3) the Galilean particle,

4) the Lorentz particle.

1. – An introduction to the formalism of the quantum physics.

In classical physics, the states of the system are described by the points of a set Ω called the phase space or sometimes the state space, and the observables are the functions defined on Ω with values in $\mathbf{R}$ or $\mathbf{R}^n$. On the other hand, in the usual quantum mechanics, the states of the system are described by the rays of a separable Hilbert space H, and the observables are the self-adjoint operators defined in H. In general, in quantum physics, the system is described by a family of separable Hilbert spaces $\{H_\alpha\}$, where the α are indices contained in a set Ω. A state is given by an index $\alpha_0 \in \Omega$ and a ray $\hat{\psi}_{\alpha_0}$ in the corresponding Hilbert space H_{α_0}. An observable is given by a family of self-adjoint operators $\{A_\alpha\}$, each A_α being defined in the corresponding H_α. The index set Ω is called the set of superselection rules. If Ω reduces to one element, we recover the usual quantum mechanics frame, inversely, if each of the H_α has dimension one, we recover the classical frame. Physically an observable is a correspondence between some properties of the apparatus and some properties of the system. To put in evidence such a correspondence let us explain the notion of property. A property of a physical system is associated to an experiment and a well-defined result. If the physical system has been prepared in such a way that we can affirm that in the event of the experiment the expected result would be certain, we say that the corresponding property is an actual property of the system in opposition to the other properties for which it is not the case, and which for this reason are only potential. In classical physics, a property is represented by a subset of the set Ω. The set $\mathscr{P}(\Omega)$ of all subsets of Ω is a complete lattice which is a Boolean algebra [1]. In the usual quantum mechanics, a property is represented by an orthogonal projector defined on the Hilbert space H. The set $\mathscr{P}(H)$ of all orthogonal projectors defined on H is a complete lattice too. In general, in quantum physics, a property is represented by a family of orthogonal projectors $\{P_\alpha\}$, each P_α being defined on the corresponding space H_α. In $\mathscr{L}$, the set of all such properties, we can define a partial-ordering relation by

$$\{P_\alpha\} < \{Q_\alpha\} \text{ if an only if } P_\alpha Q_\alpha = P_\alpha \qquad \forall \alpha$$

and an orthocomplementation by

$$\{P_\alpha\}' = \{I_\alpha - P_\alpha\} .$$

With these two structures $\mathscr{L}$ turns out to be a complete orthomodular lattice which is Boolean (distributive) if and only if each H_α has dimension one.

Given a self-adjoint operator A_α defined in the Hilbert space H_α, there exists a spectral family, *i.e.* a map E from the Borel sets in $\mathbf{R}$ into the orthogonal projectors of H_α. The Borel sets in $\mathbf{R}$ form a Boolean algebra which only is a σ-complete lattice. On the contrary, $\mathscr{B}_\alpha$, the set of classes of Borel sets in $\mathbf{R}$ modulo an element of the kernel of E, is a Boolean algebra which is a complete lattice. In conclusion, to each self-adjoint operator A_α is associated a map $\underline{A}_\alpha$ from $\mathscr{B}_\alpha$ to $\mathscr{P}(H_\alpha)$, and it is not difficult to see that this map preserves all the structures of the complete Boolean lattice $\mathscr{B}_\alpha$, *i.e.* it is a morphism in the category of complete orthomodular lattices. In the same way, if an observable is given by a family of self-adjoint operators $\{A_\alpha\}$, we can define a complete Boolean lattice $\mathscr{B}$ as the set of families of elements taken in the $\mathscr{B}_\alpha$, and a morphism $\underline{A}$ of $\mathscr{B}$ into $\mathscr{L}$ by the relation $\underline{A}\{\Delta_\alpha\} = \{\underline{A}_\alpha \Delta_\alpha\}$, where $\Delta_\alpha \in \mathscr{B}_\alpha$. This particular morphism is said to be the morphism associated to the family of self-adjoint operators. Finally, it is a theorem that every injective morphism of a complete Boolean lattice $\mathscr{B}$ into $\mathscr{L}$ can be obtained as the morphism associated to some family of self-adjoint operators.

2. – The notion of imprimitivity system.

Let us define an observable $\underline{A}$ as a morphism (in general injective) of a complete Boolean lattice $\mathscr{B}$ into $\mathscr{L}$, the lattice of the properties of the system. It is clear that the elements of $\mathscr{B}$ correspond to some of the properties of the apparatus like, for example, the position of the pointer, and that the way to distinguish one kind of observable from another is to study the group which acts on the apparatus. Mathematically, this leads to characterize an observable by an imprimitivity system (a notion due to MACKEY [2]):

$$\begin{array}{ccc} \mathscr{B} & \xrightarrow{\underline{A}} & \mathscr{L} \\ \sigma(g)\downarrow & & \downarrow S(g) \\ \mathscr{B} & \xrightarrow{\underline{A}} & \mathscr{L} \end{array} .$$

This is a commutative diagram where $\underline{A}$ is an observable, $\sigma(g)$ an automorphism of $\mathscr{B}$ and $S(g)$ an automorphism. Furthermore, as functions of g, $\sigma(g)$ and $S(g)$ are representations of a group G:

$$\begin{cases} \sigma(g_1)\sigma(g_2) = \sigma(g_1 g_2) , \\ S(g_1)S(g_2) = S(g_1 g_2) , \end{cases} \qquad \forall g_1, g_2 \in G .$$

The group G, the complete Boolean lattice $\mathscr{B}$ and the representation $\sigma(g)$ being given, if there exists a representation $S(g)$ in the automorphisms of $\mathscr{L}$ and a morphism $\underline{A}$ such that the corresponding diagram is commutative, we say that $\mathscr{L}$ admits an observable of this kind or that $\mathscr{L}$ satisfies an imprimitivity system based on G. To describe a physical system like, for example, a particle, we first construct the lattice of properties $\mathscr{L}$ and the observables satisfying the corresponding imprimitivity systems and in this way we obtain the Schrödinger picture. In the second step we construct the dynamics as a representation of the one-parameter group. We will not treat here this second step, we will just remark that in this method the construction of the dynamics is independent, the introduction of the interactions does not change the representation or the interpretation of the observables just defined before.

3. – The Galilean particle.

We define the elementary Galilean particle as a lattice of properties $\mathscr{L}$ which admits the observables momentum $\underline{\boldsymbol{p}}$ and position in space-time $q^\mu = (\underline{\boldsymbol{q}}, \underline{t})$, and no other independent observables. The group G corresponds to the different possible choices of the zeros of the scale of the instruments measuring $\underline{\boldsymbol{p}}$, $\underline{\boldsymbol{q}}$ and $\underline{t}$. This group is defined by

the momentum translations $\boldsymbol{\pi}$

$$\boldsymbol{p} \mapsto \boldsymbol{p} + \boldsymbol{\pi} , \qquad \boldsymbol{q} \mapsto \boldsymbol{q} , \qquad t \mapsto t ,$$

the space translations $\boldsymbol{a}$

$$\boldsymbol{p} \mapsto \boldsymbol{p}\,, \qquad \boldsymbol{q} \mapsto \boldsymbol{q} + \boldsymbol{a}\,, \qquad t \mapsto t\,,$$

the time translations τ

$$\boldsymbol{p} \mapsto \boldsymbol{p}\,, \qquad \boldsymbol{q} \mapsto \boldsymbol{q}\,, \qquad t \mapsto t + \tau\,,$$

the rotations R

$$\boldsymbol{p} \mapsto R\boldsymbol{p}\,, \qquad \boldsymbol{q} \mapsto R\boldsymbol{q}\,, \qquad t \mapsto t\,,$$

the Galilean transformations $\boldsymbol{v}$

$$\boldsymbol{p} \mapsto \boldsymbol{p}\,, \qquad \boldsymbol{q} \mapsto \boldsymbol{q} + \boldsymbol{v}t\,, \qquad t \mapsto t\,.$$

The translations $\boldsymbol{\pi}$, $\boldsymbol{a}$ and τ correspond to the possible choices of the origins of the scales of the instruments measuring the momentum and the position in space-time. The rotations R correspond to the possible choices of three orthogonal directions in space and the Galilean transformations $\boldsymbol{v}$ to the possible choices of a rest frame. The two imprimitivity systems are given by

$$\begin{array}{ccc} \mathscr{B}_p & \xrightarrow{\underline{\boldsymbol{p}}} & \mathscr{L} \\ \sigma_p(g)\downarrow & & \downarrow S(g) \\ \mathscr{B}_p & \xrightarrow{\underline{\boldsymbol{p}}} & \mathscr{L} \end{array}\,, \qquad \begin{array}{ccc} \mathscr{B}_q & \xrightarrow{\underline{q}^\mu} & \mathscr{L} \\ \sigma_q(g)\downarrow & & \downarrow S(g) \\ \mathscr{B}_q & \xrightarrow{\underline{q}^\mu} & \mathscr{L} \end{array}\,.$$

The complete Boolean lattice $\mathscr{B}_p$ is defined from the $\boldsymbol{p} \in \mathbf{R}^3$ and the corresponding $\mathscr{B}_q$ from the $q^\mu \in \mathbf{R}^4$. The automorphism $\sigma_p(g)$ is induced by the action of g on the $\boldsymbol{p}$ and the automorphism $\sigma_p(g)$ is induced by that of g on the q^μ. There exist two elementary Galilean particles:

a) The classical one: $\mathscr{L}$ is the lattice of the sets of $\mathbf{R}^7 = \{(\boldsymbol{p}, \boldsymbol{q}, t)\}$, $S(g)$ the canonical representation, $\mathscr{B}_p = \mathscr{P}(\mathbf{R}^3)$, $\mathscr{B}_q = \mathscr{P}(\mathbf{R}^4)$, and

$$\underline{\boldsymbol{p}}(\Delta) = \{(\boldsymbol{p}, \boldsymbol{q}, t) | \boldsymbol{p} \in \Delta,\, (\boldsymbol{p}, \boldsymbol{q}, t) \in \mathbf{R}^7\}\,,$$

$$\underline{\boldsymbol{q}}^\mu(\Delta) = \{(\boldsymbol{p}, \boldsymbol{q}, t) | (\boldsymbol{q}, t) \in \Delta,\, (\boldsymbol{p}, \boldsymbol{q}, t) \in \mathbf{R}^7\}\,.$$

These two observables can also be defined as the inverse images of the following two vector-valued functions:

$$(\boldsymbol{p}, \boldsymbol{q}, t) \mapsto \boldsymbol{p}$$

and

$$(\boldsymbol{p}, \boldsymbol{q}, t) \mapsto (\boldsymbol{q}, t)\,.$$

b) The quantal one: $\mathscr{L}$ is defined by a family of complex Hilbert spaces H_t, $t \in \mathbf{R}$, all identical to $L^2(\mathbf{R}^3, \mathrm{d}V)$. $S(g)$ is induced by the following unitary transformations:

$$(U(\boldsymbol{\pi})\varphi)_t(\boldsymbol{x}) = \exp[-i\hbar^{-1}\boldsymbol{\pi}\cdot\boldsymbol{x}]\varphi_t(\boldsymbol{x}) ,$$

$$(U(\boldsymbol{a})\varphi)_t(\boldsymbol{x}) = \varphi_t(\boldsymbol{x} - \boldsymbol{a}) ,$$

$$(U(R)\varphi)_t(\boldsymbol{x}) = \varphi_t(R^{-1}\boldsymbol{x}) ,$$

$$(U(\boldsymbol{v})\varphi)_t(\boldsymbol{x}) = \varphi_t(\boldsymbol{x} - \boldsymbol{v}t) .$$

It is easy to verify that this is a representation of G in the automorphisms of $\mathscr{L}$. Nevertheless, as transformation of $L^2(\mathbf{R}^3, \mathrm{d}V)$, $U(g)$ is only an unitary projective representation since a « phase » appears in the relations

$$U(\boldsymbol{\pi})\,U(\boldsymbol{a}) = \exp[i\hbar^{-1}\,\boldsymbol{\pi}\cdot\boldsymbol{a}]\,U(\boldsymbol{a})\,U(\boldsymbol{\pi}) .$$

These are the well-known Weyl commutation relations defining the Planck constant $\hbar$ [3].

Let us first describe $\underline{q}^\mu$, the position in space-time. The complete Boolean lattice $\mathscr{B}_q$ is defined as the lattice of families of elements Δ_t, $t \in \mathbf{R}$, which are Borel sets in $\mathbf{R}^3$ modulo the Borel sets of Lebesgue measure zero. The observable $\underline{q}^\mu$ is then given by

$$(\underline{q}^\mu(\{\Delta_t\})\varphi)_t(\boldsymbol{x}) = \chi_{\Delta_t}(\boldsymbol{x})\varphi_t(\boldsymbol{x}) ,$$

where $\chi_{\Delta_t}(\boldsymbol{x})$ is the characteristic function defined a.e. as

$$\chi_{\Delta_t}(\boldsymbol{x}) = \begin{cases} 1 & \text{if } \boldsymbol{x} \in \Delta_t , \\ 0 & \text{if } \boldsymbol{x} \notin \Delta_t . \end{cases}$$

Secondly, for the momentum $\underline{\boldsymbol{p}}$, the complete Boolean lattice $\mathscr{B}_p$ is defined as the lattice of the Borel sets in $\mathbf{R}^3$ modulo the Borel sets of Lebesgue measure zero and the momentum is given by

$$(\underline{\boldsymbol{p}}(\Delta)\,\hat{\varphi})_t(\boldsymbol{k}) = \chi_\Delta(\boldsymbol{k})\,\hat{\varphi}_t(\boldsymbol{k}) ,$$

where here $\hat{\varphi}(\boldsymbol{k})$ is obtained via the Fourier unitary transformation defined on a dense subset by

$$\hat{\varphi}(\boldsymbol{k}) = (2\pi\hbar)^{-\frac{3}{2}}\int \exp[-i\hbar^{-1}\boldsymbol{k}\cdot\boldsymbol{x}]\varphi(\boldsymbol{x})\,\mathrm{d}V .$$

Such observables can also be defined as the spectral decomposition of the families of self-adjoint operators:

$$q_t^k = x^k , \qquad t_t = tI , \qquad p_t^k = -i\hbar\partial_{x^k} .$$

4. – The Lorentz particle.

Similarly we define the elementary Lorentz particle as a lattice of properties $\mathscr{L}$ which admits the observables momentum-energy $\underline{p}^\mu = (\boldsymbol{p}, E/c^2)$ and position in space-time $\underline{q}^\mu = (\boldsymbol{q}, t)$ and no other independent observables [4]. The corresponding group G is here the semi-direct product $\mathbf{R}^8 \,\overline{\wedge}\, SO_{3,1}$ defined by

the momentum translations π^μ

$$p^\mu \mapsto p^\mu + \pi^\mu , \qquad q^\mu \mapsto q^\mu ,$$

the space-time translations a^μ

$$p^\mu \mapsto p^\mu , \qquad q^\mu \mapsto q^\mu + a^\mu ,$$

the Lorentz transformations $\Lambda^\mu{}_\nu$

$$p^\mu \mapsto \Lambda^\mu{}_\nu p^\nu , \qquad q^\mu \mapsto \Lambda^\mu{}_\nu q^\nu .$$

The Lorentz transformations $\Lambda^\mu{}_\nu$ correspond to the possible choices of the directions of the basic vectors in space-time. The two imprimitivity systems are given by

$$\begin{array}{ccc} \mathscr{B}_p & \xrightarrow{\underline{p}^\mu} & \mathscr{L} \\ \sigma_p(g)\downarrow & & \downarrow S(g) \\ \mathscr{B}_p & \xrightarrow{\underline{p}^\mu} & \mathscr{L} \end{array} , \qquad \begin{array}{ccc} \mathscr{B}_q & \xrightarrow{\underline{q}^\mu} & \mathscr{L} \\ \sigma_q(g)\downarrow & & \downarrow S(g) \\ \mathscr{B}_q & \xrightarrow{\underline{q}^\mu} & \mathscr{L} \end{array} ,$$

where $\mathscr{B}_p$ as $\mathscr{B}_q$ are here either the complete Boolean lattice of the subsets of $\mathbf{R}^4$ or the complete Boolean lattice of the Borel sets in $\mathbf{R}^4$ modulo the Borel sets of measure zero.

As for elementary Galilean particles there exist in the Lorentz case two models:

a) The classical one: $\mathscr{L}$ is the lattice of the subsets of $\mathbf{R}^8 = \{(p^\mu, q^\mu)\}$, $S(g)$ is the canonical representation and

$$\underline{p}^\mu(\Delta) = \{(p^\mu, q^\mu) | p^\mu \in \Delta,\ (p^\mu, q^\mu) \in \mathbf{R}^8\} ,$$

$$\underline{q}^\mu(\Delta) = \{(p^\mu, q^\mu) | q^\mu \in \Delta,\ (p^\mu, q^\mu) \in \mathbf{R}^8\} .$$

In the same way, these two observables can be defined by the two quadrivector functions

$$(p^\mu, q^\mu) \mapsto p^\mu$$

and

$$(p^\mu, q^\mu) \mapsto q^\mu .$$

b) The quantal one: $\mathscr{L}$ is the lattice of orthogonal projectors of $L^2(\mathbf{R}^4, \mathrm{d}V\,\mathrm{d}t)$ and $S(g)$ is induced by the following unitary transformations:

$$\begin{aligned}
\big(U(\pi^\mu)\varphi\big)(x^\mu) &= \exp\left[i\hbar^{-1}g_{\mu\nu}\pi^\mu x^\nu\right]\varphi(x^\mu)\,,\\
\big(U(a^\mu)\varphi\big)(x^\mu) &= \varphi(x^\mu - a^\mu)\,,\\
\big(U(\Lambda^\mu{}_\nu)\varphi\big)(x^\mu) &= \varphi(\Lambda^{-1\mu}{}_\nu x^\nu)\,.
\end{aligned}$$

Here also, as transformations of $L^2(\mathbf{R}^4, \mathrm{d}V\,\mathrm{d}t)$, it is an unitary projective representation, since we have the generalized Weyl commutation relations:

$$U(\pi^\mu)\,U(a^\nu) = \exp\left[+\,i\hbar^{-1}g_{\mu\nu}\pi^\mu a^\nu\right]U(a^\nu)\,U(\pi^\mu)\,,$$

where we have chosen the metric $(1, 1, 1, -c^2)$.

The observable $\underline{q}^\mu$ is defined by the projectors

$$\big(\underline{q}^\mu(\Delta)\varphi\big)(x^\mu) = \chi_\Delta(x^\mu)\varphi(x^\mu)$$

and the observable $\underline{p}^\mu$ by

$$\big(\underline{p}^\mu(\Delta)\,\hat\varphi\big)(k^\mu) = \chi_\Delta(k^\mu)\,\hat\varphi(k^\mu)\,,$$

where $\hat\varphi(k^\mu)$ is obtained via the Fourier unitary transformations defined on a dense subset by

$$\hat\varphi(k^\mu) = (2\pi\hbar)^{-2}c\int\exp\left[-\,i\hbar^{-1}g_{\mu\nu}k^\mu x^\nu\right]\varphi(x^\mu)\,\mathrm{d}V\,\mathrm{d}t\,.$$

Finally the corresponding self-adjoint operators are

$$p^\mu = -\,i\hbar g^{\mu\nu}\,\partial_{x^\nu} \qquad \text{and} \qquad q^\mu = x^\mu\,.$$

In conclusion, in this approach, the possibility of quantal particles in the Galilean case as well as in the Lorentz case comes from the existence of projective representations, *i.e.* from the existence of nontrivial extensions of the given group. The value of the « phase » which appears in the representation and which is proportional to the constant $\hbar^{-1}$ is given by a two-form defined on the Abelian subgroup of the translations. In our point of view it is the existence of such quantal particles as constituents of the macroscopic bodies which explains the part taken by the symplectic structure in analytical mechanics and not the symplectic structure which explains the, so-called, canonical commutation rules.

REFERENCES

[1] For all definitions and theorems given in this introduction see C. PIRON: *Foundations of Quantum Physics* (Reading, Mass., 1976).
[2] G. W. MACKEY: *Induced Representations of Groups and Quantum Mechanics* (Reading, Mass., and Torino, 1968).
[3] H. WEYL: *The Theory of Groups and Quantum Mechanics* (New York, N. Y., 1931), especially Chapter IV, part D.
[4] L. P. HORWITZ and C. PIRON: *Helv. Phys. Acta*, **46**, 316 (1976).

Topics in Nonequilibrium Statistical Mechanics.

I. PRIGOGINE (*) and A. P. GRECOS

Faculté des Sciences, Université Libre de Bruxelles
Campus Plaine, C.P. 231, 1050 *Brussels, Belgium*

1. – Introduction.

Is it possible to formulate a consistent microscopic theory of irreversible processes? This question is today, more than hundred years since the publication of Boltzmann's paper [1], a basic problem of theoretical physics. Perhaps, it is interesting to remember that H. POINCARÉ concludes his *Leçons de thermodynamique* [2] by the statement that no phase function exists which shares the properties of macroscopic entropy and, therefore, a dynamical interpretation of thermodynamics is impossible. This is probably one of the reasons that Boltzmann's work was considered with scepticism for a long time.

The usefulness of Boltzmann's approach is now generally accepted and the ideas and methods of kinetic theory have been proven very successful in providing a probabilistic description of irreversible phenomena in several physical systems (cf. the review articles in ref. [3], as well as several recent monographs [4-6]). Nevertheless, in spite of numerous efforts, the extension of Boltzmann's ideas to dense gases and liquids and the derivation of a general microscopic model for nonequilibrium entropy have not been accomplished till now. The relation of the (irreversible) kinetic description to the (reversible) dynamics is a subject which requires considerable clarification.

For a microscopic interpretation of entropy we need to define a functional of the state of the system which includes correlations and increases monotonically in time. The second condition, which means that entropy should be identified with a Liapounoff function, is the main difficulty that one encounters when attempting to construct a dynamical theory of irreversibility. Thus, it is not astonishing that a subjectivistic point of view has been often advocated [7, 8], which, however, tries to bypass the issue rather than resolving it.

Our point of view is that the problem of irreversibility can be studied by analysing the properties of the evolution of dynamical systems. This analysis

(*) Also at the University of Texas at Austin, Austin, Tex. 78712, U.S.A.

eventually leads to an extension of the theoretical framework, so that the concept of dissipativity can be incorporated. The following sections will be devoted to a review of the formalism which we have developed in order to deal with the question of the relation between irreversible processes and reversible dynamics.

Let us remark now that we must distinguish between two problems. The first is the justification of stationary ensembles, used in computing certain properties of systems assumed to be in thermodynamic equilibrium. In particular, for an isolated system, one would like to prove the validity of the microcanonical ensemble. The second problem is to obtain the necessary conditions for the approach to equilibrium and, eventually, to establish an « H-theorem » which would permit a microscopic interpretation of the second law of thermodynamics. It is clear, of course, that the approach to an equilibrium state does not always imply the existence of a functional which decreases monotonically in time. The possibility of defining a nonequilibrium entropy demands, as we shall see in the following, stronger conditions.

Historically the formulation of the second problem preceded that of the first [9]. It is the very success of the kinetic approach of Boltzmann and his « H-theorem » which led him and MAXWELL to the formulation of the « ergodic hypothesis » as a possible conceptual basis for equilibrium statistical mechanics, independently of any specific model [10, 11]. For classical systems, MAXWELL states this hypothesis in the form: « The only assumption which is necessary for a direct proof (of the problem of thermodynamical equilibrium) is that the system, if left to itself in its actual state of motion, will, sooner or later, pass through every phase which is consistent with the equation of energy ». We know, of course, that such a property cannot be true and the statement has been weakened by the EHRENFESTS who introduced the so-called « quasi-ergodic hypothesis » [12]. Assumptions of this type are indeed most appealing. In some sense, they imply that motion of the system is described essentially by a single trajectory in the phase space, which passes arbitrarily close to every point of the energy surface. However, the universality of this hypothesis is today rather doubtful. Recent work, initiated by KOLMOGOROFF [13], on the properties of classical Hamiltonian systems has shown that they are, in general, neither ergodic nor integrable. In the third section we will come back to this point, which implies that the dynamical theory of irreversible processes cannot be founded on the notion of ergodicity.

Since this time the dichotomy between the two problems we have mentioned has been maintained in most discussions on the foundations of statistical thermodynamics. Nonequilibrium statistical mechanics, which is the modern form of kinetic theory, has proved quite successful in elucidating the mechanism of approach to equilibrium of many systems of great physical importance. However, kinetic equations are regarded often as semi-empirical and the theoretical justification of the irreversible behaviour of dynamical systems is supposed to be found in ergodic theory.

To some extent, this situation is related to the fact that for a long time kinetic equations were derived for specific limiting cases. Typical examples are the Boltzmann equation for a classical dilute gas [14] and the Pauli master equation for a weakly coupled quantum system [15]. Thus, the generality of the kinetic approach to the description of the evolution of dynamical systems was far from being established. Moreover, the derivation was based, at least partly, on phenomenological arguments and consequently it was difficult to understand whether the predicted irreversible behaviour was due to inherent properties of the system or rather to inconsistent approximations. The objections of Zermelo [16] and Loschmidt [17], known as the *Wiederkehreinwand* and *Umkehreinwand* « paradoxes », illustrate well the conceptual problems encountered in kinetic theory and which cannot be resolved without developing a systematic approach.

An important step in this direction is the derivation of the so-called « generalized master equation » [18], which describes the evolution of some part of the state of the system and which is an exact consequence of the Liouville or von Neumann equation, *i.e.* of the dynamics of the system. It will become clear in the following that such an equation does not provide a solution to the problem of irreversibility. However, it is a convenient starting point for constructing a general form of irreversible kinetic (or master) equations and discussing their relation to dynamics.

Such a relation, as one might expect, cannot be simple. Once a kinetic equation has been established, several questions can be raised. In particular, we may inquire whether such an equation leads to an « H-theorem » and, thus, to a possible microscopic definition of entropy. We know that no H-theorem has been established for the kinetic equation derived by Choh and Uhlenbeck [19] for a moderately dense gas. The main reason for this difficulty is due to the fact that kinetic equations are, in general, incomplete in the sense that correlations are not always correctly taken into account. This problem, which will be further discussed later, is important in understanding the physical basis of the concept of « subdynamics » and the necessity of introducing nonunitary transformations in formulating a consistent (irreversible) kinetic description [20, 21].

We may summarize the point of view which has emerged through the work of our group as follows. The dynamical and thermodynamical descriptions are, in a certain sense, equivalent representations of the evolution of the system, connected via a nonunitary transformation. More precisely, there exists a microscopic description of the evolution of the dynamical system which displays explicitly « irreversibility » in the sense that we may associate to it a Liapounoff functional. Once this description is obtained from the usual formulation of dynamics through a nonunitary transformation, the macroscopic irreversible description essentially follows.

Our interest in the microscopic theory of irreversible processes was always stimulated by the hope that it may prove useful in the clarification of conceptual difficulties which are deeply rooted in the usual formulation of classical and quantum mechanics. As is well known, the theory of classical integrable systems has played an essential role in the very formulation of quantum mechanics [22]. However, such systems cannot be the correct prototype for quantum problems involving interacting elementary particles. Integrable systems give rise to well-defined, but not interacting units (a set of action variables). In general, what we need is a representation in which units and interactions are well defined simultaneously. Following Einstein's treatment of spontaneous emission [23], we suggest that « physical » particles should be defined from « bare » ones in such a way that, through their interactions, thermal equilibrium may be attained. The formalism of subdynamics and the associated theory of nonunitary transformations indicate the possibility of defining the appropriate representation.

In the next section we present some basic notions from irreversible thermodynamics. Then we review the description of dynamical systems and the derivation of generalized master equations. We make, in the fourth section, some remarks on the properties of constants of the motion which are relevant to the kinetic description. In the fifth section we present the main aspects of the theory of subdynamics and that of nonunitary transformations. To illustrate the methods that we develop, we apply them, in the sixth section, to the problem of the derivation of linearized hydrodynamics from kinetic theory in the case of a simple fluid. We conclude with a brief discussion of certain questions arising in the description of quantum systems involving unstable states and we mention certain open problems and further applications of the formalism.

2. – Irreversible thermodynamics.

Entropy plays a central role in our considerations. We believe that every dynamical theory of irreversibility should ultimately lead to a microscopic interpretation of the second law of thermodynamics and the concept of entropy. Today the importance of irreversible processes and of nonequilibrium thermodynamics is even greater. Indeed the introduction of the concept of « dissipative structures », in far from equilibrium situations [24], leads to a description of coherent processes, quite similar to those which appear in biological systems.

As we have discussed in more detail elsewhere [25], any physical theory of time-dependent phenomena must necessarily deal with irreversible processes. Such processes cannot be considered as arising from some « accident », some « improbable » initial conditions, or our ignorance of the « exact » dynamical state. Of course, there is a fundamental difference between dynamics and thermodynamics. The dynamical description, at least classically, is formulated in terms of trajectories in the phase space which never cross. On the other hand, thermodynamics is formulated in terms of processes and the existence of the entropy implies that an equilibrium state acts as an « attractor » for the evolution of the system.

Before we consider the problem of construction of a dynamical theory of irreversibility in more precise terms, we think that it is worth to summarize some elementary aspects of macroscopic thermodynamics which we will need later. More detailed discussions as well as accounts of recent developments may be found in several monographs (cf. ref. [24, 26, 27]).

The basic postulate of thermodynamics is the existence of the entropy S which is an extensive function of the state of the system and which, for isolated systems, is nondecreasing

$$\mathrm{d}S \geqslant 0 . \tag{2.1}$$

In certain cases, for nonisolated systems, other thermodynamic potentials may be used, *e.g.* the Helmholtz free energy $F = E - TS$ (E is the energy and T the temperature). However, in general, we need to split the change of entropy $\mathrm{d}S$ as a sum of two terms: the flow of entropy $\mathrm{d_e}S$ and the entropy production $\mathrm{d_i}S$

$$\mathrm{d}S = \mathrm{d_e}S + \mathrm{d_i}S . \tag{2.2}$$

Then, for a system isolated or not, the second law of thermodynamics is expressed by the inequality

$$\mathrm{d_i}S \geqslant 0 . \tag{2.3}$$

This relation permits us to classify changes taking place in the system in two categories: reversible processes ($\mathrm{d_i}S = 0$) and irreversible ones ($\mathrm{d}S_\mathrm{i} > 0$).

At this point we should insist on the qualitative character of eq. (2.3). We may conclude, of course, that an isolated system will reach equilibrium after a sufficiently long time. However, thermodynamics gives no prescription for constructing the entropy function out of equilibrium.

One case is clear, namely when the assumption of local equilibrium is valid. For an isotropic fluid, this means that the entropy per unit volume s, defined by

$$S = \int_V s \, \mathrm{d}V , \tag{2.4}$$

is the same function of the local macroscopic variables characterizing the state of the system as in equilibrium. For a simple fluid such (intensive) variables are the internal energy per unit mass $e(x)$, the mass density $n(x)$ and, possibly, the velocity field $\underline{u}(x)$. This form of entropy, together with relations familiar from equilibrium thermodynamics, such as $(\partial s/\partial e)_n = T^{-1}$, gives the correct properties in the linear regime as well as in nonlinear situations.

Let us consider some implications for the linear regime, *i.e.* assuming small deviations from absolute equilibrium. Retaining terms up to second order we have

$$s = s_{eq} + (\delta s)_{[eq]} + \tfrac{1}{2}(\delta^2 s)_{[eq]} + \dots \tag{2.5}$$

Thermodynamic stability implies that

$$(\delta^2 s)_{[eq]} < 0 . \tag{2.6}$$

Moreover, it may be shown [24] that the time derivative of this quantity gives the entropy production σ, *i.e.*

$$\tfrac{1}{2}\partial_t(\delta^2 s)_{[eq]} = \sigma \geqslant 0 . \tag{2.7}$$

In terms of thermodynamic parameters $(\delta^2 s)_{[eq]}$ takes the characteristic quadratic form

$$\delta^2 s = -\frac{1}{T}\left[\frac{c_v}{T^2}(\delta T)^2 + \frac{1}{n^2}\left(\frac{\partial P}{\partial n}\right)_T (\delta n)^2 + (\delta u)^2\right], \tag{2.8}$$

where we have included the term due to the velocity field [24].

We should note here that the evolution of the quantities δn, δu and δT is given by the equations of hydrodynamics (cf. section **6**). From eq. (2.7) we may conclude then that $\delta^2 s$ plays the role of a Liapounoff functional characterizing the approach to equilibrium. In this sense, a dynamical theory of irreversible processes amounts to the construction of a functional Ω of the microscopic state ϱ of the system, which would satisfy the relations

$$\Omega(\{\varrho\}) \geqslant 0 , \qquad d\Omega/dt \leqslant 0 \qquad (t \geqslant 0). \tag{2.9}$$

That the existence of such a functional is not a trivial matter can be seen by considering the Gibbs entropy ($\int \varrho \log \varrho \, d\Gamma$ or $\mathrm{Tr}\, \varrho \log \varrho$), which, because of the laws of dynamics, remains constant in time. Furthermore, it cannot be expected that such a functional, when it exists, is unique. Therefore the requirement that Ω leads to the macroscopic entropy imposes supplementary conditions that should be derived by investigating the dissipative processes taking place in the system. These aspects will be considered in more detail in the following sections.

3. – Dynamical evolution.

In nonequilibrium statistical mechanics, the state of a classical system is a probability density, *i.e.* it is a positive function (or a distribution) of the canonical variables, normalized to unity. Thus for a system of N particles with momenta $(p_1, p_2, ..., p_N)$ and co-ordinates $(q_1, q_2, ..., q_N)$, its state $\varrho(p, q)$ is a function such that

$$\varrho(p, q) \geqslant 0, \qquad \int_\Gamma \mathrm{d}\Gamma\, \varrho(p, q) = 1, \tag{3.1}$$

where Γ is the phase of the system and $\mathrm{d}\Gamma = \prod_{k=1}^{N} \mathrm{d}p_k \mathrm{d}q_k$. Any (real) phase function $A(p, q)$ may be considered as an observable, its expectation value being given by

$$(A, \varrho) = \int_\Gamma \mathrm{d}\Gamma\, A(p, q)\, \varrho(p, q). \tag{3.2}$$

For a Hamiltonian system, the evolution of the state ϱ is determined by the Liouville equation

$$i\,\partial_t \varrho = i\{H, \varrho\} \equiv \mathsf{L}\varrho, \tag{3.3}$$

which follows from Hamilton's equations of motion

$$\frac{\mathrm{d}p_r}{\mathrm{d}t} = -\frac{\partial H}{\partial q_r}, \qquad \frac{\mathrm{d}q_r}{\mathrm{d}t} = \frac{\partial H}{\partial p_r} \qquad (r = 1, ..., N). \tag{3.4}$$

In the quantum case, a state is represented by a density matrix ϱ, *i.e.* a positive (self-adjoint) operator on a Hilbert space $\mathscr{H}$, with trace equal to one,

$$\varrho = \varrho^+ \geqslant 0, \qquad \mathrm{Tr}\, \varrho = 1. \tag{3.5}$$

An observable A is a self-adjoint operator in $\mathscr{H}$ and its expectation value is given by the trace

$$(A, \varrho) = \mathrm{Tr}\, A\varrho. \tag{3.6}$$

The evolution of ϱ is now determined by the von Neumann equation

$$i\,\partial_t \varrho = [H, \varrho] \equiv \mathsf{L}\varrho. \tag{3.7}$$

We have used L to denote either the Poisson bracket $i\{H, \cdot\}$ (classical case) or the commutator $[H, \cdot]$ (quantum case). Formally we can develop most of the formalism without specifying whether the system is classical or quantum, and thus a notation common to both cases is quite useful.

Here we have assumed that the states are time-dependent, while the observables are kept fixed, *i.e.* a « Schrödinger picture ». It is possible of course to consider a « Heisenberg picture », keeping the states fixed and having time-dependent observables. The evolution of an observable is given by

$$i\partial_t A = -\mathsf{L}A . \tag{3.8}$$

This equation follows from the equality of the expectation values in the two pictures, namely

$$(A, \varrho(t)) = (A(t), \varrho) . \tag{3.9}$$

It is obvious that the formal solution of eq. (3.3) or eq. (3.7) is

$$\varrho(t) = \exp[-i\mathsf{L}t]\,\varrho(t=0) , \tag{3.10}$$

and that of eq. (3.8)

$$A(t) = \exp[+i\mathsf{L}t]A(t=0) . \tag{3.11}$$

If we replace in the r.h.s. of eq. (3.2) $A(p, q)$ by its complex conjugate $\bar{A}(p, q)$ and in the r.h.s. of eq. (3.6) A by its adjoint $A^\dagger$, we may consider (A, ϱ) as a scalar product. It is easily seen then that L is a symmetric operator, *i.e.*

$$(A, \mathsf{L}\varrho) = (\mathsf{L}A, \varrho) . \tag{3.12}$$

Thus the evolution is described by a unitary operator, $\exp[-i\mathsf{L}t]$, which in fact expresses the reversible character of the dynamics.

To construct a theory of dissipative processes we must analyse the contributions to eq. (3.10). This is most conveniently done by writing $\varrho(t)$ in terms of the resolvent of L:

$$\varrho(t) = \frac{1}{2\pi i}\int_{\vec{C}} \mathrm{d}z \exp[-izt]\frac{1}{\mathsf{L}-z}\,\varrho(t=0) , \tag{3.13}$$

where $\vec{C}$ is a contour parallel and above the real axis. Our problem then is to study the nature of the singularities of the integrand, having in mind that ultimately we will use $\varrho(t)$ to compute expectation values. We should note that we may proceed also by considering the evolution of the observables, cf. eq. (3.11). Clearly eq. (3.13), as well as the formulae which we will derive in the following, may be used to study $A(t)$ provided that L is replaced by $-\mathsf{L}$, an operation which we call « L-inversion ».

Irreversible (kinetic) equations, such as Boltzmann's and Pauli's, serve as a guide, and the formalism which we will develop is based on the decomposition of the state $\varrho(t)$ as a sum of two components $\varrho_0(t)$ and $\varrho_c(t)$. Their definition depends on the particular physical system. In a homogeneous classical gas $\varrho_0(t)$ may be taken as the (N-body) momentum distribution, while $\varrho_c(t)$ describes spatial correlations [18]. In a quantum system $\varrho_0(t)$ may be the diagonal part of the density matrix in some given representation, while $\varrho_c(t)$ is its off-diagonal part [18]. Other decompositions are being used as in the theory of « open systems » [28] and the theory of systems close to (absolute) equilibrium (cf. section **6**). The discussion is considerably simplified when this decomposition can be obtained by using a pair of orthogonal projections [29, 30]

$$(3.14)\qquad \mathsf{P} = \mathsf{P}^\dagger = \mathsf{P}^2, \qquad \mathsf{Q} = \mathsf{I} - \mathsf{P}; \qquad \varrho_0 = \mathsf{P}\varrho, \qquad \varrho_c = \mathsf{Q}\varrho .$$

It is then an easy matter to derive an equation for the P-component of the state.

Corresponding to the above decomposition we write the resolvent $[\mathsf{L} - z]^{-1}$ as a sum of partial resolvents, $\mathsf{P}[\mathsf{L} - z]^{-1}\mathsf{P}$, etc., which may be expressed in the form

$$(3.15)\qquad \frac{1}{\mathsf{L} - z} = [\mathsf{P} + \mathscr{C}(z)]\frac{1}{\mathsf{PLP} + \Psi(z) - z}[\mathsf{P} + \mathscr{D}(z)] + \frac{1}{\mathsf{QLQ} - z}\mathsf{Q} .$$

This equation can be proved in several ways, *e.g.* using the second resolvent identity. Here we have introduced some basic operators of the kinetic description, namely « the collision operator »

$$(3.16)\qquad \Psi(z) = -\mathsf{PLQ}\frac{1}{\mathsf{QLQ} - z}\mathsf{QLP} ,$$

the « destruction operator »

$$(3.17)\qquad \mathscr{D}(z) = -\mathsf{PLQ}\frac{1}{\mathsf{QLQ} - z}$$

and the « creation operator »

$$(3.18)\qquad \mathscr{C}(z) = -\frac{1}{\mathsf{QLQ} - z}\mathsf{QLP} .$$

This terminology stems from the fact that the « asymptotic collision operator » $\Psi(+i0) = \lim_{z\to+i0}\Psi(z)$, for specific systems, describes the effect of collision processes. An example is the case of a dilute (classical) gas, where two-particle collisions dominate. It may be shown [18] that $\Psi(+i0)$, evaluated to first order in the density, is a sum of two-particle operators which are identical to the collision term of the Boltzmann equation. Similarly, for a weakly coupled quantum system, $\Psi(+i0)$, evaluated to second order in the coupling parameter, is simply the interaction term of the Pauli equation. Furthermore, the decay of the initial correlations or the off-diagonal elements of the density matrix depends on the properties of $\mathscr{D}(z)$, and those created in the system because of the interactions are characterized by $\mathscr{C}(z)$.

Inserting eq. (3.15) into eq. (3.13), we obtain for the evolution of the P-component of the state the so-called « generalized master equation »

$$i\partial_t\varrho_0(t) + \mathsf{PLP}\varrho_0(t) = \int_0^t \mathrm{d}\tau\, \mathsf{G}\,(t-\tau)\,\varrho_0(\tau) + \frac{1}{2\pi i}\int_{\overleftarrow{C}} \mathrm{d}z \exp\left[-izt\right]\mathscr{D}(z)\,\varrho_c(0)\,, \tag{3.19}$$

where the kernel $\mathsf{G}(t)$ is given by the inverse Laplace transform of the collision operator

$$\mathsf{G}(t) = \frac{1}{2\pi}\int_{\overrightarrow{C}} \mathrm{d}z \exp\left[-izt\right]\Psi(z)\,. \tag{3.20}$$

This equation is the starting point of several investigations in nonequilibrium statistical mechanics. Specific techniques have been developed in order to obtain approximate equations for physically important systems [5, 6, 18]. However, the generalized master equation cannot be used directly to discuss questions referring to irreversibility. It follows directly from the Liouville or the von Neumann equation and, as one may easily understand, it holds for any dynamical system. Its principal merit is that it expresses the dynamics in a form which is very close to that of irreversible kinetic equations and thus a comparison between the two modes of description of the evolution becomes possible.

We do not intend to consider here the questions which arise when one attempts to formulate a mathematically rigorous theory. Most of them are related to the fact that a necessary condition for the appearance of irreversible phenomena in a dynamical system is the continuity of the spectrum of the Liouville/von Neumann operator. Although in classical mechanics this may be the case even for systems with a few degrees of freedom, in general the study of dissipative properties requires to take the limit of an « infinite » system. Then it is necessary to reconsider the definition and the meaning of the « state of the system » and that of the corresponding equations of evolution. This problem is mathematically very interesting and its investigation permits a more precise formulation of certain statements. However, in several physical problems it is sufficient to proceed formally with the Liouville/von Neumann equation keeping in mind that, eventually, reduced quantities will be computed. Also, some aspects of the problem of irreversibility may be illustrated by using quantum-mechanical models, the Hamiltonian of which has *ab initio* a continuous spectrum.

4. – Constants of motion.

We will consider in this section some asymptotic properties of the dynamical evolution for long times and, in particular, the possible stationary states that the system may attain. Our discussion will be centred on the collisional invariants and their relation to the constants of motion of the system. Such an analysis [18, 31, 32] is quite important in understanding the justification for the use of equilibrium ensembles in statistical mechanics.

As we have in mind cases where L has a continuous spectrum, it is more convenient to treat the constants of the motion, *i.e.* the solutions of the equation

$$\mathsf{L}\Phi = 0 \,, \tag{4.1}$$

as observables. For this reason, the complex variable z will appear with a minus sign in several formulae of this section.

The first step is to write eq. (4.1) in terms of the collision and the other kinetic operators which we introduced previously. Using the projections P and Q we obtain from eq. (4.1)

$$\mathsf{PLP}\Phi_0 + \mathsf{PLQ}\Phi_c = 0 \,, \tag{4.2}$$

and

$$\mathsf{QLP}\Phi_0 + (\mathsf{QLQ} + z)\,\Phi_c = z\Phi_c \,. \tag{4.3}$$

For $\operatorname{Im} z \neq 0$, assuming that the spectrum of QLQ is real, we may invert $(\mathsf{QLQ} + z)$ in the Q-subspace to get

$$\Phi_c = \mathscr{C}(-z)\,\Phi_0 + z\,\frac{1}{\mathsf{QLQ} + z}\,\Phi_c \,. \tag{4.4}$$

Then from eq. (4.2) we have

$$[\mathsf{PLP} + \Psi(-z)]\,\Phi_0 = z\mathscr{D}(-z)\,\Phi_c \,. \tag{4.5}$$

Thus any constant of the motion satisfies, for nonreal z, the above relations (and *vice versa*).

In the limit $z \to +i0$, which corresponds to $t \to \infty$, we may classify the constants of the motion as follows. When the condition

$$\lim_{z \to +i0} z\mathscr{D}(-z)\,\Phi_c = 0 \tag{4.6}$$

is satisfied, eq. (4.5) implies

$$[\mathsf{PLP} + \Psi(-i0)]\,\Phi_0 = 0 \tag{4.7}$$

(*i.e.* the P-component of Φ is a « collisional invariant »). Furthermore, if the stronger condition

$$\lim_{z\to+i0} z\frac{1}{\mathsf{QLQ}+z}\Phi_c = 0 \tag{4.8}$$

holds, the Q-component of Φ is a functional of Φ_0, namely

$$\Phi_c = \mathscr{C}(-i0)\,\Phi_0\,, \tag{4.9}$$

and the invariant is called « regular ». In this context, if eq. (4.7) and/or eq. (4.9) do not hold, we speak of a « singular » invariant.

It should be remarked that this classification of the invariants has a non-trivial meaning if the asymptotic collision operator does not vanish identically. The stationary solutions of the (irreversible) kinetic equations are determined by $\Psi(+i0)$. For most physically important systems these solutions depend only on a limited set of invariants, *e.g.* the energy, the momentum, etc. Our analysis shows that these quantities need not be the only invariants of the system. Other constants of the motion may exist which, however, are singular in the sense that we defined above.

Let us suppose now that $\Psi(+i0)\neq 0$. We want then to characterize the observables which tend to regular invariants. This is a necessary property for the class of observables to which the kinetic description might be applicable.

As we work with Laplace transforms, we must consider observables for which the Abelian limit

$$\bar{A}^t = \lim_{z\to+i0} z\frac{1}{\mathsf{L}+z}A(0) \tag{4.10}$$

exists. There are Tauberian theorems [33] which ensure, under certain conditions, the existence of a Cesaro average

$$(C,\alpha)-\lim_{t\to+\infty}A(t) = \lim_{t\to+\infty}\alpha\frac{1}{t^\alpha}\int_0^t\frac{1}{(t-\tau)^{1-\alpha}}A(\tau)\,\mathrm{d}\tau \tag{4.11}$$

and the equality of the two limits. Note that, by an Abelian theorem, eq. (4.11) implies eq. (4.10). In particular, the time average, *i.e.* the $(C,1)$-limit, of $A(t)$, if it exists, is equal to $\bar{A}^t$. Our considerations apply to observables having an Abelian limit and consequently, when the Tauberian theorem holds, to those for which the time average exists.

Taking into account eq. (3.15), we get from eq. (4.10) for $\bar{A}_0^t$

$$\bar{A}_0^t = \lim_{z\to+i0} z\frac{1}{\mathsf{PLP}+\Psi(-z)+z}[A_0+\mathscr{D}(-z)A_c]\,, \tag{4.12}$$

and for $\bar{A}_c^t$

$$\bar{A}_c^t = \lim_{z\to+i0}\left\{\mathscr{C}(-z)z\frac{1}{\mathsf{PLP}+\Psi(-z)+z}[A_0+\mathscr{D}(-z)A_c]+ + z\frac{1}{\mathsf{QLQ}+z}A_c\right\}. \tag{4.13}$$

We see then that the P-component of the observables for which

$$\lim_{z\to+i0}\mathscr{D}(-z)A_c = \text{finite} \tag{4.14}$$

tends to collisional invariants, *i.e.*

$$[\mathsf{PLP}+\Psi(-i0)]\bar{A}_0^t = 0\,. \tag{4.15}$$

Moreover, if the stronger condition

$$\lim_{z\to+i0} z\frac{1}{\mathsf{QLQ}+z}A_c = 0 \tag{4.16}$$

holds, we conclude that $\bar{A}^t$ is a regular invariant, *i.e.*

$$\bar{A}_c^t = \mathscr{C}(-i0)\bar{A}_0^t\,. \tag{4.17}$$

Thus the behaviour of the destruction operator near the origin, together with eq. (4.16), characterizes the class of observables which tend to regular invariants.

It should be remarked that from eq. (4.12) it follows that $\bar{A}_0^t$ does not depend on every collisional invariant but only on those for which $\Psi'(-i0)\Phi_0$ exists. This means that we must consider invariants such that

$$[\mathsf{PLP}+\Psi(-z)]\Phi_0 \approx -z\Psi'(-i0)\Phi_0 \qquad (|z|\to+i0), \tag{4.18}$$

a condition which is obviously more restrictive than eq. (4.7). For a system of particles interacting with central forces, collisions conserve the number of particles, the total momentum and the kinetic energy. Thus one expects corresponding collisional invariants to exist. In fact, for the number of particles and the momentum, this is trivially the case, because there is no contribution from the interactions to these quantities. On the other hand, we need some supplementary assumptions in order to ensure that the kinetic energy is a collisional invariant and that eq. (4.18) is satisfied, because of the potential energy.

In several problems, functions of the Hamiltonian are regular invariants. If the only collisional invariants are the P-components of such functions, eqs. (4.14) and (4.16) determine the class of « ergodic observables ». However, the existence of a restricted class of regular invariants is compatible with a nonergodic system, *i.e.* a system which admits constants of the motion which are not functions of H (for a more detailed discussion and references see ref. [34]).

For finite quantum systems the motion is quasi-periodic. Thus time averages may exist, but no approach to equilibrium in a strong sense may be expected. This aspect is reflected in the structure of the asymptotic collision operator $\Psi(i0)$. If it exists, or more precisely in the subspace in which the limit of $\Psi(z)$ is defined, it is a Hermitian operator. The existence of singular invariants is possible if QLQ has, in the Q-subspace, a zero eigenvalue (but this condition is not sufficient). Otherwise, all invariants are regular and $\Psi(i0)$ vanishes.

From the definition of the collision operator, eq. (3.16), it follows that

$$\Psi(x+iy)= \tag{4.19}$$

$$=-\mathsf{PLQ}\frac{\mathsf{QLQ}-x}{[\mathsf{QLQ}-x]^2+y^2}\mathsf{QLP}-i\mathsf{PLQ}\frac{y}{[\mathsf{QLQ}-x]^2+y^2}\mathsf{QLP}\,.$$

The second term in the r.h.s. has a definite sign, namely

$$-y\mathsf{PLQ}\frac{1}{[\mathsf{QLQ}-x]^2+y^2}\mathsf{QLP}\lessgtr 0 \qquad \text{for } y\gtrless 0\,. \tag{4.20}$$

In the limit $y\to 0$, it vanishes if the spectrum of L is discrete and it may not vanish if it is continuous. Then we have

$$\Psi(+i0)\neq\Psi(-i0)\,, \tag{4.21}$$

which is the so-called « dissipativity condition ». Our results apply *a fortiori* to systems for which this condition is satisfied. Moreover, in such a case there are observables for which the asymptotic limit $A(t\to\infty)$ exists, possibly in a weak sense, and no time averages are needed. Of course, the problem of determining the number of functionally independent collisional invariants cannot be solved in a general manner. Nevertheless, the number of collisional invariants cannot be larger than that of the constants of motion.

We will conclude this section with a few remarks about finite classical systems and, more specifically, interacting (nonlinear) oscillators. In terms of action-angle variables, the Hamiltonian of such a system, which is assumed periodic in the angles, is written as

$$H(J, \alpha) = H_0(J) + \lambda V(J, \alpha) = H_0(J) + \lambda \sum_m{}' V_m(J) \exp[im\cdot\alpha]. \tag{4.22}$$

To simplify the notation we have denoted by J the set $\{J_1, J_2, \ldots, J_N\}$ of action variables, by α the set $\{\alpha_1, \alpha_2, \ldots, \alpha_N\}$ of angle variables and by m the set $\{m_1, m_2, \ldots, m_N\}$ of integers. Also we define

$$m\cdot\alpha = \sum_{k=1}^{N} m_k \alpha_k, \tag{4.23}$$

a notation that we will use often when there is no danger of confusion. We have written the « interaction » $V(J, \alpha)$ as a Fourier series, taking into account its periodic dependence on the angles. The prime in the summation (which will usually be omitted) means that $m \neq 0$, because the angle independent of $V(J, \alpha)$, if any, will be included into the « free » Hamiltonian $H_0(J)$. Then the Liouville operator, cf. eq. (3.3), is a differential operator, namely

$$\mathsf{L} = \mathsf{L}_0 + \lambda \mathsf{L}_1 = \tag{4.24}$$

$$= -i \sum_k \omega_k(J) \frac{\partial}{\partial \alpha_k} - \lambda \sum_k \sum_m \exp[im\cdot\alpha] \left[i \frac{\partial V_m(J)}{\partial J_k} \frac{\partial}{\partial \alpha_k} + V_m(J) m_k \frac{\partial}{\partial J_k} \right],$$

where $\omega_k(J) = \partial H_0(J)/\partial J_k$ are the « unperturbed » frequencies.

Questions about the integrability and ergodicity of Hamiltonian systems of the type defined by eq. (4.22) have attracted mathematicians and physicists for a long time and are still of great interest for theoretical physics [35, 36]. The relatively recent work of Kolmogoroff, Arnold and Moser [13, 36, 37] has shown that for small values of λ such systems are, in general, « neither integrable nor ergodic ». Thus, for a given Hamiltonian, there are various types of motion possible, each with a nonvanishing measure. Computer calculations (see, *e.g.*, ref. [38]) illustrate this complex behaviour in a striking manner.

The Kolmogoroff-Arnold-Moser theory shows that long-time predictions for the trajectories (in phase space) of dynamical systems are impossible because of their « weak stability ». More precisely, such predictions require the knowledge of initial conditions with infinite precision. Thus, as we have discussed elsewhere [39], a statistical description becomes necessary. The nonergodic nature of the system implies that the time average of an arbitrary statistical state need not tend to a uniform distribution (for fixed energy). In this respect, the concepts of regular and singular invariants are important in studying the long-time behaviour of distribution functions.

In the context of interacting oscillators, the statistical state of the system $\varrho(J, \alpha; t)$ is periodic in the angle-variables and may be expanded in a Fourier series. The angle-independent term in the series $\varrho_0(J;t)$ may be expressed as $\mathsf{P}\varrho(J,\alpha;t)$, where P is an orthogonal projection,

$$\varrho_0(J;t) \equiv \mathsf{P}\varrho(J,\alpha;t) = \left(\frac{1}{2\pi}\right)^N \int d\alpha_1 \dots d\alpha_N \varrho(J,\alpha;t)\,. \tag{4.25}$$

Using L as given by eq. (4.24) and P by eq. (4.25), we may apply the formalism that we sketched to this problem. Provided that $\Psi(+i0) \neq 0$, eqs. (4.14) and (4.16) determine classes of states which (or their time averages) tend to functions of regular invariants (note that here $\mathsf{PLP} = 0$). It is evident that exact calculations are, in general, impossible. Some conclusions may be drawn by using perturbation expansions in terms of the « coupling » parameter λ. Although the validity of such expansions is an open problem and one must be careful in the interpretation of the results, we think that the behaviour of the collision operator calculated to second order in λ provides a qualitative illustration of the point that we want to make.

A simple calculation shows that to $O(\lambda^2)$ we have

$$\Psi_{(2)}(z) = \lambda^2 \sum_m m\cdot\frac{\partial}{\partial J}\frac{|V_m(J)|^2}{m\cdot\omega - z}\, m\cdot\frac{\partial}{\partial J}\,. \tag{4.26}$$

For nonlinear oscillators, the « resonance conditions »

$$m\cdot\omega \equiv \sum_{k=1}^{N} m_k\omega_k(J) = 0 \tag{4.27}$$

may be satisfied, and in the limit $z \to +i0$ we get

$$\Psi_{(2)}(\pm i0) = \pm i\pi\lambda^2 \sum_m m\cdot\frac{\partial}{\partial J}|V_m(J)|^2\delta(m\cdot\omega)\, m\cdot\frac{\partial}{\partial J}\,. \tag{4.28}$$

With this expression for Ψ, we look for solutions of eq. (4.7). The number of the functionally independent solutions depends on the characteristics of the Hamiltonian. Assuming that the coefficients $V_m(J)$ do not vanish ($\forall\, m$) and that eq. (4.27) is satisfied for a dense set of points in every domain of the action-variables, one may prove that

$$\Psi_{(2)}(+i0)\,\varrho_0(J) = 0 \Leftrightarrow \varrho_0(J) = f(H_0)\,. \tag{4.29}$$

Furthermore, if $\varrho(J,\alpha;t)$ has a finite number of Fourier components which are « smooth » functions of the actions, eqs. (4.14) and (4.16) are satisfied (to $O(\lambda)$!) and the average of $\varrho(J,\alpha;t)$ will tend to a function of the Hamiltonian. One should notice here that the conditions required for the validity of eq. (4.29) are those formulated by POINCARÉ [40] in proving his theorem on the « nonexistence of uniform invariants ». Let us also mention that with similar considerations one obtains approximate equations for the evolution of coupled anharmonic oscillators [41].

The physical mechanism which leads with suitable initial conditions to the microcanonical distribution is quite different from that considered for ergodic systems, as here the « resonance conditions », first introduced by POINCARÉ, play an essential role. It is the occurrence of these resonance conditions which leads to the complex microstructure of the phase space. While Poincaré's original theory introduced resonances between the unperturbed frequencies, our formal theory indicates how to generalize this conditions to arbitrary coupling parameters, given that the spectrum of QLQ near the origin determines, in general, the resonances of the perturbed system.

5. – Theory of subdynamics.

Let us now consider some problems related to the formulation of a kinetic description. In quantum theory, a typical example of an irreversible kinetic equation is, as we have already mentioned, Pauli's « master equation » [15]. It describes the evolution of the diagonal elements of the density matrix for a « weakly coupled » system, *i.e.* assuming that $H = H_0 + \lambda V$ and λ is very « small ». In terms of the collision operator evaluated to second order in λ, this equation reads

$$i\,\partial_t\,\varrho_0(t) = \Psi_{(2)}(+\,i0)\,\varrho_0(t)\,. \tag{5.1}$$

From the definition of $\Psi(z)$, cf. eq. (3.16), it may be easily shown that the r.h.s. of eq. (5.1) has the familiar gain-loss form. The dissipativity of $\Psi(+\,i0)$, cf. eqs. (4.19) and (4.20), ensures the irreversible approach to equilibrium.

There are two questions which arise. First, the connection of eq. (5.1) to eq. (3.19), *i.e.* essentially to dynamics. Second, the possibility of a generalization, and its form, of this equation to strongly coupled systems. We assume for the moment that the (initial) off-diagonal elements are such that the destruction term in the r.h.s. of eq. (3.19) decays sufficiently rapidly. Then the answer to the first question has been given by VAN HOVE [42], who has shown that eq. (5.1) is « exact » in the so-called « weak-coupling » limit: $\lambda \to 0$, $t \to \infty$, $\lambda^2 t$ finite. This asymptotic validity of the Pauli equation has been often reexamined and several of the arguments used have been made more precise and rigorous [43]. However, the « $\lambda^2 t$-approximation » does not provide an answer to what happens for finite values of the coupling parameter. Although $\Psi(+\,i0)$ determines the equilibrium states, the r.h.s. of eq. (5.1) needs to be modified. By analysing the generalized master equation, PRIGOGINE and RÉSIBOIS [18, 44] have shown that the appropriate form of an « asymptotic master equation » should be

$$i\partial_t \varrho_0(t) = \Theta \varrho_0(t)\,, \qquad \Theta = \Omega\Psi(+i0)\,, \tag{5.2}$$

where Ω is a functional of the collision operator. From this point of view eq. (5.1) should be considered as an approximation to eq. (5.2) and not directly to eq. (3.19). This approach is quite general and does not apply only to quantum systems to which we have referred in order to fix the ideas.

It is evident that the solution of eq. (5.2) represents a specific contribution, which we sometimes denote by $\tilde{\varrho}_0(t)$, to the P-component of the state $(\varrho_0(t) = \mathsf{P}\varrho(t))$. Therefore we must make precise the relation between $\tilde{\varrho}_0$ and ϱ_0 and specify the initial condition $\tilde{\varrho}_0(0)$ which should be imposed to eq. (5.2). Moreover, when $\Omega \neq I$, Θ is not necessarily dissipative (although, as we will see, its spectrum lies in the lower half-plane) and the existence of a nondecreasing functional which may be related to nonequilibrium entropy becomes questionable. As we have mentioned, this difficulty appears already in the case of a moderately dense gas where effectively Ω is nontrivial. In investigating these problems we were led to the theory of subdynamics [20] and the study of a class of invertible nonunitary transformations [21], called « star-unitary » transformations, which eventually permit the construction of a Liapounoff functional that may be used to obtain a general expression for nonequilibrium entropy.

We will review in the following the main aspects of the theory of subdynamics. The discussion is relatively simple [45-47] when one analyses the contributions to eq. (3.13) arising from isolated singularities of $\{\mathsf{PLP} + \Psi(z) - z\mathsf{P}\}^{-1}$. For this reason, here, we shall mostly refer to this case only.

When the Liouville-von Neumann operator has a discrete spectrum, the only possible singularities of this partial resolvent are real poles. This is simply another way to express the quasi-periodic nature of the time dependence. However, when the spectrum is continuous, we have a cut along the real axis and an analytic continuation, in some region of the lower half of the complex plane, might be possible. We assume here that such a continuation exists and that singularities appear off the real axis.

Thus, using the notation

$$\mathsf{M}(z) = \mathsf{PLP} + \Psi(z) - z\mathsf{P} \tag{5.3}$$

for the operator defined for $\operatorname{Im} z \geq 0$ and its analytic continuation below the real axis, we suppose that there exist values z_n ($\operatorname{Im} z_n \leq 0$) of the complex variable as well as vectors $|u_{n|\alpha}\rangle$ and $\langle v_{n|\alpha}|$ such that

$$\mathsf{M}(z_n)|u_{n|\alpha}\rangle = 0\,, \qquad \langle v_{n|\alpha}|\,\mathsf{M}(z_n) = 0\,, \qquad \alpha = 1, \ldots, k_n\,. \tag{5.4}$$

In other words we assume that the nonlinear eigenvalue problem $\mathsf{M}(z)|u\rangle = 0$

admits a set of solutions. Furthermore, we assume that this set of corresponding (left and right) eigenvectors is complete in the P-subspace. Note that certain hypotheses may be relaxed with appropriate modifications of the formalism. To each root z_n there corresponds a pole of the resolvent of L, the residue of which may be calculated in a straightforward manner. Here we should remark that, when we speak of the « analytic continuation » of the resolvent of L, we actually mean that of its matrix elements. We will come back to this point after some formal developments.

The simplest case is when the zeros of eq. (5.4) correspond to simple poles of $\mathsf{M}(z)^{-1}$. Then $\mathsf{M}(z)^{-1}$ can be written as

$$\mathsf{M}(z)^{-1} = \frac{1}{z - z_n}\,\mathsf{K}_n + \mathsf{T}_n(z)\,, \tag{5.5}$$

where $\mathsf{T}_n(z)$ has no singularity, at $z = z_n$. Here and in the following, when we speak of the inverse of an operator, we imply that this inverse is taken in a given subspace. We define now a pair of mutually orthogonal, but not necessarily Hermitian, projections F_n and G_n, where (for a given n the eigenvectors $|u_{n|\alpha}\rangle$ and $\langle v_{n|\beta}|$ can be chosen orthonormal)

$$\mathsf{F}_n = \sum_{\alpha=1}^{k_n} |u_{n|\alpha}\rangle\langle v_{n|\alpha}|\,, \qquad \mathsf{G}_n = \mathsf{P} - \mathsf{F}_n \tag{5.6}$$

in terms of which we may calculate K_n (and $\mathsf{T}_n(z_n)$). From eq. (5.5) we have

$$(z - z_n)\,\mathsf{P} = \mathsf{M}(z)\mathsf{K}_n + (z - z_n)\,\mathsf{M}(z)\,\mathsf{T}_n(z) = \mathsf{K}_n\,\mathsf{M}(z) + (z - z_n)\,\mathsf{T}_n(z)\,\mathsf{M}(z)\,. \tag{5.7}$$

Therefore the following relations

$$\mathsf{M}(z_n)\,\mathsf{K}_n = \mathsf{K}_n\,\mathsf{M}(z_n) = 0 \Rightarrow \mathsf{K}_n = \mathsf{F}_n\mathsf{K}_n\mathsf{F}_n \tag{5.8}$$

hold, where we have used that by construction F_n is the projection onto the null space of $\mathsf{M}(z_n)$. Differentiating eq. (5.7) with respect to z we obtain

$$\mathsf{P} = \mathsf{M}'(z_n)\,\mathsf{K}_n + \mathsf{M}(z_n)\,\mathsf{T}(z_n) \tag{5.9}$$

and consequently

$$\mathsf{F}_n = \mathsf{F}_n\mathsf{M}'(z_n)\,\mathsf{F}_n\mathsf{K}_n \Rightarrow \mathsf{K}_n = (\mathsf{F}_n\mathsf{M}'(z_n)\,\mathsf{F}_n)^{-1}\,. \tag{5.10}$$

The invertibility of $(\mathsf{F}_n\mathsf{M}'(z_n)\,\mathsf{F}_n)$ is of course connected to the assumption that z_n is a simple pole of $\mathsf{M}(z)^{-1}$. Note also that $\mathsf{M}'(z_n) = \Psi'(z_n) - \mathsf{P}$. Let $\tilde{\varrho}_n(t)$ be the contribution to eq. (3.13) due to z_n. With the above assumptions this is given by

$$(5.11)\qquad \begin{cases} \tilde{\varrho}_n(t) = \exp[-iz_n t]\Pi_n \varrho(0)\,, \\ \Pi_n = -[\mathsf{P} + \mathscr{C}(z_n)]\mathsf{K}_n[\mathsf{P} + \mathscr{D}(z_n)]\,. \end{cases}$$

Consequently the contribution $\tilde{\varrho}(t)$ due to the set of zeros $\{z_n\}$ is

$$(5.12)\qquad \tilde{\varrho}(t) = \sum_n \tilde{\varrho}_n(t) = \sum_n \exp[-iz_n t]\Pi_n \varrho(0)\,.$$

The results of some previous work on the theory of subdynamics [45-47] are recovered if we assume a spectral decomposition for the collision operator

$$(5.13)\qquad \mathsf{PLP} + \Psi(z) = \sum_n \sum_{\alpha=1}^{k_n} \mu_n(z)|u_{n|\alpha}(z)\rangle\langle v_{n|\alpha}(z)|$$

and assume that z_n is a solution of the dispersion relation

$$(5.14)\qquad \mu_n(z_n) - z_n = 0\,.$$

It may be shown then that

$$(5.15)\qquad \langle v_{n|\alpha}(z_n)|\mathsf{M}'(z_n)|u_{n|\beta}(z_n)\rangle = [\mu_n'(z_n) - 1]\,\delta_{\alpha\beta}\,,$$

i.e.

$$(5.16)\qquad \mathsf{K}_n = \sum_{\alpha=1}^{k_n} \frac{1}{\mu_n'(z_n) - 1}\,|u_{n|\alpha}\rangle\langle v_{n|\alpha}|\,.$$

Therefore, with eq. (5.14), we obtain directly a set of eigenvectors such that $\mathsf{M}'(z_n)$ is represented by a diagonal matrix in the F_n subspace. However, there are some technical advantages to construct $\tilde{\varrho}(t)$ using eq. (5.4) and obtain the correct set of vectors in a later stage.

Equation (5.12) may be expressed also as

$$(5.17)\qquad \tilde{\varrho}(t) = [\mathsf{P} + \mathsf{C}]\exp[-i\Theta t]\mathsf{A}[\mathsf{P} + \mathsf{D}]\varrho(0)\,.$$

This may be done by considering the sets of vectors $\{\langle\bar{v}_n|\}$ and $\{|\bar{u}_n\rangle\}$ which are bi-orthogonal to $\{|u_n\rangle\}$ and $\{\langle v_n|\}$, respectively. To simplify the notation, we have not used an index to indicate possible degeneracies in the eigenvalue problem defined by eq. (5.4). The operators which appear here are then given by

$$(5.18)\qquad \Theta = \sum_n z_n|u_n\rangle\langle\bar{v}_n|\,,$$

$$(5.19)\qquad \mathsf{A} = \sum_n \frac{1}{1 - \mu_n'(z_n)}\,|u_n\rangle\langle v_n|\,,$$

$$(5.20)\qquad \mathsf{D} = \sum_n |\bar{u}_n\rangle\langle v_n|\mathscr{D}(z_n)$$

and

(5.21) $$\mathsf{C} = \sum_n \mathscr{C}(z_n)|u_n\rangle\langle\bar{v}_n| .$$

We define also an operator Π as

(5.22) $$\Pi = \sum_n \Pi_n = [\mathsf{P} + \mathsf{C}]\mathsf{A}[\mathsf{P} + \mathsf{D}] ,$$

which is idempotent (but non-Hermitian),

(5.23) $$\Pi = \Pi^2 , \qquad \Pi \neq \Pi^\dagger \quad (\text{if } \operatorname{Im} z_n \neq 0 \text{ for some } n).$$

The proof of eq. (5.23) is based on the relation

(5.24) $$\mathsf{A}^{-1} = \mathsf{P} + \mathsf{DC} ,$$

which is readily verified by taking into account eq. (5.15) and that $\mathscr{D}(z_n)\mathscr{C}(z_n)$ must be interpreted as the analytic continuation of $-\Psi(z)'$ evaluated at $z = z_n$.

From eq. (5.17) we see that

(5.25) $$\tilde{\varrho}(0) = \Pi\varrho(0) .$$

Using eq. (5.23) we may then write

(5.26) $$\tilde{\varrho}(t) = \Sigma(t)\,\tilde{\varrho}(0) , \qquad \Sigma(t) = [\mathsf{P} + \mathsf{C}]\exp[-i\Theta t]\mathsf{A}[\mathsf{P} + \mathsf{D}] .$$

In the present formulation the concept of subdynamics may be expressed as the semi-group property which obeys formally the evolution operator $\Sigma(t)$, namely

(5.27) $$\Sigma(t)\,\Sigma(t') = \Sigma(t + t') , \qquad \Sigma(0) = \Pi \qquad (t, t' \geqslant 0) .$$

Therefore the concept of subdynamics refers to a particular class of contributions to the entire state of the state and not to its P-component alone.

If we consider $\tilde{\varrho}_0(t)$, the P-component of $\tilde{\varrho}(t)$, then from eq. (5.17) we see immediately that it satisfies eq. (5.2), with the initial condition $\tilde{\varrho}_0(0) = \mathsf{A}[\mathsf{P} + \mathsf{D}]\varrho(0)$ (if $\mathsf{Q}\varrho(0) = 0$, $\tilde{\varrho}_0(0) = \mathsf{A}\varrho(0)$). Because of our assumptions on the existence of solutions of eq. (5.4), Θ is given in eq. (5.18) in a closed form instead of the series defined by $\Omega\Psi$. One may show that, when this series converges, then its summation leads to eq. (5.4). The important point of the above discussion is that we have obtained a precise relation between $\tilde{\varrho}_0(t)$ and $\varrho(t)$ (not only $\varrho_0(t)$). Obviously there are several questions, mostly mathematical, which need further clarification. For instance, $\varrho_c(0)$ must be such that $\mathscr{D}(z)\varrho_c(0)$

can be continued analytically below the real axis and has no poles at the points $\{z_n\}$. Furthermore, we should remark that, for dissipative systems, the theory of subdynamics cannot be correctly interpreted in a normed space [48]. The reason is that the continuous spectrum of L, which is assumed to be real, is the natural boundary of its resolvent [49]. A simple Friedrichs-type model, which we have discussed briefly elsewhere [50], indicates the necessity of introducing spaces with a non-Hermitian metric, but we will not consider any further these problems here.

Although the operator Θ in eq. (5.2) has, as we have seen, a spectrum which (by construction) lies in the lower half of the complex plane, in general it is not dissipative. This means that, contrary to the asymptotic collision operator $\Psi(+i0)$, its anti-Hermitian part $(2i)^{-1}[\Theta - \Theta^\dagger]$ has no definite sign. Most of the difficulties encountered in the definition of a nonequilibrium entropy beyond the « Boltzmann approximation » may be traced to this fact. The semi-group property of $\Sigma(t)$, cf. eq. (5.27), which results when we treat simultaneously contributions to the P- and Q-component of the state, is the basis for resolving this problem. In fact, we have made the assumption that there exists a set of roots $\{z_n\}$ such that the corresponding eigenvectors in the nonlinear eigenvalue problem span the P-subspace. This implies that the projections P and Π are similar, *i.e.* that

$$\Pi = \Lambda \mathsf{P} \Lambda^{-1} . \tag{5.28}$$

Equation (5.28) is one of the determining relations of the (nonunitary) transformation Λ. It is not possible here to discuss extensively this transformation (see ref. [21]). We will indicate only its main properties. To construct it we must decompose the Q-subspace in a family of subspaces. Thus, for a quantum system, if P_ν projects onto the « ν-th » line parallel to the main diagonal (in a given representation), we write

$$\mathsf{Q} = \sum_{\nu \neq 0} \mathsf{P}_\nu \qquad (\mathsf{P}_0 = \mathsf{P}) . \tag{5.29}$$

Using similar methods as for Π, we may construct corresponding projections Π_ν, such that

$$\Pi_\nu = \Lambda\, \mathsf{P}_\nu \Lambda^{-1} , \qquad \Pi_\nu \Pi_\mu = \delta_{\nu\mu} \Pi_\nu \qquad (\Pi_0 = \Pi) . \tag{5.30}$$

There is still some indeterminacy which can be removed if we impose further conditions on Λ. Suppose, for example, that Λ depends continuously on a parameter, *e.g.* the coupling constant λ. Then we may ask that Λ satisfies the Kato equation [49]:

$$\partial_\lambda \Lambda = \tfrac{1}{2} \sum_\nu [\partial_\lambda \Pi_\nu, \Pi_\nu] \Lambda , \qquad \Lambda(\lambda = 0) = \mathsf{I} , \tag{5.31}$$

and the transformation is completely determined. However, as we shall see later, other choices are also possible.

An important aspect of the Λ-transformation is the definition of the so-called « physical representation », given by

$$\overset{(\text{p})}{\varrho} = \Lambda^{-1}\varrho \,. \tag{5.32}$$

In this representation the evolution is described by

$$i\partial_t \overset{(\text{p})}{\varrho} = \Phi \overset{(\text{p})}{\varrho} \,, \qquad \Phi = \Lambda^{-1}\mathsf{L}\Lambda \,, \tag{5.33}$$

which follows from the Liouville-von Neumann equation. When complex roots $\{z_{\nu|n}\}$ appear, Λ is not unitary and therefore Φ is not Hermitian. However, it is an accretive operator, *i.e.*, if $\Phi_{(0)}$ and $i\Phi_{(e)}$ are its Hermitian and skew-Hermitian parts, we have

$$\Phi = \Phi_{(0)} + i\Phi_{(e)} \,, \qquad \Phi_{(e)} \leqslant 0 \,. \tag{5.34}$$

The above inequality is called the « generalized dissipativity condition ». If it is satisfied, *i.e.* if $\Phi_{(e)} \neq 0$, eq. (5.33) is manifestly irreversible. Furthermore, the norm $\|\overset{(\text{p})}{\varrho}\| = (\overset{(\text{p})}{\varrho}, \overset{(\text{p})}{\varrho})^{\frac{1}{2}}$ is a decreasing function of time and it may be used to define a nonequilibrium entropy [21, 51]. It should be noticed here that through the Λ-transformation the Lt-invariance of the Liouville-von Neumann equation is broken. One may easily show that the Hermitian part of Φ is odd with respect to the L-inversion ($\Phi_{(0)}(\mathsf{L}) = -\Phi_{(0)}(-\mathsf{L})$). On the other hand, its anti-Hermitian part is even with respect to this operation ($\Phi_{(e)}(\mathsf{L}) = \Phi_{(e)}(-\mathsf{L})$). Thus Φ has the same symmetry properties as the asymptotic collision operator. In fact, we may interpret $\Phi_{(0)}$ as arising from reversible processes, while $\Phi_{(e)}$ arises from irreversible ones.

To make contact with eq. (5.2), we will consider the relation of Θ to Φ. From eqs. (5.22) and (5.28) we see that Λ must be of the form

$$\mathsf{P}\Lambda\mathsf{P} = \chi_0 \,, \qquad \mathsf{Q}\Lambda\mathsf{Q} = \chi_c \,, \qquad \mathsf{P}\Lambda\mathsf{Q} = -\mathsf{D}\chi_c \,, \qquad \mathsf{Q}\Lambda\mathsf{P} = \mathsf{C}\chi_0 \,. \tag{5.35}$$

Also that Λ^{-1} is

$$\mathsf{P}\Lambda^{-1}\mathsf{P} = \overset{*}{\chi}_0 \,, \qquad \mathsf{Q}\Lambda^{-1}\mathsf{Q} = \overset{*}{\chi}_c \,, \qquad \mathsf{P}\Lambda^{-1}\mathsf{Q} = \overset{*}{\chi}_0\mathsf{D} \,, \qquad \mathsf{Q}\Lambda^{-1}\mathsf{P} = -\overset{*}{\chi}_c\mathsf{C} \tag{5.36}$$

with

$$\chi_0\overset{*}{\chi}_0 = \mathsf{A} \,. \tag{5.37}$$

The notation takes into account that Λ is a « star-unitary » transformation, *i.e.* that, as a functional of L, it has the property

$$\Lambda^{-1}(\mathsf{L}) = \Lambda^{\dagger}(-\mathsf{L}) \equiv \Lambda^{*}(\mathsf{L}) . \tag{5.38}$$

By a direct calculation we obtain that

$$\mathsf{P}\Phi\mathsf{P} = \chi_0^{*}[\mathsf{PLP} + \mathsf{DQLP} + \mathsf{PLQC} + \mathsf{DQLQC}]\chi_0 . \tag{5.39}$$

From the definitions, cf. eqs. (5.18) to (5.21), one may deduce that C satisfies the nonlinear equation

$$\mathsf{CPLQC} - \mathsf{QLQC} + \mathsf{CPLP} = \mathsf{QLP} , \tag{5.40}$$

and that Θ is related to C by

$$\Theta = \mathsf{PLP} + \mathsf{PLQC} . \tag{5.41}$$

Thus, using these relations as well as eqs. (5.24) and (5.37), we see that eq. (5.39) becomes

$$\mathsf{P}\Phi\mathsf{P} = \chi_0^{-1}\Theta\,\chi_0 . \tag{5.42}$$

The meaning of eq. (5.42) is easily understood. From the structure of Θ, cf. eq. (5.18), we observe that Θ is similar to an accretive operator. In fact, it is sufficient to use a transformation of the form

$$\chi_0 = \sum_n \alpha_n |u_n\rangle\langle e_n| , \tag{5.43}$$

where $\{\alpha_n\}$ are arbitrary constants and $\{\langle e_n|\}$ some orthonormal basis in the P-subspace. The transformation theory, that we have indicated, leads to Λ, from which a unique χ_0 is obtained.

In eq. (5.38) we have introduced an operation called the « star-conjugation ». We may note that the L-inversion plays a central role in the investigation of the symmetries of the operators of the kinetic description. This is due to the duality of the Heisenberg and Schrödinger pictures. In general, for dissipative systems, the usual Hermitian conjugation is replaced by the star-conjugation. Thus we have

$$\mathsf{A}^* = \mathsf{A} , \qquad \mathsf{C}^* = \mathsf{D} \Rightarrow \Pi^* = \Pi . \tag{5.44}$$

These relations follow from the corresponding ones for the collision, destruction and creation operators, namely

$$\Psi^*(z) = -\Psi(-\bar{z}) , \qquad \mathscr{D}^*(z) = \mathscr{C}(-\bar{z}) , \qquad \mathscr{C}^*(z) = \mathscr{D}(-\bar{z}) , \tag{5.45}$$

which can be established easily from their definition [47]. Equation (5.44) shows that Π is « star-Hermitian » in accordance with eq. (5.28) and the star-unitarity of Λ.

We may view the problem of the existence of a nonincreasing functional of the state of the system also in a somewhat different and more general manner [52]. The Hermiticity of L implies that (ϱ, ϱ) remains constant in time. Now we may ask whether a positive operator M exists such that eqs. (2.9) are satisfied, *i.e.*

$$\Omega = (\varrho, \mathsf{M}\varrho) \geqslant 0, \qquad \frac{\mathrm{d}\Omega}{\mathrm{d}t} = i(\varrho, [\mathsf{ML} - \mathsf{LM}]\varrho) \leqslant 0 . \tag{5.46}$$

Obviously, $\mathrm{d}\Omega/\mathrm{d}t$ vanishes when M commutes with L. On the other hand, conditions on the dynamics must be imposed in order to have an inequality. In particular, L should have an absolutely continuous spectrum [53]. These ideas can be connected to the transformation theory that we have briefly mentioned. Under the assumption that Λ is invertible, M may be written as

$$\mathsf{M} = (\Lambda^{-1})^{\dagger}\Lambda^{-1} , \tag{5.47}$$

because it is a positive operator.

To conclude this section, we remark that the preceding considerations imply that, in general, the evolution of a system may be described either by a reversible equation, *i.e.* eqs. (3.3) and (3.7), or by an irreversible one, eq. (5.33). They are related by the (possibly unbounded) transformation Λ. For dissipative systems M does not commute with L and this means that the two descriptions are in some sense complementary. Note also that Poincaré's objection [2] does not apply here because the entropy is not associated to some phase function but to an operator on the state space of the system.

6. – Linearized hydrodynamics.

We will briefly discuss now the derivation of the linearized equations of hydrodynamics from kinetic theory. Our considerations will be limited to the case of a simple fluid of classical particles of mass m interacting with short-range, central forces derived from a potential V. Thus, its Hamiltonian is given by

$$H = \sum_{k=1}^{N} \tfrac{1}{2} m v_k^2 + \tfrac{1}{2} \sum_{k} \sum_{l \neq k} V(|x_k - x_l|) , \tag{6.1}$$

where x_k is the position and v_k the velocity of the k-th particle. We assume that the state of the system (for all times) is close to canonical equilibrium (T is the absolute temperature)

$$(6.2)\qquad \varrho_{\text{eq}} = \exp[-H/kT]/Z\,, \qquad Z = \int (\mathrm{d}v)^N (\mathrm{d}x)^N \exp[-H/kT]\,,$$

and that a linearized description is justified.

The microscopic theory of « hydrodynamical modes », based on a generalized kinetic equation for the one-particle distribution function $f(x, v, t)$, will be our starting point. This theory has been developed by RÉSIBOIS [54] and it is an extension of some previous work done on the level of the linearized Boltzmann equation (see, *e.g.*, ref. [55]). Moreover, similar methods have been used in the theory of correlation functions [56, 57]. In this context, « hydrodynamical modes » means certain long-wavelength contributions to the one-particle distribution function, the frequencies of which can be identified with those of the corresponding modes obtained from the macroscopic (linearized) hydrodynamical equations. However, when we consider the derivation of the macroscopic equations from kinetic theory, we have to take into account effects from two-particle correlations. In other words, we need a theory which is not limited to the one-particle distribution function. We will show (for more details see ref. [58, 59]) that the theory of subdynamics provides an adequate framework to deal with this problem.

The one-particle distribution function $f(x, v, t)$ is written as

$$(6.3)\qquad f(x, v, t) = n\varphi(v)[1 + h(x, v, t)]\,,$$

where $n = N/V$ is the mean number density and

$$\varphi(v) = (m/2\pi kT)^{\frac{3}{2}} \exp[-mv^2/kT]$$

is the Maxwellian velocity distribution. By hypothesis, the amplitude $h(x, v, t)$ of the deviation from equilibrium is small and thus the evolution of $f(x, v, t)$ can be approximated by a linearized kinetic equation. To simplify the notation we will take in the following the mass m of the particles equal to one. As we will assume an infinitely extended fluid, we will directly work with the Fourier coefficient $f_q(v, t)$ of the distribution function, as well as those of the other quantities which appear in our problem.

Formally, a linear equation for $f_q(v, t)$ may be obtained from the Liouville equation, where here

$$(6.4)\qquad \mathsf{L} = -i\sum_k v_k\cdot\frac{\partial}{\partial x_k} + i\,\frac{1}{2}\sum_k\sum_{l\neq k}\frac{\partial V(|x_k - x'_k|)}{\partial x_k}\cdot\left(\frac{\partial}{\partial v_k} - \frac{\partial}{\partial v_l}\right),$$

by using projection methods. One may show [60] that P_q defined by

$$(6.5)\qquad \langle x, v|\mathsf{P}_q|x', v'\rangle = \sum_{k=1}^{N} \varrho_{\text{eq}} \exp[iq\cdot(x_k - x'_k)]\,K_q(v_k, v'_k)$$

with

$$K_q(v, v') = V\{\delta(v - v') - B_q\varphi(v)\}\varphi(v)^{-1}, \qquad B_q = \frac{ng_q}{1 + ng_q}, \tag{6.6}$$

where g_q is the Fourier component of the radial distribution function, is a projection ($\mathsf{P}_q^2 = \mathsf{P}_q = \mathsf{P}_{-q}^\dagger$). Let us note also that in the thermodynamic limit ($N \to \infty$, $V \to \infty$, $N/V = n$ constant) B_q is simply the Fourier transform C_q of the equilibrium direct correlation function [61]. Then from eq. (3.3) we derive, as in the third section, the generalized master equation for $\mathsf{P}_q \varrho$, cf. eq. (3.19). Integrating over the positions and velocities of all particles except one, we get for $f_q(v, t)$ the equation

$$\begin{aligned}\partial_t f_q(v, t) + iq\cdot v f_q(v, t) - iq\cdot vC_q n\varphi(v)\int \mathrm{d}v' f_q(v', t) = \\ = -i\int_0^t \mathrm{d}\tau\int \mathrm{d}v'\, G_q(v, v', \tau) f_q(v', t-\tau) + F_q(t)\,.\end{aligned} \tag{6.7}$$

Here the second term in the l.h.s. is the flow term, while the third term may be interpreted as the effect of a mean field. In the r.h.s. the first term is the linearized collision term and the second one a source term due to initial correlations and nonlinear effects arising from inhomogeneities.

The time dependence of $f_q(v, t)$ is quite complicated, because all processes taking place in the system contribute to it. From processes which can eventually be identified with macroscopic transport phenomena, we will have some particular contributions. They can be studied most conveniently by analysing the Laplace transform of eq. (6.7), which reads

$$\begin{aligned} i\int \mathrm{d}v'\{q\cdot v\delta(v - v') - q\cdot vnC_q\varphi(v) + \Psi_q(v, v'|z) - z\delta(v - v')\}\hat{f}_q(v') = \\ = f_q(v, 0) + \hat{F}_q(z)\,.\end{aligned} \tag{6.8}$$

We suppose now that the initial conditions, in particular $(I - \mathsf{P}_q)\varrho(t = 0)$, are such that $\hat{F}_q(z)$ admits an analytic continuation in some domain below the real axis. Similarly we suppose that the (linearized) collision operator $\Psi_q(z)$, *i.e.* the Laplace transform of the kernel $G_q(t)$, can be continued analytically below the axis. With these assumptions hydrodynamical modes may be defined as wavelike contributions to the solution of eq. (6.7) with a (complex) eigenfrequency z_q which vanishes for $q = 0$, and they may be determined by looking for solutions of the equation

$$[q\cdot v - z_q]\hat{h}_q(v) - q\cdot vnC_q\varphi(v)\int \mathrm{d}v'\hat{h}_q(v') + \int \mathrm{d}v'\,\Psi_q(v, v'|z_q)\hat{h}_q(v') = 0\,. \tag{6.9}$$

Next we observe that, by hypothesis, for $q = 0$, $\hat{h}_0(v)$ must reduce to a collisional invariant, *i.e.*

$$\int \mathrm{d}v' \Psi_0(v, v'|+i0)\, \hat{h}_0(v') = 0\,. \tag{6.10}$$

As we deal with a one-component fluid, composed of particles without internal structure, we may suppose that eq. (6.10) admits only five solutions corresponding to the density, momentum and (kinetic) energy. Thus zero is a fivefold degenerate eigenvalue of the asymptotic collision operator.

One treats eq. (6.9) as a perturbation of eq. (6.10) with the wave vector q as a smallness parameter. Assuming that the five frequencies $z_q^{(n)}$ admit an expansion of the form

$$z_q^{(n)} = b^{(n)} q + d^{(n)} q^2 + O(q^{2+\varepsilon})\,, \tag{6.11}$$

one computes the coefficients $b^{(n)}$ and $d^{(n)}$ in terms of the matrix elements of the operators which appear in eq. (6.9). This aspect has been studied extensively by RÉSIBOIS [54]. The eigenfrequencies correspond to the two sound modes, the doubly degenerate shear mode and the thermal mode. If we now assume that we may identify the frequencies so obtained with the ones derived from the linearized equations of hydrodynamics, we get microscopic expressions for the thermodynamic derivatives and the transport coefficients of the fluid in terms of matrix elements of the first and second derivatives with respect to q and z of the collision operator. Moreover, it may be shown that expressions obtained in this manner are equivalent to the Green-Kubo [62] formulae for the transport coefficients.

To formulate the problem of the derivation of linearized hydrodynamics, we first note that the solutions of eq. (6.9), which correspond to the hydrodynamical frequencies, give rise to a contribution $\tilde{f}_q(v, t)$ to the one-particle distribution function. We may repeat part of the analysis of the previous section, to obtain for $\tilde{f}_q(v, t)$ an equation of the form

$$i\,\partial_t \tilde{f}_q(v, t) = \Theta_q \tilde{f}_q(v, t)\,, \tag{6.12}$$

where Θ_q is given by an expression similar to eq. (5.18), with eigenvalues the five frequencies z_q and eigenvectors the corresponding solutions of eq. (6.9). Now the remaining contributions to $f_q(v, t)$ have a completely different time dependence which cannot be directly linked to hydrodynamical processes. Therefore, if we take the moments of $\tilde{f}_q(v, t)$

$$\delta n_q(t) = \int \mathrm{d}v\, \tilde{f}_q(v, t)\,, \tag{6.13a}$$

$$n\, \delta u_q(t) = \int \mathrm{d}v\, v \tilde{f}_q(v, t)\,, \tag{6.13b}$$

$$n\, (3/2T)\, \delta T_q^k(t) = \int \mathrm{d}v\, (v^2/2kT - \tfrac{3}{2})\, \tilde{f}_q(v, t)\,, \tag{6.13c}$$

we may identify them with the « macroscopic » contributions to the number density, the mean velocity and the kinetic temperature. Strictly speaking, a different notation should be used to distinguish these quantities from those defined with the (total) distribution $f_q(v, t)$, but we think that it is not necessary to add more notation. The question is whether the moments of $\tilde{f}_q(v, t)$ satisfy, as in the case of the Boltzmann equation, the equations of hydrodynamics. Of course, the comparison can be carried only up to second order in the wave vector, because in macroscopic hydrodynamics it is assumed that the gradients are weak.

It turns out that to $O(q^2)$ these moments satisfy a closed set of equations, namely (we write n_q instead of δn_q etc.)

$$(6.14a)\qquad i\partial_t n_q(t) = qnu_q^{\parallel}(t)\,,$$

$$(6.14b)\qquad i\partial_t u_q^{\parallel}(t) = q\left(\frac{\partial P}{\partial n}\right)_T \frac{1}{n}\, n_q(t) + q\left(\frac{\partial P}{\partial T}\right)_n \frac{1}{n}\, T_q^k(t) - \\ - iq^2\left(\left(\frac{4}{3}\eta + \zeta + \mu\right)\Big/ n\right) u_q^{\parallel}(t)\,,$$

$$(6.14c)\qquad i\partial_t u_q^{\perp}(t) = -iq^2 \frac{\eta}{n}\, u_q^{\perp}(t)\,,$$

$$(6.14d)\qquad i\partial_t T_q^k(t) = q\left(\frac{\partial P}{\partial T}\right)_n \frac{T}{nc_v}\, u_q^{\parallel}(t) - iq^2\left(\frac{\varkappa}{nc_v} - \frac{\mu}{n}\right) T_q^k(t) + \\ + iq^2\frac{\mu}{n}\left(\frac{\partial P}{\partial n}\right)_T \left(\frac{\partial T}{\partial P}\right)_n n_q(t)\,.$$

Here $u_q^{\parallel}$ is the component of u_q parallel to the vector q and $u_q^{\perp}$ the perpendicular one.

In writing eqs. (6.14) we have used the fact that the derivatives with respect to q and z of the matrix elements

$$(6.15)\qquad \langle i|\alpha(q, z)|j\rangle = \langle i|\bar{\Psi}_q(z)|j\rangle - \langle i|\bar{\Psi}_q(z)\,\bar{Q}\,\frac{1}{\bar{Q}\Psi_q(z)\,\bar{Q} - z}\,\bar{Q}\bar{\Psi}_q(z)\,|j\rangle$$

at $q = 0$, $z = 0$ can be expressed in terms of the thermodynamic derivatives, the transport coefficients and the coefficient μ. We do not present here explicit formulae, except for μ which is given by

$$(6.16)\qquad \mu = \frac{\dot{\alpha}_{52}}{1 - \alpha'_{55}}\left(\dot{\alpha}'_{52} + \alpha''_{55}\,\frac{\dot{\alpha}_{52}}{1 - \alpha'_{55}}\right),$$

where $\langle i|$ ($|i\rangle$) for $i = 1, \ldots, 5$ denote the five invariants 1, $v_2/(kT)^{\frac{1}{2}}$, $v_3/(kT)^{\frac{1}{2}}$, $v_4/(kT)^{\frac{1}{2}}$ and $(v^2/2kT - \frac{3}{2})(\frac{2}{3})^{\frac{1}{2}}$, respectively, v_2 being parallel to q. $\bar{Q}$ is the projection on the subspace orthogonal to the one spanned by the five invariants.

$\overline{\Psi}_q(z)$ stands for the sum of $\Psi_q(z)$, the free term and the mean-field term. Finally, in eq. (6.15) we dashed the time derivatives, dotted the space ones and evaluated them at $q=0$, $z=0$.

If the coefficient μ vanishes, eqs. (6.14) reduce to the hydrodynamical ones. This is indeed the case for a dilute or a moderately dense gas, where the linearized collision operator (Boltzmann or Choh-Uhlenbeck) is independent of q and it is either frequency independent or depends linearly on z.

Clearly the difference with the equations of hydrodynamics is due to the fact that we have not included in the definition of the temperature possible contributions from potential energy. Let us consider the energy density

$$H(x)=\sum_{k=1}^{N}\left\{\tfrac{1}{2}u_k^2+\tfrac{1}{2}\sum_{l\neq k}V(|x_k-x_l|)\right\}\delta(x_k-x)\,. \tag{6.17}$$

Then, in terms of its Fourier transform H_q we may define an average energy per mode e_q, which in the linear domain is

$$e_q=\int(\mathrm{d}x)^N(\mathrm{d}v)^N H_q\,\delta\varrho\,, \tag{6.18}$$

where $\delta\varrho$ is the deviation of the N-particle state from equilibrium ($\varrho=\varrho_{\mathrm{eq}}+\delta\varrho$). In the r.h.s. of eq. (6.18) we decompose $\delta\varrho$ using the projections P_q and $\mathsf{Q}_q=\mathsf{I}-\mathsf{P}_q$:

$$e_q=\int(\mathrm{d}x)^N(\mathrm{d}v)^N H_q(\mathsf{P}_q\,\delta\varrho+\mathsf{Q}_q\,\delta\varrho)\,. \tag{6.19}$$

The kinetic temperature is obtained from the average of the kinetic energy $\left(H_q^k=\sum_{k=1}^{N}\frac{1}{2}v_k^2\exp[iq\cdot x_k]\right)$. In our analysis, the hydrodynamical modes are associated to a specific contribution, which we denote by $\widetilde{\mathsf{P}_q\,\delta\varrho}$, to the P_q-component of the state which corresponds to the contribution $\tilde{f}_q(v,t)$ of the one-particle distribution. The formalism of subdynamics, of the previous section, predicts that the contribution $\widetilde{\mathsf{Q}_q\,\delta\varrho}$ to the Q_q-component, which has the same time dependence as $\widetilde{\mathsf{P}_q\,\delta\varrho}$, is given by

$$\widetilde{\mathsf{Q}_q\,\delta\varrho}=\mathsf{C}_q\widetilde{\mathsf{P}_q\,\delta\varrho}\,. \tag{6.20}$$

Here the operator C_q is defined in a similar way as the operator C in eq. (5.21).

By a direct calculation, we get for $\tilde{e}_q$ (to first order in q)

$$\tilde{e}_q=\left(\frac{\partial e}{\partial n}\right)_T n_q+nc_v T_q\,, \tag{6.21}$$

where T_q is the kinetic temperature of eq. (6.13c) plus a potential correction which reads

$$T^p_q(t) = -iq\left(\frac{\partial T}{\partial P}\right)_n \mu u^{\parallel}_q(t)\,. \tag{6.22}$$

It should be noted that relations of this type have been derived sometime ago by ERNST and others [63]. They appear when discussing the subsidiary conditions of the Chapman-Enskog method used in the derivation of macroscopic transport theory from nonlinear kinetic equations.

It is a trivial matter to show that eqs. (6.14) and (6.22) lead to linearized hydrodynamics, *i.e.*

$$i\,\partial_t n_q(t) = qnu^{\parallel}_q(t)\,, \tag{6.23a}$$

$$i\,\partial_t u^{\parallel}_q(t) = q\left(\frac{\partial P}{\partial n}\right)_T \frac{1}{n}\, n_q(t) + q\left(\frac{\partial P}{\partial T}\right)_n \frac{1}{n}\, T_q(t) - iq^2\big(((4/3)\eta + \zeta)/n\big)\, u^{\parallel}_q(t)\,, \tag{6.23b}$$

$$i\,\partial_t u^{\perp}_q(t) = -iq^2\frac{\eta}{n}\, u^{\perp}_q(t)\,, \tag{6.23c}$$

$$i\,\partial_t T_q(t) = q\left(\frac{\partial P}{\partial T}\right)_n \frac{T}{nc_v}\, u^{\parallel}_q(t) - iq^2\frac{\varkappa}{nc_v}\, T_q(t)\,. \tag{6.23d}$$

Thus from the one-particle kinetic theory we may compute the hydrodynamical frequencies. However, to deduce the linearized transport equations we need a more general theory which accounts also for correlations. We have indicated how this question can be handled in a straightforward manner using the theory of subdynamics.

We could derive eq. (6.23) from eq. (6.14) using the theory of nonunitary transformations associated to that of subdynamics. This may be understood also from the following observation. The 5×5 matrices in the r.h.s. of eqs. (6.14) and (6.23) have, by construction, the same eigenvalues and consequently they are similar. Let χ_q be this similarity transformation, which to first order in q is

$$\chi_q = \begin{bmatrix} 1 & 0 & 0 & 0 & 0 \\ 0 & 1 & 0 & 0 & 0 \\ 0 & 0 & 1 & 0 & 0 \\ 0 & 0 & 0 & 1 & 0 \\ 0 & iq(\mu/\dot{\alpha}_{25}) & 0 & 0 & (3k/2c_v)^{\frac12} \end{bmatrix}. \tag{6.24}$$

Then we must have

$$\begin{bmatrix} n_q/n \\ u_q/(kT)^{\frac12} \\ (c_v/k)^{\frac12}T_q/T \end{bmatrix} = \chi_q^{-1} \begin{bmatrix} n_q/n \\ u_q/(kT)^{\frac12} \\ (\frac32)^{\frac12}T^k_q/T \end{bmatrix}, \tag{6.25}$$

which implies

$$T_q = T_q^k - iq\left(\frac{\partial T}{\partial P}\right)_n \mu u_q^{\parallel}\,, \tag{6.26}$$

i.e. we recover eq. (6.21). Moreover, the Gibbs formula for the entropy, cf. eq. (2.8),

$$\delta^2 s = -\frac{1}{T}\left[\left(\frac{\partial P}{\partial n}\right)_T \frac{|n_q|^2}{n^2} + |u_q|^2 + c_v \frac{|T_q|^2}{T^2}\right] \tag{6.27}$$

is a Liapounoff function for eq. (6.23). It is, up to a scale factor, proportional to the square of the « transformed » variables, as expected on general grounds. As a matter of fact, the considerations of this section lead to a microscopic derivation of Gibbs formula, which generalizes that given by one of us [64] (see also ref. [27]) some time ago, in the case of dilute gas.

7. – Concluding remarks.

We have reviewed here some basic methods used in the description of dissipative phenomena in dynamical systems. We have mainly stressed the new concepts which are necessary in order to construct the proper framework for a microscopic theory of irreversible processes. Of course, further investigations are necessary in order to get a mathematically rigorous theory. However, from a formal point of view, we think that the approach which we have presented is quite satisfactory.

The generalized master equation, which is a consequence of the decomposition of the states in two components, does not characterize the dissipativity of the system. Such a characterization requires the study of the asymptotic properties of the operators which we introduced in the third section and, in particular, those of the collision operator. In this way we may determine classes of initial conditions for which eventually the dynamical system tends to an equilibrium state. Obviously, an arbitrary initial condition may give rise to a destruction fragment—the « source » term in the r.h.s. of eq. (3.19)—which cannot be neglected for (arbitrarily) long times. This is in fact the case when we consider initial conditions resulting from an inversion of the velocities of a N-particle system. The possible existence of singular invariants illustrates vary clearly this point. Such invariants need not be excluded *ab initio*, *i.e.* a kinetic description is possible for nonergodic systems and applies to particular classes of states.

To establish the connection with irreversible thermodynamics we must impose more severe conditions on the dynamics which permit an analytic continuation of classes of matrix elements of the resolvent of the Liouville/von

Neumann operator. A necessary condition is the (absolute) continuity of the spectrum of L, which implies an « infinite » system and thus Poincaré recurrences are excluded. Nevertheless, this condition is not sufficient, because, as we have seen, we must require also that certain dispersion relations have complex solutions.

With the introduction of nonunitary transformations, we arrive at a new conceptual framework for the relation between reversible and irreversible equations. For systems which admit a nontrivial kinetic description, the Λ-transformation, which leads to the « physical representation », transforms the Hermitian operator L to the operator Φ which has a nonnegative anti-Hermitian part. Therefore, we may regard the irreversible kinetic description not as an approximation to the evolution of the system but as a different representation of it.

The linear theory of hydrodynamical modes and the microscopic derivation of the linearized hydrodynamical equations has been chosen as an example to illustrate the ideas of the theory of subdynamics. This is a subject of actual interest, because, by developing a general approach to these phenomena, one hopes to get a better understanding of the behaviour of dense systems near equilibrium. Our discussion may be extended to more complex cases, such as many-component fluids, polyatomic liquids and plasmas. It should be emphasized that, in order to obtain the Gibbs formula for the entropy, the choice of the projection P_q leading to a linear equation for the one-particle distribution is not sufficient. We need to account for contributions coming from this subspace and its complement as well.

Although in these lectures we have had in mind the application of the formalism to irreversible phenomena which appear in macroscopic systems, we may also consider the meaning of irreversibility on the microscopic level as well. As a matter of fact, these methods permit to discuss the questions connected with the formulation of a theory for decaying states in quantum systems which is a subject of continuing interest (see, *e.g.*, ref. [65, 66]). Unstable states do not correspond to eigenvalues of the Hamiltonian and thus their definition poses several difficult physical and mathematical problems. This question has been studied in the case of the Friedrichs model (for a review of quantum Hamiltonians which lead to this model, see ref. [67]). In the physical representation an equation has been obtained which describes simultaneously the decay and scattering processes. These (irreversible) processes give rise to a « microscopic » entropy which has been constructed explicitly [68].

As we have mentioned, there is a certain indeterminacy to the definition of the Λ-transformation. In the context of unstable states this indeterminacy may be removed as follows. An operator H acting on the density matrices is defined by $\mathsf{H}\varrho = \frac{1}{2}(H\varrho + \varrho H)$. We impose then the condition that $\Lambda^{-1}\mathsf{H}\Lambda$ is a diagonal form representing the energies of the states of the physical particles.

In a recent article [69] the implications of this procedure for the description of dissipative quantum systems has been analysed in detail. Here we would like to notice that it might be considered as a generalization of the Heisenberg-Born-Jordan quantization rules to which it reduces when the spectrum of the Hamiltonian is discrete.

* * *

We wish to acknowledge the collaboration of Dr. M. THEODOSOPULU in the preparation of this paper. We thank also M. DE HAAN, Cl. GEORGE, F. HENIN-JEENER, M. MARESCHAL and F. MAYNÉ for several discussions. This work has been partially supported by « Actions de Recherches Concertées » (Conv. 76/81 II 3) and the « R. Welch Foundation » (Houston, Texas).

REFERENCES

[1] L. BOTZMANN: *Wien. Ber.*, **66**, 275 (1872).
[2] H. POINCARÉ: *Thermodynamique* (Paris, 1923).
[3] E. G. D. COHEN and W. THIRRING, Editors: *The Boltzmann Equation: Theory and Applications* (Wien, 1973) (*Acta Phys. Austriaca, Suppl. X*).
[4] C. CERCIGNANI: *Mathematical Methods in Kinetic Theory* (New York, N. Y., 1969).
[5] R. BALESCU: *Equilibrium and Nonequilibrium Statistical Mechanics* (New York, N. Y., 1975).
[6] P. RÉSIBOIS and M. DE LEENER: *Classical Kinetic Theory of Fluids* (New York, N. Y., 1977).
[7] E. T. JAYNES: in *Delaware Seminar in the Foundations of Physics*, edited by M. BUNGE (Berlin, 1967), and references cited there.
[8] E. P. WIGNER: in the *Proceedings of the Robert A. Welch Foundation Conferences on Chemical Research*, XVI. *Theoretical Chemistry* (Houston, Tex., 1972).
[9] For the early history of kinetic theory see S. G. BRUSH: *The Kind of Motion We Call Heat* (two volumes) (Amsterdam, 1976).
[10] J. C. MAXWELL: *Proc. Camb. Phil. Soc.*, **12**, 547 (1879).
[11] L. BOLTZMANN: *Journ. Reine Angew. Math.*, **100**, 201 (1887).
[12] P. EHRENFEST and T. EHRENFEST: *Begriffliche Grundlagen des statistischen Auffassung in der Mechanik*, in *Encyklopädie der mathematischen Wissenschaften*, Bd. 4, Teil 32 (Leipzig, 1911) (English translation by M. J. MORAVCSIK: *The Conceptual Foundation of the Statistical Approach in Mechanics* (Ithaca, N. Y., 1959)).
[13] A. N. KOLMOGOROFF: in *Proceedings of the 1954 International Congress of Mathematics* (Amsterdam, 1957) (English translation in R. ABRAHAM: *Foundations of Mechanics* (Appendix D) (New York, N. Y., 1967)).
[14] L. BOLTZMANN: *Vorlesungen über Gastheorie*, Part I (Leipzig, 1896), Part II (Leipzig, 1898) (English translation by S. G. BRUSH: *Lectures on Gas Theory* (Berkeley, Cal., 1964)).
[15] W. PAULI: in *Probleme der modernen Physik*, edited by P. DEBYE (Leipzig, 1928).
[16] E. ZERMELO: *Ann. der Phys.*, **57**, 485 (1896).
[17] J. LOSCHMIDT: *Wien Ber.*, **73**, 128 (1876).
[18] I. PRIGOGINE: *Nonequilibrium Statistical Mechanics* (New York, N. Y., 1962).

[19] S. T. CHOH: Dissertation, University of Michigan (1958); also issued as U.S. Technical Report (P.B. 138 502) by S. T. CHOH and G. E. UHLENBECK.
[20] I. PRIGOGINE, C. GEORGE and F. HENIN: *Physica*, **45**, 418 (1969).
[21] I. PRIGOGINE, C. GEORGE, F. HENIN and L. ROSENFELD: *Chem. Scripta*, **4**, 5 (1973), and references cited there.
[22] M. JAMMER: *The Conceptual Development of Quantum Mechanics* (New York, N. Y., 1966).
[23] A. EINSTEIN: *Verh. Dtsch. Physik. Ges.*, **18**, 318 (1916); *Phys. Zeits.*, **18**, 121 (1917).
[24] P. GLANSDORFF and I. PRIGOGINE: *Thermodynamic Theory of Structure, Stability and Fluctuations* (London, 1971).
[25] I. PRIGOGINE: *From Being to Becoming*, San Francisco, to appear (1978).
[26] I. PRIGOGINE: *Introduction to Thermodynamics of Irreversible Processes* (New York, N. Y., 1961); 3rd Edition (1967).
[27] S. R. DE GROOT and P. MAZUR: *Nonequilibrium Thermodynamics* (Amsterdam, 1962).
[28] F. HAAKE: in *Springer Tracts in Modern Physics*, Vol. **66** (Berlin, 1973).
[29] R. ZWANZIG: *Physica*, **30**, 1109 (1964), and references cited there.
[30] M. BAUS: *Bull. Classe Sci., Acad. Roy. Belg.*, **53**, 1291, 1332, 1352 (1967).
[31] R. BALESCU, P. CLAVIN, P. MANDEL and J. W. TURNER: *Bull. Classe Sci., Acad. Roy. Belg.*, **55**, 1055 (1969).
[32] A. P. GRECOS: *Physica*, **51**, 50 (1971).
[33] E. HILLE and R. S. PHILLIPS: *Functional Analysis and Semi-Groups* (Providence, R. I., 1957).
[34] I. PRIGOGINE and A. P. GRECOS: in *75 Jahre Quantentheorie*, edited by W. BRAUER, H.-W. STREITWOLF and K. WERNER (Berlin, 1977).
[35] E. T. WHITTAKER: *A Treatise on Analytical Dynamics of Particles and Rigid Bodies* (London, 1964).
[36] V. I. ARNOLD and A. AVEZ: *Ergodic Problems of Classical Mechanics* (New York, N. Y., 1968).
[37] J. MOSER: *Stable and Random Motions in Dynamical Systems* (Princeton, N. J., 1973).
[38] J. FORD: *Adv. Chem. Phys.*, **24**, 155 (1973).
[39] I. PRIGOGINE, A. P. GRECOS and CL. GEORGE: *Proc. Nat. Acad. Sci.*, **73**, 1802 (1976); *Cel. Mech.*, **16**, 489 (1977).
[40] H. POINCARÉ: *Les méthodes nouvelles de la mécanique céleste* (Paris, 1892); reissue (New York, N. Y., 1957).
[41] G. M. ZASLAVSKII and B. V. CHIRIKOV: *Sov. Phys. Usp.*, **14**, 549 (1972).
[42] L. VAN HOVE: *Physica*, **21**, 517 (1955); **23**, 441 (1957).
[43] E. B. DAVIES: *Theory of Open Systems* (New York, N. Y., 1976).
[44] I. PRIGOGINE and P. RÉSIBOIS: *Physica*, **27**, 629 (1961).
[45] A. P. GRECOS and I. PRIGOGINE: in *Théories cinétiques classiques et relativistes*, edited by G. PICHON (Paris, 1975).
[46] A. P. GRECOS, T. GUO and W. GUO: *Physica*, **80** A, 421 (1975).
[47] A. P. GRECOS and M. THEODOSOPULU: *Acta Phys. Polon.*, **50** A, 749 (1976).
[48] L. LANZ, L. A. LUGIATO and G. RAMELLA: *Physica*, **54**, 94 (1971).
[49] T. KATO: *Perturbation Theory for Linear Operators* (New York, N. Y., 1966).
[50] A. P. GRECOS: in the *Proceedings of the International Conference on Frontiers of Theoretical Physics*, edited by F. C. AULUCK (New Delhi, 1978).
[51] I. PRIGOGINE: in *The Boltzmann Equation: Theory and Applications*, edited by E. G. D. COHEN and W. THIRRING (Wien, 1973) (*Acta Phys. Austriaca, Suppl. X*).
[52] I. PRIGOGINE, F. MAYNÉ, C. GEORGE and M. DE HAAN: *Proc. Nat. Acad. Sci.*, **74**, 4152 (1977).

[53] B. MISRA: *Proc. Nat. Acad. Sci.*, **75**, 1627 (1978).
[54] P. RÉSIBOIS: *Journ. Stat. Phys.*, **2**, 21 (1970).
[55] J. D. FOCH and G. W. FORD: in *Studies in Statistical Mechanics*, Vol. **5**, edited by J. DE BOER and. G. E. UHLENBECK (Amsterdam, 1970).
[56] D. FORSTER: *Hydrodynamic Fluctuations, Broken Symmetry, and Correlation Functions* (New York, N. Y., 1975).
[57] D. FORSTER: *Phys. Rev. A*, **9**, 943 (1974).
[58] M. THEODOSOPULU, A. P. GRECOS and I. PRIGOGINE: *Proc. Nat. Acad. Sci.*, **75**, 1632 (1978).
[59] M. THEODOSOPULU and A. P. GRECOS: *Physica* (1979).
[60] E. P. GROSS: *Journ. Stat. Phys.*, **11**, 503 (1974).
[61] S. A. RICE and P. GRAY: *The Statistical Mechanics of Simple Liquids* (New York, N. Y., 1965).
[62] R. KUBO: *Journ. Phys. Soc. Japan*, **12**, 1570 (1957).
[63] M. H. ERNST: Thesis, University of Amsterdam (1965); *Physica*, **32**, 252 (1966).
[64] I. PRIGOGINE: *Physica*, **15**, 272 (1949).
[65] L. P. HORWITZ and J. P. MARCHAND: *Rocky Mount. Journ. Math.*, **1**, 225 (1971).
[66] E. C. G. SUDARSHAN, C. B. CHIU and V. GORINI: *Phys. Rev. D*, to appear.
[67] A. P. GRECOS: *Adv. Chem. Phys.*, **39**, to appear (1978).
[68] M. DE HAAN, CL. GEORGE and F. MAYNÉ: *Physica*, **92** A, 584 (1978).
[69] CL. GEORGE, F. HENIN, F. MAYNÉ and I. PRIGOGINE: *Hadronic Journ.*, **1**, 520 (1978).

Foundations of Probability: A Modal Frequency Interpretation (*).

B. C. van Fraassen

University of Toronto - Toronto, Ont.
University of Southern California - Los Angeles, Cal.

1. – Introduction: probability in physics.

My topic of concern is the interpretation of probabilities in physical theories, whether quantum or classical. Since quantum mechanics is the theory in which probabilities first appeared irreducibly, many authors today concentrate on measures definable on quantum logics. These are not probabilities (nor even measures) in the usual sense, since they are not functions defined on fields of sets. On the other hand, discussions of the interpretation of probability, whether from an objective or a subjective point of view, are generally concerned with probability functions of the usual sort (additive real-valued functions on fields of sets; or normalized measures).

The view which may reconcile this tension is this: the interpretation of generalized probability measures on quantum logics (or, indeed, of any probabilities appearing in physics) will proceed through the interpretation of probabilities of the usual sort. And the reason is that the former can be explained in terms of the latter, as I shall now briefly (and sketchily) indicate.

Let M be a set of states on quantum logic L which can be associated with statistical operators on a Hilbert space in the usual manner. If μ is in M and p in L, then $\mu(p)$ is a real number between zero and one, and μ has other similarities to probability measures; but L is not Boolean. However, the association with operators on a Hilbert space is such that for each μ in M and each observable Q there is an ordinary (Kolmogoroff) probability measure $P(\mu, Q)$ on the real line, and

(1.1) for each p in L there is an observable Q and Borel set E such that $\mu(p) = P(\mu, Q)(E)$

(*) The support of the Canada Council for this research is gratefully acknowledged; also the help of Prof. R. Anderson, Department of Mathematics, McMaster University, with subsection 4.3.

and it is possible to identify proposition p with the set $\{\mu : P(\mu, Q)(E) = 1\}$ or else identify state μ with the set $\{P(\mu, Q) : Q$ is an observable$\}$, in order to provide a representation of the quantum logic. Succinctly: those aberrant quantum probabilities can all be represented in terms of « classical » probabilities, and, indeed, this is why any given calculation involving probabilities in quantum theory can be carried out through suitably chosen calculations in the classical theory of probability.

The points which will accordingly motivate my approach in these lectures (concentrating on the more usual sorts of probabilities) are these:

(1.2) The functions $P(\mu, Q)$ in terms of which states and propositions can be represented, are classical (Kolmogoroff) probabilities, formally speaking.

(1.3) These probabilities are generally interpreted as themselves conditional (« $P(\mu, Q)(E)$ is the probability that a result in E is obtained, *given* that Q is measured on a system in state μ »), and the conditions are mutually incompatible when the observables are maximal.

(1.4) Testing of the theory typically takes the form of confronting the expectation values of observable Q calculated with probability measures $P(\mu, Q)$ for states μ; hence, of comparing the probabilities $P(\mu, Q)(E)$ with the frequencies of occurrence of the corresponding events.

It seems, therefore, that even the interpretation of quantum mechanics, *in so far as it concerns what the theory says about the empirical (i.e. actual, observable) phenomena*, deals with the confrontation of classical probability measures with observable frequencies. It is exactly this confrontation which I wish to study.

I. Absolute probability and frequency.

Historically, many authors conceived of probability as basically dyadic: « the probability of A given B » was the primitive notion. This changed with KOLMOGOROFF, who made the concept monadic: « the probability of A » became basic, the event A was represented by a set of points in a sample space, and the probability by a function defined on such sets. I will refer to the monadic concept as that of *absolute* probability, and to the dyadic as that of conditional probability. In this first part, my concern will be with absolute probability, and the question whether, or to what extent, it can be interpreted in terms of frequencies of occurrence. The conclusion will be that absolute probabilities can be represented in terms of frequencies, but the representation is of the modal (« many world ») sort.

2. – The axiomatic basis.

Most writers today agree on the axioms laid down by KOLMOGOROFF, which make probability theory a part of measure theory. There is one point, however, on which there remains a dispute: should probability be asserted to be *countably* [1] *additive*, or merely *finitely additive*? The main proponent of the latter is DE FINETTI, and he is followed in this by various statisticians and philosophers of the « Bayesian » school. For the working statistician, who applies probability to economic and industrial problems, the issue is no doubt academic. But in physics, where more abstract mathematics is needed, and accordingly in philosophy where questions of interpretations of physics are a main concern, the issue is of more immediate interest.

I shall accordingly distinguish *probability functions*, which are finitely additive, and *probability measures*, which are countably additive. After defining these, I shall discuss the reasons for and against imposing the strong requirement of countable additivity.

A *probability function* is a real-valued function P defined on a field F of subsets of a given (nonempty) set K, such that

$$0 \leqslant P(A) \leqslant P(K) = 1\,, \tag{2.1}$$

$$P(A \cup B) = P(A) + P(B) \qquad \text{if } A \perp B\,, \tag{2.2}$$

for all A and B in F, where « $A \perp B$ » stands for « $A \cap B = \wedge$ ». The triple $S = \langle K, F, P \rangle$ I shall call a *probability-function space*. The couple $\langle K, F \rangle$ may be called a *sample space*.

A *probability measure* is a real-valued function P defined on a Borel field (also called σ-field: a field closed under countable union) F of subsets of a given (nonempty) set K, such that (2.1) and

$$P(\bigcup X) = \sum \{P(A) : A \in X\} \quad \text{if } X \text{ is a countable disjoint subfamily of } F, \tag{2.3}$$

where a disjoint family is a family X such that $A \perp B$ whenever A and B are in X. Because of the upper bound of P noted in (2.1), the countable sum in (2.3) exists. It will be convenient often to index a family of sets, and write « $X = \{A_i\}$, $i = 1, 2, \ldots$ » and « $\sum_{i=1}^{\infty} P(A_i)$ » accordingly. When P is a probability measure with K and F as indicated, I shall call the triple $S = \langle K, F, P \rangle$ a *probability space*. Note that all probability spaces are probability-function spaces.

In a few cases it will be convenient to talk of a space with P unspecified: let us call $M = \langle K, F \rangle$ a *sample space* provided F is a field of subsets of K, and a *Borel sample space* if F is a Borel field on K.

2·1. *The question of additivity.* – DE FINETTI has asserted that, in practical cases, it is generally not needed to assume countable additivity; and also that much of measure theory can be re-developed for finitely additive functions. While he has provided some basis for these assertions, in part through some admirably ingenious mathematics, they are not sufficiently substantiated to provide us with an immediately available alternative to the standard theory.

However, DE FINETTI has also argued that the imposition of countable additivity itself precipitates some mathematical difficulties. The family of probability measures on a given sample space, he argues, is in general not topologically closed, while the family of probability functions is. Let us first make this precise.

If $M = \langle K, F\rangle$ is a sample space, let $PF(M)$ be the family of probability functions defined on F. If furthermore F is a Borel field, let $P(M)$ be the family of probability measures defined on F—note that $P(M) \subseteq PF(M)$. If $\{P_i\}$, $i = 1, 2, \ldots$, is a countable sequence of members of $PF(M)$, let us call this sequence *convergent* exactly if for each set A in F there is a number $p(A)$ such that

$$p(A) = \operatorname*{limit}_{i\to\infty} P_i(A)$$

and call P the *limit* of $\{P_i\}$ exactly if $P(A) = p(A)$ for all A in F. It is easily shown that P will be a member of $PF(M)$ if all the P_i are, and, in this sense, $PF(M)$ is topologically closed.

However, if all the functions P_i are probability measures, there is no guarantee that P is a measure. For let K be the natural numbers, and F the Borel field of *all* its subsets. Define

$$P_n(A) = 1/n\,\#\,(A \cap \{1, \ldots, n\}) \,,$$

where « $\# X$ » denotes the cardinality of X. It is clear that

$$P_n(K) = 1\,, \qquad \text{so} \qquad \operatorname*{limit}_{n\to\infty} P_n(K) = 1\,,$$

$$\operatorname*{limit}_{n\to\infty} P_n(\{q\}) = 0 \qquad \text{for all } q \text{ in } K\,,$$

$$K = \bigcup \{\{q\} : q \in K\}\text{—a countable union}\,.$$

Hence the defined limit is not a countably additive function. (This example of additive functions defined on the natural numbers will reappear in later contexts as a touchstone for conclusions about probability.) Therefore, as DE FINETTI concludes, $P(M)$ is not in general topologically closed in the indicated sense [2].

Without disputing the conclusion as stated, we must still point out that the lack of closure has only been shown for the case of functions defined on a denumerable set. Whether the result is symptomatic or peculiar therefore depends on whether we may not be able to embed sample spaces generally in larger, more manageable mathematical structures. It should be added that in the cases in which the sample space M is a separable metric space, then $P(M)$ is closed, topologically complete, and also compact (in a naturally defined topology) exactly if M itself is [3].

I conclude tentatively that the lack of closure remarked upon by DE FINETTI, which certainly appears for certain sorts of sample spaces, may after all not prove to cause any difficulties. We should, therefore, turn to the requirement of countable additivity and see if it brings any benefits.

2·2. *Equivalents of countable additivity.* – KOLMOGOROFF already pointed out that countable additivity is, in effect, a continuity postulate. In his discussion in ref. [2] DE FINETTI provided two further equivalents, which concern the *conditional probability* defined by

$$P(A/B) = P(A \cap B)/P(B) \text{—provided } P(B) \neq 0 .$$

The equivalents are these: let P be a probability function on $\langle K, F \rangle$ whose domain F of definition is in fact a Borel field. Then P is a probability measure exactly if one of the following conditions holds:

(2.4) $P\left(\bigcup^{\infty} A_i\right) = \sum^{\infty} P(A_i)$ when $\{A_i\}$, $i = 1, 2, \ldots$, is a disjoint family;

(2.5) $\operatorname{limit}_{n\to\infty} P(E_n) = 0$ if $\{E_n\}\downarrow\Lambda$ [*continuity*];

(2.6) if $\{E_n\}$ is a countable measurable partition of K and $p_1 \leqslant P(A/E_n) \leqslant p_2$ for all n, then $p_1 \leqslant P(A) \leqslant p_2$ [*conglomerability*];

(2.7) if $\{E_n\}\downarrow E$ in F and $P(E) \neq 0$, then $\operatorname{limit}_{n\to\infty} P(A/E_n) = P(A/E)$ [*condition continuity*].

Here I write « $\{E_n\}\downarrow X$ » for the assertion that $E_n \supseteq E_{n+1}$ for all $n = 1, 2, \ldots$ and $\bigcap\{E_n : n = 1, 2, \ldots\} = X$; we also say then that the series $\{E_n\}$ converges to X. A measurable partition of K is a disjoint subfamily of F whose union equals K.

I shall not discuss (2.6) and (2.7) further, but the equivalence of countable additivity to continuity in the sense of (2.5) is instructive. Let us first suppose that P is countably additive and that $\{E_n\}\downarrow\Lambda$. In that case $K - \bigcap^{\infty} E_n = K$, so $\bigcup^{\infty} (K - E_n) = K$. We see, thererefore, that our problem is equivalent to proving continuity from below at K:

(2.8) if $F_n \subseteq F_{n+1}$ for $n = 1, 2, \ldots$ and $\bigcup^{\infty} F_n = K$, then $\operatorname{limit}_{n\to\infty} P(F_n) = 1$.

But now we can argue

$$K = \bigcup^{\infty} F_n = F_1 \cup \left[\bigcup^{\infty} (F_{n+1} - F_n)\right],$$

So

$$1 = P(F_1) + P\left[\bigcup^{\infty} (F_{n+1} - F_n)\right] = P(F_1) + \sum^{\infty} P(F_{n+1} - F_n) =$$

$$= P(F_1) + \operatorname*{limit}_{n\to\infty} \sum^{n} P(F_{n+1} - F_n) = P(F_1) + \operatorname*{limit}_{n\to\infty} P(F_{n+1} - F_1) =$$

$$= \operatorname*{limit}_{n\to\infty} P(F_1) + P(F_{n+1} - F_1) = \operatorname*{limit}_{n\to\infty} P(F_n) .$$

The converse argument is simpler; for if $\bigcup^{\infty} A_i = X$, then the series $(K - X) \cup \bigcup^{n} \{A_i\}$ converges to K from below.

It will be clear from the method of proof that not only is (2.5) equivalent to (2.8), but that it is moreover equivalent to assertions of continuity, both from above and from below (when applicable) at all measurable sets.

2·3. *The Radon-Nikodym theorem.* – We have seen that, if we concentrate attention on a single probability function, countable additivity is a continuity condition. The advantage we can expect from this is that the familiar techniques of differentiation and integration will be found applicable. That this is so—and how—is the content of the Radon-Nikodym theorem, which can be proved on the assumption of countable additivity.

Before giving a precise account of the content of this theorem, let me give a brief summary (the terms in the summary will be defined below): If f is a random variable, we can define its *expectation value* on a measurable set X as $Ef(X) = \int_X f \, \mathrm{d}P$.

The theorem says, in part, that every absolutely continuous countably additive set function is the *expectation* Ef of some random variable f.

I turn now to a more detailed exposition. If $S = \langle K, F, P \rangle$ is a probability space, we call the members of F the *measurable* sets. A *measurable function* or *random variable* (r.v.) on S is a real-valued function h defined on K such that $h^{-1}(E)$ is in F for every Borel set E. This means that

$$P(h \in E) = P(\{x \in K : h(x) \in E\})$$

is well defined for each Borel set E on the real line.

A *simple* r.v. is any finite linear combination of characteristic functions IA of measurable sets A. These are integrable, and so are certain other r.v., by the definition of the integral below.

(2.9) If $h = \sum_{i=1}^{n} a_i IA_i$, then $\int_K h \, \mathrm{d}P = \sum_{i=1}^{n} a_i P(A_i)$.

(2.10) If h is a nonnegative r.v. and h_n a nondecreasing sequence of nonnegative simple functions converging to h, then

$$\int_K h \, dP = \underset{n \to \infty}{\text{limit}} \int_K h_n kP .$$

(2.11) If h is a r.v. equal to $h_1 - h_2$, where h_1 and h_2 are nonnegative integrable r.v., then

$$\int_K h \, dP = \int_K h_1 \, dP - \int_K h_2 \, dP .$$

In (2.11) the positive and negative parts h_1 and h_2 of h can be chosen as $hI\{h \geqslant 0\}$ and $hI\{h < 0\}$. We call h integrable only if the integral is defined and finite; the notion is trivially extended to any r.v. which equals an integrable r.v. *almost everywhere*, *i.e.* outside a null set (set of measure zero). We also extend it to integration over specific measurable sets by

(2.12) $$\int_A h \, dP = \int_K hIA \, dP \qquad \text{for } A \in F.$$

We note that the integral over A exists when the integral over K does.

As defined, the integral is a linear function on the r.v. and also additive for integration over disjoint sets, and order preserving. We note specifically for future reference:

(2.13) the *expectation* Eh of a r.v. h is its integral:

$$Eh(A) = \int_A h \, dP \qquad \text{for } A \text{ in } F.$$

(2.14) $Eh = 0$ if and only if $h = 0$ almost everywhere $\big(P(\{h = 0\}) = 1\big)$.

(2.15) $Eh > Eh'$ implies $h > h'$ a.e. (« almost everywhere »).

(2.16) If f_n, f be integrable r.v. with $|f_n| \leqslant f$, and $f_n \to f$ a.e., then f is integrable,

$$Ef_n(K) \to Ef(K) , \qquad \text{and} \qquad E|f_n - f|(K) \to 0 .$$

(2.17) If f_n, g are integrable and $\left|\sum_{n=1}^{r} f_n\right| \leqslant g$ for all r, then $E \sum_{n=1}^{\infty} f_n = \sum_{n=1}^{\infty} Ef_n$.

Result (2.17) is a useful convergence result, and shows that the expectation

Eh is itself a countably additive function on the probability space:

(2.18) $Eh\left(\bigcup_{i=1}^{\infty} A_i\right) = \sum_{i=1}^{\infty} Eh(A_i)$ when $\{A_i\}$ is a countable disjoint family of measurable sets.

For, we reason, for the case of nonnegative r.v. h,

$$Eh\left(\bigcup_{i=1}^{\infty} A_i\right) = \int_K hI \bigcup_{i=1}^{\infty} A_i \,\mathrm{d}P = \int_K h \sum_{i=1}^{\infty} IA_i \,\mathrm{d}P = \int_K \sum^{\infty} hIA_i \,\mathrm{d}P = \sum_{i=1}^{\infty} \int_K hIA_i \,\mathrm{d}P\,,$$

which is justified by (2.17) because we know from the preceding line that $\sum_{i=1}^{\infty} hIA_i$ is integrable, and of course $\sum_{i=1}^{r} hIA_i \leqslant \sum_{i=1}^{\infty} hIA_i$ for all r so that (2.17) is applicable. This proof is simply extended to r.v. which are not nonnegative.

(2.19) Eh is *absolutely continuous* with respect to P; *i.e.* if $P(A) = 0$, then $Eh(A) = 0$, for all measurable sets A.

We have now found that the expectation of a r.v. is a countably additive, absolutely continuous set function. The Radon-Nikodym theorem now says that the converse is also true:

(2.20) Radon-Nikodym. Let $S = \langle K, F, P\rangle$ be a probability space, and g any countably additive real-valued function defined on the family F. Then there exists an integrable r.v. h and a set function q such that, for A in F,

a) $g(A) = Eh(A) + q(A)$,

b) $q(A) = q(A \cap Z)$ for a fixed null set Z,

c) $q = 0$ if g is absolutely continuous with respect to P.

We call q the *singular component* of g; if $q = 0$, then we call h the *Radon-Nikodym derivative* of g. This derivative is unique up to a null function; that is, if $Eh = Eh'$, then $h = h'$ a.e.; see statement (2.15).

Remark on the proof. The heart of the proof is an application of Zorn's lemma, which establishes the result for the special case in which g is nonnegative and absolutely continuous. Take the family G of all nonnegative integrable r.v. f such that $Ef(A) \leqslant g(A)$ for all measurable sets A. This family is partially ordered by $\leqslant$ a.e.; a chain in this family is shown to have an upper bound in it; any maximal element in the family, which exists now by Zorn's lemma, is then shown to have an expectation equal to g on any measurable set.

Remark on the uses. If X is a region in n-dimensional Euclidean space, and μ the Lebesgue measure, then $\mu(X \cap A)/\mu(X)$ is a probability measure. We note accordingly that, if P is another probability measure with $P(X) = 1$ and $P(Y) = 0$ whenever $\mu(Y) = 0$, then the theorem implies that $P = Ef$, for a certain r.v. f—the *probability density* of P—and hence definable in terms of Lebesgue integration. Such probabilities are called *geometric probabilities*.

We have now seen that if P is countably additive, then we can define an integral $\int dP$, and we can also differentiate certain set functions with respect to P—the derivative (« Radon-Nikodym derivative ») being a random variable. And finally, these P-differentiable set functions will include further probability measures, which are therefore definable in terms of P in a suitably broad sense. These are the advantages that accrue from the countable additivity postulate.

Remark on terminology. Note that « the expectation of r.v. h » sometimes refers to the function Eh and sometimes to the value $Eh(K)$. Also, if more than one probability measure P is discussed, one should write « EPh » rather than « Eh ».

3. – The strict frequency interpretation.

Many empirical propositions are constantly being put to the test. Every day provides a test of the assertion that the sun comes up every day, and every time you make coffee you test the assertion that water boils at a temperature of 100 °C. When the assertion is one of probability, a test will confront that probability with an actual frequency. This led to the interpretation, most strictly defended by REICHENBACH, that a probability assertion *is* an assertion about an actual frequency.

Although most of Reichenbach's theory concerns limits of relative frequency in an infinite long run, no references need be made to infinity in the initial statement of the interpretation. The reference class may be finite or infinite. However, a theoretical assertion of probability must very often be understood as conditional on the reference class being « large enough ». There will still be implications for frequencies in smaller classes, but these will be much more approximate, or less reliable. This sort of consideration automatically leads us to the infinite case as the only « pure » case, the only one for which we can make precise statements. For imagine I say:

(3.1) The proportion of A's in a class of B's will be $\frac{1}{2}$ provided the class of B's is large enough .

One attempt to make this precise yields

(3.2) There is a size x such that, in any class of B's of size $\geqslant x$, the proportion of A's equals $\frac{1}{2}$.

This is self-contradictory, since either x or $x+1$ will be odd, and the number of A's is an integer. Clearly the size of the class of B's itself places a limitation of the exactness of the assertion. So we must say something like

(3.3) There is a size x such that for any $n \geqslant x$, if a class of B's has size n, then the proportion of A's therein equals $\frac{1}{2} \pm 1/n$.

This entails that, as the size of the class of B's approaches infinity, the proportion of A's tends to the limit $\frac{1}{2}$. True; what (3.3) says is stronger than this limit assertion. But just in the way that (3.3) is stronger, it is unacceptable: there is no reason to take (3.1) to rule out fluctuations around $\frac{1}{2}$ that are larger than $1/n$ in classes of size n, provided only these fluctuations become smaller and smaller as the size increases. Hence, although, according to REICHENBACH, a probability statement is literally no more and no less than a statement of relative frequency in the *actual* reference class (however large or small it may be), the probability assertions of physical science should be taken as relating to ideally extended, infinite long runs, and applying only approximately to smaller cases.

One main problem for the interpretation is how we can tease out implications for actual, finite reference classes. Should these be regarded as random samples from nonactual infinite series? If so, *which* nonactual (possible) infinite series? But this speculation and question already introduces an element of *modality* which is foreign to the strict extensionalist interpretation with which REICHENBACH began.

Instead, I want to concentrate here on the pure case of an actual long run. Perhaps in the « basic » case of interactions among elementary particles, say, this infinity is in any case actual! But have we here a tenable interpretation of probability?

In the preceding section we were looking at a purely mathematical interpretation: the axioms describe the class of mathematical structures called *probability spaces*. We have both *soundness* (the axioms are true in all probability spaces) and *completeness* (anything which does not follow from the axioms will be false in some probability space). Now REICHENBACH offers another sort of structure: a long run (that is a countable sequence of events $\sigma = \langle\sigma(1), \sigma(2), \ldots\rangle$) plus a set function *relative frequency* defined on certain sets of members of that sequence, which I will quickly make precise: if A is a set, let a be its characteristic function, and define

(3.4)
$$+\, a(\sigma, n) = \sum\{a(\sigma(m)) : m \leqslant n\},$$
$$\text{rel}\,(a, \sigma, n) = +\, a(\sigma, n)/n,$$
$$\text{relf}\,(A, \sigma) = \lim_{n\to\infty} \text{rel}\,(a, \sigma, n),$$

which definitions will be useful later. Note that relf (A, σ) is not guaranteed to exist; it exists if and only if rel (a, σ, n) tends to a limit as n increases indefinitely. The domain of definition of the set function relf $(-, \sigma)$ is given only in this way.

We now turn to the question whether soundness or completeness could be claimed for the probability axioms thus interpreted; *and more important*, whether every application of probability theory (every use of some probability space or other) can be viewed as an application of a theory of relative frequencies. It is very important not to attend only to the logical requirements—this is, I think, one factor that led REICHENBACH astray.

3·1. *Failure of the probability axioms.* – Let the actual long run be counted in days: 1 (today), 2 (tomorrow), and so on. Let $A(n)$ be an event that happens only on the n-th day. Then the limit of the relative frequency of the occurrence of $A(n)$ in the first $n+q$ days, as q goes to infinity, equals zero. The sum of all these zeros, for $n = 1, 2, 3, \ldots$, equals zero again. But the union of the events $A(n)$—that is $A(1)$-or-$A(2)$-or-$A(3)$-or...; symbolically $\bigcup \{A(n): n \in N\}$—has relative frequency 1. It is an event that happens every day.

So relative frequency is not countably additive [4]. Indeed, its domain of definition is not closed under countable unions, and so is not a Borel field. For let B be an event whose relative frequency does not tend to a limit at all. Let the events $B(n)$ be as follows: $B(n) = A(n)$ if B happens on the n-th day, while $B(n) = \wedge$, the « empty » event, if B does not occur on the n-th day. The limit of the relative frequency of $B(n)$ exists, and equals zero, for each number n. But B happens on the same days as the union of the events $B(n)$, and the limit of its relative frequency does not exist.

A somewhat more complicated argument, due independently to DE FINETTI and to RUBIN, and reported by SUPPES, establishes that the domain of relative frequency in the long run is not a field either [5]. Let us divide the long run of days into segments: $X(n)$ is the segment which stretches from segment $X(n-1)$ to day 2^n inclusive ($X(1)$ is just the first and second day). We note that $X(n)$ is as long as the sum of all preceding segments. Call $X(n)$ an odd segment if n is odd, an even segment otherwise. Let A be an event that happens every day in every odd segment, but on no other days. In that case, the relative frequency of A has no limit in the long run. A is not in the domain of definition of relative frequency.

We let B and C be two events that overlap inside A, in a regular way: Let B happen on all the even dates on which A happens, and on all the odd dates when A does not happen. And let C also happen on all the even dates on which A happens, and, in addition, on all even dates when A does not happen. About B and C it is true to say that, in each segment, each of them happens every other day. So each has relative frequency $\frac{1}{2}$.

But the intersection $B \cap C$ is curious. Up to the end 2^n of segment $X(n)$, there have been exactly half as many ($B \cap C$) days as there were A days. So if $B \cap C$ had a relative frequency in the long run, so would A. And A does not.

Let me sum up the findings so far. The domain of relative frequency is not closed under countable unions, nor under finite intersections. But still, countable additivity fares worse than finite additivity. For when the relative frequency of a countable union of disjoint events exists, it need not be the sum of the relative frequencies of those components. But, if the relative frequencies of B, C and $B \cup C$ all exist, while B is disjoint from C, then the relative frequency of $B \cup C$ is the sum of those of B and C.

We cannot say therefore that relative frequencies are probabilities. But we have not ruled out yet that all probabilities are relative frequencies (of specially selected families of events). For this question it is necessary to look at « large » probability spaces; specifically, at geometric probabilities.

It is sometimes said that a finite or countable sample space is generally just a crude partition of reality—reality is continuous. Of course, you cannot have countable additivity in a sensible function on a countable space! But the problem infects continuous sample spaces as well. In the case of a mechanical system, we would like to correlate the probability of a state with the proportion of time the system will spend in that state in the limit, that is in the fullness of time. But take a particle travelling forever in a straight line. There will be a unique region $R(n)$ filled by the trajectory of the particle on day n. The proportion of time spent in $R(n)$ tends to zero in the long run; but the particle always stays in the union of all the regions $R(n)$.

Geometric probabilities such as these have further problems. Take Reichenbach's favorite machine gun example, shooting at a circular target. The probability that a given part of the target is hit in a given interval is proportional to its area, we say. But idealize a bit: let the bullets be point-particles. One region with area zero is hit every time: the set of points *actually* hit in the long run. Its complement, though of area equal to the whole target, is hit with relative frequency zero.

3·2. *Implications of the laws of large numbers.* – The difficulties we have just found in the idea of relative frequency in the long run may be somewhat surprising in view of Bernoulli's law of large numbers. Does not the theory of probability itself establish its own frequency interpretation? For that law says that, if we examine more and more events, the relative frequency of A's found will tend to agree more and more with the probability of A.

As was pointed out by KYBURG, however, the law really says only that such agreement is to be expected with ever greater certainty. Hence it is compatible with the *actual* long-run relative frequency being quite different from the probability. So Bernoulli's law cannot convince us of any actual connection between frequency and probability, and does not establish even the consistency of Reichenbach's interpretation.

However, when Cantelli's lemma is added to Bernoulli's law, it becomes possible to deduce certain consequences that show Reichenbach's claims to be tenable at least in a limited way. These consequences are known as strong laws of large numbers, and were appealed to by CANTELLI himself in this connection [6].

The specific large-number law we need is that for identically distributed random variables. We call random variables g and h independent if $P(g\leqslant x/h\leqslant y) = P(g\leqslant x)$ for all numbers x and y, and say that g and h have the same distribution if $P(g\leqslant x) = P(h\leqslant x)$ for all numbers x. The expectation we have already defined as $\int_K g\,\mathrm{d}P$. The result is then:

(3.5) *Strong law of large numbers of identically distributed variables*: Let g_n be a countable sequence of r.v. which are mutually independent, all have the same distribution and the same finite expectation m. Let $h_n = \sum_{i=1}^{n} g_i$. Then $h_n/n \to m$ almost everywhere.

This last statement, made precise, means that

$$P\Big(\Big\{x\in K : \operatorname*{limit}_{n\to\infty} h_n(x)/n = m\Big\}\Big) = 1\,.$$

I will now apply this to the space of possible « infinite long runs » of events.

We begin with a probability space $S = \langle K, F, P\rangle$. The elements of K stand for simple events, like possible outcomes of a given experiment. Thus a long run in this space is a countable sequence of elements of K. These can themselves be regarded as possible outcomes of a complex experiment, and so we look at the *product space* S^* with

$K^* = \{\sigma : \sigma(1), \sigma(2), \ldots \in K\}$,

F^* = the least Borel field containing all the sets $\{\sigma\in K : \sigma(1)\in A_1, \ldots, \sigma(n)\in A_n\}$—briefly designated as $A_1\times\ldots\times A_n\times K$ formed from sets $A_1, \ldots, A_n$ in F.

As far as the new probability P^* is concerned, there is a very natural one:

$$P^*(A_1\times\ldots\times A_n\times K) = P(A_1)\ldots P(A_n)\,.$$

In other words, consecutive events are treated as mutually independent. Let me say at once that P^* is not being interpreted as correctly representing any probability judgment in the present context. It only plays the role of a normed measure.

Let A be in F, that is an event in the old space S. I want to assert that there is a long run σ—namely the actual one!—such that $P(A) = \text{relf}\,(A, \sigma)$. To establish this, I focus on the product space, and define the r.v. g_i on S^* by

$$g_i(\sigma) = \begin{cases} 1, & \text{if } \sigma(i) \in A\,, \\ 0, & \text{otherwise}\,. \end{cases}$$

These variables g_i are mutually independent in S^* because the measure P^* is specially designed to ensure this. That the g_i are identically distributed is also clear for $P^*(g_i \leqslant 1) = 1$ and $P^*(g_i \leqslant 0) = 1 - P(A)$. Now h_n/n measures the proportion of events in A which are found in the first n events $\sigma(1), \ldots, \sigma(n)$ in a sequence σ. Hence h_n/n is in fact rel (a, σ, n) and so

$$\operatorname*{limit}_{n\to\infty} h_n(\sigma)/n = \text{relf}\,(A, \sigma)\,.$$

Finally, the expectation m of g_i equals $\int g_i \,\mathrm{d}P^*$, but g_i is a simple r.v. being the characteristic function of $K \times \ldots \times A \times K$, where A appears in the i-th place; so that expectation is $P^*(K \times \ldots \times A \times K) = P(A)$. We now apply the theorem to obtain

$$P^*(\{\sigma : \text{relf}\,(A, \sigma) = P(A)\}) = P^*\Big(\Big\{\sigma : \operatorname*{limit}_{n\to\infty} h_n(\sigma)/n = m\Big\}\Big) = 1\,.$$

A set which receives measure 1 from P^* cannot be empty, so we conclude

(3.6) there is at least one possible long run σ such that $\text{relf}\,(A, \sigma) = P(A)\,.$

This establishes that REICHENBACH was as it were « point wise » consistent: for each event A, the claim that $P(A)$ *is* the relative frequency of A in the long run can *tenably* be made.

We can strengthen this conclusion, by noting the simple fact that a countable intersection of sets of measure 1 still has measure 1. Hence,

(3.7) if G is any countable family of measurable sets, then there is a long run σ such that $\text{relf}\,(A, \sigma) = P(A)$ for all A in G.

Therefore, REICHENBACH can tenably claim, *for any countable family of events*, that their probabilities *are* their relative frequencies in the long run.

Indeed, we can add one bit more. Since fields are formed by finitary operations, the same can be asserted *for all countably generated fields of events* (« fields », not « Borel fields » !).

This is certainly as far as REICHENBACH ever looked (being misled in this by his concentration on logical syntax). But in probability theory we have to go on to consider uncountable families of events and Borel fields.

3.3. *Polya's proof: the separable atomistic Borel field.* – There is an ingenious proof by POLYA, reported by VON MISES [7], to the effect that the strict frequency interpretation can be maintained for the probabilities on a certain class of countably generated Borel fields. Before giving the proof we must identify this class.

A Borel field B is called *separable* exactly if it is countably generated; that is there is a countable class G of sets such that B is the least Borel field which contains G. An *atom* of the Borel field is an element E such that $\wedge \subseteq X \subseteq E$ implies $X = E$ for X in B.

If B is separable, then it has atoms, and, indeed, every event is a union of atoms. For let G be the countable class of generators $\{X_i\}$, and let $\bigcap_i X_i^*$ be the intersections in which X_i^* is X_i or $K - X_i$. In that case it is at once clear that these intersections are in B, are atoms of B, and that $X_i = \bigcup \{\bigcap_j X_j^* : X_i^* = X_i\}$. But this union need not be a countable union. The class of atoms may have any cardinality up to and including that of the continuum.

(3.8) A separable Borel field has a class of atoms, of cardinality at most that of the continuum, and each element is a union (though perhaps not a countable union) of these atoms.

We call a Borel field *atomistic* if its set of atoms is also a set of generators, *i.e.* each element is a countable union of atoms. The atoms clearly form a partition of the space, but not necessarily a countable partition.

(3.9) A Borel field is both separable and atomistic exactly if it is generated by a countable partition of the space.

The sufficiency is obvious; suppose then that B has a countable class G of generators, and is atomistic. Then for each X in G we have a countable class $A(X)$ of atoms such that $\bigcup A(X) = X$. But $K = \bigcup \{X : X \in G\}$. Since the union of all the atoms is also K, the only atoms are the ones found in the countable family of sets $A(X)$, each of which is countable—so there are countably many atoms.

Polya's result is that, if the Borel field of events is separable and atomistic, there will be a long run σ such that relf $(A, \sigma) = P(A)$ for all events A.

Let $S = \langle K, F, P \rangle$ be a probability space in which F is indeed separable and atomistic, generated by the measurable countable partition $\{A_i\}$, which means that F is simply the null set plus all countable unions of these sets A_i.

We choose a long run σ: $\sigma(1), \sigma(2), \ldots$ such that $P(A_i) =$ relf (A_i, σ) for $i = 1, 2, \ldots$. This is possible as we have seen in the preceding section. It now remains to prove that $P(X) =$ relf (X, σ) for all X in F. The facts to be used are that

$a)$ $0 \leqslant P(A_i) \leqslant 1$,

$b)$ $\sum_{i=1}^{\infty} P(A_i) = 1$,

$c)$ $P(A_i) = \text{relf}(A_i, \sigma)$,

$d)$ $X = \bigcup \{A_i : i \in I\}$, $I \subseteq \{1, 2, \ldots\}$.

However, I shall state Polya's result in the general form reported by VON MISES, as an independent result about infinite sums.

To facilitate the statement and proof of the result, let the index set I be a subset of $\{1, 2, 3, \ldots\}$ and $I_k = I \cap \{1, \ldots, k\}$.

(3.10) *Polya's result*: Let $\{p_i\}$, $i = 1, 2, \ldots$, be a set of real numbers such that

$a)$ $0 \leqslant p_i \leqslant 1$ and $\sum_{i=1}^{\infty} p_i = 1$,

$b)$ $p_i = \text{limit}_{m\to\infty} m_i/m$,

$c)$ $m = \sum_{i=1}^{\infty} m_i$ and m, m_i are nonnegative integers.

Then $\sum \{p_i : i \in I\}$ exists and equals $\text{limit}_{m\to\infty} \sum \{m_i/m : i \in I\}$.

This is a commutation-of-limits result. To see this, note that the infinite sum $\sum_{i \in I}$ is *defined* as the limit of the partial finite sums $\sum_{i\in I_k}$. Hence the lemma states

$$\text{limit}_{k\to\infty} \sum_{i\in I_k} p_i = \text{limit}_{m\to\infty}\ \text{limit}_{k\to\infty} \sum_{i\in I_k} m_i/m,$$

which, by clause $b)$ in the statement of the result, means that

$$\text{limit}_{k\to\infty} \sum_{i\in I_k} \left[\text{limit}_{m\to\infty} m_i/m\right] = \text{limit}_{m\to\infty}\ \text{limit}_{k\to\infty} \sum_{i\in I_k} m_i/m,$$

hence, since $\sum_{i\in I_k}$ is a finite summation,

$$\text{limit}_{k\to\infty}\ \text{limit}_{m\to\infty} \sum_{i\in I_k} m_i/m = \text{limit}_{m\to\infty}\ \text{limit}_{k\to\infty} \sum_{i\in I_k} m_i/m.$$

Finally the use of this lemma will be clear if we set $m_i/m = \text{rel}(a_i, \sigma, m)$. For then $\text{relf}(A_i, \sigma) = p_i$ and $\text{relf}\left(\bigcup \{A_i : i \in I\}, \sigma\right) = \text{limit}_{m\to\infty} \sum_{i\in I} m_i/m$, by the definitions.

Exposition of Polya's result. The result follows from four simple lemmas.

Lemma 1 (finite case). $\sum_{i\in I_k} p_i = \text{limit}_{m\to\infty} \sum_{i\in I_k} m_i/m$.

This follows immediately from the fact that I_k is finite and $\underset{m\to\infty}{\text{limit}}$ is a finitely additive function.

Lemma 2. $\sum_{i\in I} p_i$ exists (is finite).

By definition $\sum_{i\in I} p_i = \underset{k\to\infty}{\text{limit}} \sum_{i\in I_k} p_i$. This is the limit of an increasing series bounded from above by 1 $\left(\text{because } I_1 \subseteq I_2 \subseteq \dots \{1, 2, \dots\} \text{ and } \sum_{i=1}^{\infty} p_i = 1\right)$, hence that limit does exist.

For the next two lemmas we need preliminary definitions of the inferior and superior limits of a sequence, and how these are related to the ordinary limit if it exists.

(3.11) *a*) $\limsup_{n\to\infty} x_n = \inf\{\sup\{x_k : k \geqslant n\} : n = 1, 2, \dots\}$,

b) $\liminf_{n\to\infty} x_n = \sup\{\inf\{x_k : k \geqslant n\} : n = 1, 2, \dots\}$,

c) $\lim_{n\to\infty} x_n$ exists and equals y if and only if

$$\liminf_{n\to\infty} x_n = \limsup_{n\to\infty} x_n = y .$$

It will be obvious from the definitions that the inferior limit must be less than or equal to the superior limit, and that, if $x_n \leqslant y_n$ for all $n = 1, 2, \dots$, then their inferior and superior limits are also related in this way. The two sorts of limits do not coincide if the sequence oscillates. For example, let $a_n = (-1)^n(1 + 1/n)$. This is the sequence -2, $+1+\frac{1}{2}$, $-1+\frac{1}{3}$, $+1+\frac{1}{4}$, ... Here $\liminf_{n\to\infty} a_n = -1$ and $\limsup_{n\to\infty} a_n = +1$. Such oscillating sequences can easily be constructed by mixing two convergent sequences. If the infinum or supremum mentioned do not exist in the real numbers, we shall set them equal to $-\infty$ and $+\infty$, respectively, so that *these* limits will always exist.

Lemma 3. $\sum_{i\in I} p_i \leqslant \liminf_{m\to\infty} \sum_{i\in I} m_i/m$.

We note that $m > \sum_{i\in I} m_i \geqslant \sum_{i\in I_k} m_i$ because $m = \sum_{i=1}^{\infty} m_i$ (which is really a finite sum, all but finitely many factors being zero). Because of this inequality,

$$\liminf_{m\to\infty} \sum_{i\in I} m_i/m \geqslant \liminf_{m\to\infty} \sum_{i\in I_k} m_i/m . \tag{3.12}$$

We know from lemma 1 and line (3.11) *c*) that the right-hand side is an ordinary limit, so we have established

$$\text{for all } k = 1, 2, \dots,\ \liminf_{m\to\infty} \sum_{i\in I} m_i/m \geqslant \sum_{i\in I_k} p_i . \tag{3.13}$$

But $\sum_{i\in I} p_i = \underset{k\to\infty}{\text{limit}} \sum_{i\in I_k} p_i$; hence the lemma follows.

Lemma 4. $\limsup_{m\to\infty} \sum_{i\in I} m_i/m \leqslant \sum_{i\in I} p_i$.

Here we concentrate on the complements $I' = \{1, 2, \ldots\} - I$ and $I'_k = \{1, \ldots, k\} - I_k$. Clearly,

$$\sum_{i \in I} m_i = m - \sum_{i \in I} m_i \leqslant m - \sum_{i \in I'_k} m_i \,. \tag{3.14}$$

Hence

$$\limsup_{m \to \infty} \sum_{i \in I} m_i/m \leqslant \limsup_{m \to \infty} \Big[1 - \sum_{i \in I'_k} m_i/m\Big] \leqslant \tag{3.15}$$

$$\leqslant 1 - \limsup_{m \to \infty} \sum_{i \in I'_k} m_i/m \leqslant 1 - \sum_{i \in I'_k} p_i \,.$$

This last step follows by noting that lemma 1 holds for all finite index sets, and hence for I'_k as well as for I_k. Thus we conclude

$$\text{for all } k = 1, 2, \ldots, \qquad \limsup_{n \to \infty} \sum_{i \in I} m_i/m \leqslant 1 - \sum_{i \in I'_k} p_i \,, \tag{3.16}$$

so we can continue (3.15) still further:

$$\leqslant \operatorname*{limit}_{k \to \infty} 1 - \sum_{i \in I'_k} p_i \leqslant 1 - \operatorname*{limit}_{k \to \infty} \sum_{i \in I'_k} p_i \leqslant 1 - \sum_{i \in I'} p_i \,,$$

which latter step follows by the definition of the infinite sum $\sum_{i \in I'}$ because I' is clearly the union of the increasing series $I'_1 \cup I'_2 \ldots$. Finally, since $I \cup I' = \{1, 2, \ldots\}$ and $\sum_{i=1}^{\infty} p_i = 1$, we know that $\sum_{i \in I} p_i$ equals $1 - \sum_{i \in I'} p_i$; and so the lemma follows.

Polya's result has now beeen established because lim inf $\geqslant$ the sum $\geqslant$ lim sup; but lim inf $\leqslant$ lim sup; so all three are equal, and hence equal the ordinary limit.

(3.17) *Corollary.* Let $S = \langle K, F, P \rangle$ be a probability space and G the Borel field of measurable sets generated by a countable measurable partition $\{A_n\} \subseteq F$. Then there is a countable sequence σ of members of K such that $P(A) = \text{relf}\,(A, \sigma)$ for each A in G.

For if X is in G, then $X = \bigcup \{A_i : i \in I\}$ for some index set I. By the continuity of P, we know that

$$P(X) = \operatorname*{limit}_{k \to \infty} P(X_k)\,, \qquad X_k = \bigcup \{A_i : i \in I_k\}\,.$$

But there is a sequence σ such that, for each member A_i of the partition, $P(A_i) = \text{relf}\,(A_i, \sigma) = p_i$; moreover, $\sum p_i = 1$.

Letting m_i, m be with respect to σ as in the proof, m_i being the number of A_i in $\sigma(1), \ldots, \sigma(m)$, we reason:

$$P(X) = \operatorname*{limit}_{k\to\infty}\left(\operatorname*{limit}_{m\to\infty} \sum \{m_i/m : i \in I_k\}\right) = \operatorname*{limit}_{m\to\infty}\left(\operatorname*{limit}_{k\to\infty} \sum \{m_i/m : i \in I_k\}\right),$$

this equality having been established by Polya's result. By definition, $\operatorname*{limit}_{k\to\infty} \sum \{m_i/m : i \in I_k\} = \sum_{i\in I} m_i/m$. Thus $P(X) = \operatorname*{limit}_{m\to\infty} \sum_{i\in I} m_i/m = \text{relf}\,(X, \sigma)$.

3·4. *Geometric probability.* – Among geometric probabilities we find structures much «larger» than those discussed in the preceding subsection; these structures are in fact the norm in physics. In this subsection I shall describe these to some extent, aiming mainly to exhibit the central place taken by the structure of the unit interval with Lebesgue measure.

Let $S_0 = \langle [0, 1], B_0, \mu \rangle$, where B_0 is the class of Borel sets on $[0, 1]$ and μ is Lebesgue measure restricted to B_0. This field clearly has atoms, namely the unit sets $\{r\}$, $0 \leqslant r \leqslant 1$; and each element of B_0 is a union of these atoms; but B_0 is not atomic for its elements are not all countable unions of atoms.

However, B_0 is separable. For B_0 is defined as the least Borel field containing all the intervals on $[0, 1]$. It suffices to take one sort of intervals only, for example the open ones provided one adds the unit sets $\{0\}$ and $\{1\}$; for instance,

$$[a, b] = \bigcap \{(a - 1/n, b + 1/n),\ n = 1, 2, \ldots\}$$

is a countable intersection of open intervals. Indeed it suffices to take open intervals with rational end points. For example,

$$(0, b) = \bigcup \{(0, r) \colon\ r \text{ is rational and } r < b\}\,,$$

because the set of rationals is dense in $[0, 1]$. But, of course, the family of open intervals with rational end points is countable; so B_0 is countably generated, *i.e.* separable.

We can generalize this reasoning to all Euclidean n-spaces, and, indeed, all separable complete metric spaces. In any topological space, the *Borel sets* are defined to be the members of the Borel field generated by the closed (or, equivalently, open) sets; the space is called *separable* exactly if it has a countable dense subset; if it is separable, then it has a countable base (spheres with rational radii around the points in the dense subset). We can go farther: in a *metric space*, the open sets are exactly the spheres of positive radius; these are countable unions of spheres with positive rational radius. Let A be a dense set; this means that the closure of A is the whole space, and implies, in a *complete metric space*, that each point x is the limit of a converging sequence x_n of members of A. Take the sphere $S(r, x)$ of radius r and center x, and construct the sequence of spheres $S(r + d(x_n, x), x_n)$ which decreases uniformly to $S(r, x)$, its infimum. Thus the open sets around any point are countable intersections of open sets around members of A, which in turn are countable unions of spheres around members of A with rational radius. We have thus found a countable family that generates all the Borel sets:

(3.18) In any complete separable metric space, the Borel field of Borel sets is countably generated (separable).

This result can be extended, in a corollary to a result due to KURATOWSKI [8]:

(3.19) Let X be any uncountable Borel subset of a complete separable metric space. Then the Borel field of all Borel subsets of X is isomorphic to B_0.

As far as the Borel fields are concerned, therefore, we shall not expect too many surprises when the higher cardinalities are utilized in physics. We must now look at the probability measures that can be defined there.

We saw already that any separable Borel field has atoms. But this is meant only set-theoretically: these atoms may not generate the space (by countable operations). In the Borel sets on $[0, 1]$, the atoms are the unit sets $\{x\}$; each of these has zero probability, and the family *they* generate by union, that is the family of countable point sets, consists entirely of null sets. We need a new, probabilistic sense of « atoms », to designate something which has minimal positive probability as well.

To do this, we switch our attention to *Borel algebras*, by which I mean an algebra which is the homomorphic image of a Borel field. The obvious way to produce such algebras is to reduce a Borel field modulo a suitable equivalence relation, and the natural instance of that procedure here is identification modulo differences of measure zero in a probability space.

To be precise, let $S = \langle K, F, P \rangle$ be a probability space, and define

(3.20) $$A \equiv B \text{ if and only if } P(A - B \cdot\cup\cdot B - A) = 0\,, \qquad A, B \text{ in } F\,.$$

Call F/P the result of reducing G modulo $\equiv$. Then F/P is a Borel algebra, with the obvious definitions

$$[A] = \{B : A \equiv B\}\,,$$

$$[A] \cap [B] = [A \cap B]\,,$$

$[A] \subseteq [B]$ if and only if there is a C such that $A \subseteq C$ and C is in $[B]$ and so on, and we can define

$$p[A] = P(A)\,,$$

$$d(x, y) = p(x \cup y) - p(x \cap y)\,, \qquad x, y \text{ in } F/P\,,$$

to produce the *probability algebra* $\langle F/P, p\rangle$ with the metric d. We note that we have a natural minimal element $0 = [\wedge]$; and that $d(0, x) = 0$ if and only if $p(x) = 0$ if and only if $x = 0$.

Now we can characterize atoms in the usual way: x in F/P is an atom if $0 \subseteq y \subseteq x$ implies that $y = x$, while $0 \neq x$. Every atom has, therefore, positive measure and positive distance from 0.

When the Borel sets on $[0, 1]$ are identified modulo differences of Lebesgue measure zero, we obtain an algebra which has no atoms. The reason is simple: if $p[A]$ is positive, so is $P(A)$, and we can then find a $B \subseteq A$ such that $0 < P(B) < P(A)$. This gives us the central example of the structure of geometric probability, according to a theorem proved by BIRKHOFF (essentially as a corollary to a result due to KURATOWSKI) [9]. Let $S_0 = \langle [0, 1], B_0, \mu\rangle$ be the probability space in which B_0 is the family of Borel sets on the unit interval and μ Lebesgue measure, and let $\langle B_0/\mu, \mu\rangle$ be the corresponding probability algebra. Then

(3.21) (Birkhoff). Let M be a probability algebra which has no atoms, and has a countable dense subset. Then M is isometrically isomorphic to $\langle B_0/\mu, \mu\rangle$.

This result has been extended to a complete classification of probability algebras in work by HORN, TARSKI and MAHARAM (see [9]).

In conclusion I wish to say something about attempts to approximate geometric probability by means of relative frequency, which were made by REICHENBACH, VON MISES and others. The idea is quite simple. Recall that the intervals with rational end points on $[0, 1]$ form a countable set of generators for B_0. We know from before that there will be a countable sequence σ in which the relative frequency agrees with the measure on that countable family. In addition, familiar extension results show that there is a unique extension of any measure restricted to these generators, to a measure on the whole Borel field. Hence, it is said, there is a sense in which the relative frequency approximates the geometric probability.

This is a very strange sense indeed. For the relative frequency is defined on many Borel sets outside that countable family, and *conflicts* with the measure found through the extension theorem.

The argument can be made to look as if it exhibits a proper approximation procedure. Let $\{A_n^1\} \subseteq \{A_n^2\} \subseteq \ldots$ be a series of ever finer countable measurable partitions of the space $\langle K, F, P\rangle$. We know that there are sequences

$$\sigma^m : P(A_n^m) = \text{relf}\,(A_n^m, \sigma^m) \qquad \text{for all } n.$$

Moreover, if we are concerned with a metric space, these partitions can be chosen in such a way that

$$Z = \bigcup\{\{A_n^m\}: m, n = 1, 2, ...\} \qquad \text{generates Borel field } F$$

and the extension theorem shows therefore that the assignment $p(A_n^m) = \text{relf}(A_n^m, \sigma^m)$ can be extended to exactly one measure on F, namely P. Since the Borel fields generated by these successive finer partitions « approximate » F, it is then asserted that the successive assignments of relative frequency approximate P.

That there is a certain amount of window-dressing going on here is immediately clear when we reflect that Z is itself a countable family. Hence we know that there is a sequence

$$\sigma: P(A_n^m) = \text{relf}(A_n^m, \sigma) \qquad \text{for all } m, n,$$

that is for all members A_n^m of Z. The machine gun example shows that relf $(-, \sigma)$ and P conflict on members of F. Hence the sequence $\sigma^1, \sigma^2, ...$ of sequences yields an approximation to relative frequencies in σ, which is the wrong thing: it conflicts with P. (And, in any case, the single sequence σ gives us all the information that $\sigma^1, \sigma^2, ...$ gave, so the latter play no real role at all.)

Not only do those chosen long-run sequences approximate the wrong thing, they can also be exhibited and discussed before Polya's result is known. Hence, far from extending our knowledge of the relation between probability and frequency, they fall short of it.

4. – The modal frequency interpretation.

We have now seen that there are insuperable obstacles to the *identification* of probabilities and relative frequencies. As the same time, we have found much reason to believe that probability statements can be explained in some fashion in terms of relative frequencies. Only, the frequency view of probability cannot consist in a naive identification. Accordingly, we turn to Peirce's famous suggestion that probability statements are about frequencies, but involve a « would-be »—something like possibilities or necessities or perhaps even a « many worlds » approach.

4·1. *Popper: the virtual sequence.* – In 1957, POPPER argued that probability is not a property of an actual course of events (such as experimental outcomes) but of the conditions in which these events occur (such as the experimental arrangement or set-up). He explicitly presented this as a more sophisticated version of the frequency interpretation, the difference lying in an appeal to what *would happen*, if the conditions are realized sufficiently often:

> Every experimental arrangement is liable to produce, if we repeat the experiment very often, a sequence with frequencies which depend upon this particular experimental arrangement. These virtual frequencies may be called probabilities. But since the probabilities turn out to depend upon the experimental arrangement, they may be looked upon as *properties of this arrangement. They characterize the disposition, or the propensity*, of the experimental arrangement to give rise to certain characteristic frequencies *when the experiment is often repeated* [10].

We can put it this way: when we say that outcome A has a probability $\frac{1}{2}$ in experiment C under present conditions, we mean that A *would* occur half the time, if experiment C *were* repeated sufficiently often. Thus we envisage a sequence σ: $\sigma(1)$, $\sigma(2)$, ... of outcomes, which is *the sequence* of outcomes *which would occur* if C were repeated indefinitely under the same conditions. The probability of A is the relative frequency of A in that sequence σ; not its frequency in the actual course of events (which may end too soon, or in which the conditions of the experiment vary with time etc.). Hence POPPER speaks of a « virtual frequency »—the frequency in the « virtual sequence ».

How is that probability assertion—*i.e.* the assertion about frequency in the virtual sequence—to be tested? POPPER says explicitly that it is tested by looking at what happens in the actual sequence of events (ref. [10]). But how can the actual sequence (which has, say, 1000 members) provide a test of what the virtual frequency is? Only, as far as I can see, by being regarded as a random part of the virtual sequence. For we can begin with the premise that in sequence σ the relative frequency of A is $\frac{1}{2}$; then use a χ^2 table or similar test to see what we should expect to find in a random sample of 1000 members—and compare this with what we find in the actual 1000 outcomes. The comparison would make no sense if we could not regard the actual sequence as a random selection from the virtual one.

Let us go a bit further. If the virtual sequence has been completely identified by what POPPER says, then he must be saying that, if in the actual world the relevant conditions *do* occur infinitely many times (as Nature does continually test our propositions!), then the actual frequency equals the virtual frequency.

I claim that he must be saying that; for otherwise he would be subject to the following objection. I could imagine doing the experiment once every minute, or once every two minutes, and so on. But when you say « the sequence of outcomes that would occur » you do not specify the rate at which the experiments are performed. Hence your words do not pick out one definite frequency. The answer POPPER must give, it seems to me, is that the sequence in question is performed at the rate at which it would be performed if we actually did it. Hence the words « if C were ... », unlike the word « could », pick out a single possible course of events which is determinate in all respects—

and which is *indeed* realized if the conditions are realized. (How else could he speak of *the* virtual sequence?)

So now we know two things about Popper's virtual sequence: *a*) the actual sequence is a random selection from it, and *b*) if the actual sequence is infinite, the two are identical. This would seem to be the correct explication of Popper's rather vague account.

POPPER called his new interpretation the *propensity interpretation*; subsequent advocates of this interpretation, however, do not all explain propensity as a frequency in what *would* happen under certain conditions. Some, like GIERE, follow him only in saying that probability is an objective quality distinct from actual frequency, and refuse to say even that, in an infinite sequence of experiments under correct fixed conditions, the relative frequency and the probability would coincide [11].

4'2. *Kyburg: the many-world view.* – KYBURG does not advocate the propensity interpretation of probability. But in a provocative article, he outlined a way in which the propensity interpretation can be made precise [12]. To begin, he notes the sorts of differences between POPPER and other propensity advocates which I described at the end of the last subsection. Some propensity theorists, like GIERE, insist on a complete logical separation between probability and frequency (though asserting that beliefs about probabilities are the rational guide to expectations about relative frequencies). Others, like POPPER, talk of probabilities as being the frequencies there would be under suitable conditions.

What KYBURG outlines is a view he calls the *hypothetical frequency interpretation*, which may satisfy propensity theorists of the second sort. The virtue of Kyburg's proposal is that it is stated in terms of mathematical models of experimental situations. I will explain it here in simplified form.

Suppose I measure a table leg three times, and find the lengths 100 cm, 100.3 cm, 100.5 cm. I assert that the real length is 100.3 cm $\pm$ 0.3 cm. This indicates my confidence that, if I were to keep repeating these measurements, the sequence of numbers obtained would have superior and inferior limits inside [100, 100.6]. There is no assertion that none of these measurements would yield values outside the interval, only that, in the long run, those anomalous values would play a negligible role.

Experiments designed to test or establish probabilities are very similar. Suppose I toss a coin 100 times and obtain 49 heads. I say the coin is fair, indicating my confidence about a regularity that would be observed if this coin were to be tested by other people, say, tossing it respectively 10^3, 10^4, 10^5 times, and so on. So I am envisaging many other experiments which are not actually done. My model of the actual situation involves reference to many possible situations.

This way of modelling situations has been used with varying success elsewhere. In logic and linguistics, it has been used to develop hypotheses about such qualifiers as « necessary », « possible »; such adverbial modifiers as « slowly », « deliberately »; verbs such as « believe », « want », « ought »; and tense and mood. In foundations of physics, the many-world view of Everett and De Witt is well known, and such models have also been used for quantum logic, the measurement problem and the quantum-mechanical paradoxes, and also for the axiomatization of classical and relativistic mechanics [13].

The elements of a many-world model are referred to variously as « set-ups », « possible situations », « possible worlds », or just « worlds ». I shall use the last term, mainly for its brevity. In a Kyburg model, each world is or has a sequence of events in an outcome space; most or all of these series are finite. A *maximal* world is one which is not part of another; or is infinite. Precisely (but in simplified form):

1) A model is a triple $M = \langle S, W, \tau \rangle$, where S and W are disjoint non-empty sets (the *outcome space* and the set of *worlds*), and τ is a function which associates with each element α of W a sequence $\tau(\alpha)$—finite or countably infinite—of members of S.

2) World α in model M is *maximal* if and only if there is no other world β in M such that $\tau(\alpha)$ is an initial segment of $\tau(\beta)$.

3) If $A \subseteq S$, then probability$_M$ $(A) = r$ if and only if the relative frequency of A in $\tau(\alpha)$ equals r for every maximal world α in model M; and is not defined if there is no such number.

This view differs from Popper's in that there is in general no single virtual sequence, but many; and the probability of an event equals r exactly if, in all maximally extended series of possible experiments, the relative frequency equals r. This notion of possible is clearly not *logical possibility*; on the contrary, such assertions of possibility as are present here are themselves contingent, and are meant to reflect empirical facts. In this way we can similarly say that, if a body were weighed twice in quick succession, we would get approximately the same values for its weight; though from a *logical* point of view this is not necessarily so.

We can regard Popper's view as a special case: his models are Kyburg's, with the stipulation that the set of worlds has only one maximal element.

However, as they stand, both these views fall prey to all the difficulties that beset the strict frequency interpretation. For, clearly, there are models containing but a single world (a single long run). In that case we find that the domain of the function probability$_M$ may not be a Borel field, and indeed not even a field; and even where defined, not countably additive.

Given these difficulties, which propensity theorists *can* avoid by the simple expedient of denying logical connections between frequency and probability, the models will not be acceptable. But an improvement of Kyburg's proposal may be possible.

4·3. *A modal frequency representation.* – In an actual experiment we generally find numbers attaching to large but finite ensembles (whether of similar coexistent systems, or of the same system brought repeatedly into the same state) which we compare with theoretical expectation values.

But this comparison is based on the idea that the number found depends essentially on a (finite) frequency in just the way that the theoretical expectation value depends on a probability. Moreover, we demand closer agreement between the number found and the expectation value as the size of the ensemble is increased.

Hence it is reasonable to take as an *ideal* (*repeated*) *experiment* an experiment performed infinitely many times under identical conditions or on systems in identical states. The relation between the ideal and the actual should then be this: the actual experiment is thought about in terms of its possible extensions to ideal repeated experiments (its ideal extension). If this is correct, we compare an actual experiment with a conceptual model, consisting in a family of ideal repeated experiments. In that model there should be an intimate relation between frequencies and probability, so that the model can be directly compared with the theory under consideration. Secondly, the theory of statistical testing should be capable of being regarded as specifying the extent to which the actual « fits » or « approximates » that model of the experimental situation.

Let us then inquire how an ideal experiment is to be described. In any actual experiment we can only make finitely many discriminations. We can, for instance, determine whether a given spot appears with an x–co-ordinate, in a given frame of reference, of $(y \pm 1)$ cm, with $-10^5 \leqslant y \leqslant 10^5$. The possible values of the x–co-ordinate are all real numbers, but we focus attention on a finite partition of this outcome space. The *first idealization* is to allow that partition to be countable, that is finite or countably infinite. Secondly, in the actual experiment we note down successively a finite number of spots; the *second idealization* is to allow this recorded sequence of outcomes also to be countable.

Note that theory directs the experimental report at least to the extent that the outcome space (or sample space) is given; in different experiments of the same sort we focus attention on different partitions of that space. So we begin with a space $\langle K, F\rangle$, F a Borel field, and describe one ideal experiment by means of a countable partition $\{A_n\} \subseteq F$ and a countable sequence σ of members of K.

We also note that, in the idealization, we could have proceeded in different ways; for instance, we could have merely lifted the lower and upper limits to y, or we could also have made the discrimination accurate to within 1 mm as opposed to 1 cm. So our model will incorporate many ideal experiments with partitions $\{A_n^\alpha\}$, all of which are finer than the partition in the actual experiment. These represent different experimental set-ups, and we cannot expect the series of outcomes to be the same, partly because the change in the set-up may affect the outcomes, and partly because we must allow for chance in the outcomes: if many ideal experiments *were* performed, just as in the case of actual ones, we would expect a spread in the series of outcomes.

But we are building a model here of what would happen in ideal experiments, and this model building is guided either by a theory we assume, or by the expectations we have formed after learning the results of actual experiments. Hence we expect not only a spread in the outcomes, but also a certain agreement, coherence. In actuality, such agreement would again be approximate, but here comes the *third idealization*: we assume that the agreement will be exact.

To make this precise, let us call *significant events* in an ideal experiment with partition $\{A_n\} = G \subseteq F$ exactly the members of the Borel field $BG \subseteq F$ generated by G. We now *stipulate first* that, if that ideal experiment has outcome sequence σ, then relf (A_n, σ) is well defined for each A_n in G and that $\sum_{n=1}^{\infty} \text{relf}\,(A_n, \sigma) = 1$. By Polya's result it follows then that relf $(-, \sigma)$ is well defined and countably additive on BG. Only the relative frequencies of those significant events generated by the partition will be considered in the use or appraisal of any ideal experiment. We *stipulated secondly* that, if A is a significant event in several ideal experiments, or the countable union of significant events in other such experiments, then the frequencies agree as required. To answer the question how many ideal experiments the model should contain, we *stipulate thirdly* that the significant events together form a Borel field on the set K of possible outcomes. This must include, for instance, that, if we consider experiments with partitions $\{A_n^m\}$, $m = 1, 2, \ldots$, and $A_1^m \downarrow A$, then there must be some experiment in the model in which A is a significant event.

Together, these three stipulations give the exact content of the third idealization. And the three idealizations together yield the notion of a model of an experimental situation by means of a family of ideal experiments (*a good family*), which I shall now state precisely.

(4.1) *A good family* (*of ideal experiments*) is a couple $Q = \langle K, E\rangle$ in which K is a nonempty set (the « possible outcomes » or « worlds ») and E is a set of couples $\alpha = \langle G_\alpha, \sigma_\alpha\rangle$ (the « possible experiments ») such that

i) G_α is a countable partition of K and σ_α a countable sequence of members of K (the « outcome sequence » of experiment α);

ii) if $A_1, A_2, \ldots$ are in $BG_{\alpha_1}, BG_{\alpha_2}, \ldots$, the Borel fields generated by $G_{\alpha_1}, G_{\alpha_2}, \ldots$ (the Borel fields of « significant events » of $\alpha_1, \alpha_2, \ldots$), then there is an experiment β in E such that $B = \bigcup^{\infty} A_i$ is in β, and relf $(B, \beta) = \sum \{\text{relf}\,(A_i, \alpha_i) : i = 1, 2, 3, \ldots\}$ if the A_i are disjoint events;

iii) relf (A, σ_α) is well defined for each A in G_α;

iv) $\sum\{\text{relf}(A, \sigma_\alpha): A \in G_\alpha\} = 1$;

v) if $A \in BG_\alpha \cap BG_\beta$, then relf $(A, \sigma_\alpha) =$ relf (A, σ_β).

As noted, in the experiment $\langle G_\alpha, \sigma_\alpha \rangle$ we call σ_α the *outcome sequence* and the members of BG_α the *significant events*. Polya's result guarantees, from iii) and iv), that relf (A, σ_α) is defined for each significant event A. Finally, we call $\langle K, F \rangle$, where $F = \bigcup\{BG_\alpha: \alpha \in E\}$, the *sample space of* good family Q. It will be noted from condition ii) that F is itself a Borel field, for, if $A_1, A_2, \ldots$ are significant events of $\alpha_1, \alpha_2, \ldots$, their union is in BG_β and hence in F.

(4.2) If $Q = \langle K, E \rangle$ is a good family with sample space $\langle K, F \rangle$, we define $PQ(A) = r$ if and only if relf $(A, \sigma_\alpha) = r$ for all α in E such that A is a significant event in α.

(4.3) *Result* I: if $Q = \langle K, E \rangle$ is a good family with sample space $\langle K, F \rangle$, then $\langle K, F, PQ \rangle$ is a probability space.

For let $A_1, A_2, \ldots$ be disjoint members of F. Then, by condition ii) of (4.1), there is an experiment β such that their union is a significant event in β. Their relative frequencies are the same in all experiments in which they are significant events, by condition v) of (4.1), hence PQ is countably additive.

(4.4) *Result* II: If $\langle K, F, P \rangle$ is a probability space, then there is a good family $Q = \langle K, E \rangle$ with sample space $\langle K, F \rangle$ such that $P = PQ$.

This is easy to prove: let E be the family of all couples $\langle G, \sigma \rangle$ such that $G \subseteq F$ is a countable measurable partition of K and σ a sequence of members of K such that relf $(-, \sigma)$ restricted to G is exactly P restricted to G. That such a sequence σ exists we know from result (3.7) in subsect. **3**.2. Condition ii) of (4.1) is satisfied simply because the measurable sets are closed under countable union, and every measurable set is a member of some countable (that is finite or denumerable) measurable partition.

The other conditions are satisfied because P is a probability measure.

We now have the desired representation result: probability spaces bear a natural one-to-one correspondence to good families of ideal experiments. It is possible, therefore, to say:

(4.5) the probability of event A equals the relative frequency with which it would occur, were a suitably designed experiment performed often enough under suitable conditions.

This is the modal frequency interpretation of probability.

II. Conditional probability.

Before Kolmogoroff's axiomatization became orthodoxy, and the mathematical theory of probability part of measure theory, ***conditional probability*** was taken as the more basic notion. KOLMOGOROFF defined $P(A/B)$ as $P(A \cap B)/P(B)$, for those *antecedent events* B for which $P(B) \neq 0$ and left it undefined elsewhere. The principal approach before him was to define $P(A)$ as $P(A/K)$, where K is the necessary event.

I shall discuss first standard (Kolmogoroff) conditionalization; second, conditional relative frequencies; third, the extension of conditional probability to antecedent events with probability zero. Such extensions have been proposed by POPPER and RENYI.

The conclusions reached are: 1) if there must be a general rule for calculating probabilities of events conditional on antecedent events with positive probability, applicable to all probability spaces, then Kolmogoroff's rule is the only one (Teller); 2) conditional relative frequencies are defined for some antecedent events with relative frequency zero, and do not conform to Kolmogoroff's, nor to Popper's or Renyi's concepts of conditional probability; 3) there are reasonable motives for a concept of conditional probability wider than the standard one; but this can be done by a construction in terms of standard probability measures.

5. – Standard conditionalization.

Let $S = \langle K, F, P \rangle$ be a probability space; we also say that P is a probability measure on $\langle K, F \rangle$. It is easily verified that if $P(B) \neq 0$, and we define

$$P_B(X) = P(X \cap B)/P(B)\,, \qquad X \text{ in } F, \tag{5.1}$$

then P_B is again a probability measure on $\langle K, F \rangle$. We call P_B the ***conditionalization of P on B***, and also write

$$P(X/B) = P_B(X)\,. \tag{5.2}$$

The main question addressed in this section is whether the definition of P_B given here is mathematically reasonable and/or unique. To answer it we shall have to relate conditionalization to orthogonal decomposition; in a sense they are the same.

5·1. *Orthogonal and full measures.* – If $M = \langle K, F \rangle$ is a sample space, let $P(M)$ be the family of probability measures on M (on the Borel field F)—we encountered this family in de Finetti's discussion of additivity. This family has a certain amount of structure; it is at least a *convex* set, for

(5.3) if P, P' are in $P(M)$, and $0 \leqslant r \leqslant 1$, then $rP + (1-r)P'$ is in $P(M)$,

mainly since countable additivity is clearly preserved under linear combination. There is also a natural *orthogonality* relation on it:

(5.4) $P \perp P'$ in $P(M)$ exactly if there is a measurable set X such that $P(X) = 1$ and $P'(X) = 0$,

(5.5) $P \perp P'$ if and only if there are disjoint measurable sets X and X' such that $P(X) = P'(X') = 1$

are equivalent characterizations of this relation.

For future use, I introduce a further relation, which holds between measures and measurable sets:

(5.6) P is *full on* measurable set X (equivalently: X is a *full set* in $\langle K, F, P\rangle$) if and only if $P|F(X)$ takes every value between zero and $P(X)$.

Here $F(X) = \{Y \cap X : Y \text{ in } F\}$, itself a Borel field; and the upright bar denotes restriction of the function P to the domain $F(X)$.

A probability measure may fail to be full on any set at all; consider the case $P(X) = 1$ if and only if $x \in X$ (« P is concentrated on point x ») [14]. However, we are always able to study a probability space by switching attention to another space in which all relevant sets are full. For let $S = \langle K, F, P\rangle$ be a probability space and $S_0 = \langle [0, 1], B_0, \mu\rangle$ the unit interval with Lebesgue measure. We form the product space $S \times S_0$ in which the set of points is the Cartesian product $K \times [0, 1]$, the Borel field that generated by the sets $A \times E$: A in F and E in B_0, and the probability measure the one found by extension, according to Caratheodory's extension theorem, from the function

$$P^*(A \times E) = P(A)\mu(E).$$

Then we have a natural embedding $h: h(A) = A \times [0, 1]$, which is measure and set-operation preserving; and which takes each measurable set in S into a full set in the product space. To sum up:

(5.7) If $S = \langle K, F, P\rangle$ is a probability space, then there is a probability space $S^* = \langle K^*, F^*, P^*\rangle$ and a function $h: F \to F^*$ such that

i) h is a measure-preserving set-isomorphism onto a Borel subfield of F^*;

ii) $h(A)$ is full in S^* for each A in F.

It will be obvious that, if Y is in $F(B)$, and P full on Y, then P_B is full on Y also.

Henceforth I shall call a function a *measure embedding* if it maps measurable sets of one space one-to-one into measurable sets of another, while preserving measure and countable set operations in both directions (as h does in (5.7)).

5.2. *Partition and orthogonal decomposition.* – If M is a sample space and $P_1, P_2, \ldots$ are probability measures on M, then so is

$$P = \sum_{n=1}^{\infty} r_n P_n , \qquad 0 \leqslant r_n \leqslant 1, \quad \sum_{n=1}^{\infty} r_n = 1. \tag{5.8}$$

I shall call eq. (5.8) a *decomposition* of P, and $P_1, P_2, \ldots$ the factors in that decomposition. If all the measures $P_1, P_2, \ldots$ are mutually orthogonal, (5.8) is an *orthogonal decomposition.*

Suppose for instance that a draw is made from one of two urns, U_1 and U_2. The probability that a black ball is drawn (B) is clearly given by

$$P(B) = aP_1(B) + bP_2(B) , \tag{5.9}$$

where a is the probability that the ball is drawn from U_1, $P_1(B)$ the probability of drawing a black ball from that first urn, and so on. Obviously, P can contain all that information, and we can say

$$P(B) = P(U_1)P(B/U_1) + P(U_2)P(B/U_2) . \tag{5.10}$$

This is a clear case of orthogonal decomposition, since U_1 and U_2 are disjoint.

On the other hand, let A be the event of drawing a rubber ball, while half of all the balls are neither rubber nor black. In that case

$$P' = aP_A + bP_B \tag{5.11}$$

is also a probability measure, but not P, for $P(A \cup B) = \frac{1}{2}$ but $P'(A \cup B) = 1$. Nor is (5.11) an orthogonal decomposition of P', unless $P_A(B) = 0$.

Clearly we obtain an orthogonal decomposition of P if we choose a countable measurable partition $G = \{A_n\}$ and set $P_n = P(-/A_n)$, $r_n = P(A_n)$. I will now show that every orthogonal decomposition is formed in exactly this way. To begin, a lemma:

(5.12) If $\{P_m\}$ is a countable orthogonal family of measures on sample space M, then there is a countable disjoint family of measurable sets $\{X_n\}$ such that

$$P_m(X_n) = \delta_{mn} \qquad (m, n = 1, 2, \ldots).$$

Since $\{P_m\}$ is an orthogonal family, there are sets X_{mn} such that $P_m(X_{mn}) = 1$ and $P_n(X_{mn}) = 0$ when $m \neq n$. Define

$$X_m^1 = \bigcap_{\substack{n=1 \\ m\neq n}}^{\infty} X_{mn} ,$$

$$X_m^0 = \bigcup_{\substack{k=1 \\ k\neq m}}^{\infty} X_{km} ,$$

$$X_m = X_m^1 - X_m^0 .$$

First, the sets X_m are mutually disjoint, for $X_m \subseteq X_m^1 \subseteq X_{mn} \subseteq X_n^0 \perp X_n$. (Indeed, we can take $X_{mn} = K - X_{nm}$.)

Secondly, $P_m(X_m^0) = \sum P_m(X_{km}) = 0$, so $P_m(X_m) = P_m\Big(\bigcap_{\substack{n=1 \\ m\neq n}}^{\infty} X_{mn}\Big) = 1.$

Thirdly, $P_m(X_n) \leqslant P_m(X_n^1) \leqslant P_m(X_{nm}) = 0$, hence $P_m(X_n) = 0$, when $m \neq n$.

(5.13) $P = \sum_{n=1}^{\infty} r_n P_n$ with $r_n > 0$ is an orthogonal decomposition of P if and only if there is a countable measurable partition $G = \{A_m : m = 0, 1, 2, \ldots\}$ such that $P(A_0) = 0$, $r_n = P(A_n)$, and $P_n = P(-/A_n)$, for $n = 1, 2, \ldots$.

The proof is now straightforward. First if G is as stipulated, then $\sum P(A_n) = 1$ and the measures $P_n = P(-/A_n)$ are orthogonal, so

$$P = \sum_{n=1}^{\infty} P(A_n) P_n$$

is an orthogonal decomposition.

Secondly, let $P = \sum_{n=1}^{\infty} r_n P_n$ be an orthogonal decomposition, then by (5.12) there is a disjoint measurable family $\{X_m\}$ such that $P_n(X_m) = \delta_{mn}$. We can turn this into a partition $G = \{A_m\}$ by setting $A_0 = K - \bigcup \{X_m\}$ and $A_n = X_n$ for $n = 1, 2, \ldots$.

Now $P(A_0) = P(K) - \sum P(X_m) = 1 - \sum \sum P_n(X_m) = 0$ since $P_m(X_m) = 1$. Also, $P(X_m) = \sum r_n \delta_{mn} P_n(X_m) = r_m P_m(X_m) = r_m$, which fixes the values of the coefficients. So

$$P(A/X_m) = \frac{P(A \cap X_m)}{P(X_m)} = \frac{\sum r_n \delta_{mn} P_m(A \cap X_m)}{r_m} = P_m(A \cap X_m) = P_m(A)$$

because $P_m(X_m) = 1$; which finishes the proof.

5·3. *Teller's proof: conditionalization is unique.* – We have seen that standard conditionalization is essentially the same thing as orthogonal decomposition. But why should this be regarded as the correct way to conditionalize probabilities? There are two sorts of arguments: the first looks at the interpretation of probability (and, accordingly, we shall look in sect. **6** at conditional frequencies), the second at the mathematical relationships. Following this second approach, TELLER has proved that, in effect, there can be only one rule for calculating conditional probabilities (namely the standard one) if we assume that they can be calculated from the absolute probabilities (in pointwise fashion) *and* that rule is the same for all probability spaces [15]. The way in which I shall present the result is much indebted to improvements made by A. FINE.

Let me make this precise. If A is a nonnull measurable set in $S = \langle K, F, P\rangle$, we wish to construct a new probability measure P' such that $P'(A) = 1$. But we wish more. There must be « essential agreement » between P and P' on subsets of A. If that requirement is phrased as $P(B)/P(C) = P'(B)/P'(C)$ for all nonzero subsets B and C of A, we obtain the result that $P' = P_A$ immediately. Here is the weakest form of the requirement:

(5.14) There is a function f of $[0, P(A)]$ into $[0, 1]$ such that $P'(B) = f(P(B))$ for all measurable subsets B of A in S.

Since $P'(A) = 1$, and $P'(B) = P'(A \cap B) + P'(B - A)$, it follows that $P'(B) = P'(A \cap B)$. Hence the required property of f is equivalent phrased as

(5.15) $P'(B) = f(P(A \cap B))$ for all measurable sets B in S.

We must now add that this function can be specified in the same way regardless of the probability space S and measurable set A considered. I propose:

(5.16) If h is a measure embedding of S into $S^* = \langle K^*, F^*, P^*\rangle$, then $P^{*\prime}(E) = f\big(P^*(E \cap h(A))\big)$ is also a probability measure on F^* with $P^{*\prime}(h(A)) = 1$.

Let us check first that standard conditionalization meets these criteria. The function f is simply division by $P(A)$. If h is a measure embedding, then $P^*(h(A)) = P(A)$, so, equivalently, to apply f is to divide by $P^*(h(A))$, which is standard conditionalization in the space S^*, hence as required.

We now deduce first that f is an additive function. For this we consider as h and S^* the construction for result (5.7), namely the product of S with $S_0 = \langle [0, 1], B_0, \mu\rangle$. Let $x + y \leqslant P(A)$, then $B = A \times [0, x/P(A))$ has P^*

measure x, and $C = A \times [x/P(A), (x+y)/P(A))$ has P^* measure y. But then

$$f(x) + f(y) = P^{*\prime}(B) + P^{*\prime}(C) = P^{*\prime}(B \cup C) = f(P^*(B \cup C)) = f(x+y)\,,$$

since these are disjoint subsets of $h(A)$ in the domain of P^*.

A nonnegative additive function on $[0, P(A))$ is continuous and has a constant derivative. By integration we deduce

$$f(x) = xk + m\,;$$

$$f(0) = P'(\wedge) = 0\,, \qquad \text{so} \qquad m = 0\,;$$

$$f(P(A)) = 1\,, \qquad \text{so} \qquad k = 1/P(A)\,;$$

that is application of f is division by $P(A)$.

There are obvious limitations to this result: we could dispute that conditional probabilities must be calculable from absolute ones; and we can also doubt that the rule must be the same when we shift attention from one probability space to another. If, for instance, I ask what the probability is of a measurement result $y \geqslant 3$ given that y must be positive, we naturally think of so modifying the apparatus that it cannot register values $\leqslant 0$. This modification might affect the probability ratios however. It can be objected, on the other hand, that we are now considering the probability of $y \geqslant 3$ given that modification, which is not the same question. However this may be, Teller's proof shows that there is very little leeway for alternative mathematical formulations of conditional probability.

6. – Conditional relative frequencies [16].

The standard (Kolmogoroff) theory leaves $P(A/B)$ undefined when $P(B)=0$. This is in contrast with earlier theories of probability, and also somewhat in contrast with common intuitions. The probability of finding the center of mass of a given airplane *exactly* on the parallel 42 °N is surely zero; but *given* that it is on this parallel, and that it is equally likely to be any part thereof, the probability that it lies between 0 °W and 15 °W surely equals 1/24. We shall see that, in relative frequencies, conditionalization on such « zero antecedents » makes sense and is possible.

6'1. *Informal discussion.* – There certainly are examples of conditional relative frequency, and their general construction is easy. For instance, let σ be the sequence of natural numbers 1, 2, 3, ... The relative frequency of even numbers equals $\frac{1}{2}$. If we now look at the even numbers only, we see that in that series (2, 4, 6, 8, ...) the relative frequency of multiples of 4 equals again $\frac{1}{2}$. We can call that relf (*multiples of* 4/*even*, σ).

Consider now the integral powers of $3: 3^1, 3^2, 3^3, \ldots$. The distances between them are ever increasing: there are 2 in the first 10, 2 more in the next 100, 2 more in the next 1000, and so on. Therefore, relf (*power of* 3, σ) $= 0$.

But now look at the series $3^1, 3^2, 3^3, \ldots$ itself. Every other one is an even power of 3 (3^x, x even). Hence relf (*even power of* 3, σ) $= \frac{1}{2}$. Thus we have an example of a well-defined relative frequency conditional on a « zero antecedent », that is on an event which itself has relative frequency zero.

There have been several efforts to develop theories of probability with $P(A/B)$ well-defined in at least some cases when $P(B) = 0$. In a later section I shall discuss such theories given by POPPER and RENYI. Unfortunately these theories make an assumption which is not valid for relative frequencies. They assume:

> If $P(-/A)$ is well defined, its domain of definition is the same as that of $P(-/K)$

for an initial sample space $\langle K, F\rangle$. Suppose, however, that A is the set of powers of 3, and C a subset of A such that relf $(C/A, \sigma)$ does not exist. This means that relf (C, σ'), where $\sigma' = 3, 3^2, 3^3, \ldots$ is not well defined. Then relf $(A, \sigma) = 0$ and hence relf $(C, \sigma) = 0$ also. Therefore C is in the domain of definition of relf $(-, \sigma)$ but not in the domain of relf $(-/A, \sigma)$.

It is necessary therefore to study conditional relative frequencies directly. I shall show that conditional relative frequencies are, properly regarded, absolute relative frequencies after all. (More precisely: the conditional relative frequencies in a given sequence are obtained by a conditionalizing operation on the events.) Secondly, conditional relative frequencies have a good deal of algebraic structure in themselves.

6·2. *The natural frequency space.* – Let us define a *frequency space* as a couple $\langle \sigma, F\rangle$ for which there exists a set K such that

(6.1) σ is a countable sequence of members of K,
F is a family of subsets of K,
relf (A, σ) exists for each A in F.

We can easily map any such structure into what I shall call the natural frequency space $\langle \omega, F(\omega)\rangle$, where

(6.2) ω is the natural number series $(1, 2, 3, \ldots)$,
$F(\omega)$ is the family of sets A of natural numbers such that
relf (A, ω) exists.

What exactly is in the family $F(\omega)$ is very mysterious. The map $A \to \{i : \sigma(i) \in A\}$ will relate $\langle \sigma, F\rangle$ to $\langle \omega, F(\omega)\rangle$ in a natural way, preserving relative frequencies. It also preserves the set operations on the members of F which are sets of members of σ.

If K has in it members foreign to σ, these will of course play no role in the determination of relative frequencies, and are ignored in this map. If A is in $F(\omega)$, let a be its characteristic function: $a(i) = 1$ if i is in A, and $a(i) = 0$ otherwise. In that case, define

$$+\, a(n) = \sum \{a(m) : m \leqslant n\}, \tag{6.3}$$

$$\operatorname{rel}(a, n) = \frac{+\, a(n)}{n}, \tag{6.4}$$

$$\operatorname{relf}(A) = \operatorname*{limit}_{n\to\infty} \operatorname{rel}(a, n), \tag{6.5}$$

$$\operatorname{relf}(B/A) = \operatorname*{limit}_{n\to\infty} \frac{\operatorname{rel}(ba, n)}{\operatorname{rel}(a, n)} = \operatorname*{limit}_{n\to\infty} \frac{+\, ba(n)}{+\, a(n)}, \tag{6.6}$$

where ba is the characteristic function of AB, namely $ba(n) = b(n)a(n)$.

What I would like to show now is that the natural frequency space is *already* closed under conditionalization (insofar as it can be). That is, there is for each pair of sets B and A in $F(\omega)$ another set $A \rightharpoonup B$ such that

> relf $(A \rightharpoonup B)$ exists if an only if relf (B/A) exists; and if they exist, they are equal.

We find this set by constructing its characteristic function $(a \rightharpoonup b)$. As a first approximation, I propose

$$(a \rightharpoonup b)(i) = \begin{cases} 1 \text{ if the } i\text{-th member of } A \text{ is in } B, \\ 0 \text{ otherwise.} \end{cases}$$

The members of A have of course a natural order: if A is $\{1, 3, 5\}$, then 3 is its *second* and 5 its *third* member. If A is infinite, the above will do very well, for we shall have

$$\operatorname{rel}(a \rightharpoonup b, n) = \frac{+\,(a \rightharpoonup b)(n)}{n} =$$

$$= (\text{the number of } B\text{'s among the first } n\ A\text{'s})/n = \frac{+\, ab(m)}{+\, a(m)},$$

where m is the number at which the n-th A occurs. The variables n and m go to infinity together, and the left-hand limit equals the right-hand limit.

If A is finite, we need a slight emendation. For example, the relative frequency of the even numbers among the first 10 should be $\frac{1}{2}$. If we kept the above definition, it would be zero. So we call k the *index* of A if its characteristic function a takes the value 1 exactly k times. Each number i is, of course, a multiple of k plus a remainder r (ranging from one to k; this slightly unordinary usage of « remainder » makes the sums simpler). We let $a \rightharpoonup b$ take value 1 at i exactly if the (r)-th A is a B. In that way,

$$\operatorname{rel}(a \rightharpoonup b)(k) = \frac{\text{the number of } B\text{'s among the first } k\ A\text{'s}}{k},$$

$$\operatorname{rel}(a \rightharpoonup b)(km) = \frac{m \times \text{the number of } B\text{'s among the first } k\ a\text{'s}}{mk} =$$

$$= \frac{+\,(a \rightharpoonup b)(k)}{k},$$

which, if there are only k A's, is just correct. If we call that number x/k, we also see that

$$\frac{+\,(a \rightharpoonup b)(km+r)}{km+r} \leqslant \frac{mx+(k-1)}{km+(k-1)},$$

which series converges to x/k as m goes to infinity. So this is indeed the correct characteristic function.

Let us make this precise. Let 0 and 1 stand also for those characteristic functions which belong to $\wedge$ and N, respectively, that is take constant values 0 and 1.

(6.7) $\operatorname{index}(a) = k$ if and only if $\sum \{a(n) : n \in N\} = k$ [let k here be any natural number, or ∞].

(6.8) $i \dot{-} k = r$ if and only if *either* $k = \infty$ and $i = r$, or $(\exists\, m)(mk + r = i)$, where $0 < r \leqslant k$.

(6.9) $\#\, a(r) = i$ if and only if $a(i) = 1$ and $+\, a(i) = r \dot{-} \operatorname{index}(a)$.

(6.10) $$(a \rightharpoonup b) = \begin{cases} b \,\#\, a \text{ if } a \neq 0, \\ 1 \text{ otherwise}. \end{cases}$$

Here (6.9) must be read as defining « the r-th (modulo the index of A!) A occurs at i ». Thus if A is $1, \ldots, 10$, then the 5th A occurs at 5; also the 25th (modulo the index 10) occurs at 5. So $\#\, a(5) = \#\, a(25) = 5$ in this example. If this « 25th » A occurs at 5, then $a \rightharpoonup b$ should take the value 1 at 25 exactly if b takes the value 1 at 5. Thus $(a \rightharpoonup b)(25) = b(\#\, a(25)) = b(5)$. This is what (6.10) says. In the case where a never takes the value 1, I have arbitrarily given $a \rightharpoonup b$ the value 1 everywhere. This is only a trick to keep the object well defined.

The class $A \rightharpoonup B$ is the one which has characteristic function $a \rightharpoonup b$. While the operation $\rightharpoonup$ is very « linear », it is certainly not a conditional in the sense of logical implication, because even the analogue to *modus ponens* would not hold. It serves the purpose of showing that the natural frequency space already contains all the *conditional relative frequencies.*

6·3. *A partial algebra of questions.* – I shall now continue the general discussion of relative frequencies, taking the conditionality for granted. That is, we know at this point that relf $(B/A, \sigma)$ can be reduced to an assertion of « absolute » form relf (X, σ). Henceforth, the conditional concept relf $(B/A, \sigma)$ will therefore be used without comment.

Let us view suitable relative frequencies in the long run as answers to questions of the form « What is the chance that an A is a B? » The terms A and B I take to stand for subsets of a large set K; the set of possible situations or states or events. The question I shall reify as the couple $\langle B, A\rangle$. Since the answer would be exactly the same if we replaced « a B » in the question by « an A which is a B », I shall simplify the matter by requiring that $B \subseteq A$. This couple is a *question on* K, and I shall call B the *Yes-set* and A the *domain* of the question. The *answer* relative to long run σ (*a* countable sequence of members of K) will be relf $(B/A, \sigma)$, if indeed relf $(B/A, \sigma)$ exists; otherwise the question is mistaken relative to σ.

The word « chance » was proposed by HACKING as a neutral term. I use it here, but mean of course exactly what REICHENBACH thought we should mean with « probability ». The occurrence of *actual* long run σ in the determination of the answer makes the question empirical.

i) *Questions.* A *question on* set K is a couple $q = \langle qY, qD\rangle$ with $qY \subseteq \subseteq qD \subseteq K$. Call questions q and q' *comparable* exactly if $qD = q'D$. The set $[q]$ of questions comparable to q is clearly ordered through the relations on its first members (*Yes-sets*), and we could define

$$q \cap q' = \langle qY \cap q'Y, qD\rangle\,, \qquad -q = \langle qD - qY, dD\rangle\,,$$

and so on, to show that $[q]$ is a family isomorphic by a natural mapping to the power set of qD.

The operation of *conditionalization* which I shall now define takes us outside $[q]$, unless of course $qY = qD$:

$$q \to q' = \langle qY \cap q'Y, qY\rangle\,.$$

This leads us to the next topic.

ii) *Unit questions.* If q is a question and $qY = qD$, I shall call it a *unit question.* Its answer must always be 1; and in $[q]$ it plays the role of unit, that is supremum of the natural partial ordering.

Henceforth, let Q be the set of questions on K, and U the set of unit questions thereon; let u, v, ... always stand for unit questions.

The unit questions are not comparable to each other. But they are easily related nevertheless; they form a structure isomorphic to $P(K)$ under the natural map: A to $\langle A, A\rangle$. Let us use the symbols $\neg$, $\wedge$, $\vee$, $\leqslant$ in this context, to maintain a distinction between the natural ordering of the unit questions and that of the questions comparable to a given one:

$$\neg u = \langle K - uY, K - uY\rangle ,$$

$$u \wedge v = \langle uY \cap vY, uY \cap vY\rangle ,$$

$$u \leqslant v \text{ if and only if } uY \subseteq vY ,$$

where of course $u = \langle uY, uD\rangle$ and so on. Every question is a conditionalization of unit questions:

$$q = \langle qY, qD\rangle = \langle qD, qD\rangle \to \langle qY, qY\rangle .$$

Therefore define

$$uq = \langle qD, qD\rangle ,$$

$$vq = \langle qY, qY\rangle ,$$

$$q = (uq \to vq) .$$

We note that, if q and q' are *comparable*, the operations on them are definable in terms of those on unit questions:

$$-q = uq \to \neg vq ,$$

$$(q \cap q') = uq \to (vq \wedge vq') ,$$

$$(q \cup q') = uq \to (vq \vee vq') ,$$

$$q \subseteq q' \text{ if and only if } vq \leqslant vq' .$$

It would not make much sense to generalize these except for $-$ which is of course defined for all questions. However, I want to add one operation on all questions

$$q \cdot q' = uq \to (vq \wedge vq') ,$$

which reduces to $\cap$ when q and q' are comparable, but is otherwise not commutative. However, it gives the comparable question to q that comes closest to being its conjunction with q'.

iii) *The logic of questions.* Since I have not given a partial ordering of all questions, it may seem difficult to speak of a logic at all. Should we say that q *implies* q' if the Yes-set of q is part of the Yes-set of q'? Or should we require in addition that their domains are equal? Or that the No-set (domain minus Yes-set) of q' be part of the No-set of q? The minimal relation is certainly that the Yes-set of q be part of the Yes-set of q'. We may think of a yes-no question q as related to a proposition which is *true* at x in K if $x \in qY$, *false* at x if $x \in (qD - qY)$, and *neither true nor false* in the other cases. In that context the relation of semantic entailment is just that minimal relation (corresponding to valid arguments) of « if true, then true ». So let us define

$$q \Vdash q' \text{ if and only if } qY \subseteq q'Y$$
$$\text{if and only if } q \to q' \text{ is a unit question}.$$

Then we note that the analogue to *modus ponens* holds, but only because something stronger does

$$q \cdot (q \to q') \Vdash q \to q' \Vdash q,$$

which should not be surprising, because these conditionals are just like Belnap conditionals (a Belnap conditional says something only if its antecedent is true; in that case it says that its consequent is true) [17]

(6.11) $\quad q \to q' = vq \to vq',$

(6.12) $\quad q_0 \to (q \to q') = q_0 \to (vq \to vq') =$

$$= q_0 \to \langle vqY \cap vq'Y, vqD\rangle = q_0 \to (vq \wedge vq') = q_0 \to (q \cdot q').$$

Corollary: $u \to (v \to v') = u \to (v \wedge v')$.

(6.13) $\quad (q_0 \to q) \to q' = \langle q_0 Y \cap qY, q_0 Y\rangle \to q' =$

$$= \langle q_0 Y \cap qY \cap q'Y, q_0 Y \cap qY\rangle = (vq_0 \wedge vq) \to vq' = (q_0 \cdot q) \to q'.$$

Corollary: $(u \to v) \to v' = (u \wedge v) \to v'$.

(6.14) If $u_0 \leqslant u \leqslant u'$, then $u \to u_0 = (u' \to u) \to (u' \to u_0)$.

The second and third show that iteration is trivial; the fourth is a trivial corollary which I mention because of the way it recalls the « multiplication axiom ». Which brings us to the next topic.

iv) *The multiplication axiom.* Reichenbach's fourth axiom was the « theorem of multiplication ». In our symbolism, it states:

(6.15) If $\text{relf}(B/A, \sigma) = p$ and $\text{relf}(C/A \cap B, \sigma) = r$ exist, then $\text{relf}(B \cap C/A, \sigma)$ also exists and equals $p \cdot r$.

The answer to the question q, relative to σ, is

$$\text{relf}\,(qY/qD, \sigma) = m(q)\,.$$

Just for the moment, let A, B, C stand equally for the unit questions, $\langle A, A\rangle$, $\langle B, B\rangle$, etc. Then the axiom clearly says:

> If $m(A \to B) = p$ and $m(A \cap B \cdot \to C) = r$ exist, then $m(A \to \cdot B \cap C)$ also exists and equals $p \cdot r$.

Because $X \to Y = X \to X \cap Y$, there are only three sets really operative here: $A \cap B \cap C$, $A \cap B$, A. So we can phrase this also as follows:

> If $u_0 \leqslant u \leqslant u'$, and $m(u' \to u) = p$ and $m(u \to u_0) = r$ exist, then so does $m(u' \to u_0)$ and equals $p \cdot r$.

Thus in the favorable case of $p \neq 0$, which here means only that a certain conditional probability is not zero,

$$m(u \to u_0) = \frac{m(u' \to u_0)}{m(u' \to u)} \qquad \text{if } u_0 \leqslant u \leqslant u'\,. \tag{6.16}$$

But at the end of subsection iii) we just saw that, with $u_0 \leqslant u \leqslant u'$ given, $u \to u_0 = (u' \to u) \to (u' \to u_0)$. So we have

$$m[(u' \to u) \to (u' \to u_0)] = \frac{m(u' \to u_0)}{m(u' \to u)}\,.$$

Let us now generalize this to conditionals for which the antecedent and consequent are not specially related:

$$\begin{aligned} m(q \to q') &= m(vq \to (vq \wedge vq')) = \\ &= \frac{m(uq \to (vq \wedge vq'))}{m(uq \to vq)} = \\ &= \frac{m(q \cdot q')}{m(q)} \qquad \text{provided } m(q) \neq 0\,, \end{aligned} \tag{6.17}$$

where I went from the first line to the second by the reflection that $uq \geqslant vq \geqslant vq \wedge vq'$ always, and applying (6.16).

We have seen that in the general theory of questions « What is the chance that an A is a B? » there is a logically reasonable conditionalizing operation; relative frequency of the conditional object looks quite familiar in the relevant special cases.

7. – Extended conditional probabilities [18].

In the standard (Kolmogoroff) treatment, $P(A/B)$ is defined only if $P(B) > 0$. Apart from the intuitive counterexamples furnished by conditional frequencies, there is a clear boundary to the use of conditional probabilities, indicated by the following elementary observation:

(7.1) If G is a disjoint measurable family in $S = \langle K, F, P\rangle$ and $P(A)$ is positive for each A in G, then G is countable.

For consider the series $1, \frac{1}{2}, \frac{1}{3}, \ldots$. Only finitely many members of G can have measure greater than or equal to $1/n$; indeed there must be no more than n such members, for the sum cannot exceed 1. But each event A in G has measure $\geqslant 1/n$ for some n; hence there are only countably many.

We can, therefore, conditionalize, in standard fashion, upon only countably many incompatible antecedents. This boundary is relaxed somewhat by conditional expectations which exist by the Radon-Nikodym theorem. These, however, are defined within the limits of the standard theory and do not extend it. In this section I shall discuss the treatments of conditional probability by POPPER and RENYI which do allow the domain of definition to be extended to uncountable disjoint families.

7'1. *Popper: axioms.* – POPPER defined a conditional probability function to be a map from a domain of form $F\times F$ into $[0, 1]$. The structure of F was not assumed, but defined in terms of P; it was seen that P could not satisfy Popper's axioms without imposing on F the structure of a Boolean algebra [19].

This procedure can easily be adapted to present purposes by assuming F to be a Borel field on a given set K, and requiring countable additivity where POPPER requires merely finite additivity. Moreover, discussions by STALNAKER and HARPER have simplified Popper's presentation considerably, and I shall further simplify it here [20].

(7.2) A ***Popper (probability) measure*** on a Borel field F on set K is a map of $F\times F$ into $[0, 1]$ such that

A1) $0 \leqslant P(A/B) \leqslant P(B/B) = 1$;

A2) if $P(K - B/B) \neq 1$, then $P(-/B)$ is a probability measure on $\langle K, F\rangle$;

A3) $P(A \cap B/C) = P(A/C)P(B/A \cap C)$;

A4) if $P(A/B) = P(B/A) = 1$, then $P(-/A) = P(-/B)$.

Even this simple formulation is heavily beset by «don't cares», namely the sets B such that $P(\bar{B}/B) = 1$. The following results show how these can be discarded. Let us call A *normal* exactly if $P(\bar{A}/A) \neq 1$ and ***abnormal*** otherwise.

(7.3) A is abnormal if and only if $P(B/A) = 1$ for all B in F.

(7.4) A is abnormal if and only if $P(A/B) = 0$ for all normal B in F.

For suppose that A is abnormal and B normal. Then $0 = P(A \cap \bar{A}/B) = P(A/B)P(\bar{A}/A \cap B)$ by $A2$) and $A3$); so either $P(A/B)$ or $P(\bar{A}/A \cap B)$ equals zero. But $P(B \cap \bar{A}/A) = 1$ by (7.3), so $P(B/A)P(\bar{A}/B \cap A) = 1$ by $A3$); and $P(B/A) = 1$ by (7.3) again; hence $P(\bar{A}/B \cap A) = 1$. We conclude that $P(A/B) = 0$.

We can now discard $A4$) from the axioms. For suppose that $P(A/B) = P(B/A) = 1$. Then if either A or B is abnormal, so is the other, by (7.4). But if A and B are abnormal, $P(-/A) = P(-/B) =$ the constant map into 1, by (7.3). If A and B are both normal, we have $P(B/A) = 1$ so $P(X/A) = P(X \cap B/A)$ by $A1$) and $A2$), which equals $P(B/A)P(X/B \cap A) = P(X/B \cap A)$; *mutatis mutandis* for $P(X/B)$; hence $P(-/A) = P(-/B) = P(-/A \cap B)$. Hence $A4$) is redundant. This leads us to the following reformulation of the theory:

(7.5) A *Popper space* is a quadruple $T = \langle K, F, G, P \rangle$, with K a nonempty set, F a Borel field on K, G a nonempty subclass of F (the « normal events »), and P a map of $F \times G$ into $[0, 1]$ such that

I) $P(B/B) = 1$,

II) $P(-/B)$ is a probability measure on $\langle K, F \rangle$,

III) $P(A \cap B/C) = P(A/C)P(B/A \cap C)$,

IV) if $P(A/B) \neq 0$ for some B in G, then A is in G; all wherever defined.

There is a natural one-to-one relation between Popper spaces and Popper measures, in that we can extend P by setting $P(A/B) = 1$ for A in F and B in $F - G$ to obtain the correlate Popper measure. We note some simple but useful consequences; the use of « $P(A/B)$ » will carry the presumption that B is normal.

(7.6) If $A \subseteq B$ and $P(A/B) \neq 0$, then $P(X/A) = \dfrac{P(X \cap A/B)}{P(A/B)}$.

(7.7) $P(X/A) = P(X \cap A/A)$.

(7.8) If $A \subseteq B \subseteq C$ then $P(A/C) \leqslant P(A/B)$, provided A is normal.

Result (7.6) is a useful equivalent formulation of axiom $A3$) in (7.2) or principle III) in (7.5), and (7.8) is proved by noting that $P(A/C) = P(A \cap B/C) = P(A/B)P(B/A \cap C) = P(A/B)P(B/A)$ because $A = A \cap B = A \cap C$ in this case; but $0 \leqslant P(B/A) \leqslant 1$ so $P(A/C)$ is a fraction of $P(A/B)$.

(7.9) If $P(A/C) \neq 0$ and $A \subseteq B$, then $P(B/C \cup B) \neq 0$.

Note first that by IV) of (7.5), if C is normal, so is $C \cup B$. Assume then that $P(A/C)$ is positive and $A \subseteq B$. Suppose first that $P(C/C \cup B) = 0$, then $P(C \cup B/C \cup B) = 1 = P(C/C \cup B) + P(B - C/C \cup B)$, so $P(B - C/C \cup B) = 1$, and hence $P(B/C \cup B) = 1$.

Suppose second that $P(C/C \cup B) \neq 0$. In that case, since $A \subseteq B$, it suffices to prove that $P(A/C \cup B)$ is positive. But $P(A/C \cup B) \geqslant P(A \cap C/C \cup B) = P(C/C \cup B)P(A/C \cap \cdot C \cup B) = P(C/C \cup B)P(A/C)$, the product of two positive numbers and hence positive.

This last result is a useful lemma for later purposes.

7.2. *Renyi: quotients of measures.* – Renyi's theory of conditional probability is formulable by means of the definition of a *conditional probability space*; and this is just the definition of a Popper space with condition IV) omitted [21]. RENYI calls one of his spaces *additive* if the family G is closed under finite unions. Thus Popper spaces are a special case of Renyi's additive conditional probability spaces.

RENYI envisaged the construction of conditional probabilities as quotients in the usual way, but by measures which need not be probability measures, not even bounded. He gave the following example. Let $G_1, \ldots, G_n, \ldots$ be a series of subfamilies of F which are mutually disjoint, and $\mu_1, \ldots, \mu_n, \ldots$ a series of measures defined on F which may take any value up to $+\infty$ inclusive. Let G be the union of the series $G_1, \ldots, G_n \ldots$ and define

$$P(X/B) = \mu_m(XB)/\mu_m(B)$$

for the unique number m such that B is in G_m. Provided $\mu_m(B)$ is positive and finite when B is in G_m, the quadruple $\langle K, F, G, P \rangle$ is a conditional probability space.

CSASZAR proved quite easily that all conditional probability spaces can be so constructed. We use a separate measure for each B in G defined as

$$\mu_B(X) = P(X/B) \text{ if } X \subseteq B, \text{ and } +\infty \text{ otherwise.}$$

But RENYI had suggested a condition that would make the representation interesting: that the measures be *dimensionally ordered*: if $m < n$ and $\mu_m(X)$ is finite, then $\mu_m(X) = 0$. (The indices m and n are linearly ordered, and the family of measures need not be countable.)

Here we may note the origin of the terminology. Figures on a line are also figures on a plane, which in turn are figures in space. Length, area and volume are the appropriate measures; and line figures have zero area, while plane figures have zero volume. If we further stipulate that two- and three-dimensional figures have infinite length, and the latter infinite area, then we have here a dimensionally ordered series of measures.

RENYI then proved a result, which was reported and generalized by CSASZAR, namely that every *additive* conditional probability space can be generated by a series of measures which are dimensionally ordered [22].

Thus all Popper spaces can be so generated.

7·3. *Representation of extended conditional probabilities.* – To what extent can the extended conditional probabilities be represented in terms of (Kolmogoroff) absolute probabilities? And among the cases so representable, is there a significant subclass corresponding to what RENYI calls dimensional order? To begin there is a simple negative result:

(7.10) If $T = \langle K, F, G, P \rangle$ is a Popper space and $S = \langle K, F, P' \rangle$ a probability space such that $P(-/A) = P'_A$ for each set A in G, then every disjoint subfamily of G is countable.

This follows from (7.1) stated at the beginning. This suggests strongly that probability spaces can be regarded as a special proper subclass of Popper spaces, and this will turn out to be so in our representation result. Each Popper space will be formed by « pasting together » families of probability spaces.

(7.11) The set $I = \{P_\alpha : \alpha < \lambda\}$ is an *ordinal family* (*of probability measures*) (indexed by the ordinal λ) on set K exactly if K is a nonempty set, λ an ordinal, and I such that

a) P_α is a probability measure on a Borel field F_α on a subset K_α of K;

b) $\bigcup \{F_\alpha : \alpha < \lambda\}$ is a Borel field F on K;

c) if A is in F and B in F_α, then $A \cap B$ is in F_α;

d) if A is in F and $P_\alpha(A \cap B) \neq 0$, then $P_\alpha(A)$ is defined and positive for some $\beta < \lambda$;

e) if $\alpha < \beta < \lambda$, $\mu_\alpha(X)$ and $\mu_\beta(X)$ positive, and β the first ordinal such that $\mu_\beta(X \cup Y)$ is positive, then $\mu_\alpha(X \cap B)/\mu_\alpha(X) = \mu_\alpha(X \cap B)/\mu_\beta(X)$.

Note that clause *e*) is vacuously satisfied if all members of I have the same domain of definition; and also if there are disjoint sets K_α such that $P_\alpha(K_\alpha) = 1$ for $\alpha < \lambda$. Suppose that I is an ordinal family on K; define G to be $\{B \in F : P_\alpha(B)$ is defined and positive for some $\alpha < \lambda\}$, and define

(7.12) $P(A/B) = P_\alpha(A \cap B)/P_\alpha(B)$ for the first ordinal $\alpha < \lambda$ such that $P_\alpha(B)$ is defined and positive.

Call $T = \langle K, F, G, P \rangle$ so defined *the space generated by I.*

In the simplest example, let $\{X_\alpha : \alpha < \lambda\}$ be a partition of K and $P_\alpha(X_\alpha) = 1$, so that $I = \{P_\alpha : \alpha < \lambda\}$ is an orthogonal family of probability measures, all defined on the same Borel field F. In that case clause e) of definition (7.11) is vacuously satisfied. Looking now at

$$P(B \cap A/C) = P(A/C)P(B/A \cap C)$$

let α be the first ordinal such that $P_\alpha(C)$ is positive. If $P_\alpha(A \cap C)$ is positive, this equation holds because then α is also the first ordinal such that $P_\alpha(A \cap C)$ is positive. If, on the other hand, $P_\alpha(A \cap C) = 0$, then both sides of the equation are zero. In addition, it is easily seen that, if λ is not countable, then there is no single probability measure such that $P_\alpha = P(-/X_\alpha)$. So if the space generated by an ordinal family is indeed a Popper space, then several of our questions will be answered.

(7.13) If I is an ordinal family of probability measures, then the space it generates is a Popper space.

It is clear that if $I = \{\mu_\alpha : \alpha < \lambda\}$ generates $T = \langle K, F, G, P\rangle$, then F is a Borel field on K, G contains at least K, and is a subset of F, and $P(-/B)$ is a probability measure on F by clause c). Concerning IV) in (7.5), if $P(A/B)$ is positive, then A is in G by clause d) in (7.11). Hence there remains III) of (7.5) to verify.

We note by d) again that, if $A \cap C$ is in G, so is C. Let α be the first ordinal such that $\mu_\alpha(A \cap C)$ is positive and β the first such that $\mu_\beta(C)$ is positive. By c), $\mu_\alpha(A \cap C)$ is defined.

Suppose first that $\mu_\beta(A \cap C) = 0$. In that case $\mu_\beta(A \cap B \cap C) = \mu_\beta(A \cap C) = 0$, so $P(A \cap B/C) = P(A/C) = 0$ and III) holds.

Suppose second that $\mu_\beta(A \cap C)$ is positive. If $\beta \leqslant \alpha$, then $\beta = \alpha$, and III) holds. If $\alpha < \beta$, then clause e) of (7.11) applies with $X = A \cap C$, $Y = C = X \cup Y$, and $\mu_\alpha(A \cap B \cap C)/\mu_\alpha(A \cap C) = \mu_\beta(A \cap B \cap C)/\mu_\beta(A \cap C)$. Hence

$$\begin{aligned} P(A/C)P(B/A \cap C) &= \mu_\beta(A \cap C)/\mu_\beta(C) \cdot \mu_\alpha(A \cap B \cap C)/\mu_\alpha(A \cap C) = \\ &= \mu_\beta(A/C)\mu_\beta(A \cap B \cap C)/\mu_\beta(A \cap C) = \mu_\beta(A/C)\mu_\beta(B/A \cap C) = \\ &= \mu_\beta(A \cap B/C) = P(A \cap B/C), \end{aligned}$$

where $\mu_\beta(X/Y)$ abbreviates $\mu_\beta(X \cap Y)/\mu_\beta(Y)$. Hence III) holds in all possible cases.

(7.14) Every Popper space is generated by some ordinal family of probability measures.

If $T = \langle K, F, G, P\rangle$ is a Popper space, define for each set A in G

$$F(A) = \{A \cap B : B \in F\},$$

$$P_A : P_A(B) = P(B/A) \qquad \text{for } B \text{ in } F(A),$$

and well-order G in any way you like, denoting the ordinal attached to A in G by the same symbol as A itself.

We note first that, if $P_A(E)$ and $P_B(E)$ are both positive, then $E \subseteq A$ and $E \subseteq B$, so I claim *a fortiori* that $P(X/E)$ equals both

$$\frac{P(X \cap E/A)}{P(E/A)} \quad \text{and} \quad \frac{P(X \cap E/B)}{P(E/B)},$$

hence $P_A(X \cap E)/P_A(E) = P_B(X \cap E)/P_B(E)$. For that reason it does not matter whether A or B comes first in the well-ordering. But that claim follows from (7.6).

The family $I = \{P_A : A \in G\}$ is now our ordinal family, and the method of definition guarantees (7.11) clauses *a*)-*c*). Also IV) of (7.5) entails here *d*) of (7.11). Finally, for *e*) we show that, if $P_A(X)$ and $P_E(X)$ are positive, and $P_E(X \cup Y)$ also positive, then

$$P_A(X \cap B/P_A(X)) = \frac{P(X \cap B/A)}{P(X/A)} = P(B/X) = \frac{P(X \cap B/E)}{P(X/E)}$$

again by (7.6) because $X \subseteq A \cap E$ for $P_A(X)$, $P_E(X)$ to be defined.

In this representation a large number of measures were used; and clearly, if $A \subseteq B$ and $P(A/B) \neq 0$, then P_A can be dropped from the ordinal family without affecting the space generated. The question of more economical representations is similar to the « dimensional order » question broached by RENYI. Define

(7.15) $I \leqslant I'$ if and only if I and I' generate the same space and all measures in I are in I'; I is *minimal* if $I \leqslant I'$ implies $I = I'$.

(7.16) Ordinal family I is minimal if and only if there is for each set B in G the generated space $T = \langle K, F, G, P\rangle$ one and only one member μ of I such that $\mu(B)$ is defined and positive.

If the generated space is as described, deletion of μ_α takes its universal set K_α away from F, but $F - \{K_\alpha\}$ cannot be a Borel field, so then the generating family is minimal.

Conversely, suppose that $\mu_\alpha(B)$ and $\mu_\beta(B)$ are positive while $\alpha \neq \beta$. Then $B \subseteq K_\alpha \subseteq K_\alpha \cup K_\beta$, and B is in G, so all these sets are in G. But also, by (7.9), we see that $P(K_\alpha/K_\alpha \cup K_\beta)$ and $P(K_\beta/K_\alpha \cup K_\beta)$ are positive. Finally by (7.6) se see that $P(X/K_\alpha) = P(X \cap K_\alpha/K_\alpha \cup K_\beta)/P(K_\alpha/K_\alpha \cup K_\beta) = \mu_\gamma(X \cap K_\alpha)/\mu_\gamma(K_\alpha)$, where γ is the first ordinal such that $\mu_\gamma(K_\alpha \cup K_\beta)$ is positive; similarly for $P(X/K_\beta)$. The conclusion is that, if we replace μ_α and μ_β by μ_γ, the conditional probabilities will remain the same. (For suppose that $X \subseteq K_\alpha \cup K_\beta$ and δ the first ordinal such that $\mu_\delta(X)$ is positive, and $\alpha < \beta < \delta < \gamma$. The replacement of μ_α by μ_γ will not affect $P(-/X)$, for if $\mu_\gamma(X)$ is also positive, then clause *e*) of (7.11) with $K_\alpha \cup K_\beta = X \cup Y$ guarantees that $\mu_\delta(Y \cap X)/\mu_\delta(X) = \mu_\gamma(Y \cap X)/\mu_\gamma(X)$.) So we see that, if the generated space does not satisfy the stated condition, then the ordinal family is not minimal.

Those Popper spaces which are generated by minimal ordinal families have an especially simple structure. To begin, there is a natural compatibility relation among normal sets:

(7.17) If A, B are normal sets in $\langle K, F, G, P\rangle$, A is *compatible* with B exactly if, for some X in G, $A \cup B \subseteq X$ and $P(A/X)$, $P(B/X)$ are positive.

It is easily shown that A and B are compatible exactly if $P(A/A \cup B) \neq 0 \neq P(B/A \cup B)$. Secondly, compatibility is an equivalence relation on G. It is necessary only to prove transitivity: so suppose that $A \cup B \subseteq X$ and $B \cup C \subseteq Y$ and $P(A/X)$, $P(B/X)$, $P(B/Y)$, $P(C/Y)$ are all positive. I argue that $P(A/X \cup Y)$ and $P(C/X \cup Y)$ are also positive:

a) $X \cap Y \subseteq Y$ and $P(X \cap Y/Y) \geqslant P(B/Y)$ is positive, so by (7.9) $P(X \cap Y/X \cup Y)$ is positive; hence also $P(X/X \cup Y)$ and $P(Y/X \cup Y)$ are positive.

b) $P(A/X \cup Y) = P(A \cap X/X \cup Y) = P(X/X \cup Y)P(A/X)$ and therefore positive.

c) $P(C/X \cup Y) = P(C \cap Y/X \cup Y) = P(Y/X \cup Y)P(C/Y)$ and therefore positive.

Using this equivalence relation we characterize a family of simply structured Popper spaces:

(7.18) A Popper space $\langle K, F, G, P\rangle$ is *dimensional* exactly if for each A in G the union of the sets compatible with A is in G, and is compatible with A.

(7.19) A Popper space is dimensional exactly if it is generated by a minimal ordinal family.

First let $\langle K, F, G, P\rangle$ be dimensional and let $R = \{\bigcup[A] : A \in G\}$, where $[A]$ is the set of members of G compatible with A. Hence $R \subseteq G$. For each set

X in R, let P_X be the function defined by $P_X(B) = P(B/X)$ on the subsets B of X which are in F. Well-order R in any way and denote the ordinal attached to X by the same symbol as X. Thus $K_X = X$ is the « universal set » of P_X. Since compatibility is an equivalence relation, the sets $[A]$ are disjoint. Therefore, if $P_X(B)$ is defined and positive, then B and X are compatible, and so $X = \bigcup [B]$. Hence $P_X(B)$ is both defined and positive for at most one X in R. The well-ordering chosen can make no difference therefore; and clause e) of (7.11) is vacuously satisfied. Hence $\{P_X : X \in G\}$ is minimal ordinal family generating this space.

Conversely, suppose that minimal ordinal family $I = \{\mu_\alpha : \alpha < \lambda\}$ generates the space. If A and B are compatible, then $A \cup B$ is compatible with both, and β is the first ordinal such that $\mu_\beta(A \cup B)$ is positive, then $\mu_\beta(A)$ and $\mu_\beta(B)$ are defined and positive. It follows that, if I is minimal and $\mu_\beta(A)$ is defined and positive, then $\mu_\beta(B)$ is also thus for all B compatible with A; hence all these sets are compatible with K_β, of which they are subsets, and which is their union since it is compatible with A as well. Hence the space is dimensional.

(7.20) Ordinal family $I = \{\mu_\alpha : \alpha < \lambda\}$ is minimal if and only if the sets K_α are linearly ordered by set inclusion, $K_\alpha = K_\beta$ implies $\alpha = \beta$, and $\mu_\alpha(K_\beta) = 0$ when $K_\alpha \supset K_\beta$.

Suppose first that the sets K_α satisfy the three stated conditions. In that case, if $\mu_\alpha(B)$ and $\mu_\beta(B)$ are positive, let $K_\alpha \supset K_\beta$; then $\beta \subseteq K_\beta$ and $0 < \mu_\alpha(B) \leqslant \leqslant \mu_\alpha(K) = 0$ which is impossible. So $\mu_\alpha(B)$ is positive for at most one ordinal α, and I is minimal by (7.16).

Conversely, suppose that I is minimal. Clearly $K_\alpha = K_\beta$ if $\alpha = \beta$; and also, $\mu_\alpha(K_\beta)$ cannot be defined and positive if $\alpha \neq \beta$, so if $K_\alpha \subset K_\beta$, in which case K_β is in the domain of μ_β by c) of (7.11), then $\mu_\alpha(K_\beta) = 0$.

To show that the sets K_α are linearly ordered by set inclusion we must show that $K_\alpha \supset K_\beta$ or $K_\beta \supset K_\alpha$ when $\alpha \neq \beta$. Define

A is *superior* to B if and only if $P(B/A \cup B) = 0$

in the generated space. Then

a) K_α is superior to K_β if and only if $K_\alpha \supset K_\beta$,

b) if $\alpha \neq \beta$ then one of K_α, K_β is superior to the other,

from which we conclude that set inclusion orders them linearly.

For a) we have just seen that $K_\alpha \supset K_\beta$ implies $P(K_\beta/K_\alpha) = 0$, and hence superiority; if K_α is superior to β, then $P(K_\beta/K_\alpha \cup K_\beta) = 0$, in which case $P(K_\alpha/K_\alpha \cup K_\beta) \neq 0$, so K_α is compatible with $K_\alpha \cup K_\beta$. But a minimal ordinal family generates a dimensional space, so we conclude that $K_\alpha \cup K_\beta \subseteq K_\alpha$, and so $K_\alpha \supset K_\beta$.

For $b)$ we note that, if $P(K_\alpha/K_\alpha \cup K_\beta) \neq 0 \neq P(K_\beta/K_\alpha \cup K_\beta)$, then all three sets are compatible; but then $\alpha = \beta$, because there can be only one measure giving a positive value to any one of them.

We see therefore that a dimensional space can be pictured by means of a chain of sets

$$K = K_1 \supset K_2 \supset \ldots \supset K_\alpha \supset \ldots$$

such that $P(K_{\alpha+1}/K_\alpha) = 0$ and $P(-/B)$ is definable as $P(- \cap B/K_\alpha)/P(B/K_\alpha)$ for the one and only ordinal α such that $P(B/K_\alpha) \neq 0$.

REFERENCES

[1] In this paper, « countable » will mean « finite or countably (denumerably) infinite ».

[2] B. DE FINETTI: *Probability, Induction and Statistics* (New York, N. Y., 1972), subsect. **5**·22 (translation of a paper which appeared in Italian in 1949).

[3] K. R. PATHASARATHY: *Probability Measures on Metric Spaces*, Chapt. II, theor. 6.4 and 6.5 (New York, N. Y., 1967). A set $\{\mu_\alpha\}$ in $P(M)$ is said to converge to a measure μ exactly if $\mu(A) = \text{limit}_\alpha\, \mu_\alpha(A)$ for every Borel set A in M whose boundary has μ-measure zero; this will indicate the sense of the completeness asserted.

[4] This was pointed out by G. BIRKHOFF: *Lattice Theory* (Providence, R. I., 1940), XII, 5; B. DE FINETTI: *Probability, Induction and Statistics* (New York, N. Y., 1972), subsect. **5**·17 (translation of a paper which appeared in Italian in 1949); but seems to have been ignored in philosophical discussions.

[5] B. DE FINETTI: *Probability, Induction and Statistics* (New York, N. Y., 1972), subsect. **5**·8 (translation of a paper which appeared in Italian in 1949); P. SUPPES: *Set-Theoretical Structures in Science* (Stanford, Cal., 1967).

[6] F. P. CANTELLI: *Rend. Accad. Lincei*, **26**, 39 (1917).

[7] R. VON MISES: *Mathematical Theory of Probability and Statistics* (New York, N. Y., 1964), p. 18.

[8] See G. BIRKHOFF: *Lattice Theory*, 3rd edition (Providence, R. I., 1967), XI, 3; C. KURATOWSKI: *Topologie*, 2nd edition (Warsaw, 1948), p. 358.

[9] G. BIRKHOFF: *Lattice Theory*, 3rd edition (Providence, R. I., 1967), XI, 5; D. A. KAPPOS: *Probability Algebras and Stochastic Spaces* (New York, N. Y., 1969), II, 4 and III, 3.

[10] K. POPPER: *The propensity interpretation of the calculus of probability, and the quantum theory*, in *Observation and Interpretation*, edited by S. KÖRNER (New York, N. Y., 1957), p. 67.

[11] R. GIERE: *Objective single case probabilities and the foundations of statistics*, in *Proceedings of the Fourth International Congress on Logic, Methodology and Philosophy of Science* (Bucharest, 1971).

[12] K. E. KYBURG jr.: *Brit. Journ. Phil. Sci.*, **25**, 358 (1974).

[13] H. EVERETT: *Rev. Mod. Phys.*, **29**, 454 (1957); B. S. DEWITT: *Phys. Today*, **23**, No. 9, 30 (1970); B. VAN FRAASSEN: *A formal approach to philosophy of science*, in *Paradigms and Paradoxes*, edited by R. COLODNY (Pittsburgh, Pa., 1972), p. 303; *Semantic analysis of quantum logic*, in *Contemporary Research in the Foundations*

and Philosophy of Quantum Theory, edited by C. A. HOOKER (Dordrecht, 1973), p. 80.

[14] If the probability algebra $\langle F/P, P\rangle$ has no atoms, every set in $\langle K, F, P\rangle$ is full; see M. LOÈVE: *Probability Theory*, 2nd edition (New York, N. Y., 1960), p. 100, exercise 7.

[15] P. TELLER: *Conditionalization, observation and change of preference*, in *Foundations of Probability Theory, Statistical Inference and Statistical Theories of Science*, edited by W. L. HARPER and C. A. HOOKER (Dordrecht, 1976), p. 205.

[16] This section is an abbreviated version of sect. **2** of B. C. VAN FRAASSEN: *Synthese*, **34**, 133 (1977).

[17] N. D. BELNAP jr.: *Nous*, **4**, 1 (1970).

[18] This section is an improved version of B. C. VAN FRAASSEN: *Journ. Philos. Logic*, **5**, 417 (1976) (see also erratum, **6**, issue 3 (1977)).

[19] K. POPPER: *Brit. Journ. Phil. Sci.*, **6**, 51 (1955-1956); *The Logic of Scientific Discovery*, revised edition (London, 1968), Appendices (*iv) and (v), p. 326.

[20] W. L. HARPER: *Synthese*, **30**, 221 (1975); R. STALNAKER: *Phil. Sci.*, **37**, 64 (1970).

[21] A. RENYI: *Acta Math. Hungar.*, **6**, 285 (1955).

[22] A. CSASZAR: *Acta Math. Hungar.*, **6**, 337 (1955).

Frontiers of Time.

J. A. WHEELER

Center for Theoretical Physics, Department of Physics, The University of Texas
Austin, Tex. 78712

1. – Law without law.

> *species will never vary, and have remained the same since the creation of each species.*
>
> Charles LYELL [1], writing almost three decades before *The Origin of Species*

> [The astronomer Sir John Frederick William] *Herschel says my book is 'the law of higgledy-piggledy'.*
>
> Charles DARWIN [2], 18 days after the November 24, 1859 publication of *The Origin of Species*

Are the laws of physics eternal and immutable? or are these laws, like species, mutable [3] and of « higgledy-piggledy » origin?

The hierarchical speciation of plant and animal life, we now know, arises out of the blind accidents of genetic mutation and natural selection [5, 6]. Likewise the gas laws, the pressure-volume-temperature relation for water and for other substances, and the laws of thermodynamics take their origin in the chaos of molecular collisions. But as for the molecules themselves, the particles of which they are made and the fields of force that couple them, is it conceivable that they too derive their way of action, their structure and even their existence from multitudinous accidents?

Such questions about the « plan » of physics we would hardly raise if we had the skeleton of it in hand. But we don't. Now and then we meet a colleague in another realm of thought who still thinks physics is in possession of this plan. He cites the words of Laplace [7] and reiterates the Laplacean vision as he understands it: the laws are definite, the initial co-ordinates and momenta are definite, and therefore the future is definite. The Universe is a machine.

No, we have to tell him; that is a cracked paradigm. Quantum mechanics allows us to know a co-ordinate, or a momentum, but not both. Of the initial-value data that LAPLACE needed, the principle of complementarity [8] or indeterminacy [9] says half do not and cannot exist.

It is no use to warn our colleague of the grip of determinism, nor to compare it for him with « the hold of astrology on the Renaissance mind; neither education nor enlightenment (Jacob BURCKHARDT insisted [10]) could do anything against this delusion... because it was supported by the authority of the ancients and satisfied passionate fantasies and the fervent wish to know and determine the future » [11]. He reads more physics and comes back convinced that the plan of the world is still determinism. No one can deny, he insists, that the Schrödinger equation foreordains in every detail the time development of the wave function. Yes, there is a probability element in the physics that shows up at the instant of an observation, he admits. However, that element of chance is not at all in contradiction with determinism, he tells us, but evidence that we have failed to include the observer in our bookkeeping. In support of this contention he cites the thesis of Everett [12-14], that the measurement postulate of quantum mechanics [15] can be derived out of the wave equation itself, rather than being added from outside as a mysterious and foreign element.

On this view, our colleague reminds us, the relevant dynamical system is the system under study augmented by the observer system. The wave function for this larger system lends itself to being written as the sum of products. Each product contains one factor referring to the system under study, multiplied by a second factor referring to the observer. Measurement is described in terms of the correlation within these two factors. The first factor describes the system under study as being in a specific quantum state. The second, according to Everett's analysis, represents the observer as aware that the system under study is in that quantum state. The « coexistence » in one overall wave function of these alternative states of observer-plus-observed-system has given rise to such phrases as « branching histories » and « the many-world interpretation of quantum mechanics ». That one can get all this, our friend concludes, out of the deterministic Schrödinger equation shows more clearly than any argument ever advanced that Nature is at bottom deterministic.

Imaginative Everett's thesis is, and instructive, we agree. We once subscribed to it [16]. In retrospect, however, it looks like the wrong track. First, this formulation of quantum mechanics denigrates the quantum. It denies from the start that the quantum character of Nature is any clue to the plan of physics. Take this Hamiltonian for the world, that Hamiltonian, or any other Hamiltonian, this formulation says. I am in principle too lordly to care which, or why there should be any Hamiltonian at all. You give me whatever world you please, and in return I give you back many worlds. Don't look to me for help in understanding this universe.

Second, its infinitely many unobservable worlds make a heavy load of metaphysical baggage. They would seem to defy Mendeléev's demand of any proper scientific theory, that it should « expose itself to destruction ».

WIGNER [17] (see also [18-20]), WEIZSÄCKER [21] and WHEELER [22] have made objections in more detail, but also in quite contrasting terms, to the relative-state or many-world interpretation of quantum mechanics. It is hard to name anyone who conceives of it as a way to uphold determinism.

You tell me what isn't the plan of physics, our friend rejoins. If you understand quantum mechanics so well, why don't you tell we what *is* the plan of physics?

No one knows, we reply. We have clues, clues most of all in the writings of Bohr [23-25], but no answer. That he did not propose an answer, not philosophize, not go an inch beyond the soundest fullest statement of the inescapable lessons of quantum mechanics, was his way to build a clean pier for some later day's bridge to the future.

What kind of a « plan of physics » do you think BOHR had in mind, our colleague asks. I know Einstein's words [26], « Physics is an attempt to grasp reality as it is thought independently of its being observed ». I know Bohr's reply [28], « These conditions [of measurement] constitute an inherent element of any phenomenon to which the term ' physical reality ' can be attached .... [This requires] a final renunciation of the classical ideal of causality and a radical revision of our attitude towards the problem of physical reality ». But if I could have asked BOHR, how did he think the Universe came into being, and what is its substance, what would he have said?

It is too late to ask. The plan is up to us to find.

The Universe can't be Laplacean. It may be higgledy-piggledy. But have hope. Surely someday we will see the necessity of the quantum in its construction. Would you like a little story along this line?

Of course! About what?

About the game of twenty questions. You recall how it goes—one of the after-dinner party sent out of the living room, the others agreeing on a word, the one fated to be questioner returning and starting his questions. « Is it a living object? » « No. » « Is it here on earth? » « Yes. » So the questions go from respondent to respondent around the room until at length the word emerges: victory if in twenty tries or less; otherwise, defeat.

Then comes the moment when we are fourth to be sent from the room. We are locked out unbelievably long. On finally being readmitted, we find a smile on everyone's face, sign of a joke or a plot. We innocently start our questions. At first the answers come quickly. Then each question begins to take longer in the answering—strange, when the answer itself is only a simple « yes » or « no ». At length, feeling hot on the trail, we ask, « Is the word ' cloud '? » « Yes », comes the reply, and everyone bursts out laughing. When we were out of the room, they explain, they had agreed not to agree in advance on any word at all. Each one around the circle could respond « yes » or « no » as he pleased to whatever question we put to him. But however he replied he had to have a word in mind compatible with his own reply—and with all the replies that went before. No wonder some of those decisions between « yes » and « no » proved so hard!

And the point of your story?

Compare the game in its two versions with physics in its two formulations, classical and quantum. First, we thought the word already existed « out there » as physics once thought that the position and momentum of the electron existed « out there », independent of any act of observation. Second, in actuality the information about the word was brought into being step by step through the questions we raised, as the information about the electron is brought into being, step by step, by the experiments that the observer chooses to make. Third, if we had chosen to ask different questions we would have ended up with a different word—as the experimenter would have ended up with a different story for the doings of the electron if he had measured different quantities or the same quantities in a different order. Fourth, whatever power we had in bringing the particular word « cloud » into being was partial only. A major part of the selection—unknowing selection—lay in the « yes » or « no » replies of the colleagues around the room. Similarly, the experimenter has some substantial influence on what will happen to the electron by the choice of experiments he will do on it; but he knows there is much impredictability about what any given one of his measurements will disclose. Fifth, there was a « rule of the game » that required of every participator that his choice of yes or no should be compatible with *some* word. Similarly, there is a consistency about the observations made in physics. One person must be able to tell another in plain language what he finds and the second person must be able to verify the observation.

Go on!

That is difficult! Interesting though our comparison is between the world of physics and the world of the game, there is an important point of difference. The game has few participants and terminates after a few steps. In contrast, the making of observations is a continuing process. Moreover, it is extraordinarily difficult to state sharply and clearly where the community of observer-participators begins and where it ends.

This comparison between the world of quantum observations and the game of twenty questions misses much, but it makes the vital central point. In the real world of quantum physics, *no elementary phenomenon is a phenomenon until it is an observed phenomenon*. In the surprise version of the game no word is a word until that word is promoted to reality by the choice of questions asked and answers given. « Cloud » sitting there waiting to be found as we entered the room? Pure delusion! Momentum, $p_x = 1.4 \cdot 10^{-19}$ gcm/s, or position, $x = 0.31 \cdot 10^{-8}$ cm, of the electron waiting to be found as we start to probe the atom? Pure fantasy! MANN may be going too far when he suggests [29] that « ... we are actually bringing about what seems to be happening to us ». However, it is undeniable that each of us, as observer, is also *one* of the participators in bringing « reality » into being.

Until I heard your story I had never grasped what a strange and fascinating quality the Universe has, and never understood how absolutely indefensible determinism is. Won't you go on? What do *you* think the quantum is trying to tell us about the structure of physics?

Nobody wants conjecture!

But how can anybody even begin to ask the right questions if he doesn't have at least some thought in his mind about how the answers look? I can see you have some suspicions about the shape of things. What are they?

Little though we know, I agree we owe it to each other to talk as frankly as we can.

Please do.

« Law without law »: it is difficult to see what else than that can be the « plan » of physics. It is preposterous to think of the laws of physics as installed by a Swiss watchmaker to endure from everlasting to everlasting when we know that the Universe began with a big bang.

The laws must have come into being [3, 30]. Therefore they could not have been always a hundred percent accurate. That means that they are derivative, not primary. Also derivative, also not primary is the statistical law of distribution of the molecules of a dilute gas between the two interconnecting portions, V_1 and V_2, of a total volume $V = V_1 + V_2$,

$$N_1 = V_1(N/V)\,, \qquad N_2 = V_2(N/V)\,. \tag{1}$$

This law is always violated and yet also always upheld. The individual molecules laugh at it; yet as they laugh they find themselves obeying it. The statistical fluctuations about the predicted values,

$$(\delta\bar{N}_1)_{\text{RMS}} = (\delta\bar{N}_2)_{\text{RMS}} = (\bar{N}_1\bar{N}_2/N)^{\frac{1}{2}}\,, \tag{2}$$

in every normal circumstance are absolutely negligible.

Are the laws of physics of a similar statistical character? And if so, statistics of what? Of billions and billions of acts of observer-participancy which individually defy all law?

The only thing harder to understand than a law of statistical origin would be a law that is not of statistical origin, for then there would be no way for it—or its progenitor principles—to come into being. On the other hand, when we view each of the laws of physics—and no laws are more magnificent in scope or better tested—as at bottom statistical in character, then we are at last able to forego the idea of a law that endures from everlasting to everlasting.

Individual events. Events beyond law. Events so numerous and so uncoordinated that, flaunting their freedom from formula, they yet fabricate firm form.

« Fabricate form »? Do you suggest that even the 4-dimensional space-time manifold is only a fabrication, only a theory—irreplaceable convenience though that theory is?

Yes! Compare space-time with cloth. Each it is useful under everyday circumstances to call a manifold. Yet each is exactly then most obviously not a manifold where it comes to an end, whether in the selvedge made by the loom, or in the geodesic terminations made by one of the « gates of time »—big bang or big crunch [31, 32] or black hole [33]. Nowhere more clearly than in the ending of space-time are we warned that time is not an ultimate category in the description of Nature [34].

Aren't you being extreme? I see the lesson of the game of twenty questions. I begin to believe with you that no elementary phenomenon is a phenomenon until it is an observed phenomenon. I accept that events of observer-participancy, as you call them, occupy a special place in the scheme of things. I agree that that word « cloud » was brought into being entirely through such elementary events. But that such events, however numerous, should be the *sole* blocks for building the laws of physics—and space and time themselves—seems to me preposterous. You surely have been involved enough in times past with nuts-and-bolts physics to know the difference between science and poetry; yet if I appreciate the drift of what you say, you might as well be quoting SHAKESPEARE [35],

> ... These our actors,
> As I foretold you, were all spirits and
> Are melted into air, into thin air:
> And, like the baseless fabric of this vision,
> The cloud-capp'd towers, the gorgeous palaces,
> The solemn temples, the great globe itself,
> Yea, all which it inherit, shall dissolve
> And, like this insubstantial pageant faded,
> Leave not a rack behind. We are such stuff
> As dreams are made on ...

I can't believe any such dreamlike vision of the physical world. As Samuel JOHNSON used to say, I have only to kick a stone to find it real enough.

Why do you say « preposterous »? Perhaps SHAKESPEARE understood this universe of ours better than we do ourselves! You have known for years that the atom is more than 99.99 percent emptiness. If matter turns out in the end to be altogether ephemeral, what difference can that make in the pain you feel when you kick the rock? And how can matter—and space-time—be anything but mutable, coming into being at one gate of time and fading out of existence at the other? No physics before the big bang, or after the big crunch? No! The lesson of Einstein's standard closed-space cosmology is different and stronger. It denies all meaning to such terms as « before the big bang » and « after the big crunch ».

Particles or fields or mathematics won't do for ultimate building blocks. They can't come into being or fade out of existence [30].

Yes, I appreciate the reasons given [36] against believing in any « magic particle » or any « magic field » or [37] any « magic mathematics » as the foundation of physics; but isn't it even more difficult to think of acts of observer-participancy as the magic ingredient?

Difficult, yes; inconceivable, no.

Go on!

No, we have to stop here. It is beyond the power of today to fit together the pieces of the puzzle.

Don't stop! You've carried me halfway into an exciting mystery story. You can't leave me without the traditional half-way-point review of the important clues and first try at a working hypothesis.

Review? A proper review would be impossibly ambitious. And how can one advance a working hypothesis that will not be wrong tomorrow and ridiculous the day after?

I appeal to you to go on. You have told me more than once that science advances only by making all possible mistakes; that the main thing is to make the mistakes as fast as possible—and recognize them. You like to quote the motto of that engine inventor, John Kris: « Start her up and see why she don't run ». You point to Einstein's definition of a scientist, « An unscrupulous opportunist ». If you believe all this, and are a true colleague of mine, you must go on.

You leave no escape!

Good!

Then let us agree to go on; but let us replace the comprehensive review of clues that you wanted by something more modest. How would it do, for example, to survey some of the lessons we have learned from the study of time, and how those lessons bear on « observer-participancy »?

I accept, and with many thanks. But first tell me the central point as you see it.

The absolute central point would seem to be this: The Universe had to have a way to come into being out of nothingness, with no prior laws, no Swiss watchworks, no nucleus of crystallization to help it—as on a more modest level, we believe, life came into being out of lifeless matter with no prior life to guide the process [5, 6, 38].

When we say « out of nothingness » we do not mean out of the vacuum of physics. The vacuum of physics is loaded with geometrical structure and vacuum fluctuations and virtual pairs of particles. The Universe is already in existence when we have such a vacuum. No, when we speak of nothingness we mean nothingness: neither structure, nor law, nor plan.

A conception more clearly impossible I never heard!

Preposterous we have to agree is the idea that everything is produced out of nothing—as preposterous, but perhaps also as inescapable, as the view that life had its origin in lifeless matter.

But how?

« Omnibus ex nihil ducendis sufficit unum », LEIBNIZ told us [39]; for producing everything out of nothing one principle is enough. Of all principles that might meet this requirement of Leibniz nothing stands out more strikingly in this era of the quantum than the necessity to draw a line between the observer-participator and the system under view. Without that demarcation it would make no sense to do quantum mechanics, no sense to speak of quantum theory of measurement, no sense to say that « No elementary phenomenon is a phenomenon until it is an observed phenomenon ». The necessity for that line of separation is the most mysterious feature of the quantum. We take that demarcation as being, if not the central principle, the clue to the central principle in constructing out of nothing everything.

Let me ask if your reasoning couldn't be turned around. You talk of the observer-participancy of quantum theory as the mechanism for the Universe to come into being. If that is a proper way of speaking, would the converse not also hold: The strange necessity of the quantum as we see it everywhere in the scheme of physics comes from the requirement that—via observer-participancy—the Universe should have a way to come into being?

Your point is exciting indeed. If true—and it is attractive—it should provide someday a means to *derive* quantum mechanics from the requirement that the Universe must have a way to come into being [40].

I know that in that empty courtyard many a game cannot be a game until a line has been drawn—it does not matter where—to separate one side from the other. I know that no Gaussian flux integral can be a flux integral until the 2-surface over which it runs—bumpy and rippled though we make it and deform it as we will—has been extended to closure. But how much arbitrariness is there in this more ethereal kind of demarcation, the line between « system » and « observing device »?

Much arbitrariness! BOHR stresses [42] that the stick we hold can itself be an object of investigation, as when we run our fingers over its surface. The same stick, when grasped firmly and used to explore something else, becomes an extension of the observer or—when we depersonalize—a part of the measuring equipment. As we withdraw the stick from the one role, and recast it in the other role, we transpose the line of demarcation from one end of it to the other. The distinction between the probed and the probe, so evident at this scale of the everyday, is the without-which-nothing of every elementary phenomenon, of every « closed » quantum process.

Do we possess today any mathematical or legalistic formula for what the line is or where it is to be drawn?

No.

Then what is important about this demarcation?

Existence, yes; position, no. It is the mark of an observation to leave an « indelible » record, according to BELINFANTE [43]. WIGNER argues that an observation is only then an observation when it becomes part of « the consciousness of the observer » [44] and points to « the impressions which the observer receives as the basic entities between which quantum mechanics postulates correlations » [45]. For BOHR the central point is not « consciousness », not even an « observer », but an experimental device—grain of silver bromide, Geiger counter, retina of the eye—capable of an « irreversible act of amplification » [47]. This act brings the measuring process to a « close » [48]. Only then, he emphasized, is one person able « to describe the result of the measurement to another in plain language » [49]. He adds that « all departures from common language and ordinary logic are entirely avoided by reserving the word 'phenomenon' solely for reference to unambiguously communicable information » [50].

I would have felt very uncomfortable if BOHR had used the term « consciousness » in defining the elemental act of observation. I would not have known what he meant. However, I am beginning to understand and accept the terms he actually adopts, « brought to a close by an irreversible act of amplification » and « communicable in plain language ». What *was* his position on consciousness?

We have asked Jørgen KALCKAR, who collaborated with BOHR in his last months, and he has kindly replied [51], « During work on the preparation of some lecture, to define the phenomenon of consciousness, BOHR used a phrase somewhat like this: a behaviour so complex that an adequate account would require references to the organism's 'self-awareness'. I objected jokingly that with this definition he would soon have to ascribe a consciousness to the highly developed electronic computers. This did not worry BOHR. 'I am absolutely prepared', said he, 'to talk of the spiritual life of an electronic computer; to state that it is reflecting or that it is in a bad mood .... The question whether the machine *really* feels or ponders, or whether it merely looks as though it did, is of course absolutely meaningless'. »

Other outstanding thinkers have argued otherwise. For them « consciousness » makes an unclimbable difference of principle between even the most powerful imaginable computer and the brain [52].

Do you agree with that argument?

How can we possibly accept such a difference of principle?

Do we not believe that brain function itself will someday be explained entirely in terms of physical chemistry and electrochemical potentials? What escape is there from the reasoning of von Neumann [53] and Bohr and many active present-day investigators? When one of the three discoverers of the mechanism of superconductivity today gives us, chapter by chapter and verse by verse, an entirely cellular account of the mechanism of memory [54-56], who can dismiss it?

When a distinguished computer expert and student of the structure of society details, one by one, the distinctions proposed in times past between « consciousness » and the computer, and painstakingly analyzes each down to nothingness [57], what case can anyone possibly maintain for *any* distinction of principle between the computer and the brain?

I am happy not to have to delve today into the term « consciousness ». I find it hard enough to know what to make of « irreversible act of amplification ». Never have I heard of an act of amplification that was not characterized by an amplification factor, or an equivalent quantity; and never an amplification factor that was not a finite number.

Between infinity and a finite number there may be a difference of principle; but between one finite number and another there is only a difference of degree. How big does the grain of silver bromide have to be, or the avalanche of electrons in the Geiger counter, before we count the measuring process as brought to a close by an irreversible act of amplification?

According as I specify one or another number as the critical level of amplification, don't I make all the difference between rating or not rating a given process as an « elementary phenomenon »?

According as the closed Gaussian surface encloses a given elementary charge or not, we find an unmistakable difference in the surface integral of the electric flux. Nevertheless, we know enough about the relevant invariance principle never to question the correctness of always identifying flux with enclosed charge. About « elementary quantum phenomenon » we have not today learned, but have a deep obligation someday to learn, enough to display a similar covariance with respect to where we draw the line. That is what « complementarity » is all about.

Even if neither you nor I know how to define that line, I like the idea that the « game » in the empty courtyard is only then possible when *a* line is drawn. May I question you now about the game itself? How would you describe it if forced to commit yourself?

Let us try to squeeze an answer [36] into three sentences and a picture (fig. 1). The Universe is a self-excited circuit. As it expands, cools and develops, it gives rise to observer-participancy. Observer-participancy in turn gives what we call « tangible reality » to the Universe.

Thank you for the brevity and challenge of that working hypothesis. Forgive me if I respond with an immediate objection. Surely the Universe existed long before any acts of observation were going on. Doesn't that mean that the Universe cannot possibly owe either its structure or its existence to those elementary acts however numerous they are in the more recent history of the Universe?

Agreed we all certainly are that the big bang occurred some $10 \cdot 10^9$ a ($10 \cdot 10^9$ years [58]) ago. We also confess that we know more about the radiation physics and the building of the elements that went on in those long ago days than we do about the organization of the synapses in our own brains

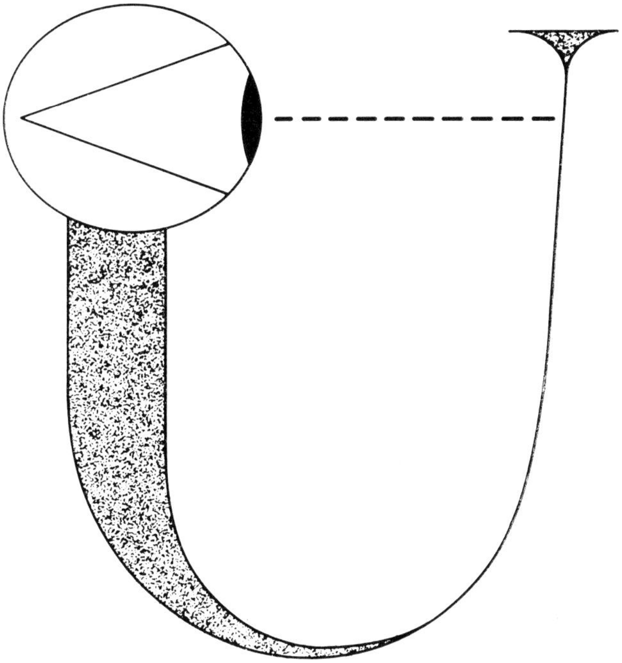

Fig. 1. – The Universe (big U) viewed as a self-excited circuit grows and in time gives rise to observer-participancy which in turn gives tangible reality to the Universe.

right now. But how do we get that information? From the primordial cosmic fireball radiation, and by way of photons from far away stars. Moreover, no photon is counted as a photon until it is an observed photon—any more than that word « cloud » counted as a word until it had been conjured out of nothingness by question and answer. There is a real sense in which we in the here and now play a part in giving tangible reality to that which had supposedly already happened in the remote past.

Isn't what you say directly contradictory to what we know about the direction of time? How can an observation made now have any influence whatsoever on what has already happened?

Ah, but « what has already happened » is not so easy to say. Perhaps here is where we should start to outline that review that you made me promise to give. Let us take up in our next encounter (Review II) « The 'Past' and the 'Delayed-Choice' Double-Slit Experiment ». Here we shall see in what sense, after an electron or photon has *already* traversed a screen with two holes in it, we can choose whether it *shall have* gone through only one of the two holes (triggering one or the other of two distinct counters) or through both of them (contributing to building up an interference fringe) [59]. A decision in the present thus makes a striking difference in what we can rightfully say about the past.

To say « no elementary phenomenon is a phenomenon until it is an observed phenomenon » is to make no small change in our traditional view that something has « already happened » before we observe it. The word « cloud », we mistakenly thought, already existed in the room before we « uncovered » it. The photons of the primordial cosmic fireball radiation that enter our telescope today, we customarily assume, already had an existence in the very earliest days of the Universe, long before life evolved. However, not until we catch a particular one of those photons in a particular state with particular parameters, not until the elementary phenomenon is an observed phenomenon, do we have the right even to call it a phenomenon. This is the sense, the limited sense, but the inescapable sense, in which we, here, now, have a part in bringing about that which « had already happened » at a time when no observers existed.

But what about the unbelievably more numerous relict photons that escape our telescope? Surely you do not deny them « reality »?

Of course not; but their « reality » is of a paler and more theoretic hue. The vision of the Universe that is so vivid in our minds is framed by a few iron posts of true observation—themselves also resting on theory for their meaning—but most of the walls and towers in the vision are of papier-maché, plastered in between those posts by an immense labor of imagination and theory. In this labor, « ... we can never neatly separate what we see from what we know ... what we call seeing is invariably coloured and shaped by our knowledge (or belief) of what we see » [61]. « Without some initial system, without a first guess to which we can stick unless it is disproved, we could ... make no 'sense' of the milliards of ambiguous stimuli that reach us from our environment. In order to learn, we must make mistakes ... the simplicity hypothesis cannot be learned. It is ... the only condition under which we could learn at all » [62]. « ... our mind will still react to the challenge of this conundrum [of what we 'see'] by throwing out a random answer, making ready to test it in terms of consistent possible worlds. It is these answers that will transform the ambiguous stimulus pattern into the image of something 'out there' » [63].

What keeps these images of something « out there » from degenerating into separate and private universes: one observer, one universe; another observer, another universe?

That is prevented by the very solidity of those iron posts, the elementary acts of observership-participancy. That is the importance of Bohr's point that no observation is an observation unless we can communicate the results of that observation to others in plain language [49].

I have the impression that texts on quantum mechanics deal with the case where one observer is involved; research papers on the Einstein-Podolsky-Rosen experiment [64] address the situation where two observers are making measurements; and nobody deals with the richness of the case where many observers are at work on the system. How then is the limiting case to be analyzed that you have in mind? As I understand it, you propose that the statistics of billions upon billions of elementary acts of observation gives rise all by itself, without law or plan, to all the structure and laws of physics. Ever to check that proposal would seem an impossibility.

All of us—we agree—will have to try long and hard before we learn how to do that kind of statistics; but have hope that we will! It is good fortune for this enterprise that some of the necessary ground work has already been laid in a paper by Houtappel, Van Dam and Wigner [65]. Why don't we make it the subject of Review III, « 'Development in Time' Gives Way to 'Correlation in Time' »?

Good! There will surely come a day when the concept of « law without law »—out of the statistics of acts of observation-participancy—can be tested, and either fleshed out or disproved. Until then, however, I shall be one of the many who will persist in considering Maxwell's electromagnetism, Einstein's geometric theory of gravity and the Yang-Mills theory [66] of the quark-binding field as among the truly great achievements of science. They take an enormous range of experience, and measurements made over many years by gifted experimenters, and by way of a few simple principles bring these results into beautiful order. Those laws, in my view, mark our deepest penetration to date into the working of Nature.

Beautiful, yes; marvellous in their summarizing power, yes; but depth of penetration— is that so clear? Isn't the quantum, the fact that no elementary phenomenon is a phenomenon until it is an observed phenomenon—the clear evidence that this universe of ours is in some strange sense a participatory universe—a far deeper discovery?

For me what counts is not words, but equations: we have Nature boiled down into three laws, each with its own definite equation.

Then perhaps we should devote Review IV to the theme of « Many-Fingered Time, 'Imbeddability', and the Laws of Physics ». Those equations that took the efforts of so many investigators so many years to work out can be derived today, all three of them, and in a few minutes, from the utterly simple requirement of Hojman, Kuchař and Teitelboim [67-70] that the physics of the field should be « imbeddable » in space-time. Never has one got so much physics from so little.

This is news to me—and exciting, too. I realize we can't get into technical details now. But can you at least give me the flavor of the idea—and appraise it?

Compare the three field theories with the theory of elasticity. For a homogeneous isotropic solid subject to a small strain—symmetry considerations tell us—there are only two ways to form an expression for the stored energy of second order in the strain. Either take the trace of the strain tensor and square it, or square the strain tensor and take the trace of it. A linear combination of the two quantities with appropriately chosen coefficients gives the most general acceptable expression. By this simple line of reasoning we conclude that the elasticity of a homogeneous isotropic substance is characterized by exactly two elastic constants. If this example shows how much can be obtained from arguments of symmetry, it also illustrates that those symmetry considerations conceal from sight any view of the underlying machinery [71, 72]. A hundred years of the study of elasticity would never have revealed that those elastic constants are formed by adding the second derivatives of hundreds of complicated molecular potential-energy curves, each multiplied by the appropriate direction cosines. And a hundred years of the chemistry of interatomic forces would never have revealed that these forces—and the « hundred laws » of chemistry—have their origin in something so fantastically simple as a system of positively and negatively charged masses moving in accordance with the Schrödinger wave equation. Symmetry is the quick road to the mathematics of law, but no road at all to the machinery behind law. That's why the work of Hojman, Kuchař and Teitelboim makes those three deepest laws of physics today no longer look so deep; rather, as « superficial » as elasticity is superficial. Deeper we must look if we would know in what soil those laws are rooted.

To constitute that soil, to compose that substrate, to serve as that primordial building « substance », what can we possibly propose today except the totality of elementary quantum acts of observer-participancy?

When you speak of the machinery underlying the great laws, are you suggesting that all of the important field equations have already been discovered?

Quite the opposite. Never more rapidly than today is progress being made in unravelling spinor field equations, as seen especially in that beautiful development of our times known as « supersymmetry » [74-77]. Moreover, as bombarding energies go up, and distances probed in collision experiments go down, new effects will come to light which will provide—so distinguished colleagues in elementary particle physics suggest—evidence for still other fields of force, world without end.

I view that as an almost hopeless prospect.

Not at all! Cheer yourself up by remembering what it is to do a harmonic analysis of the tides. The more components we include in our Fourier analysis, the better fit we get to the past, and the better predictions of the future. No matter how many terms we include, however, they won't do a thing to forecast the splash of tomorrow morning's ship launching, or the tsunami from next week's earthquake. On the contrary, the more terms there are for dealing with the expected, the more prominently those features will show up which belong to the unexpected. Even to begin to include them in the bookkeeping requires one to go to a far more comprehensive form of analysis that altogether transcends the traditional treatment of the tides. When the number of Ptolemaic epicycles becomes too great, or the number of « elementary fields » too large, we have a compelling motive to look for a new paradigm.

In the topic you proposed for our Meeting IV you mentioned « many-fingered time », in keeping with your overall theme of « Frontiers of Time ». But what about time itself, and space-time? Are they primordial concepts? Or are they secondary and approximate?

Space-time is a classical and approximate concept that utterly contradicts the uncertainty principle. Give up one or give up the other; you can't keep both.

I can't see how space-time can possibly relate to the uncertainty principle, let alone violate it or be « approximate ».

Then let us look at another, and simpler, classical concept that also violates the uncertainty principle: a world-line. At every point along its length the world-line attributes to the particle in question both a position and a velocity, or momentum. That degree of definiteness violates the uncertainty principle in the most evident way, would you not agree?

Of course. And I know what we do about it. We give up altogether the idea of a deterministic world-line. In its place we speak of a wave function or probability amplitude.

You are right; but you have to go further and say more about what is right and what is wrong with the idea of a world-line if later on you expect to see what is right and what is wrong with the idea of space-time.

I don't see anything right about the idea of a world-line.

Let us recall two features of a world-line which you surely know and accept. First, the classical world-line is a useful approximation to replace the quantum wave function you spoke about when the particle wavelength is small compared to all other relevant physical dimensions. In this limit—we agree—the predictions of « geometrical optics » closely model the predictions of « physical optics ». Second, and regardless of these relative dimensions, the individual world-line or history, H, of the particle's motion is the elementary building block in Feynman's prescription for the wave function [78, 79]. He emphasizes what we may call « the democratic equality of all histories ». The probability amplitude for one history is as great in magnitude as the probability amplitude for any other history. One of these probability amplitudes is differentiated from another only by its phase. The phase is given by the action integral for that history:

$$\text{phase} = I_H/\hbar = (1/\hbar)\int_{x',t'}^{x'',t''} \text{Lagrangian}\,[x_H(t), \dot{x}_H(t), t]\,\mathrm{d}t\,. \qquad (3)$$

The total probability amplitude to transit from the original position x', at the original time t', to the final position x'', at the final time t'', is given by summing the elementary probabilities over all conceivable histories with equal weight for each; thus

$$\langle x'', t'' | x', t' \rangle = \int \exp[iI_H/\hbar] \mathscr{D}H \,, \tag{4}$$

where $\mathscr{D}H$ is a suitably normalized measure over the « space » of all histories. This Feynman « sum over histories » or over world-lines, and the principle of correspondence between world-line and wave, are the two « rights » about a classical world-line. To see what is « wrong » about a classical history, we have only to note the overwhelming preponderance of « unruly » histories [80] over smooth ones in the Feynman sum. The shorter the intervals between the times t_1, t_2, t_3, ..., t_i, t_n at which we specify the position, x_i to $x_i + dx_i$, of the particle, the greater are the zig-zags in velocity in the histories which contribute most to the sum over histories. The more numerous and wilder the histories are that we are forced to consider, the clearer we become that the classical concept of « one history » is wrong.

What has all this to do with « space-time »?

« Space-time » is the history of space geometry changing with time. « World-line » is the history of particle position changing with time. What we have just said of the history of the particle inescapably applies to the history of space geometry. « History » is right, « space-time » is right, for an approximate and semi-classical description of space geometry changing with time [81]. It is right too as a building block in the Feynman sum over histories to give the quantum description of the dynamics of geometry [82-84]. But a single classical history of space geometry, a single space-time, is wrong; it is incompatible with any proper quantum description of the dynamics of geometry.

Why worry?

Because « space-time » violates the uncertainty principle. Take any deterministic classical space-time, such as the Friedman universe, the Schwarzschild geometry, or the Taub universe [85]. Make any spacelike slice whatsoever through it [86]. That slice assigns to space a definite 3-geometry. However, it also assigns to space a definite curvature with respect to the enveloping 4-geometry. That means a definite « extrinsic curvature » [87]; or, in the language of Hamiltonian field theory, a definite « field momentum » [88]. Moreover, when we ascribe to space both a field co-ordinate—a definite intrinsic 3-geometry—and a field momentum—a definite extrinsic curvature,—we collide head-on with the uncertainty principle. We can talk of « space-time », as we talk of a « world-line », but both are classical, antiquantum concepts. You agree that a « world-line » ascribes to a particle both a co-ordinate and a momentum?

Yes.

And you agree that that is incompatible with the uncertainty relation?

I do.

And you concede that it is equally wrong to assign to a field both a coordinate and a momentum?

I must.

Then what escape is there from ruling out « space-time » as a deterministic classical concept, applicable only at the level of approximation theory?

I see no escape. But I would like to understand this matter better. Why not make it the subject of Meeting V?

Agreed. Let's give that discussion the title « Transcending Time ». In giving up « space-time » as a basic idea in the description of Nature we have to give up « time », too; and with time gone, even the concepts of « before » and « after » lose all meaning [89].

I feel completely lost. I have never heard anything in philosophy or logic that did not rest in the end, explicitly or implicitly, on the distinction between what comes first and what follows. How can I or anyone hope to make sense out of a nature in which the terms « before » and « after » have « lost all meaning »?

You have to recognize that we are discussing questions of principle. In all every-day situations, and even in radiative processes and in the collisions of GeV particles, the relevant distances and times are enormous compared to the Planck distance [90, 91],

$$L^* = (\hbar G/c^3)^{\frac{1}{2}} = 1.6\cdot 10^{-33}\ \text{cm}\,, \tag{5}$$

and the corresponding Planck time. Only at such small distances in present-day geometry, or in the extreme geometry of big bang or collapse, do we expect to have to give up the idea of before and after. And why shouldn't we accept that limitation on our customary presuppositions, even welcome it? How else are we to come to terms with what Einstein's theory tells us? How else can we begin to understand that there is no such thing as a « before » before the big bang? no « after » after gravitational collapse? The only thing worse than having to give up « before » and « after » would be not having to give them up.

I can't understand how we can arrive at such limitations on our customary ideas of time starting from a theory—Einstein's geometrical theory of gravity—which accepts from the start the familiar local special-relativity theory distinction between past and future.

Be happy that we have sure and simple guides through these questions. We have not only Einstein's standard theory of the dynamics of geometry. We also have the standard principles of Hamiltonian dynamics, interpreted as we interpret them today in the light of quantum theory [92]. We shall need no more to see how and why « time » is transcended as a primary category in the description of Nature.

Now I feel better prepared for Meeting V. What do you propose for Meeting VI?

« Initial Conditions and the Asymmetry in Time of Radiative Reaction ». There we can come back to the idealization of flat space-time and classical theory. There we may summarize the account of radiative reaction given by WHEELER and FEYNMAN in 1945 [93, 94]. In it every charged particle is envisaged as coupled to every other charged particle by a field that is symmetric in time: half advanced, half retarded.

Interconnections run forward and backward in time in such numbers as to make an unbelievable maze. That weaving together of past and future seems to contradict every normal idea of causality. However, when the number of particles is great enough to absorb completely the signal starting out from any source, then this myriad of couplings adds up to a simple result: the familiar retarded actions of everyday experience, plus the familiar force of radiative reaction with its familiar sign.

How can couplings symmetric in time add up to a result so obviously asymmetric in time?

Asymmetry in the boundary-value data provide the explanation. The particles of the absorber are either at rest or in random motion before the acceleration of the source. They are correlated with it in velocity after that acceleration. Thus radiation and radiative reaction are understood in terms, not of pure electrodynamics, but of statistical mechanics [95].

Don't I also understand why heat always flows from hot to cold and why entropy increases in terms of asymmetry in time in the boundary-value data? In that reasoning don't I dispense with interactions propagated in time? Don't I idealize—and idealize with good results—to instantaneous couplings? Then why so much emphasis on half-advanced-plus-half-retarded interactions?

Our emphasis is on the directly opposite point: We need asymmetry in time of boundary-value data to understand the asymmetry in time that we see in Nature, regardless whether the elementary time-symmetric interactions are idealized as instantaneous or are propagated in time.

I am happy with the perspective you give me; happy, that is, with all except one point, and I fear it is an absolutely central point. Why should it be initial-value data that are specified in statistical mechanics? Why not final-value data?

Your question couldn't be more appropriate. You put your finger on one of the great mysteries. It is even conceivable that we can't make any headway in answering your question until we finally begin doing statistical mechanics in a proper cosmological setting. Why then don't we make « Asymmetry in Time and the Expansion of the Universe » the topic of Meeting VII? It will suggest some observations and measurements.

I look forward to that topic.

And while we are on mysteries let us discuss in Meeting VIII another:

« Memory ». How does it come about that we remember the past but not the future? Is this asymmetry in time a consequence of and witness to the « observer-participancy » that we would make the underpinning for all the laws of physics? It is not necessary for us to have answers to raise questions.

Cosmological issues are so central to all you have to say that before you end I would like to hear more about the big bang, the big crunch and the black hole—what you call [96] « The Gates of Time ».

Then let's make that the topic for a final Meeting IX. That will bring to a natural close our survey of some of the « Frontiers of Time ». Nothing indicates more clearly than those gates of time that the Universe did not exist forever. No evidence gives more incentive to conceive of the laws of physics as having come into being. None suggests more forcefully that proud unbending immutability is a mistaken ideal for physics; that this science now shares, and must forever share, the more modest mutability of its sister sciences, biology and geology.

A new species of bird may appear unbelievable. The upended strata of a mountain slope may look incredible. Yet both biology and geology find their explanation in the accumulative consequences of many individual small effects. Today we do not abandon reason when we regard the kingdom of life, rich though it is, or when we look up at the Himalayas, tall though they stand. How these wonders came about we now understand in outline, and count on someday being able to describe in detail. Have equal confidence that we shall find out how the laws of physics—and the Universe—came into being, incredibly remote though they today seem from being also the accumulative consequences of many individual small effects.

Small effects? Accidents? Accidents like mutations, or like the rainstorms that wear away mountains? Blind accidents?

We have to be careful with that word « blind accident ». « Blind » implies blind towards future consequences. It suggests a happening that is rooted in the past and heedless of the future. Such a conception implies that an order in time is already in being. The direct opposite is lesson number one of our survey of « Frontiers of Time ».

Time, we discover, is not a primordial concept in the structure of Nature. It is secondary and derived. So too, it would appear, is the asymmetry between past and future that shows up so strikingly in radiative reaction, in the flow of heat from hot to cold, in biological evolution, and even in the mechanism of the memory. In contrast, how can any elementary building process be an elementary process for building existence and law unless it transcends the category of time?

To identify an elementary building process that transcends time: is that why you put « The 'Past' and the 'Delayed-Choice' Double-Slit Experiment » ahead of all other topics on our list of meetings?

Yes. The act of observer-participancy in such an experiment, right now, irretrievably alters what we have the right to say about « the past ». In that sense, that carefully restricted sense, that act is an inescapable part of the actual building of « the past ».

I begin to realize that not only topic II, but all the topics on our list of meetings make time their central concern. How did this come about, when the original focus of our discussion was « How did the Universe come into being, and what is its substance? »

The answer is simple. We don't understand genesis and we never will until we rise to an outlook that transcends time. That is why a review of frontiers of time is precondition for any proper analysis of the ultimate issue.

Thanks! With your permission I plan to bring colleagues to our further meetings, even if that forecloses most of the questions I would like to ask along the way. However, I worry lest they miss the bearing of « time » on « genesis ». Therefore may I ask if for them—and me—you would please boil down into a few lines that I can copy the gist of what you've said today?

We would do better to have no summary at all than a summary so short we cannot analyze for each point the evidence, whether weak or strong.

I understand your concern. Nevertheless, please put everything in a dozen brief points. Leave it to me to supply later, in the light of what you have said today, the qualifications and caveats I know you would want.

Then let us try.

As surely as we now know how tangible water forms out of invisible vapor, so surely we shall someday know how the Universe came into being. We will first understand how simple the Universe is when we recognize how strange it is.

The simplicity of that strangeness, Everest summit, so well directs the eye that the feet can afford to toil up and down many a wrong mountain valley, certain stage by stage to reach someday the goal.

Of all strange features of the Universe, none are stranger than these: time is transcended, laws are mutable, and observer-participancy matters.

« Before » and « after » don't rule everywhere, as witness quantum fluctuations in the geometry of space at the scale of the Planck distance. Therefore « before » and « after » cannot legalistically rule anywhere. Even at the classical level, Einstein's standard closed-space cosmology denies all meaning to « before the big bang » and « after the big crunch ». Time cannot be an ultimate category in the description of Nature. We cannot expect to understand genesis until we rise to an outloook that transcends time.

There never was a law of physics that did not require space and time for its statement. With collapse the framework falls down for everything one ever called a law. The laws of physics were not installed in advance by a Swiss watchmaker, nor can they endure from everlasting to everlasting. They must have come into being. They could not always have been accurate. They are derivative and superficial, not primary and revelatory.

Quantum physics teaches that no elementary phenomenon is a phenomenon until it is an observed phenomenon. The « delayed-choice experiment » shows that an act of observer-participancy in the present has an irretrievable consequence for what one can say—with the help of theory—about the past.

Conformant to these three strangenesses, how else can the Universe come into being except as a « self-excited circuit »? As it expands, cools and develops, it gives rise to observer-participancy. Observer-participancy in turn gives what we call « tangible reality » to the Universe.

« Omnia ex nihil ducendis sufficit unum »—one principle suffices to build everything from nothing.

From what kind of nothingness?

« Nothingness » is not the vacuum of physics, loaded with geometry and field fluctuations; it is a nothingness devoid of structure, law or plan; it is the zeroness of existence of that word « cloud » at the beginning of the surprise-version game of twenty questions.

Build how much out of nothingness?

Law; and space-time as part of law; and out of law substance. Build law out of the statistics of billions upon billions of acts of observer-participancy each of which by itself it utterly random. Recognize law as the accumulative consequence of many individual small effects. How else could law come into being?

No test of these views looks more like being someday doable, nor more interesting and more instructive, than a derivation of the structure of quantum theory from the requirement that everything have a way to come into being out of nothing.

If you would have an epitome of this summary, let it be this: Nothing. No time. The line. Acts. Statistics. Law. Space-time. Substance. Observer-participator. Closed circuit. Test. But all our further meetings, we have agreed, will focus on one part of this larger theme: on time, and what it means to transcend time.

2. – The « past » and the « delayed-choice » double-slit experiment.

Reality is theory.

Torgny Segerstedt

... the past has no existence except as it is recorded in the present [59].

(The following is abbreviated from ref. [59].)

Partway down the optic axis of the traditional double-slit experiment stands the central element, the doubly slit screen. Can one choose whether the photon (or electron) *shall have* come through both of the slits, or only one of them, after it has *already* transversed this screen? That is the new question raised and analyzed here.

Known since the days of Young is the possibility to use the receptor at the end of the apparatus to record well-defined interference fringes. How can they be formed unless the electromagnetic energy has come through both slits? In later times EINSTEIN noted that in principle one can determine the lateral kick given to the receptor by each arriving quantum. How can this kick be understood unless the energy came through only a single slit?

Einstein's further reasoning as reported by BOHR [97] is familiar. Record both the kicks and the fringes. Conclude from the kicks that each quantum of energy comes through a single slit alone; from the fringes, that it nevertheless also comes through both slits. But this conclusion is self-contradictory. Therefore quantum theory destroys itself by internal inconsistency.

Bohr's reply [97] has become by now a central lesson of quantum physics. One can record the fringes or the kicks but not both.

The arrangement for the recording of the one automatically rules out the recording of the other. The quantum has momentum p, de Broglie wavelength $\lambda = h/p$, and reduced wavelength $\lambda\!\!\!^{-} = \hbar/p$. To record for it well-defined interference fringes, one must fix the location of the receptor within a latitude

$$\Delta y < (\text{fringe spacing})/2\pi = (L/2S)\,\lambda\!\!\!^{-}\,. \tag{6}$$

To tell from which slit the quantum of energy arrives, one must register the transverse kick it gives to the receptor within a latitude small enough to distinguish clearly between a momentum $p = \hbar/\lambda\!\!\!^{-}$ coming from below, at the inclination S/L, and a momentum coming from above at a like inclination; thus,

$$\Delta p_y < (S/L)(\hbar/\lambda\!\!\!^{-})\,. \tag{7}$$

However, for the receptor simultaneously to serve both functions would be incompatible with what the principle of indeterminacy has to say about receptor dynamics in the y-direction,

$$\Delta y\,\Delta p_y > \hbar/2\,. \tag{8}$$

Not being able to observe simultaneously the two complementary features of the radiation, it is natural to focus on the one and forego examination of the other. Either one will insert the pin through the hole shown in fig. 2. It will couple the receptor to the rest of the device. It will give the receptor a well-defined location. Then one will be able to check on the predicted pattern of interference fringes. Or one will remove the pin. Then one can measure the through-the-slot component of momentum of the receptor before and after the impact of the quantum. Then one will say that one knows through which slit the energy came.

Pin in or pin out: when may the choice be made? Must it be made before the quantum of energy passes through the doubly slit screen? Or may it be

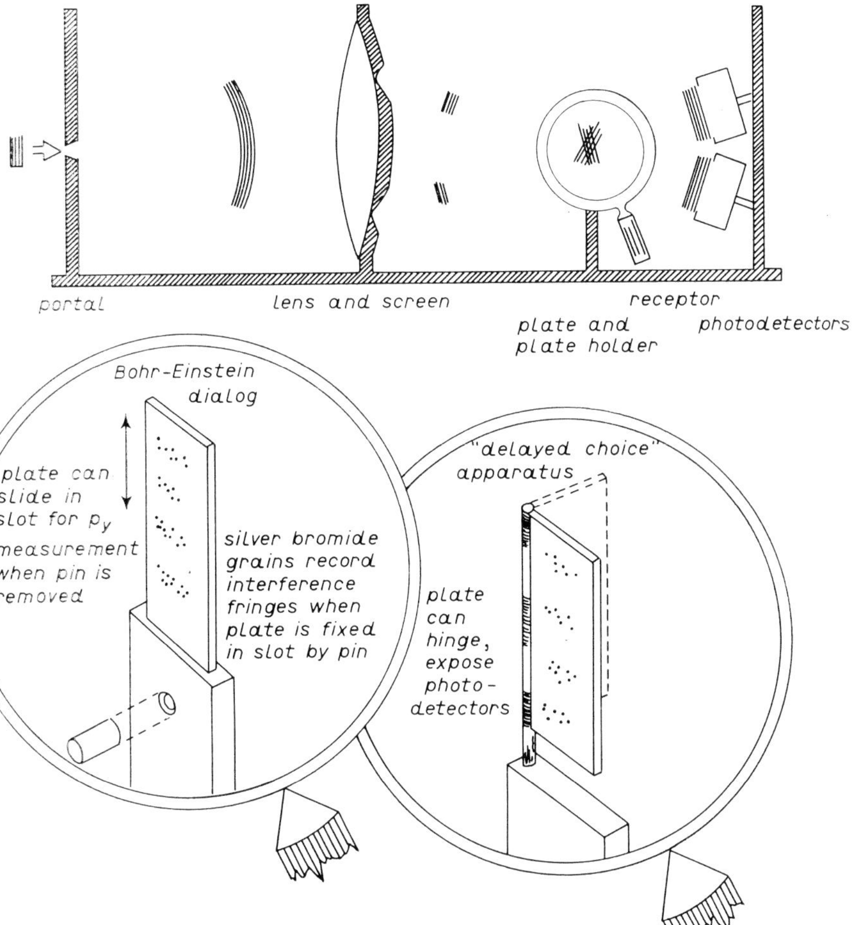

Fig. 2. – Top: Idealized double-slit experiment. Distance of each slit from optic axis, S; from photographic plate, L. For simplicity, details of the plate and plate holder are omitted from the circle encompassed by the magnifying glass and are presented below, magnified and in perspective. Lower left: the version of the Bohr-Einstein dialog. The plate catches every photon. It registers precisely the y–co-ordinate of impact or the y-component of impulse delivered—but does not and cannot do both. Omit the photodetectors. Lower right: The present « delayed choice » version. Include the photodetectors. One or other of them is sure to catch the quantum of energy when the plate is swung aside. Whether to expose the plate or expose the photodetectors, whether thus to infer that a single quantum of energy *shall have* gone through both slits in the screen or only one, is subject to the free choice of the observer after the energy has *already* traversed the screen.

made after? That is the central question in this paper as that question first seems to impose itself. However, a closer look shows that the measurement of transverse-momentum kick, in principle conceivable, is practically almost out of the question (because of the large mass of the photographic plate; details of analysis omitted in this abbreviated account). Therefore it is appropriate to alter the idealized experiment before taking up the question of « before » *versus* « after ». The difficulty is overcome by a simple change (fig. 2, lower right):

1) Give up measuring the y-component of the momentum of the photographic plate.

2) Hold its y–co-ordinate fixed.

3) By means of a hinge parallel to the y-axis arrange that this high narrow plate can be swung out of the way of the incident light—at the last-minute option of the observer—quicker than the flight of light from screen to plate. (Switch from « operative » to « open » position.)

4) Sufficiently far beyond the region of the plate, the beams from upper and lower slits cease to overlap and become well separated. There place photodetectors. Let each have an opening such that it records with essentially 100 percent probability a quantum of energy arriving in its own beam, and with essentially zero probability a quantum arriving in the other beam.

Now the choice is clear; and the objective, too. We today cannot argue, and Einstein in his later years would not even have wanted to argue, his erstwhile case of logical inconsistency against quantum theory: the photon goes through both slits, as evidenced in interference fringes, and yet simultaneously through only one, as evidenced in lateral momentum kick. Choose we know we must between the two complementary features open to study; and choose we do, by putting the plate athwart the light or turning it out of the line of fire. In the one case the quantum will transform a grain of silver bromide and contribute to the record of a two-slit interference fringe. In the other case one of the two counters will go off and signal in which beam—and therefore from which slit—the photon has arrived.

In our arrangement the photographic plate registers only the point of impact of a photon. In the earlier idealized experiment it could additionally (Einstein) or alternatively (Bohr) record the transverse momentum delivered by the impact. We have assigned the two distinct kinds of measurement to two distinct kinds of register. We have demoted the plate from a privileged status. That demotion is irrelevant to any question now at issue. Equally irrelevant is the different distance—and time of flight—from entry portal to plate, or photodetector, according as the one or other register is exposed. But the essential new point is the timing of the *choice*—between observing a two-slit effect and a one-slit one—until after the single quantum of energy in question has *already* passed through the screen.

Let the reasoning be passed in review that leads to this at first sight strange inversion of the normal order of time. Then let the general lesson of this apparent time inversion be drawn: « No [elementary] phenomenon is a phenomenon until it is an observed phenomenon ». In other words, it is not a paradox that we choose what *shall* have happened after « it has *already* happened ». It has not really happened, it is not a phenomenon, until it is an observed phenomenon.

Whatever we now do to spell out the otherwise idealized experiment, we will leave idealized its most unusual features: the « swinging door photographic plate ». That term includes the arrangement, whatever it may be,

1) for a last-minute choice, to swing the plate aside or leave it athwart the beam, after the arriving energy has already traversed the doubly slit screen, and

2) for completion of that movement before the energy arrives at the plate.

In practice it will be more reasonable to swing the beam than swing the plate. Fix the plate. Halfway from screen to it, position a Kerr cell. Apply to it a positive or a negative voltage [98] according as one wishes to record fringes on the plate, or register « which beam » on a counter. Or, still better, Manfred FINK suggests, replace the experiment with the photon by an experiment with an electron. Then the last-minute deflection of the electron beam can be accomplished by a localized magnetic field centered between screen and plate. One or another of these arrangements to swing the beam will be understood hereafter to apply in practice when in principle we speak of swinging the plate.

(Other requirements and presuppositions of the idealized experiment are analyzed in the original publication but these details are omitted here. Also omitted here are six other types of « delayed choice » experiments.)

. . . .

The double-slit experiment, like the other six idealized experiments (microscope, split beam, tilt-teeth, radiation pattern, one-photon polarization and polarization of paired photons), imposes a choice between complementary modes of observation. In each experiment we have found a way to delay that choice of type of phenomenon to be looked for up to the very final stage of development of the phenomenon, *whichever* type we then fix upon. That delay makes no difference in the experimental predictions. On this score everything we find was foreshadowed in that solitary and pregnant sentence of Bohr [99], « ...it ... can make no difference, as regards observable effects obtainable by a definite experimental arrangement, whether our plans for constructing or handling the instruments are fixed beforehand or whether we prefer to postpone the completion of our planning until a later moment when the particle is already on its way from one instrument to another ».

Not one of the seven delayed-choice experiments has yet been done. There can hardly be one that the student of physics would not like to see done. In none is any justification whatsoever evident for doubting the obvious predictions.

We search here, not for new experiments or new predictions, but for new insight. Experiments dramatize and predictions spell out the quantum's consequences; but what is its central idea? A pedant of Copernican times could have calculated planetary positions from the equations of Copernicus as well as COPERNICUS himself; but what would we think of him if his eyes were closed to the main point, that the « Earth goes around the Sun »?

. . . .

[No analysis] in recent times moved our understanding forward more than the Einstein-Bohr dialog [97]. Out of that dialog no concept emerged of greater fruitfulness than « phenomenon » [100]: « ... [In my discussions with EINSTEIN] I advocated the application of the word *phenomenon* exclusively to refer to the observations obtained under specified circumstances, including an account of the whole experimental arrangement » [101]. No other point does the present analysis of idealized delayed-choice experiments have but to investigate what « phenomenon » means as applied to the « past ».

After the quantum of energy has *already* gone through the doubly slit screen, a last-instant free choice on our part—we have found—gives at will a double-slit-interference record or a one-slit-beam count. Does this result mean that present choice influences past dynamics, in contravention of every formulation of causality? Or does it mean, calculate pedantically and don't ask questions? Neither; the lesson presents itself rather as this, that the past has no existence except as it is recorded in the present. It has no sense to speak of what the quantum of electromagnetic energy was doing except as it is observed or calculable from what is observed. More generally, we would seem forced to say that no [elementary] phenomenon is a phenomenon until—by observation, or some proper combination of theory and observation—it is an observed phenomenon.

. . . .

3. – « Development in time » gives way to « correlation in time ».

> *... it appears that our theory denies*
> *the existence of absolute reality—a denial*
> *which is unacceptable to many ... I do not know*
> *how one could define operationally*
> *the reality of anything.*
>
> E. P. WIGNER [102]

Most instructive of all the idealized experiments considered in the great dialog between BOHR and EINSTEIN was the Einstein-Podolsky-Rosen experiment [104, 105], later simplified by BOHM [106] to the version illustrated at the middle of fig. 3. The very light isotope of hydrogen composed of one positive and one negative electron is allowed to cascade down to its ground state of 0 angular momentum. There it sits until it undergoes annihilation. Two photons come off with equal and opposite momenta, as illustrated by the two wavy lines in the diagram. An observer on the right determines whether the photon travelling to the right is circularly polarized to the right or to the left. Whatever the result, he is assured that a measurement of the circular polarization of the left-hand photon will give exactly that result, right-handed or left-handed, which is required for conservation of angular momentum.

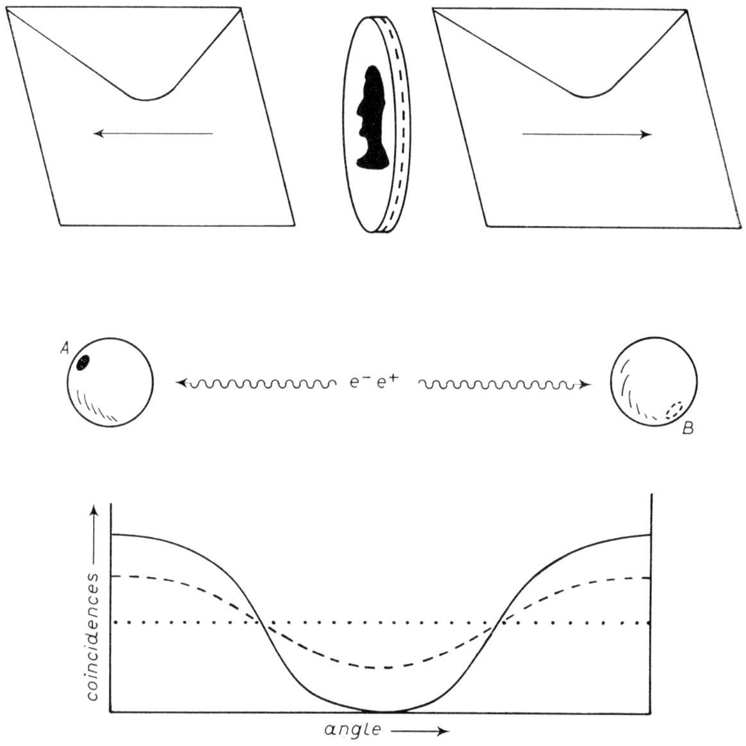

Fig. 3. – In contrast to splitting a coin, putting the two slices into envelopes, shuffling them, and sending them to two remote observers (upper part of diagram), the Einstein-Podolsky-Rosen experiment in the version of Bohm (middle part of diagram) permits a double infinity of choices (point on right-hand Stokes sphere) for the polarization to be looked for in the right-hand (e^+e^-) annihilation photon, with corresponding consequences [108] for the polarization (point on left-hand Stokes sphere) that will be found for the left-hand photon. If the polarizations were determined in the act of emission (« hidden variables ») the coincidences between the two photons would show only half the dependence on relative orientation of the two polarizations (dashed curve in lower diagram) that is predicted by quantum mechanics—and observed (full curve).

Alternatively, he may choose to study the photon going to the right with the help of an analyzer of linear polarization. Then he makes a clean measurement as to whether the polarization lies in the y-direction (or the z-direction). Then he is assured that a study by similar equipment of the photon travelling to the left will show it to be vibrating with 100 % certainty in the z-direction (or the y-direction).

At first sight there is nothing very startling about the correlation in polarization between the two photons.

What difference in principle is there, one might well ask, with respect to the old game in which a coin is sawed in half? The two halves are put in separate envelopes and sealed and dispatched to observers far away on the left and the right. If the observer on the right opens his envelope and finds the head in it, then he knows that the other far away observer will find the tail of the coin when he opens his envelope. There is no paradox involved. There is no possibility of using the arrangement of envelopes to send a signal in excess of the speed of light.

In the e^+e^- annihilation, the new feature is this: The polarization of a photon is a more sophisticated quantity than the differences between the two faces of a coin.

According to Stokes' parametrization of polarization (see, for example, [107]) each of the many alternative ways to specify a well-defined polarization can be set into one-to-one correspondence with a point on the surface of the unit sphere. The observer on the right can delay until the very last picosecond before the arrival of his photon the determination of which kind of polarization he will look for, as symbolized by point B on the right-hand sphere. Whatever the choice—and he cannot, of course, know whether the photon that arrives will have the polarization B or the polarization « anti-B »—he is assured that the running of the corresponding experiment on the other photon will give for it the uniquely mated polarization (see [108] for the coincidence rate for the general case of arbitrary polarization).

What is to be said of the polarizations of the two photons in the course of all their long travel from the site of the e^+, e^- annihilation to their respective points of reception? Nothing. Nothing until the experiment is over. No elementary phenomenon is a phenomenon until it is an observed phenomenon.

Instead of accepting this lesson of the quantum one can try to quarrel with it. Why not assign a probability amplitude to the state of the two photons, as for example $[\alpha(1)\beta(2)-\alpha(2)\beta(1)]/2^{\frac{1}{2}}$? Why not go further and view the process of measurement in a Lorentz frame in which the right-hand photon arrives first at its analyzer-detector? Is the right-hand photon not suddenly recorded as having, for example, the polarization $\alpha(1)$? Does it not follow that the left-hand photon, still en route from the site of annihilation to the left-hand analyzer-detector, must suddenly in mid-course be redescribed as having the mated polarization $\beta(2)$?

Does not this redoing of the state function imply the existence of an effect propagated from right to left in excess of the speed of light? Then look at the whole process all over again in a Lorentz frame in which the left-hand photon arrives first at the left-hand analyzer-detector. By the same reasoning is one not led to speak of an effect propagated this time from left to right in excess of the speed of light? What a confusion! What a warning not to identify these pencil-and-paper readjustments, these pencil-and-paper supra-light-velocity effects, with anything real. What an indication that the wave function is not itself real, but a purely formalistic device, and within the present incomplete marriage of quantum theory and relativity not a very happy device, for calculating the probability of real coincidences. Only when the counters have gone off has the reality of the situation declared itself. No elementary phenomenon is a phenomenon until it is an observed phenomenon.

Continue to contest this lesson of the quantum. Argue that a reality *is* to be attributed to the polarizations of the two photons on their way towards the two detectors, regardless of the settings of the two analyzers. Declare that the chance for the first detector to go off is

$\cos^2$ (angle between the « true direction » of the polarization of photon 1 and the setting of analyzer 1),

and the chance for the second detector to go off is

$\cos^2$ (angle between the « true direction » of the polarization of photon 2 and the setting of analyzer 2).

Argue that the chance for a coincidence is the product of these two expressions averaged over the random direction of the polarization of one of these two photons—the second being of necessity orthogonal to the first. In this way end up with the dashed curve of fig. 3 for number of coincidences as a function of the relative setting of the analyzers on left and right. For the difference in rate between the least favorable and the most favorable setting one gets only half what is predicted by quantum mechanics and only half of what is observed (polarization of e^+e^- annihilation photons: theory, WHEELER (1946) [109], PRYCE and WARD (1947) [110], SNYDER, PASTERNACK and HORNBOSTEL (1948) [111]; observations by BLEULER and BRADT (1948) [112], HANNA (1948) [113], VLASOV and DZELJEPOV (1949) [114], WU and SHAKHOV (1950) [115], HEREFORD (1951) [116], BERTOLINI, BETTONI and LAZZARINI (1955) [117], LANGHOFF (1960) [118], KASDAY, ULLMAN and WU (1970) [119], KASDAY (1971) [120], FARACI, GUTKOWSKI, NOTARRIGO and PENNISI (1974) [121], WILSON, LOWE and BUTT (1976) [122]; polarization of the photons given out by an atom in a 2-step transition, KOCHER and COMMINS (1967) [123], FREEDMAN and CLAUSER (1972) [124]; HOLT (1973) [125] result in contradiction with quantum predictions but not confirmed by CLAUSER (1976) [126]; FREEDMAN and HOLT (1975) [127], FRY and THOMPSON (1976) [128] and CLAUSER (1976) [126]). Quantum mechanics thus « exposes itself to destruction » in numerous decisive tests—and stands up to these tests. It is a central point of this quantum mechanics that it denies to photons any « real » polarizations merely in virtue of their being « on their way » and in default of any actual act of observation. In other words, an elementary phenomenon is a phenomenon only when it is an observed phenomenon.

If « development in time » of « the wave function » is not a happy way to describe the state of the two photons in the EPR experiment, how should one describe this and more complex situations, in which observations are made at several or even many, locations in space-time? Correlation of observations: this is the appropriate concept, according to HOUTAPPEL, VAN DAM and WIGNER [65] and WIGNER [102].

In this approach one makes a conceptual reformulation of the equations of quantum mechanics, « eliminating explicit reference to the equations of motion and to state vectors. According to this [philosophy], the function of quantum mechanics is to give statistical correlations between the outcomes of successive observations » [102].

As an example, WIGNER considers the correlation between two measurements. In the first measurement the physical quantity under examination is described by some operator Q. The various possible outcomes for this measurement are labelled by an index j. Associated with the j-th outcome is a projection operator P_j, with $P_j^2 = P_j$. Both Q and the P_j are envisaged to vary with time in accordance with the same law,

$$Q(t) = \exp[iHt/\hbar]\, Q(o) \exp[-iHt/\hbar]\,, \tag{9}$$

$$P_j(t) = \exp[iHt/\hbar]\, P_j(o) \exp[-iHt/\hbar]\,, \tag{10}$$

in which H is the Hamiltonian operator. As WIGNER notes, we attribute « the operator $Q(t)$ to the same measurement, carried out at time t, to which we attribute the operator $Q(o)$ if carried out at time o », and $P_j(t)$ is « the projection operator which leaves the state vectors of outcome j unchanged, annihilates the state vectors of all other measurement outcomes ». A measurement of quite a different physical quantity at quite a different time has associated with many quite different possible outcomes, of which the k-th is associated with the projection operator P'_k. « The probability that the second measurement yields the result k if the first one's outcome was i is then given by [the ratio of traces]

$$\text{probability} = \operatorname{Tr}(P'_k P_j)/\operatorname{Tr} P_j = \operatorname{Tr}(P_j P'_k P_j)/\operatorname{Tr} P_j\,;$$

and similar expressions (see [65], eqs. (4.4)-(4.7)) can be given for the probabilities of the different outcomes of several successive measurements. »

This formalism replaces the older view of dynamics as development in time by a proper quantum concept of correlation as it depends upon time. It does not go the whole way towards what is eventually envisaged under the heading of « law without law ». First, the correlation treatment does not analyze, nor was it aimed at analyzing, down to the substrate of observation the ultimate make-up of the particle or field under study. It takes the existence of the dynamic entity under study as for granted. Second, it takes the concept of time for granted. It does not transcend the concept of time nor was it intended to. However, it has the great merit of providing a first framework for working towards an ultimate statistics of « billions upon billions of acts of observer-participancy ». At that point the idea will come closer to testability, « Is everything—including time—built from nothingness by acts of observer-participancy? »

4. – Many-fingered time, « imbeddability », and the laws of physics.

... according to the remark
which I formerly made on the occasion
of an optical law ... final cause ...
even in physics ... serves to find and to discover
hidden truths.

G. W. LEIBNIZ [129]

The more one learns about the laws of physics, the more one learns how little one has learned. Maxwell electrodynamics, Einstein geometrodynamics and the chromodynamics of the Yang-Mills quark-binding field, laws won through decades of effort, summarizing an unbelievable richness of experience, and representing our deepest penetration to date into the machinery of Nature, spring out, all three, full-bodied, at a single simple Alladin-like command, « let final values be independent of the choice of many-fingered time ».

It seems at first sight almost unbelievable that so much hard-won experience should be deducible from a demand so simple. Therefore it may be appropriate to spell out a bit more fully this beautiful discovery of Hojman, Kuckař and Teitelboim (hereafter abbreviated as HKT) [67, 68]. Specify the initial value of the field in question and its conjugate momentum on an arbitrary smooth initial spacelike hypersurface σ_1. From these initial data and from the Hamiltonian, H, of the field calculate the final value of the field and its conjugate momentum on an arbitrary smooth final spacelike hypersurface σ_2. Do the calculation over and over again moving forward from σ_1 to σ_2 by all arbitrary choices of many-fingered time (fig. 4). Demand that the final values on σ_2 shall be the same for all these different ways of marching forward from σ_1 to σ_2. HKT, recognizers and exploiters of this requirement, call it the demand for « imbeddability ». If it were violated, the dynamics could not be imbedded in a single space-time manifold. « Imbeddability » is the magic word that summarizes and delivers forth almost all we know of the laws of Nature.

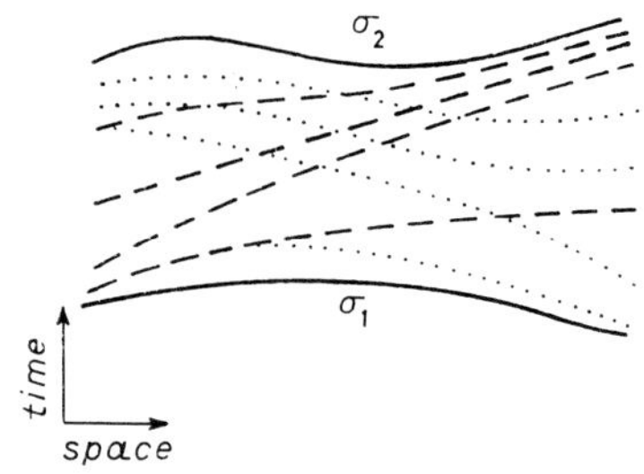

Fig. 4. – The « history of deformation » indicated by the dashed hypersurface leads from initial-value hypersurface σ_1 to final-value hypersurface σ_2. So does the history indicated by the dotted hypersurfaces. The physics on σ_2 resulting from a complete specification of the initial-value data on σ_1 must be independent of the history one chooses to integrate along in passing from σ_1 to σ_2 via the Hamiltonian equations of motion. This heavy but simple requirement suffices to fix the form of the Hamiltonian for the dynamics of a vector field (giving Maxwell theory), for the dynamics of the 3-geometry itself (giving Einstein's geometrodynamics) (HOJMAN, KUCHAŘ and TEITELBOIM, 1973) and for chromodynamics [69].

Following this brief overview, it is appropriate to come back for a little more detail, especially on « time ». « Unfolding in time » being the essence of dynamics, it is natural that changing views of time have led to changing concepts of dynamics. The progress from Newtonian time to the time of special relativity, and from that to the many-fingered time of general relativity, was essential precondition for the discovery of HKT.

Time is absolute and universal in Newtonian physics. In special relativity successive times correspond to successive slices through space-time. These slices are parallel to one another and normal to the time axis of the particular observer in question. For another observer there is another time axis and normal to it a different set of parallel flat spacelike hypersurfaces. Thus for each observer, each inertial frame, there is a globally defined time. Dynamics may be described in one of these global Lorentz frames, or another, or another; but in any one frame it is described with respect to a single time variable.

In 1932 DIRAC, FOCK and PODOLSKY [130] introduced a way of analyzing particle dynamics in which there are as many time parameters as there are particles. These time parameters are at the disposition, not of the particles, but of the analyst. He picks up business-machine printout cards telling what each particle has been doing. Nobody can keep him from placing on his desk the card that tells what particle 1 was doing at a particular time t_1, and alongside it a card that tells what particle 2 was doing at a particular time t_2, and correspondingly other cards from other times t_3, t_4, ..., t_n in the lives of the other particles. Not to have any of these particles in the zone of influence of any other of these particles it is useful to impose the requirement that the chosen « events » on the several world-lines should have a spacelike relation, each to all the others.

This concept of what we may call a « many-fingered time » TOMONAGA [131] generalized in 1946 from the dynamics of n particles in flat space-time to the dynamics of the electromagnetic field in flat space-time. The field is conceived to be studied in its dependence, not on one time parameter, not on n time parameters, but on as many time parameters,

$$t = t(x, y, z)\,, \tag{11}$$

as there are points in space—which is to say, a continuous infinity of time parameters. There is a simple way to visualize this collection of parameters. It constitutes a hypersurface, a slice through space-time, what LANDAU and LIFSHITZ [132] call a « simultaneity ». Generalizing from an arbitrarily curved, bent, or wiggly slice through flat space-time to an arbitrary slice through the curved space-time of general relativity, we impose the same kind of requirement that DIRAC, FOCK and PODOLSKY did. We require each point on this hypersurface to stand in a spacelike relationship to all the other points. None is to be able to send a signal to, or exert a force at, the others. In this sense the hypersurface is « spacelike ». To demand the existence of such a global spacelike hypersurface is a powerful condition. Gödel's model of a rotating universe with closed timelike lines [133] does not satisfy this requirement. On this account that space-time is generally regarded as nonphysical. It we therefore exclude from consideration along with every space-time that does not admit global spacelike hypersurfaces.

In classical physics the electromagnetic field has a deterministic evolution in space-time. What does this mean for the description of this field in terms of many-fingered time? Nothing startling. Pick the spacelike hypersurface. Pick one point in the 3-dimensional space thus defined. Erect at that point the unique timelike unit vector normal to the local tangent hypersurface. With respect to that vector and that local tangent 3-space the electromagnetic field falls apart into the magnetic field, $\boldsymbol{B}$, and the electric field, $\boldsymbol{E}$, both 3-vectors located in the local tangent 3-space. The magnetic field $\boldsymbol{B} = \nabla \times \boldsymbol{A}$ or, better, the vector potential $\boldsymbol{A}$ from which $\boldsymbol{B}$ lets itself be derived, thus specified from point to point throughout the spacelike hypersurface, may be regarded as the electromagnetic field co-ordinate; the electric field, divided by 4π, as the electromagnetic field momentum, $\boldsymbol{\pi} = -\boldsymbol{E}/4\pi$, in a canonical Hamiltonian description of the electromagnetic field.

It is enough to give $\boldsymbol{B}$ and $\boldsymbol{E}$ as initial data on an initial spacelike hypersurface, and to know the Hamiltonian density for Maxwell's field,

$$H = (\boldsymbol{B}^2 + \boldsymbol{E}^2)/8\pi = (1/8\pi)(\nabla \times \boldsymbol{A})^2 + [(4\pi)^2/8\pi]\boldsymbol{\pi}^2\,, \tag{12}$$

to be able to predict how the field changes with changes in many-fingered time as the hypersurface is pushed slowly forward (or backward) in time. We will not write down the necessary Hamiltonian equations; they can be imagined.

Why the particular Hamiltonian (12)? Why not some other Hamiltonian, some other law of physics? It provides only a partial answer to this question to turn back to Hilbert's famous paper of 1915 [134]. He derived electrodynamics and general relativity, or vacuum geometrodynamics—and the combined theory of the two fields together—by postulating the simplest action principle that depends on a 4-dimensional vector field

$$A_\alpha \tag{13}$$

or on a 4-dimensional metric field

$$g_{\mu\nu} \tag{14}$$

or on the combination of the two. But why the simplest action principle? Why not some one of the thousand and one alternative action principles that contain these two fields in some other invariant combination?

« Imbeddability » is the new and magic and beautiful answer that HOJMAN, KUCHAŘ and TEITELBOIM [67, 68] (see also [69, 135-139]) give to this old question. They envisage a 3-vector field

$$A_i (i = 1, 2, 3) \text{ (and its conjugate momentum)}, \tag{15}$$

$$g_{jk} (j, k = 1, 2, 3) \text{ (and its conjugate momentum)} \tag{16}$$

or both (and their conjugate momenta). Whatever the dynamic law that governs the evolution of these fields with time, as many-fingered time is pushed forward from the spacelike hypersurface σ_1 to the spacelike hypersurface σ_2 in fig. 4, that law must give the same result for the dynamic variables whether this hypersurface is pushed forward first more rapidly on the « right » and then more rapidly on the « left »; or first more rapidly on the « left » and then more rapidly on the « right ». If the conditions obtained at σ_2—by step-by-step forward integration of the Hamiltonian field equations on an electronic computer—depended upon the choice of history adopted in proceeding from σ_1 to σ_2, then the history of the fields could not be imbedded in any single space-time manifold. With « independence of history » lost, imbeddability would also be lost.

No local second-order law for the development with many-fingered time of a vector field (15) will satisfy this condition of imbeddability except Maxwell's theory. No local Hamiltonian for the development with time of space geometry (16) will give a history-independent result except Einstein's general relativity. No local Hamiltonian for a vector field with an « internal-spin » degree of freedom

$$A_i^{(S)} \qquad (i = 1, 2, 3) \tag{17}$$

is compatible with imbeddability except the Yang-Mills theory [66], today's standard and widely accepted theory of the quark-binding field. Thus simply derived from almost nothing are electromagnetism, gravitation and the current theory of the forces that hold elementary particles together, theories that summarize an unbelievable wealth of experience, years of experimentation and the life work of some of the most gifted men of the last two centuries. No one has ever seen a simpler or more compelling theme than « imbeddability » to summarize the requirements that lead to physics as we know it.

Not only the Hamiltonian, but the gauge features of field theory follow from the argument of imbeddability. As TEITELBOIM showed [137], « $\boldsymbol{A}$ is not observable but only its curl $\boldsymbol{B} = \nabla \times \boldsymbol{A}$ is, because the propagation of $\boldsymbol{A}$ itself will not be in general integrable. Thus the gauge transformation

$$A_i \rightarrow A_i + \partial \Lambda / \partial x^i \tag{18}$$

must have no physical effect: Our efforts for preserving path-[history-]independence have led us to gauge invariance ». He finds a similar result for the dynamics of the metric: No dynamics of the g_{jk} is imbeddable in space-time unless the theory is left unchanged by a co-ordinate transformation.

The significance of gauge invariance is familiar. The physically meaningful quantity is not the vector field A_i $(i = 1, 2, 3)$, but its curl, a quantity that rises above gauge, $\boldsymbol{B} = \nabla \times \boldsymbol{A}$; not the tensor field g_{jk}, but a quantity that rises above co-ordinates, the 3-geometry, ${}^{(3)}G$, about which sect. **5** has more to say; not the Yang-Mills field $A_i^{(S)}$, but a new geometrical entity that once more rises above gauge.

A scalar field, φ, departs in two ways from the pattern of the Yang-Mills field, gravitation and electromagnetism. First, the requirement that its dynamics be imbeddable does not introduce gauge. If such a field existed in Nature, it would be directly observable. Second, imbeddability does not determine a unique local Hamiltonian. The function $f(\varphi)$ in

$$H = (\nabla\varphi)^2/8\pi + f(\varphi) \tag{19}$$

is arbitrary.

Is this arbitrariness of the Hamiltonian for a scalar the reason why no scalar field has ever been found in Nature? Is this omission a clue to how Nature may build law without law? No one can rest happy with « history-independence of dynamics » as the foundation of physics, simple guide though it is to the great laws. It does not explain how it comes about that the dynamics must be imbedded in a manifold of 1 time and 3 space dimensions in the first place, nor why Nature drops the scalar field.

No questions bring us closer than these to the frontiers of time. No way seems reasonable for deriving the dimensionality of $3+1$ which does not start from a viewpoint that transcends dimensionality. No building blocks offer but elementary acts of observer-participancy. No method of construction that has been seen at work in other contexts looks more applicable than Feynman's sum over histories [78, 79], applied however here to the higgledy-piggledy of yes-no-decision observations. No feature of such a sum over histories would seem more immediately susceptible to test than this, that it should kill out by destructive interference any contribution that looks like a scalar field. For the other three fields there is a uniqueness of contribution that can be imagined to lead to a constructive interference of elementary Feynman amplitudes, and therefore a nonzero representation of such fields in the physics. For the scalar field, however, does the very wealth of Hamiltonians acceptable at the classical level mean a wealth of values for the classical action I_H, and therefore wide-ranging values for the phase, $I_H/\hbar$ (see eq. (4)), of the elementary Feynman amplitude? Does this feature of the phase in turn imply destructive interference, and, therefore, finally zero representation for the scalar field in the scheme of physics?

Whatever the next steps may be towards deriving « everything out of nothing », the HKT result would seem to mark one of the largest leaps of recent times. Their way of analysis, starting from the « group » of deformations of a spacelike hypersurface, reminds us again of the power of symmetry considerations to simplify the content of physical law, and their impotence in revealing the machinery behind law. No one would dream of studying the laws of elasticity to uncover the principles of quantum mechanics. Neither would anyone investigate the work-hardening of a metal to learn about atomic physics. The order of understanding ran not

(20) work-hardening (1 cm) $\rightarrow$ dislocations (10^{-4} cm) $\rightarrow$ atoms (10^{-8} cm),

but the other way,

(21) atoms (10^{-8} cm) $\rightarrow$ dislocations (10^{-4} cm) $\rightarrow$ work-hardening (1 cm).

One had to know about atoms to conceive of dislocations, and had to know about dislocations to understand work-hardening. Is it not likewise hopeless to go from laws of physics to underlying machinery? Must the order of progress not be the direct opposite? If so, what course offers itself except to try « acts of observer-participancy » as the underlying « machinery », and see if out of them one can derive the laws of physics? Nothing does more to give a little encouragement in such an enterprise than the HKT achievement of deriving so much from so little, with the help of the concepts of « many-fingered time » and « imbeddability ».

5. – Transcending time.

... space and time are orders of things and not things.

G. W. LEIBNIZ [140]

... time and space are modes by which we think and not conditions in which we live.

A. EINSTEIN [141]

There is no such thing as space-time, quantum mechanics tells us. Space-time is a purely classical concept. It is a classical history of space geometry changing with the progress of time. What is meant by a « classical history of space changing with time »? How does it come about that quantum mechanics forbids this way of speaking? And what does it offer instead as acceptable way of describing the dynamics of space? But first, before any of these questions, why focus on 3-geometry at all when a casual impression might have made it seem that space-*time* is the « without-which-nothing » ingredient of modern theoretical physics?

How can one accept going back from four dimensions to three when one knows that going from three dimensions to four marked one of the great steps forward in the history of science? Not putting the fourth dimension into his curved-space geometry accounts more reasonably than any other circumstance that one can easily name for Riemann's failure to discover general relativity. Already at the age of 27, in his habilitation lecture of June 10, 1854 on entry into the philosophical faculty of the University of Göttingen, he had set forth the mathematical tools to describe curvature in any number of dimensions; and he had declared that « the properties which distinguish space from other conceivable triply-extended magnitudes are only to be deduced from experience .... At every point the three-directional measure of curvature can have an arbitrary value if only the effective curvature of every measurable region of space does not differ noticeably from zero » [142]. EINSTEIN speaks of the inspiration he derived from this lecture of Riemann in developing his own geometrical theory of gravity, « But ... physicists were still far removed from such a way of thinking; space was still, for them, a rigid, homogeneous something, susceptible of no change or conditions. Only the genius of Riemann, solitary and uncomprehended, had already won its way by the middle of the last century to a new conception of space, in which space was deprived of its rigidity, and in which its power to take part in physical events was recognized as possible » [143]. Dying of tuberculosis at Selasca on Lake Maggiore July 20, 1866, twelve years later, in his final days achieving with BETTI a system for characterizing multiply connected topologies, RIEMANN failed in the other great enterprise to which he gave his last measure of devotion: to provide a unified explanation of gravitation and electromagnetism. It took 1905, EINSTEIN and special relativity to provide the missing concept: four dimensions, not three. With that recognized, it took only a decade to achieve general relativity and a fully geometrical theory of gravity in the spirit of Riemann.

It took much longer to recognize the dynamic structure of Einstein's geometrodynamics. The point that was most central, and took the longest to grasp, was also the simplest: The dynamic object is not space-time. It is space. The geometric configuration of space changes with time. But it is space, three-dimensional space, that does the changing.

That 3-space is the dynamic object would have been recognized much sooner had the work and results [144, 145] of Élie Cartan been more widely appreciated, whose deep insights into the theory of partial differential equations gave him a hold on many of the essential ideas. However, physics already had a standard machinery for dealing with dynamic problems, and it seemed natural to lay out general relativity in the Hamiltonian pattern without further thought. If the basic theory is 4-dimensional, should not the Hamiltonian be 4-dimensional, and was it not, therefore, reasonable to think of the dynamic object itself as also being 4-dimensional (space-time)? No wonder that the resulting equations persisted in yielding up zero quantities, statements that « zero equal zero » and deeper difficulties. These difficulties clouded the subject for several decades until DIRAC on the one hand [146, 147] and ARNOWITT, DESER and MISNER on the other [148] moved from a 4-dimensional treatment to a 3-plus-1-dimensional analysis. Still further down the road one began to see the larger pattern of subject in all its basic simplicity.

The central concept lends itself to statement in a single sentence: *A 3-geometry describes the momentary configuration of space as it undergoes its dynamic change with time* [34].

« 3-geometry » is a co-ordinate–free concept. One does not have to use co-ordinates to speak of « a 2-sphere of radius a », nor co-ordinates to define « a 3-sphere of radius a », nor co-ordinates to describe the deformation of a 3-sphere of radius a into a 3-ellipsoid of principal dimensions a, b, c. But neither do co-ordinates hurt—nor the combination of co-ordinates and metric that gives the square of the element of distance,

$$\mathrm{d}s^2 = g_{ij}\,\mathrm{d}x^i\,\mathrm{d}x^j\,. \tag{22}$$

For the 2-sphere one choice of co-ordinates gives

$$\mathrm{d}s^2 = a^2(\mathrm{d}\theta^2 + \sin^2\theta\,\mathrm{d}\varphi^2)\,; \tag{23}$$

another choice of co-ordinates on the same 2-sphere gives

$$\mathrm{d}s^2 = \frac{\mathrm{d}x^2 + \mathrm{d}y^2}{[1 + (x^2 + y^2)/4a^2]^2}\,; \tag{24}$$

and there are similar options, infinite in number, for the co-ordinates on the 3-sphere and the 3-ellipsoid. What counts in these options is not the name given to the co-ordinates. The names for those co-ordinates one can standardize so that they always read x^1, x^2, x^3. What counts rather than name is the dependence on these co-ordinates of the metric coefficients. How is one to know that the metric of (23)

$$\begin{Vmatrix} g_{11} & g_{12} \\ g_{21} & g_{22} \end{Vmatrix} = \begin{Vmatrix} a^2 & 0 \\ 0 & a^2 \sin^2 x^1 \end{Vmatrix} \tag{25}$$

and the metric of (24)

$$\begin{Vmatrix} g_{11} & g_{12} \\ g_{21} & g_{22} \end{Vmatrix} = \{1 + [(x^1)^2 + (x^2)^2]/4a^2\}^2 \begin{Vmatrix} 1 & 0 \\ 0 & 1 \end{Vmatrix} \tag{26}$$

describe the same ${}^{(2)}\mathscr{G}$—in this case, the same radius-a 2-sphere—whereas the metric

$$\begin{Vmatrix} g_{11} & g_{12} \\ g_{21} & g_{22} \end{Vmatrix} = \begin{Vmatrix} a^2 & 0 \\ 0 & a^2 \sin^2 x^1 + \varepsilon^2 \sin^4 x^1 \end{Vmatrix} \tag{27}$$

describes a figure with an equatorial bulge?

An alteration in metric coefficients that marks a real change in 3-geometry is distinguished most easily at the infinitesimal level from an alteration in the g_{ij} that arises from a mere change of co-ordinates.

Let a certain definite ${}^{(3)}\mathscr{G}$, with all its lumps, bumps and ripples, be expressed throughout one «local co-ordinate patch» [149] in terms of one set of co-ordinates x^i by one set of metric coefficients g_{ij}. Reexpress that 3-geometry in terms of new co-ordinates $\bar{x}^i$ shifted by the small amount ξ^i,

$$\bar{x}^i = x^i - \xi^i . \tag{28}$$

As a picture at the back of one's mind of what is going on, envisage the 3-geometry in question as the right-hand front fender of a Ford automobile, short though it is by one dimension of measuring up to a proper mental image. It is distinct in shape from the right-hand front fender of cars of a hundred other kinds. The difference is clear cut. There is no need of co-ordinates to see the difference. But co-ordinates provide a useful means to express the difference. To supply co-ordinates, take a sufficiently large and transparent rubber sheet. Mark on it an intersecting grid of lines, $x^1 = ..., 13, 14, 15, ...$; $x^2 = ..., 77, 78, 79, 80, ...$. Apply it to the fender, stretching it so it fits snugly. Then every point on the fender acquires a pair of co-ordinates, (x^1, x^2). Therefore, a measurement of the distances from a given point to several nearby points provides a straightforward way to determine the several metric coefficients g_{ij}. The distinction is clear between a mere change of co-ordinates and a true alteration in the shape of the fender, as for example in a collision. To illustrate a mere change in co-ordinates, slip the marked rubber sheet over the surface of the fender a little way in the direction of increasing x^1. In that process a given scratch mark on the fender acquires a slightly decreased co-ordinate, $\bar{x}^1 = x^1 - \xi^1$; hence the minus sign in (28).

The distance from one scratch mark on the fender to the several nearby scratch marks naturally is not changed by the movement of the rubber sheet over the surface. In other words, $(ds)^2$ is co-ordinate independent. More concretely, we have for each pair of nearby scratch marks,

$$(29)\qquad \mathrm{d}\bar{s}^2 = \bar{g}_{ij}(\bar{x})\,\mathrm{d}\bar{x}^i\,\mathrm{d}\bar{x}^j = \mathrm{d}s^2 = g_{mn}(x)\,\mathrm{d}x^m\,\mathrm{d}x^n =$$
$$= g_{mn}(\bar{x} + \xi)(\mathrm{d}\bar{x}^m + \mathrm{d}\xi^m)(\mathrm{d}\bar{x}^n + \mathrm{d}\xi^n) =$$
$$= [g_{mn}(\bar{x}) + (\partial g_{mn}/\partial\bar{x}^s)\xi^s]\times[\mathrm{d}\bar{x}^m + (\partial\xi^m/\partial\bar{x}^p)\,\mathrm{d}\bar{x}^p][\mathrm{d}\bar{x}^n + (\partial\xi^n/\partial\bar{x}^q)\,\mathrm{d}\bar{x}^q],$$

an expression reminiscent of how one analyzes strain in the theory of elasticity. Comparing the coefficient of $\mathrm{d}\bar{x}^i\,\mathrm{d}\bar{x}^j$ on left and right, and taking account of the symmetry of metric coefficients in their two indices, we arrive at an expression showing how metric coefficients change as a result of the infinitesimal slippage of the sheet:

$$(30)\qquad \bar{g}_{ij} = g_{ij} + \xi_{i|j} + \xi_{j|i}\,,$$

or

$$(31)\qquad \delta g_{ij} = \xi_{i|j} + \xi_{j|i}\,.$$

Here

$$(32)\qquad \xi_{i|j} = \partial\xi_i/\partial x^j - \Gamma_{ij}{}^m\,\xi_m$$

is an abbreviation for the covariant derivative of the i-th component of the displacement of the « rubber sheet » with respect to the j-th co-ordinate. Moreover

$$(33)\qquad \Gamma_{ij}{}^n = g^{mn}\Gamma_{ijm} = (g^{mn}/2)(\partial g_{jm}/\partial x^i + \partial g_{im}/\partial x^j - \partial g_{ij}/\partial x^m)\,,$$

expressed in terms of the rate of change of the metric coefficients and the elements g^{mn} of the matrix reciprocal to the metric tensor, is the typical « connection coefficient », having to do with the way the co-ordinate grid turns and swells or shrinks as one moves from point to point *on* the rubber sheet, regarded as fixed.

In brief, an infinitesimal change in metric coefficients, δg_{ij}, that lets itself be expressed in the form $\xi_{i|j} + \xi_{j|i}$ betokens no change in 3-geometry at all, only a change in co-ordinates, otherwise known as a « gauge change ». In contrast, an infinitesimal change in metric coefficients that does not let itself be represented in this form is the sign of a real change in 3-geometry:

(34) $$\delta g_{ij} = \xi_{i|j} + \xi_{j|i} + (\delta g_{ij})_{\text{real change}} \, .$$

This latter type of change, like the slow crumpling of an automobile fender, is what one means when one talks about the dynamics of geometry.

In what arena does the dynamics of geometry unroll?

Superspace [34, 150], $\mathscr{S}$. Superspace, with suitable mathematical amendments [151], is the manifold made up by the totality of all 3-geometries. This manifold contains infinitely many points. Each point represents one and only one 3-geometry. A collection of these points makes up the dynamic history of space evolving with time.

How does one co-ordinatize superspace, and how does one describe that movement from one point to a nearby point which is the essence of dynamics?

First ever to consider superspace was RIEMANN himself [152], though not in the context of relativity, of course. His superspace was composed of the totality of all conformally equivalent closed Riemannian 2-geometries of the same topology. Such a superspace is known today as Teichmüller space. For more on Riemann's contribution to such superspaces and the subsequent development of the relevant theory, reference may be made to the literature [150, 153]. For 2-geometries of genus g the superspace in question has dimension $6g - 6$ for $g \geqslant 2$ ($g = 0$, 2-sphere, dimensionality 0; $g = 1$, 2-torus, dimensionality 2; $g = 2$, figure eight shape, dimension 6); it is a manifold of a very limited dimensionality. In contrast the superspace built of 3-geometries requires an infinity of parameters for its representation.

For mathematical simplicity limit attention here and hereafter to closed, or in mathematical terms « compact », 3-geometries. The physics associated with such a restriction is briefly recapitulated in sect. **9**. Among compact 3-geometries the easiest to consider is a 3-sphere. LIFSHITZ and KHALATNIKOV have given a complete classification of the small deformations of a 3-sphere into tensorial harmonics [154], analogous to the scalar harmonics that one finds so useful in electrostatics. The coefficients in this expansion provide countable and convenient co-ordinates to describe the small deformations of the geometry of the 3-sphere. In the language of superspace, they allow one to « reach out » a little way in every conceivable direction from one chosen point in the ∞-dimensional arena, $\mathscr{S}$. Similar ways have been discussed [150] for parametrizing, not only the small deformations of other 3-geometries, but also the general finite deformation—and thus co-ordinatizing superspace in its entirety.

An alternative approach to mathematizing superspace contents itself with an approximation that provides additional insight. As a smooth auditorium roof can be approximated arbitrarily closely by a geodesic dome constructed of sufficiently many sufficiently small flat triangles, so a smooth 3-geometry can be approximated arbitrarily closely by a locked-together assembly of sufficiently many sufficiently small Euclidean tetrahedrons. This scheme of approximation, devised by REGGE [155], has received the name of « Regge calculus » in a subsequent review [156].

The triangles that meet at a common vertex on the geodesic dome there ordinarily have angles that fall short by some small amount δ of adding up to $2\pi = 360°$ (fig. 5). This « deficit angle » provides a measure of the curvature that is concentrated at that point of the dome. Moreover, that angle and that curvature, and the analogous angles at all the other vertices of the dome— and, therefore, the « shape » or « 2-geometry » of the dome as a whole,— are all determined by a finite number of parameters, the edge lengths, $l_1, l_2, l_3, \ldots, l_N$, of these triangles. Therefore, it might seem reasonable directly to adopt these N lengths as co-ordinates to single out and specify the one 2-geometry in question in contrast to all the other 2-geometries available in the « truncated

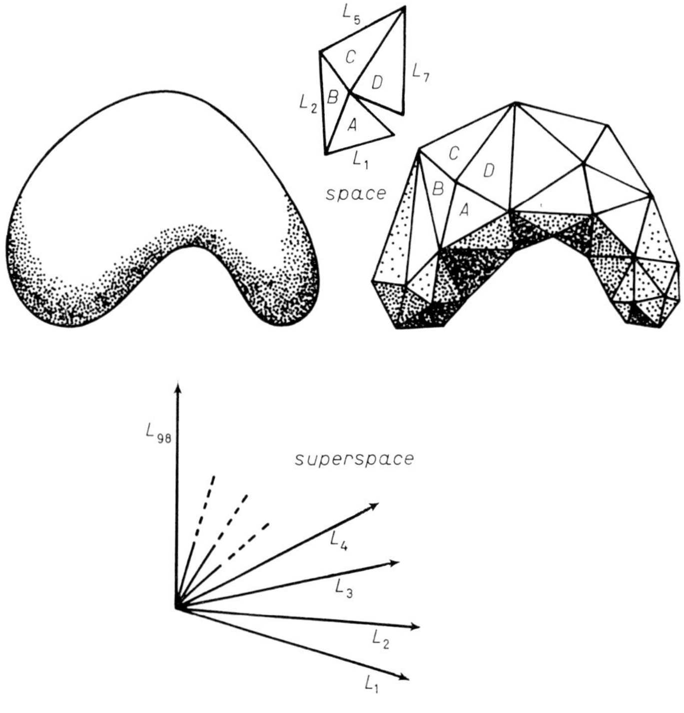

Fig. 5. – A 2-geometry (upper left) is approximated by a skeleton 2-geometry (upper right). All the details of the shape of this skeleton 2-geometry are completely specified by giving (in this example) all 98 edge lengths, $L_1, L_2, \ldots, L_{98}$. This information is represented by a single point (lower diagram) in a 98-dimensional « truncated superspace ».

N-dimensional superspace » of the l_i. However, some changes in the l_i amount in effect to mere reexpression of essentially the same 2-geometry in terms of triangles of slightly altered sizes and locations. Excluding such uninteresting alterations by appropriate supplementary conditions, one reduces the number of independent parameters from N to some lesser number N', which has to be regarded as the proper dimensionality of the « truncated superspace » built of « skeleton 2-geometries » with the given number of vertices. The larger the number of vertices, the more closely one expects to be able to reproduce the results of an analysis based on the full ∞-dimensional superspace of 2-geometries.

When one turns from a skeleton 2-geometry to a skeleton 3-geometry, the locus of curvature shifts from the vertex common to a set of triangles to the edge common to a set of tetrahedra, and other details alter, but the end result is similar. The curvature concentrated on each locus, and the shape of the entire 3-manifold, is fully fixed by a finite number, N, of edge lengths, on which one imposes certain supplementary conditions (having to do with « evenness of zoning »), leaving over a number $N' < N$ of parameters. With their help one describes each 3-geometry as a point in an N'-dimensional « truncated superspace ». So much for illustrations of superspace!

What is the relation among space, space-time and superspace? Figure 6 tells the story in brief. At the right is space-time, the deterministic classical history of space geometry evolving with time. Space-time is the history of space in this sense, that any spacelike slice through it, such as A, is a 3-geometry, a simultaneity in the sense of Landau and Lifshitz, a momentary configuration of space. That momentary 3-geometry, conceived here for definiteness as « closed » or « compact », and endowed with the topology of a 3-sphere, is illustrated schematically at the upper left—for want of dimensions on the paper—by a small deformed 2-sphere. In it are two bumps. They symbolize the local curvature of space produced by two large agglomerations of mass-energy at an early stage in the history of the Universe when the dimensions of space were much smaller than they are today and galaxies were closer together. That entire 3-geometry A, with all its curves and bumps, is represented by a single point A in the infinite-dimensional superspace at the bottom of fig. 6.

Another slice B through the same space-time at the upper right provides another 3-geometry, another momentary configuration for space in its dynamical evolution with time. The Universe in this case is larger, but the two great clouds of mass-energy, because they happen to have started off moving towards each other, are now closer than they were in moving-picture-frame A. In the superspace description of the dynamics at the bottom of fig. 6 this configuration of the Universe is described by a single point, B.

A one-parameter family of spacelike slices through a given space-time thus evidently « generates » a one-parameter family of points running through superspace: a line or curve. However, time in general relativity has a many-

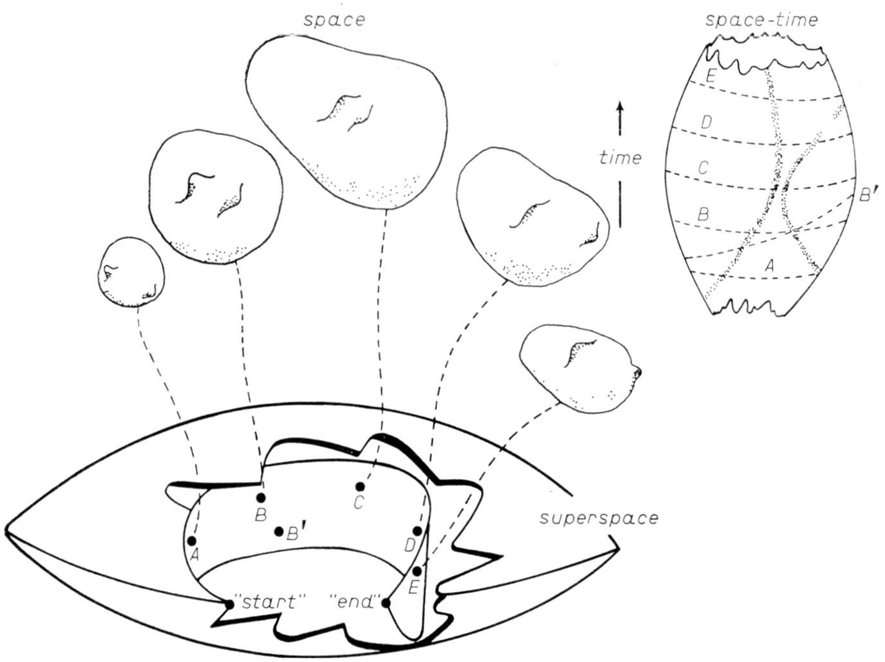

Fig. 6. – Space (upper left), space-time (upper right) and superspace (below). The *leaf of history* that curves through superspace includes all the configurations $(A, B, B', ...)$ achieved by space in its classical dynamical evolution in time; that is all spacelike slices through the given space-time. A different space-time (not shown), that is a classical history of space when the dynamics of space is started off with different initial conditions, corresponds to a different leaf of history (also not shown) cutting through superspace.

fingered character. It bursts the bounds of anything so narrow as a one-parameter family of spacelike slices. The explorers of space-time have full liberty to push ahead their exploration faster in one place than another. They have perfect freedom to measure up the 3-geometry of the spacelike slice B'. This 3-geometry is represented by another point, B', in superspace. No one simple line in superspace can accommodate all the points A, B, B', ... all the 3-geometries, that one gets by making spacelike slices in all conceivable ways through a given space-time. The region of superspace occupied by all these points is not a line; it is a leaf.

A *leaf of history* (illustrated schematically by the bent leaf visible through the cut-away part of the lower diagram in fig. 6) cuts through superspace. It describes the deterministic dynamical development of space with time.

To be more specific, consider one of the 3-geometries, say C, that is met with in the history of space, changing its shape with time. At each space point of this 3-dimensional manifold there are three independent and meaningful alterations that can be conceived in this 3-geometry (6 freely variably metric coefficients g_{ik}, diminished by the 3 types of change that arise out of mere changes in co-ordinates as in eq. (34), giving a net of 3 « adjustable parameters » or « real degrees of freedom » per space point). One of these three modifications amounts to pushing the hypersurface ahead in time a small amount in the given space-time or 4-geometry: so much here, so much there, so much at each of the points in 3-space. It describes that freedom of exploration of a given space-time which we subsume under the title of « many-fingered time ». It describes a movement in superspace that leaves the representative point on the given leaf of history. It provides the tool for analyzing the given dynamics of geometry; for reaching, bit by bit, every conceivable spacelike slice that one can think of making through the given space-time, illustrated at the upper right in fig. 6.

Conversely, given all details of the space-time geometry in question, and given all details of some particular 3-geometry that lies on that leaf of history in superspace, say C, then—apart from nongeneric symmetries or degeneracies—one can say exactly where that particular spacelike slice is located, and must necessarily be located, in that particular space-time. In other words, in this sense the specification of a 3-geometry compatible with the given 4-geometry is entirely equivalent to the complete specification of many-fingered time. This is what one means by speaking of a 3-geometry as a « carrier of information about time » [157]. « Time » conceived in these terms means nothing more or less than the location of the ${}^{(3)}\mathscr{G}$ in the ${}^{(4)}\mathscr{G}$.

Put in still other language, time is the relation between the ${}^{(3)}\mathscr{G}$'s that lie on the given leaf of history in superspace and the description of that same history as a ${}^{(4)}\mathscr{G}$. The child's toy can be removed from its box only to reveal another box and—that taken away—another box, and so on, until eventually there are dozens of boxes scattered over the floor. Or conversely the boxes can be put back together, nested one inside the other, to reconstitute the original package. The packaging of ${}^{(3)}\mathscr{G}$'s into a ${}^{(4)}\mathscr{G}$ is much more sophisticated. Nature provides no monotonic ordering of the ${}^{(3)}\mathscr{G}$'s. Two of the dynamically allowed ${}^{(3)}\mathscr{G}$'s taken at random will often cross each other one or more times. When one shakes the ${}^{(4)}\mathscr{G}$ apart, he therefore gets enormously more ${}^{(3)}\mathscr{G}$'s « spread out over the floor » than he might otherwise have imagined. Conversely, when one puts back together all of the ${}^{(3)}\mathscr{G}$'s allowed by the condition of constructive interference, he gets a structure with a rigidity that he might not otherwise have foreseen. This rigidity arises from the infinitely rich interleaving and intercrossing of clear-cut well-defined ${}^{(3)}\mathscr{G}$'s one with another.

How different from the textbook concept of space-time! There the geometry of space-time is conceived as constructed out of elementary objects, or points, known as « events ». Here, by contrast, the primary concept is 3-geometry, and the event is secondary: 1) The events lies at the « intersection » of such and such ${}^{(3)}\mathscr{G}$'s. 2) Its timelike relation to some other ${}^{(3)}\mathscr{G}$

is determined by the structure of the ${}^{(4)}\mathscr{G}$, which in turn derives from the intercrossings of all the other ${}^{(3)}\mathscr{G}$'s.

Whether one starts with ${}^{(3)}\mathscr{G}$'s as primary and regards the « event » as a derived concept, or *vice versa*, might make little difference, if one were to remain in the domain of classical geometrodynamics. It makes all the difference when one turns to quantum geometrodynamics.

There is no such thing as a 4-geometry in quantum geometrodynamics, and for a simple reason. No probability amplitude function $\psi({}^{(3)}\mathscr{G})$ can propagate through superspace as an indefinitely sharp wave packet. It spreads. It has a finite probability amplitude in a domain of superspace of finite measure. This domain encompasses a set of ${}^{(3)}\mathscr{G}$'s for too numerous to accommodate in any one ${}^{(4)}\mathscr{G}$. One can express this situation in various terms. One can say that propagation takes place in superspace, not by following any one classical history of space, not by following any one ${}^{(4)}\mathscr{G}$, but by summation of contributions from an infinite variety of such histories. In whatever way one states the matter, however, the facts are clear. The ${}^{(3)}\mathscr{G}$'s that occur with significant probability amplitude do not fit and cannot be fitted into any single ${}^{(4)}\mathscr{G}$. That « magic structure » of classical geometrodynamics simply does not exist. Without that building plan to organize the ${}^{(3)}\mathscr{G}$'s of significance into a definite relationship, one to another, even the « time ordering of events » is a notion devoid of all meaning.

These considerations reveal that the concepts of space-time and time itself are not primary but secondary ideas in the structure of physical theory. These concepts are valid in the classical approximation. However, they have neither meaning nor application under circumstances when quantum-geometrodynamical effects become important. Then one has to forgo that view of Nature in which every event, past, present, or future, occupies its preordained position in a grand catalog called « space-time ». There is no space-time, there is no time, there is no before, there is no after. The question what happens « next » is without meaning.

How does one see these lessons of the quantum in more detail? And how close to being inescapable are they? [158-161].

Is geometry measurable anyway? Especially is it measurable in principle down to distances comparable to the Planck length of eq. (5), where the concepts of « before » and « after » are predicted to lose all applicability?

Consider first geometry at the classical level. Compare the space-time interval PQ anywhere in space-time with a fiducial interval MN at a particular location in space-time [162-166]. Use any and all routes of intercomparison one pleases. Get in every case without exception the same value for the ratio [163]

$$r = \overline{PQ}/\overline{MN}\,. \tag{35}$$

That is the central point and prediction of Riemannian geometry. It exposes itself to destruction on a hundred fronts. Were it not true, then, for example, electrons brought by different routes to the same iron atom at the center of the Earth would be expected to have different properties. Then the Pauli exclusion principle would not apply. The electrons would all fall to the K-orbit. The iron atom—and the center of the Earth—would collapse, contrary to observation [163].

When one turns from the classical to the quantum dynamics of geometry, then field co-ordinate and field momentum have to be accepted as complementary, conjugate, and not simultaneously measurable quantities, the reciprocal uncertainty relations between which are given by the theory itself. Into these relations enters only one physical quantity, the Planck length of eq. (5). WIGNER and SALECKAR [162], looking at possible methods to measure the geometry compatible with the quantum principle, conclude that any determination of substantial precision is limited, not by the Planck distance, but by a distance many powers of ten greater. If this conclusion were to be upheld, one would have to accept that the quantum theory of the dynamics of geometry is incomplete or incorrect or both. A similar incompleteness or incorrectness in what quantum electrodynamics has to say about the possibilities for field measurements was claimed by LANDAU and PEIERLS [167]. It took the famous papers of Bohr and Rosenfeld [168, 169] to show that the possibilities for making measurements had been too narrowly conceived, and that the precision predicted by theory could be attained in principle by idealized measuring equipment when one looked apart from limitations imposed by the atomic constitution of matter. In brief, devising measuring equipment that won't work is easier than devising equipment that will. In the end field theory itself would seem to be the safest guide—in the absence of other evidence—on the reciprocal uncertainties of the field quantities, and on the precision attainable in measurements of the 3-geometry intrinsic to a spacelike hypersurface or the extrinsic curvature of that geometry relative to the enveloping space-time. We adopt this point of view pending further analysis and assessment of the conclusions of Wigner and Saleckar.

The plain straightforward conclusions of quantum geometrodynamics about uncertainties in space-time geometry follow from an elementary line of reasoning as familiar in the physics of the simple harmonic oscillator as in the analysis of the electromagnetic field.

The essential ideas show up already in such an elementary system as a single harmonic oscillator. There we write the wave function of a typical state, for example the ground state, in the form

$$\psi(x) = N \exp[-m\omega x^2/2\hbar]\,, \tag{36}$$

where N is a normalization factor. We proceed similarly with a collection of harmonic oscillators; and with suitably normalized displacement co-ordinates

$\xi_1, \xi_2, \ldots$, we have for the ground state probability amplitude function the expression

$$\psi(\xi_1, \xi_2, \ldots) = N \exp[-\xi_1^2 - \xi_2^2 - \ldots]. \tag{37}$$

More familiar in the case of the electromagnetic field than this description in terms of oscillator amplitudes is the so-called occupation number representation; but a third, *spacelike*, representation prepares the way for situations, as in general relativity, where Fourier analysis is not appropriate. The magnetic field at the point x, y, z, expressed in terms of normal modes and the amplitudes $\xi_1, \xi_2, \xi_3, \xi_4, \ldots$ of these normal modes, has the form

$$\boldsymbol{B}(x, y, z) = \sum_{n=1}^{\infty} \xi_n \boldsymbol{B}^{(n)}(x, y, z). \tag{38}$$

To specify the amplitudes is to specify the magnetic field; but, conversely, to specify the magnetic field everywhere is to have all the information required to determine the ξ's and therefore the wave function (37); thus,

$$\psi = \psi(B(x, y, z)) = N \exp[-1/16\pi^4\hbar c]\int\int r_{12}^{-2} \boldsymbol{B}(1)\cdot\boldsymbol{B}(2)\, d^3x_1\, d^3x_2. \tag{39}$$

In expression (39) one has the probability amplitude for a given, global, configuration of the magnetic field. For example, a configuration in which $\boldsymbol{B}$ is zero everywhere, except for a nonzero value $\Delta\boldsymbol{B}$ in a region of extension $\sim L$, has a probability amplitude in which the exponent in (39) is of the order

$$L^4(\Delta B)^2/\hbar c. \tag{40}$$

In this sense a field fluctuation $\Delta\boldsymbol{B}$ has a negligible probability unless its magnitude is of the order

$$\Delta B \sim (\hbar c)^{\frac{1}{2}}/L^2 \tag{41}$$

or less.

In a fuller description the appropriate wave function depends on the time t as well as on the entire configuration of the magnetic field at that time. However, in a curved space-time one generalizes from a time co-ordinate t to an arbitrary spacelike hypersurface σ. The probability amplitude depends as well on σ as on the configuration of the magnetic field *upon* this hypersurface:

$$\psi = \psi(\boldsymbol{B}, \sigma). \tag{42}$$

This wave function in this spacelike representation satisfies Tomonaga's wave equation, with its « bubble time » functional differentiation [131],

$$i\hbar\, \delta\psi/\delta\sigma = (\boldsymbol{B}^2/8\pi)\,\psi + (1/8\pi)[(4\pi\hbar/i)\,\delta/\delta\boldsymbol{A}]^2\,\psi. \tag{43}$$

The wave function ostensibly depends on all three components of the vector potential $\boldsymbol{A}$; thus,

$$\psi = \psi(\boldsymbol{A}, \sigma)\,. \tag{44}$$

However, the change in these components induced by the arbitrary infinitesimal « change in gauge » λ,

$$A_k \to A_k + \partial\lambda/\partial x^k\,, \tag{45}$$

produces no change in what alone counts physically, the magnetic field $\boldsymbol{B}$ (eq. (42)). Therefore the change in ψ resulting from the transformation (45) must vanish for arbitrary choice of the gauge function λ; that is, in the last integral below,

$$\delta\psi = \int (\delta\psi/\delta A_k)\,\delta A_k\,\mathrm{d}^3x = -\int \lambda[(\partial/\partial x^k)(\delta\psi/\delta A_k)]\,\mathrm{d}^3x\,, \tag{46}$$

the expression in square brackets must vanish everywhere. This is the condition that the divergence of the electric field should vanish, expressed in operator language. It is also the condition that ψ, ostensibly dependent upon the potential $\boldsymbol{A}$, with its three independent components per space point, should really depend only on the divergence-free field $\boldsymbol{B}$, with its two independent components per space point.

In a similar way the superspace formulation of general relativity (here taken for simplicity to be source-free) expresses the state functional as ostensibly dependent on the six independent g_{ik} of the metric upon a spacelike hypersurface, but in reality as dependent only on the co-ordinate–independent 3-geometry ${}^{(3)}\mathcal{G}$ described by this metric. This 3-geometry is not at all affected by the arbitrary infinitesimal co-ordinate transformation

$$\begin{cases} \bar{x}^r = x^p - \xi^p\,, \\ \bar{g}_{pq} = g_{pq} + \xi_{p|q} + \xi_{q|p}\,, \end{cases} \tag{47}$$

where the vertical slash stands for covariant differentiation. Therefore, the change in ψ resulting from (47), calculated in exact analogy to (46), must vanish for arbitrary choice of the three infinitesimal co-ordinate shifts ξ^p, from which one concludes that the three conditions

$$(\delta\psi/\delta g_{pq})_{|q} = 0 \tag{48}$$

must be fulfilled everywhere. Thus ψ, instead of depending upon 6 quantities g_{ij} per space point, depends only on the three quantities per space point that are carried in ${}^{(3)}\mathscr{G}$:

$$\psi = \psi({}^{(3)}\mathscr{G}) \, . \tag{49}$$

Of these three « informations », two have to do with gravitational wave amplitudes and one with time. In the case of electromagnetism these two kinds of data are cleanly separated in (42). In the case of a 3-geometry no such clean separation of wave amplitude from time is possible. The 3-geometry as a whole is a « carrier of information about time » [157]. Each 3-geometry requires for its specification an infinite number of parameters and can be represented as a point in an infinite-dimensional manifold, superspace [150].

The propagation of the probability amplitude, ψ, in the superspace of geometrodynamics requires a propagation law analogous to the Tomonaga equation (43) of electrodynamics; symbolically,

$$\delta^2\psi/(\delta^{(3)}\mathscr{G})^2 + {}^{(3)}R = 0 \, , \tag{50}$$

where ${}^{(3)}R$ is the local value of the curvature scalar of the 3-geometry. In the WKB approximation, where ψ is represented as a slowly varying amplitude factor times a rapidly varying phase factor,

$$\psi \sim A \exp[iS/h] \, , \tag{51}$$

the « dynamical phase », or Hamilton-Jacobi function $S({}^{(3)}\mathscr{G})$, satisfies the Einstein-Hamilton-Jacobi equation of Peres [170], the « dispersion relation »

$$(16\pi)^2 g^{-\frac{1}{2}}[\tfrac{1}{2} g_{pq} g_{rs} - g_{pr} g_{qs}](\delta S/\delta g_{pq})(\delta S/\delta g_{rs}) + g^{\frac{1}{2}\,(3)}R = 0 \, . \tag{52}$$

All of (source-free) classical general relativity follows from this one equation [171].

Consider a classical history H_{class} of 3-geometry developing deterministically in time in accordance with Einstein's field equation. Consider the « leaf of history » in superspace that describes this dynamics. Consider one of the ${}^{(3)}\mathscr{G}$'s that is met on this leaf of history. Per space point of this 3-dimensional manifold there are three independent modifications that can be conceived in this 3-geometry ($6 - 3$ arbitrary co-ordinates $g_{ik} = 3$ real degrees of freedom per space point). On of these modifications amounts to pushing the hypersurface ahead in time a small amount in the given 4-geometry. The other two modifications change gravitational wave degrees of freedom, therefore change the space-time, and therefore carry the representative point in superspace off the given leaf of history. In other words, the infinite-dimensional space of small deformations away from the given point ${}^{(3)}\mathscr{G}$ on the leaf of history (« local tangent space of superspace ») breaks down into the product of two subspaces, each also infinite-dimensional. One has one-third the dimensionality of the original space. It is the subspace of deformations that leave ${}^{(3)}\mathscr{G}$ on the leaf of history. The other has two-thirds the dimensionality of the full tangent space. It is the subspace of deformations that move ${}^{(3)}\mathscr{G}$ off the leaf of history. Quantum geometrodynamics makes no such sharp distinction. It assigns a finite probability amplitude $\psi({}^{(3)}\mathscr{G})$ to 3-geometries off the classical leaf. This spread of the state function in superspace is the superspace description of the quantum fluctuations in geometry. A closer analysis [172, 173] tells us that, in a probe region of extension L, the quantum fluctuations in the normal metric coefficients $(-1, 1, 1, 1)$ are of the order

$$\Delta g \sim L^*/L \,, \tag{53}$$

where L^* is the Planck length.

To summarize, the sharp division of superspace by a classical history into « Yes » and « No » ${}^{(3)}\mathscr{G}$'s is denied by the quantum principle, which assigns a probability amplitude $\psi({}^{(3)}\mathscr{G})$ to every 3-geometry. The ${}^{(3)}\mathscr{G}$'s with appreciable probability amplitude are too numerous to be accommodated into any one space-time. Thus the uncertainty principle declares that space-time is only an approximate and classical concept. In reality there is no such thing as space-time. « Time » itself loses its meaning, and the words « before » and « after » are without application. These long-known considerations are of importance only at the Planck scale of distances. They all flow out of the straightforward analysis of the dynamics of geometry in the arena of superspace, inescapable conceptual adjunct of general relativity.

(The above ten paragraphs are from ref. [174].)

6. – Causal order without causal order.

Time is not primordial. It, like every concept that man works with, is secondary and derived. How time, and space-time, force themselves upon us in our efforts to organize our observations is a question on which only a miniscule beginning has so far been made despite the impressive pioneer work of Mach [175] and Piaget [176]. Not one bit of further headway into this enterprise do the present lectures intend to make. Their purpose is much less courageous. Don't try to « take time apart » into the elementary quantum acts of observer-participancy out of which we conceive it—and everything—to be built. Instead, sticking to the solid ground of physics as we know it, identify domains where familiar concepts of time and causality come to the limit of applicability and have to be modified. We have just finished exploring one such frontier. We have seen how both time and space-time, according to existing theory, lose all application at the Planck distance and the Planck time;

but how out of a description that transcends time—out of superspace—we come back in the appropriate correspondence principle limit to familiar views of time. We now turn to another question. Can we similarly arrive at the familiar ordering of cause and effect from a description that transcends that order?

We consider a system of point charges coupled with each other by elementary electromagnetic actions-at-a-distance, individually time symmetric, in the sense that the force exerted on particle b by particle a is given by half the retarded field of a, as usually calculated, plus half the advanced field. Of the motives for considering such a coupling—that it should be derivable from an action principle, that it should be compatible with a principle of action and reaction, that it should reproduce all the familiar physics of electrostatics and of electromagnetism—we shall say nothing here, for the subject is discussed at length in two papers written with Richard FEYNMAN [93, 94]. For simplicity the interactions are treated in the context of classical theory and a pre-existing flat space-time. The point of interest is the field created by one of these particles when it undergoes a sudden acceleration. Experience says that the effect produced will be confined to the future light-cone of the acceleration. With this observation the model seems absolutely incompatible. It links past and future in a maze of backward and forward running light rays. Nowhere can the slightest change be made without altering motions everywhere into the indefinite past and future.

Why should we be interested in trying to derive causality out of an apparently so preposterous model? Because we want to establish in this one example a point of more general application: The apparent inability of an action taken now to influence the past by no means rules out a direct influence of the present in « bringing about that which we call the past ». It is in no way suggested here that this is the actual mechanism by which acts of observer-participancy in the present bring about that which we call the tangible or communicable reality of the Universe at an era when no observers existed. That is a deeper question with which physics is not yet prepared to deal. However, one is open to believe that the kind of considerations that elucidate the one may clarify the other.

The concrete issue then is this: How is one to reconcile the $(1/2)R + (1/2)A$ field that the accelerated particle produces in the model with the $1R$ field that the particle in actuality produces? The answer is easily summarized. The far away absorber driven by the source produces a field which in the neighborhood of the source, though source-free there, looks as if it were a field for which the source is directly responsible, $(1/2)R - (1/2)A$. Combined with the field due to the source, this field generated in the absorber gives rise in the vicinity of a to total field, R, in full agreement with experience. Thus the familiar ordering in time of cause and effect is upheld in a model which at the beginning violated that ordering as outrageously as one could well imagine.

The idea thus so briefly stated raises several questions. How can the « superposition of the advanced fields of a large number of particles ... give the appearance of both retarded and advanced fields due to the source itself [93] [?] The advanced field of a single charge of the absorber can be symbolized as a sphere which is converging towards the particle and which will collapse upon the source. But at the moment when the source particle itself was accelerated, the sphere in question had a substantial radius. One point on it touched, or nearly touched, the source. The shrinking sphere, therefore, appears to the source as a nearly plane wave which passes over it headed towards one of the particles of the absorber. When we consider the effect of all the absorbing charges, we have to visualize an array of approximately plane waves, all marching towards the source and passing over it in step. The resultant of these individual effects is an spherical wave, the envelope of the many nearly plane waves. The sphere converges, collapses on the source, and then pours out again as a divergent sphere. An observer in the neighborhood will gain the impression that this divergent wave originated from the source. »

« Why does radiation have [an] irreversible character even in a formulation of electrodynamics which is from the beginning symmetrical with respect to the interchange of past and future? ... We have to conclude with Einstein [95] that the irreversibility of the emission process is a phenomenon of statistical mechanics connected with the asymmetry of the initial conditions with respect to time. In our example the particles of the absorber were either at rest or in random motion before the time at which the impulse was given to the source. »

« That it is solely the nature of the initial conditions which governs the direction of the radiation process can be seen by imagining a reversal of the direction of time .... We have then a solution of the equations of motion just as consistent as the original solution. However, our interpretation of the solution is different. As the result of chaotic motion going on in the absorber, we see each one of the particles receiving at the proper moment just the right impulse to generate a disturbance which converges upon the source at the precise instant when it is accelerated. The source receives energy and the particles of the absorber are left with diminished velocity. No electrodynamic objection can be raised against this solution of the equations of motion. Small *a priori* probability of the given initial conditions provides our only basis on which to exclude such phenomena. »

What is the effect on the source particle of the $(1/2)R - (1/2)A$ field produced by the absorber? It gives rise to the familiar and well-tested force of radiative reaction [93]. What for our present purpose is the central lesson of this study in electrodynamics? That an order in time, ostensibly causal, can originate from an underlying machinery that is very far from causal.

Why is this point relevant to our larger theme (sect. **1**) of « law without law »? Because we see here a sample law, causality, emerging from a description of Nature that contains no such law.

7. – Asymmetry in time and the expansion of the Universe.

Rhenium-187 has a half-life of $40\cdot10^9$ a (a = year), as measured today. In other words, of ^{187}Re atoms now present, the fraction

$$(54)\qquad -\,\mathrm{d}N/N = \lambda_{\text{apparent}}\,\mathrm{d}t$$

will disappear, on the average, in the time $\mathrm{d}t$, where the familiar constant for radioactive decay has the value

$$(55)\qquad \lambda_{\text{apparent}} = 0.693/40 \text{ in units of } (10^9\,\text{a})^{-1}\,.$$

Therefore, it has often seemed natural to suppose that the number N of these atoms has been, is now and will continue falling off as $\exp[-\lambda_{\text{app}}t]$. This assumption is a special case of the belief of older times that the Universe will endure forever but that all activity in it will eventually slow down and end in a « heat death ». In that final condition, it was imagined, temperature differences, net outflow of particles from radioactive nuclei and all other measures of a departure from statistical equilibrium will have sunk to zero, and « the entropy of the Universe » will have attained « the absolute maximum » of which N. L. Sadi CARNOT was already writing in 1824, inspiration for the phrase « the heat-death of the Universe » that CLAUSIUS first set down on paper in 1865 [177]; and that Bertrand RUSSELL much later took as gospel truth when he wrote [178]: « The second law of thermodynamics makes it scarcely possible to doubt that the Universe is running down, and that, ultimately, nothing of the slightest interest will be possible anywhere ».

Will the amount of ^{187}Re really fall off exponentially with time? Will temperature differences really sink exponentially to zero? Is perpetual approach to equilibrium guaranteed? It is impossible to face up to such questions in our own time without encountering issues of cosmology and without having to ask, is there a connection between statistical mechanics and cosmology? An exponential can only be brought to zero in an infinite time; but a finite time is all that is available in the familiar Friedmann model of a closed universe. If the Universe is to end out of equilibrium, who knows enough to say that it should not end as much out of equilibrium as it started? How then can one properly predict the amount of ^{187}Re over a cosmological range of time without first coming to grips with this question of « double-ended » statistical mechanics [179]?

The first line of the first page of a recent and distinguished book [180] by two leading mathematicians declares that « The foundamental problem of mechanics is computing, or studying qualitatively, the evolution of a dynamical system with prescribed initial data ». Moreover, thus to focus on an initial time and at that time to specify all co-ordinates and momenta is often the most useful way to apply dynamics to a given problem and sometimes the only way. However, one states the data in a quite different and thoroughly time-symmetric, « double-ended », way when one derives dynamics in the first place from either

1) the Euler-Lagrange variation principle of point mechanics,

2) the Hamilton variation principle of point mechanics,

3) the Hilbert variation principle of electrodynamics,

4) the Hilbert variation principle of general relativity,

5) the Hojman-Kuchař-Teitelboim imbeddability argument of sect. **3**, or

6) the Feynman sum-over-histories.

One deals with the co-ordinates of particles or fields, and co-ordinates only, but at two times, or on two spacelike hypersurfaces.

If one thus plumbs some of the deepest issues of dynamics in terms of « double-ended data », can one escape from asking what statistical mechanics looks like when it too is stated in terms of double-ended data? No more quickly than by this route is one led—if one is ever led—to question [181-188] the automatic presupposition that departure from equilibrium will necessarily decrease and entropy will inescapably increase in the Einstein-Friedmann-predicted phase of contraction of the Universe.

A recent review [188] puts the issue in these terms: « As dynamic time marches forward, what will happen then [in the phase of contraction] to [the arrows of] statistical and biological time? Will they continue to point in the same direction or will they point in opposite directions? In the one case, to a person alive in the second phase of the Universe, the Universe will appear to be contracting. In the other case, it will appear to be expanding, simply because a moving picture of contraction run backwards looks like expansion. Many colleagues agree that the question is open and that the answer is one of the great puzzles of our day; but others are strongly convinced that the one answer or the other is the only right answer and that the answer is perfectly obvious and should be accepted without question. This is the insanity of the subject [of the arrow of time]. » To paraphrase, it is not a question of accepting a solution; it is a question of accepting a problem [189].

Is there any real doubt that each revolution of the Earth around the Sun will see a greater statistical-mechanical disorder in the Universe, down to the end of time? Doubt shows clear in the works of leading figures in statistical mechanics from BOLTZMANN to today. Presuppose order in the initial conditions, and randomness otherwise? That assumption, all recognize, will reproduce the evidence of experience that entropy increases. But has so cosmic an assumption any deeper foundation? Doubt begins when it is asked whether entropy will increase *forever*. Doubt grows when it is asked, why order in the *initial* conditions? Doubt takes a new turn with the advances of relativistic astrophysics of recent years. How can a *cosmological* requirement on initial conditions possibly be imagined to be well grounded when it presupposes the out-of-date cosmological model of a universe that endures from everlasting to everlasting?

That this doubt about the right end-point conditions for statistical mechanics has a long history one can forbear from reminding oneself anew by skipping the next few pages of brief quotes, extracted for the most part from the collection of reprints and translations of reprints edited by Stephen BRUSH [190]. Nothing stands out more strikingly from this quick oversight of the last hundred years than the « foreverness » of the cosmology taken as forgranted in all the discussions.

GIBBS [191] (1875): « The impossibility of an uncompensated decrease in entropy seems to be reduced to an improbability. »

BOLTZMANN [192] (1877): « ... The fact that this integral [$\int dQ/T$] is actually $\leqslant 0$ for all processes in the world in which we live (as experience shows) is not due to the nature of the forces, but rather to the initial conditions. » « It is only because there are many more uniform distributions than nonuniform ones that the distribution of states will become uniform in the course of time. One, therefore, cannot prove that, whatever may be the positions and velocities of the spheres at the beginning, the distribution must become uniform after a long time; rather one can only prove that infinitely many more initial states will lead to a uniform one after a definite length of time than to a nonuniform one. » « ... When we follow the state of the world into the infinitely distant past [here BOLTZMANN is speaking without benefit of the present-day evidence for big-bang cosmology, and is tacitly assuming that the Universe endures from everlasting to everlasting], we are actually just as correct in taking it to be very probable that we would reach a state in which all temperature differences have disappeared, as we would be in following the state of the world into the distant future ... if we know that in a gas at a certain time there is a nonuniform distribution of states, and that the gas has been in the same container without external disturbance for a very long time, then we must conclude that much earlier the distribution of states was uniform and that the rare case occurred that it gradually became nonuniform. » « If perhaps this reduction of the second law to the realm of probability makes its application to the entire Universe appear dubious, yet the laws of probability theory are confirmed by all experiments carried out in the laboratory. »

POINCARÉ [193] (1893): « A theorem, easy to prove, tells us that a bounded world, governed only by the laws of mechanics, will always pass through a state very close to its initial state. On the other hand, according to accepted experimental laws ..., the Universe tends towards a certain final state [of uniform temperature], from which it will never depart .... I do not know if it has been remarked that the English kinetic theories can extricate themselves from this contradiction. The world, according to them, tends at first toward a stage where it remains for a long time without apparent change; and this is consistent with experience; but it does not remain that way forever, if the theorem cited above is not violated; it merely stays there for an enormously long time, a time which is longer the more numerous are the molecules. This state will not be the final death of the Universe, but a sort of slumber, from which it will awake after millions of millions of centuries. According to this theory, to see heat pass from a cold body to a warm one, it will not be necessary to have the acute vision, the intelligence and the dexterity of Maxwell's demon; it will suffice to have a little patience. »

ZERMELO [194] (1896): « Poincaré's theorem [195] says that in a system of mass-points under the influence of forces that depend only on position in space, in general any state of motion (characterized by configurations and velocities) must recur arbitrarily often, at least to any arbitrary degree of approximation even if not exactly, provided that the co-ordinates and velocities cannot increase to infinity. Hence, in such a system irreversible processes are impossible ... (aside from singular initial states). »

« Suppose we have a gas enclosed in a solid container with elastic sides that are impermeable to heat. In general there will indeed be an infinite manifold of states of the molecules for which the gas will undergo *permanent* changes of state, such as viscosity, heat conduction, or diffusion. However, there will also be a much larger number of possible initial states, which can be reached by arbitrarily small displacements from the former states, and these states, instead of undergoing irreversible changes, will come back periodically to their initial states as closely as one likes .... »

BOLTZMANN [196] (1896): « Poincaré's theorem, which ZERMELO explains at the beginning of his paper, is clearly correct, but his application of it to the theory of heat is not. » « ... according to the laws of probability a certain quantity H (which is some kind of measure of the deviation of the prevailing state from Maxwell's) can only decrease for a stationary gas in a stationary container. ... [Thereafter the H-curve] almost always runs very close to the abscissa [time] axis. Only very rarely does it rise up above this axis; we call this a peak, and indeed the probability of a peak [significant deviation from Maxwell's distribution] decreases very rapidly as the height of the peak increases .... It is just for certain singular initial states that the Maxwell distribution is never reached, for example when all the molecules are initially moving in a line perpendicular to two sides of the container. ... Whereas ZERMELO says that the number of states that finally lead to the Maxwellian state is small compared to all possible states, I assert on the contrary that by far the largest number of possible states are 'Maxwellian' and that the number that deviate from the Maxwellian state is vanishingly small. ... According to the molecular-kinetic view, this [second] law [of thermodynamics] is merely a theorem of probability theory. According to this view, it cannot be proved from the equations of motion that all phenomena must evolve in a certain direction in time. »

« An answer to the question—how does it happen that at present the bodies surrounding us are in a very improbable state—cannot be given, any more than one can expect science to tell us why phenomena occur at all and take place according to certain laws. »

« ... One may say that according to Poincaré's theorem the entire Universe must return to its initial state after a sufficiently long time, and hence there must be times when all processes take place in the opposite direction. How shall we decide, when we leave the domain of the observable, whether the age of the Universe, or the number of centers of force which it contains, is infinite? »

ZERMELO [197] (1896): « ... as long as one cannot make comprehensible the *physical origin* of the initial state, one must merely assume what one wants to prove; instead of an explanation one has a renunciation of any explanation. »

BOLTZMANN [198] (1897): « The second law will be explained mechanically by means of assumption A (which is of course unprovable) that the Universe, considered as a mechanical system—or at least a very large part of it which surrounds us—started from a very improbable state, and is still in an improbable state. »

« ... Poincaré's theorem does not contradict the applicability of probability theory, but rather supports it, since it shows that in eons of time there will occur a relatively short period during which the state probability and the entropy of the gas will significantly decrease, and that a more ordered state similar to the initial state will occur. » « One has the choice of two kinds of pictures. One can assume that the entire Universe finds itself at present in a very improbable state. [Or one can assume that for] the Universe as a whole the two directions of time are indistinguishable ... [with, however,] here and there relatively small regions of the size of our galaxy (which we call worlds), which during the relatively short time of eons deviate significantly from thermal equilibrium ... a living being that finds itself in such a world at a certain period of time can define the time direction as going from less probable to more probable states (the former will be the 'past' and the latter the 'future') and by virtue of this definition he will find that this small region, isolated from the rest of the Universe, is 'initially' always in an improbable state. »

EHRENFEST-AFANASSJEWA [199, 200] (1959): « Although BOLTZMANN did not fully succeed in proving the tendency of the world to go to a final equilibrium state, there remain after all criticisms the following valuable results. First, the derivation of the Maxwell-Boltzmann distribution for equilibrium states, then the kinetic interpretation of the entropy by the H-function, and finally the *explanation* of the existence of an integrating factor for $\mathrm{d}U + \mathrm{d}A$ .... »

« The so important irreversibility of all observable processes can be fitted into the picture in the following way. The period of time in which we live happens to be a period in which the H-function of the part of the world accessible to observation decreases. This coincidence is really not an accident, since the existence and functioning of our organisms, as they are now, would not be possible in any other period. To try to explain this coincidence by any kind of probability considerations will in my opinion necessarily fail. The expectation that the irreversible behaviour will not stop suddenly is in harmony with the mechanical foundations of the kinetic theory. »

UHLENBECK [201] (1968): « ... one then can conclude that, if the system is *not* in thermal equilibrium, it almost always will go into that state; and if the system is in thermal equilibrium, it almost always will stay in that state although fluctuations away from equilibrium will and must occur because of the quasi-periodic nature of the motion of the Γ-point. This is the Boltzmann picture; it clearly reconciles the reversibility of the mechanical motion as expressed by the Poincaré theorem with the approach to equilibrium as required by the zeroth law of thermodynamics ... how is it possible that a contracted description can be *closed* and *causal* [?]. In a bona-fide macroscopic theory it should of course *not* be necessary to go back to the microscopic, molecular picture (in this sense the theory must be closed), and it *should* be possible to make predictions, that is the theory must be causal. This is the macroscopic causality problem and, although it is in my opinion still far from clarified, one begins to see some light thanks to the basic papers of Bogoliubov [203]. BOGOLIUBOV pointed out that in any macroscopic theory the macroscopic variables must be in some sense *secular variables*, that is they must vary in time much slower than all the remaining variables needed to describe the molecular system. »

COHEN [205] (1973): « It is the Boltzmann Ansatz, the statistical Ansatz of molecular chaos, which introduces the arrow of time or ... the approach to equilibrium. It is the assumption of the factorization of the s-particle distribution at time $t = 0$, which is a generalization of the statistical Ansatz, which introduces the irreversibility. »

For more on the history and the issues, reference may be made to a review article of Prigogine [206], Klein's biography of *Paul Ehrenfest* [207], and especially the books of Gold [186], Reichenbach [208] and Davies [209].

In summary, after a century and more, half the battle has been won to understand the direction of time in heat flow and other statistical processes; but the other half looks like being a long struggle. Evidently it is generally accepted that the elementary molecular interaction is time symmetric in thermal conduction, in viscosity and in other irreversible processes of everyday interest; and that the observed macroscopic irreversibility takes its origin in two circumstances: the enormous number of molecules involved and the asymmetry in time of the initial conditions. In other words, conditions were ordered before the relevant observations were made and disordered afterwards. In the rare case in which conditions are guaranteed instead to be disordered before, and ordered after one measures—say every five minutes for an hour—the temperature difference between a hot block of metal and a cold one in contact with it, then the same reasoning tells us that the temperature difference, rather than falling exponentially with time, should rise exponentially. This reasoning about exponential rise has been confirmed observationally so far only at the level of small fluctuations. For the temperature difference to increase exponentially by chance fluctuations to any truly macroscopic—and macroscopically observable—level would require a time so fantastically long as to put a test at this level utterly beyond reach. All this is not only understandable, but also well understood, as the quotes indicate. Different investigations use different words to make the same by now generally agreed points: all the elementary processes normally taken into consideration are reversible in time at the microscopic level; and the macroscopic resultant of large numbers of such processes is shown to go according to the usual sense of the arrow of time only by appealing to boundary-value conditions on the microscopic motions that presuppose order in the past, disorder in the future. Here consensus ends. Shall one or shall one not impose boundary conditions near the big crunch similar to those that one imagines imposing near the big bang? Or is it even without sense, as some would suggest, to raise such issues? To flee the abstractness of these question, let us turn to a concrete model.

No model ever illustrated approach to equilibrium more instructively than the Ehrenfest double-urn model [209-213] (fig. 7). No model allows one to see more vividly what it means to prescribe a departure from equilibrium at a «final time» $t = +T$ as great as the departure at an «initial time» $t = -T$. The idea is due to COCKE [179]. One starts as did the EHRENFESTS with the 100 balls divided $75 = 50 + 25 = 50 + n = 50 +$ «surplus» in the left-hand urn and $25 = 50 - 25 = 50 - n$ in the right-hand one. Each spin of the roulette wheel brings up a number between 1 and 100. The ball with that number painted on it thereupon jumps from whichever urn it's in to the other urn. To start with there are three times as many balls on the left. Therefore, it is three times more probable that a given number will turn up on a ball on the left. Consequently the most probable course of events is a gradual decrease in the number of balls on the left:

$$A = \text{chance that roulette wheel causes ball on left to jump} = \frac{50+n}{100},$$

$$B = \text{change of } n \text{ in such a jump} = -1,$$

$$C = \text{chance that roulette wheel causes ball on right to jump} = \frac{50-n}{100},$$

$$D = \text{change of } n \text{ in such a jump} = +1$$

$$\text{expectation value of change in } n = AB + CD = -n/50\,. \tag{56}$$

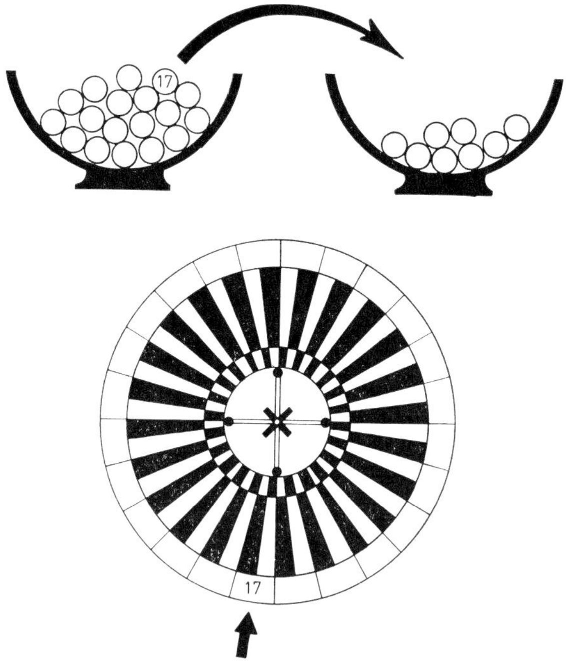

Fig. 7. – The Ehrenfest double urn in a 1978 rendering. When number 17 comes up on the roulette wheel, the ball carrying that number is transferred from whichever urn it happens to be in to the other urn. Thus 100 balls, 75 of them initially in the left-hand urn and 25 in the right-hand one, gradually approach (see fig. 8) a 50-50 distribution as « time increases » (more spins of the roulette wheel).

Taking for the unit of time the interval between spins of the roulette wheel, and dealing only with averages or expectation values, one thus analyzing events at the simplest level finds a differential equation for approach to equilibrium

$$\mathrm{d}n/\mathrm{d}t = -\,n/50\,. \tag{57}$$

The solution of this equation shows the familiar feature of exponential approach to equilibrium

$$n = 25\,\exp[-\,\mathrm{time}/50]\,, \tag{58}$$

in agreement with the standard « law of cooling ».

Equation (58) predicts that the expectation value of the « surplus number », n, in the left-hand urn will drop to $25/2.718 = 9.2$ after 50 spins of the roulette wheel; to $\overline{n} = 3.4$ after 100 spins; to $\overline{n} = 1.25$ after 150 spins; and to $\overline{n} = 0.46$ after 200 spins. However, superposed on this regular fall-off—to be seen only by averaging over many independent runs, each starting with $n = 25$—will be the fluctuations about this average unique to any one individual run. These random variations quickly grow to a magnitude given to a good approximation by the familiar formula

$$(59) \qquad \overline{(N_{\text{left}} - \bar{N}_{\text{left}})^2} = \bar{N}_{\text{left}} \bar{N}_{\text{right}}/N\,,$$

implying a root-mean-square fluctuation in the « surplus », $n = N_{\text{left}} - 50$, given by

$$(60) \qquad (\Delta n)_{\text{RMS}} = (\bar{N}_{\text{left}}/2)^{\frac{1}{2}} = 5\,.$$

These ups and downs drown out the tail of the exponential.

The fluctuations in the « surplus », n, in the left-hand urn are not limited to the magnitude $(\overline{n^2})^{\frac{1}{2}}$. From time to time a larger variation occurs; and, very rarely, a much larger one.

What is to be said about a much larger than average fluctuation as at the « Boltzmann peak » or point P in fig. 8? What about time asymmetry? There is none. The behavior prior to P is dominated by exponential rise as that later than P is dominated by exponential fall-off. In other words the one-sidedness in time of the exponential law of fall-off of the « surplus » has to be understood, not as an indicator of any asymmetry in time of the elementary process itself, but as a consequence of the special initial conditions (« order » at P). We see

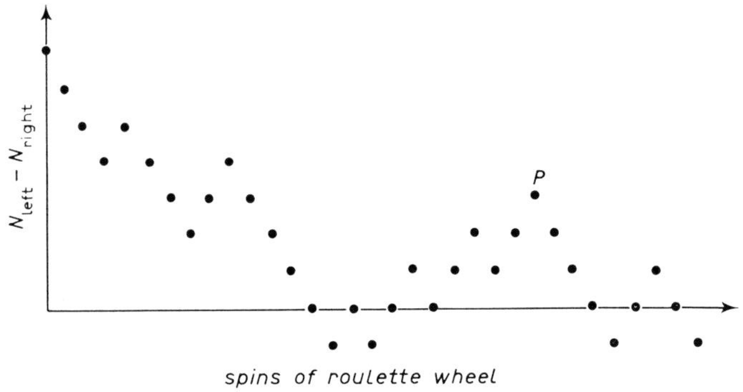

Fig. 8. – Approach to equilibrium, and fluctuations about equilibrium, as they show up in a typical « run » of the Ehrenfest double-urn experiment. The point P marks a larger-than-average fluctuation away from equilibrium. When one makes statistical run after statistical run, each run containing for example 300 spins of the roulette wheel in fig. 7, one will find some runs in which statistical fluctuation brings $N_{\text{left}} - N_{\text{right}}$ at the end of the run back to its original value. When one averages over sufficiently many of the runs that satisfy these special end-point conditions one washes out the statistical fluctuations and arrives at a cosh-curve (fig. 9). If we rule out all the other—and much more numerous—runs as « might-have-been » but « never-were » runs, we have a model for what is meant by a universe ruled by « double-ended statistics ».

here in an example as elementary as anyone has ever devised what is also apparent in the phenomenon of heat conduction and in the coupling of a complete absorber with an accelerated source by half-advanced, half-retarded potentials (sect. **6**).

All three processes—spins of the roulette wheel, molecular collisions, radiative coupling—convert ordered into disordered states, although every elementary interaction is microscopically reversible. All three are epitomized by heat conduction. WHEELER and FEYNMAN remark [93], « A portion of matter observed at the present moment to be warmer than its surroundings will cool off in the future with a probability overwhelmingly greater than the chance for it to grow hotter. About the past of the same portion of matter Boltzmann's H-theorem however also predicts an enormously greater likelihood that the body warmed up to its present state rather than cooled down to it. In other words, we are asked to understand the present temperature of the body as the result of a simple statistical fluctuation in the distribution of energy through the entire system. This deduction is based on the premise that the system was isolated before observation. However, common experience tells us that the given portion of matter probably acquired its abnormal temperature, not via an internal statistical fluctuation, but because it had earlier not been isolated from the outside. For the radiative analogy of this example of heat conduction, conceive a charged particle bound to a position of equilibrium by a quasi-elastic force. Furthermore suppose its energy at the moment of observation is large in comparison with the agitation of the surrounding absorber particles. There is then as overwhelming probability that the oscillator will lose energy to the absorber at a rate in close accord with the law of radiative damping. What can be said of the particle prior to the moment of acceleration? In an ideal absorbing system completely free of special disturbances, there is an equally overwhelming chance that the energy of the charge was then increasing at a rate given approximately by the inverse of the law of radiative damping. In this case as in heat conduction the abnormally high energy of the object is to be interpreted as the result of a statistical fluctuation. However, that the Sun at some past age acquired its energy by such a fluctuation no one now would seriously propose. Obviously the Universe is a special system with respect to the origin of which probability considerations cannot freely be applied ».

How do these considerations bear on the Ehrenfest double urn? The first part of our response is immediate. We identify the point P with a statistical fluctuation. The dominant feature of $n(t)$ before P is exponential rise, and after P an exponential fall-off. However, when we turn from the point P to the start of play, we do not suggest that n acquired the value $n = 25$ as a consequence of a prior and very large statistical fluctuation. On the contrary we understand $n = 25$ as an initial condition. That initial condition in the double-urn problem symbolizes the quiescence of the absorber *before* the acceleration of the source charge and—in the problem of heat flow—the initial condition of energy disequilibrium in the early universe. In other words the initial surplus, $n = 25$, symbolizes a cosmological boundary condition at the start of time.

Are any cosmological boundary conditions complete that do not deal with the end of time as completely as with the beginning of time? If the Universe collapses to a big crunch as it begins with a big bang, is it not as natural for requirements on particle motions and field configurations to be imposed at the end as at the beginning? Why not then see what are the consequences for the Ehrenfest double urn of imposing a final $n=n''$ at a final time $t=t''=T$ on the same footing as the initial $n = n'$ at the initial time $t = t' = -T$? W. J. COCKE [179] was the first to ask and analyze this question for the Ehrenfest double urn. That analysis is carried further here.

What does it mean to impose on the double-urn problem a final condition, $n=n''=25$, symmetric to the initial condition, $n = n' = 25$? For definiteness let the length of play, $2T$, be limited always to $2T = 200$ spins of the roulette wheel. By that time the initial condition will be almost forgotten ($\bar{n} = 25 \exp[-200/50] = 0.46$) and fluctuations ($\Delta n_{\text{RMS}} = 5$) will dominate. Thus, let the 200-spin play be repeated over and over 10^9 times, each time starting with a surplus of balls on the left, $n = n' = 25$. The play will end with $n = n''$ sometimes equal zero, sometimes $+6$, sometimes -10 and, very rarely, but occasionally, $+25$, the initial value. The probability that exactly this value is attained at the time $t = t'' = T = 100$ is estimated most easily by neglecting altogether any « memory of the past » (surplus $\bar{n} = 0.46 \simeq 0$) and treating each of the 100 balls as having a probability (1/2) to be in the left-hand urn and an equal probability to be in the right-hand one.

All possible ways to distribute the balls between the two urns are contained—each with its characteristic probability—in the binomial expansion

$$(61) \qquad 1 = \left[\left(\frac{1}{2}\right)_{\text{left}} + \left(\frac{1}{2}\right)_{\text{right}}\right]^{100} = \sum_{N_{\text{left}}+N_{\text{right}}=100} \frac{100!}{N_{\text{left}}!\,N_{\text{right}}!}\left(\frac{1}{2}\right)^{N_{\text{left}}}\left(\frac{1}{2}\right)^{N_{\text{right}}}.$$

Therefore, the desired probability to return at $t = T = 100$ to the surplus $n = 25$ on the left—in the stated approximation (low by not quite a factor 2, but uncorrected here)—is

$$(62) \qquad w(n = 25) = \frac{100!}{75!\,25!\,2} \sim (2\pi)^{-\frac{1}{2}} \frac{100^{100.5}}{75^{75.5}\,25^{25.5}\,2^{100}} = 192 \cdot 10^{-9}.$$

In other words, out of the 10^9 repetitions of a 200-spin run, of the order of 192 will end up with $N_{\text{left}} = 50 + n = 75$. Let these ~ 192 histories be called « acceptable ». Let existence be denied to all the others; let them be ruled out as « unacceptable », as « might-have-been » but « never born » universes. This is what we shall mean by speaking of « double-ended statistics ».

What are the features of the typical history that is allowed by double-ended statistics? It is marked by an almost exponential decay of n at the beginning and an almost exponential rise of n at the end. Superposed on this general trend are the inevitable fluctuations. To iron them out we turn attention from the individual history to the average of all 192 acceptable histories. Better, increase the number of tries from 10^9 to 10^{12} and the number of acceptable histories from ~ 192 to $\sim 192 \cdot 10^3$. Or multiply the number of trials by still further powers of ten. In this way reduce below any preassigned level the effect of the fluctuations which show so clearly in any one acceptable run and which still show a little when one takes the average of 192 acceptable runs.

The « ideal average run » in the sense just described follows a simple mathematical formula. There is a quick way to this formula: a differential equation. The appropriate differential equation is not the usual law of cooling

$$\mathrm{d}n/\mathrm{d}t = -n/50 = -n\,. \tag{63}$$

That is asymmetric in time. The new law must treat the two directions of time symmetrically. It must make no reference to the initial time or the final time. It must make no reference to initial n' or final n''. Those boundary-value data must go into the final formula for n only *as* boundary-value data. The only law that meets the physical requirements of the problem is one that treats exponentially rising and exponentially falling functions on the same footing,

$$\mathrm{d}^2n/\mathrm{d}t^2 = \lambda^2 n\,. \tag{64}$$

This is the law of change of n with time in double-ended statistics.

The general solution of (64) is a linear combination of $\exp[-\lambda t]$ and $\exp[\lambda t]$; or a linear combination of $\sinh \lambda t$ and $\cosh \lambda t$.

That solution which takes on the value n' at time t' and n'' at time t'' is given by the expression

$$n = [n' \sinh \lambda(t''-t) + n'' \sinh \lambda(t-t')]/\sinh \lambda(t''-t')\,. \tag{65}$$

This solution is characterized, for positive n' and n'', by 3 regions (fig. 9). The first is a region of nearly exponential fall-off near t'. The last is a region of nearly exponential rise near t''. Between is a region of transition from fall-off to increase.

These considerations make a little clearer what it means to ask whether there is any correlation between statistics and cosmology. In further pursuance of this point, let « the turning of the tide » refer to the phase in the dynamics of the Universe where expansion gives way to contraction, and let the term « the statistical turn of the tide » refer to the minimum in « the departure from equilibrium », as represented in fig. 9. Even if there is any correlation between

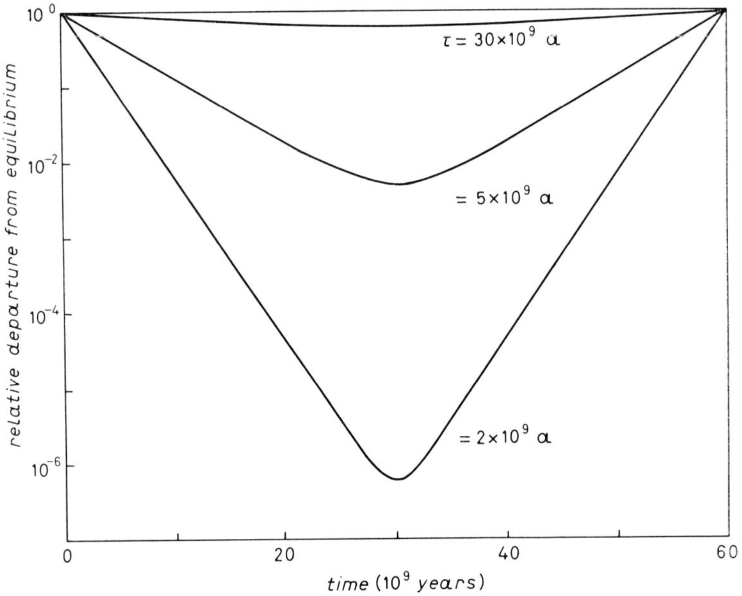

Fig. 9. – Relative departure from equilibrium of temperature or number of radioactive atoms as a function of time calculated under quite schematized assumptions for one illustrative scenario (see eq. (65)) out of many equally conceivable alternatives; specifically: 1) initial and final departures from equilibrium identical, 2) total time available from start to end $60\cdot10^9$ a, 3) symmetry in time, 4) no reaction chains; only one characteristic time relevant, $\lambda^{-1}=\tau=30$, 5, or 2, in units of 10^9 a for the three cases illustrated. These are gross and highly arbitrary simplifications. The departure from standard exponential decay in the first half of time shows up strongly only in the last e-folding time before « turnaround ».

statistics and cosmology, it is not a necessary consequence of the reasoning that the statistical tide should turn at the same time as the cosmological tide, nor is it necessary that either time occur exactly half-way between start and stop. There are few model universes easier to analyze in all detail than the Taub model universe [214-217]. For the extreme time-asymmetric (large m') case of this model the volume varies with proper time in accordance with a relation which, written parametrically, is [217]

$$\begin{cases} V = 32\pi^2 l^3 (m')^2 \sin f(1-\cos f)\,, \\ \tau = lm'(f-\sin f)\,. \end{cases} \tag{66}$$

Only for the special choice of the parameter, $m'=0$, is the dynamics time-symmetric. Moreover, there is no obvious reason why the final value of $n=n''$ in the double-urn experiment should be identified with the initial value; or, to spell out the analogy, no obvious reason why the conditions at the big crunch should be in every way identical to those at the big bang. In addition, the inescapable fluctuations that occur in any given history and that produce deviations from any idealized statistical law will normally be quite distinct in the descending and ascending phases of the curve of fig. 9. Despite all these provisos and caveats, the simplest model makes the greatest appeal in any first sketch of the possibilities. In it the turning of the tide for the statistics is identical in its timing with the transition from expansion to contraction. Also both are mirror symmetric with respect to that common time. The « homogeneity and isotropy » of the Friedmann model, if it applies to the Universe at all, applies in the large, not in detail. Likewise « mirror symmetry in time », if it applies to the Universe at all, applies in the large, not in detail. If it applied in detail, the configuration of every part of the Universe at the time $t_{\text{mirror}} + t$ would have to be identical with its configuration at the time $t_{\text{mirror}} - t$. That would mean that every motion would come to a halt at t_{mirror} itself. So detailed a requirement would plainly be incompatible with the motion of the planets around the Sun and the Moon around the Earth.

To accept double-ended statistics for investigation is to deal with no small change in familiar ideas of time and causality. What is the observer of the roulette wheel to think as he watches the end of the play approaching? Spin after spin the wheel turns up predominantly the identifying numbers of the balls that lie in the right-hand and less occupied urn. Against all normal odds the smaller number N_{right} grows still smaller. And with the final spin of the wheel the numbers of balls in the two urns are restored as if by magic to their initial values $N_{\text{left}} = N'_{\text{left}} = 75$ and $N_{\text{right}} = N'_{\text{right}} = 25$. He would find this outcome utterly beyond understanding, if he did not know that every history had been thrown out as impossible which did not end as it began with prescribed conditions.

With what words will one describe the biased probabilities spun out by the roulette wheel? « Bias » or « providence factor » are the only terms that immediately suggest themselves. A factor is at work that pushes the probabilities ever more strongly toward the predetermined end as the final time of reckoning approaches. A providence factor defines itself naturally in the context of the smooth average number $n(t)$ (average over many repetitions of a 200-spin run) in its dependence on time. It also shows up in quite another way in the biasing for or against certain numbers on the roulette wheel according as the balls so-numbered lie in the right-hand or left-hand urn.

Let us turn to the continuum description first as the simpler way to analyze this bias. We want to say that the surplus, $n = N_{\text{left}} - 50$, decreases in time in accordance with the normal law of cooling except as modified by a bias of « unknown origin » that will see to it that the predetermined end is brought about.

Thus we write

$$dn/dt = -\lambda n + \text{« bias term »}. \tag{67}$$

We compare this expression with what we get by taking the general solution (65), differentiating it once, and eliminating from the two expressions dn/dt and n the initial value n' (and, simultaneously, t'); thus,

$$(dn/dt)\sinh\lambda(t''-t)+\lambda n\cosh\lambda(t''-t)=\lambda n''\,; \tag{68}$$

or

$$dn/dt=-\lambda n+\underbrace{\frac{\lambda[n''-n\exp[-\lambda(t''-t)]]}{\sinh\lambda(t''-t)}}_{\text{« bias term »}}\,. \tag{69}$$

In other words, we have defined the bias term in such a way that it should make reference to present value and final value alone and no reference at all to the initial value, n'. As solution to this requirement we find one and only one answer, the second term on the right-hand side of (69).

The meaning of (69) is clear. Final requirements have no influence on present happenings so long as the time of reckoning lies many relaxation times in the future. However, as the time available for the final adjustments becomes of the order of a couple of relaxation times or less, the predestined end impresses itself on the game in an ever heavier bias. In the very last stage, only a few spins before the game must end, the normal decay rate is essentially without effect. The development proceeds practically deterministically to its end. In mathematical terms, the differential equation (69) reduces in this « last-moment limit » to the form

$$dn/dt=\frac{n''-n}{t''-t}\,. \tag{70}$$

The solution is

$$n=n''-\text{constant}\,(t''-t)\,; \tag{71}$$

in other words, single-minded straight-line progression towards the final goal, unmoderated by any influence of the relaxation constant, λ.

It is characteristic of « double-ended statistics » that one direction of time, t, is as good as the other, $\bar{t}=-t$, for describing it. The two very different looking equations,

$$dn/dt=-\lambda n+\text{bias} \tag{72}$$

and

$$dn/d\bar{t}=-\lambda n+\overline{\text{bias}}\,, \tag{73}$$

or

$$dn/dt=\lambda n-\overline{\text{bias}}\,, \tag{74}$$

deal with two completely equivalent ways of describing the same time dependence, $n = n(t)$, of the surplus in the left-hand urn. When the total length of the run, $t'' - t'$, amounts to many relaxation times, then one equation is « useful » near one limit, and the other equation is « useful » near the other limit. Here « useful » means that the term $\pm \lambda n$ dominates, and the bias term is negligible by comparison. But either equation, and both, are valid for the entire stretch of time from $t = t'$ to $t = t''$. To say that the « providence factor » or « bias » is important or is negligible at such and such an epoch is therefore not a statement that is invariant with respect to the change of description,

$$t \to \bar{t} = - t . \tag{75}$$

This existence of covariance but not of invariance under time reversal is reminiscent in some respects of the alterative descriptions of approach to equilibrium developed by PRIGOGINE [206, 218, 219]; but the statement of boundary conditions at both ends of time is unique to « double-ended statistics ».

It would be possible to go to the next step beyond the continuum description of eqs. (64) and (69) and deal with fluctuations about the continuum description. Thus the number $n(t)$ dealt with so far does not refer to any individual history. Rather it is the average over many acceptable histories, not 192 histories, not $192 \cdot 10^3$ histories, but $192 \cdot 10^p$ histories, where the power p can be made large enough to guarantee approach to a continuum description to any preassigned degree of precision. When we turn to the characterization of individual histories in all their fluctuations about the continuum, the relevant quantity is the probability, w_n, that any given surplus of balls, n, will be found in the left-hand urn. This probability will vary with time according to the equation

$$\begin{aligned} \mathrm{d}w_n/\mathrm{d}t = [50 + (n+1)](w_{n+1}/100) - w_n + \\ + [50 - (n-1)](w_{n-1}/100) + \text{« (bias term)}_n \text{»} . \end{aligned} \tag{76}$$

The first three terms follow from the elementary probabilities of eq. (69). They will suffice to account for what goes on when many relaxation times intervene between « now » and the end. They give for the average value of the surplus on the left at any given time,

$$\bar{n} = \sum_{n=-50}^{50} n w_n , \tag{77}$$

the familiar cooling equation

$$\mathrm{d}\bar{n}/\mathrm{d}t = - \lambda \bar{n} \tag{78}$$

with $\lambda = 1/50$. However, as the end comes nearer, the fourth or bias term begins to become effective. If at this stage the number of balls on the left does not measure up to the prescribed final number, this term sees to it that in the spinning of the roulette wheel 1) all those numbers show up with greater probability which belong to balls to be moved from right to left and 2) all those numbers show up with decreased probability which represent balls to be moved from left to right. To state and derive the explicit formula for this bias term, to discuss the ostensible upper limit on the rate of change of $\bar{n}$ with time (one unit per spin of the roulette wheel) and to examine what it would mean to try to circumvent this limit by allowing negative values for jump probabilities are all interesting questions; but they deflect attention from the main point: The double-urn model of Cocke, as analyzed here, provides the simplest model that one can well imagine for what it means to speak of double-ended statistics.

Incentive though the double-urn model is for asking new questions about the Universe, it is inadequate for answering them. One shortcoming is evident from the start. The double-urn model is characterized by a single transition rate, the λ of (78). In contrast, the Universe is characterized by almost as many transition rates as there are physical processes, from elementary-particle decay rates to the rates of thermonuclear processes in stars, and from the rate of dynamical evolution in star clusters to the rate of decay of turbulence. Nowhere does this limitation of the double-urn model show more conspicuously than in the difficulties it makes for predictions about β-decay of ^{187}Re. Which is relevant, the $40 \cdot 10^9$ a half-life for expulsion of the β-particle or the 10^{-12} s time for reducing the expelled β-particle to thermal equilibrium with its surrounding? Or a complex resultant of these two and many other characteristic times? The predictions of the double-urn model, if one can call them predictions, are utterly different according as one correlates the characteristic decay constant, λ, of that model with the short time or the long time (fig. 9), let alone some unknown third «resultant time constant». In the one case the transition from exponential decay of ^{187}Re to exponential increase takes place within an extremely short interval of the turning of the tide. To hope to see any evidence of that transition today, at a time when the Universe is still expanding, would seem preposterous. However, if the long time of the β-decay itself is the relevant quantity, then the transition from fall to rise takes place gradually over the whole range from start to end (top curve in fig. 9). In this case a significant difference in the effective half-life of ^{187}Re might be expected as between today and $4.5 \cdot 10^9$ a ago, when certain stony meteorites were formed.

Consider first the customary hypothesis that the decay has been exponential ever since the time, t_{form}, of the formation of the meteorite, and has continued to have a decay rate, the $\lambda_{\text{apparent}}$ of eq. (54), equal to that found today [220],

$$N_{\text{form}}(\text{Re}) = N_{\text{now}}(\text{Re}) \exp[\lambda_{\text{ap}}(t_{\text{now}} - t_{\text{form}})] . \tag{79}$$

In this event the number of daughter ^{187}Os atoms that should have accumulated in the meteorite is

$$N_{\text{now}}(\text{Os}) = N_{\text{form}}(\text{Re}) - N_{\text{now}}(\text{Re})\,. \tag{80}$$

Thus correcting for any primordial ^{187}Os present in the relevant granules of the meteorite, or verifying that the amount of primordial ^{187}Os was negligible, we have

$$R = \frac{N_{\text{now}}(\text{daughter Os})}{N_{\text{now}}(\text{surviving Re})} = \exp[\lambda_{\text{app}}\Delta t] - 1\,, \tag{81}$$

where we use the abbreviation

$$\Delta t = t_{\text{now}} - t_{\text{form}}\,. \tag{82}$$

Now ask how the situation will differ if ultimately there is to be a turnabout in statistics, a turnabout that already produces today a premonitory effect. Adopt a simple illustrative cosmology (sect. **9**, table II), with a big bang $10\cdot10^9$ a in the past and a maximum in the expansion, or a turning of the tide, $20\cdot10^9$ a into our future. Make further the purely illustrative assumption that the number of ^{187}Re atoms in an undisturbed meteorite is symmetric in time with respect to that same time, t_{tt}, of the turning of the tide. Then we have

$$N_{\text{form}}(\text{Re}) = N_{\text{tt}}(\text{Re})\cosh\lambda(t_{\text{tt}} - t_{\text{form}})\,, \tag{83}$$

$$N_{\text{now}}(\text{Re}) = N_{\text{tt}}(\text{Re})\cosh\lambda(t_{\text{tt}} - t_{\text{now}})\,. \tag{84}$$

Neither the number of ^{187}Re atoms at turnabout, N_{tt}, nor the true transformation constant, λ, is directly observable. The observable quantities are the apparent decay rate today,

$$\lambda_{\text{app}} = \lambda\tanh\lambda(t_{\text{tt}} - t_{\text{now}})\,, \tag{85}$$

and the ratio of accumulated ^{187}Os to surviving ^{187}Re,

$$R = \frac{N_{\text{now}}(\text{daughter Os})}{N_{\text{now}}(\text{surviving Re})} = \frac{\cosh\lambda(t_{\text{tt}} - t_{\text{form}})}{\cosh\lambda(t_{\text{tt}} - t_{\text{now}})} - 1\,. \tag{86}$$

In the limit where the time of turning of the tide is sufficiently far into the future ($t_{\text{tt}} \to \infty$), then statistical turnabout is destined never to arrive, and expressions (85), (86) reduce to the familiar result (81). However, for a value of $t_{\text{tt}} - t_{\text{now}} = 20\cdot10^9$ a—a cosmologically reasonable order of magnitude—and a specimen that has been undisturbed for $t_{\text{now}} - t_{\text{form}} = 4\cdot10^9$ a since formation, the calculated accumulation of ^{187}Os (table I) is about 8 percent greater than one would have expected from the standard straightforward Rutherford-Soddy theory of radioactivity.

TABLE I. – *Calculated effect of future « turning of the tide of statistics » on amount of daughter* ^{187}Os *accumulated up to now in an ancient rock or meteorite containing* ^{187}Re (*present-day apparent decay constant* $\lambda_{app} = -dN/N\,dt = 0.693/40$ *billion years*). There is a 7.8 percent difference between the two numbers marked in the table by arrows.

Time from now to « turning of tide » $t_{tt} - t_{now}$	λ (true) required to make $T_{\frac{1}{2}}$ (apparent, today) equal $40\cdot10^9$ a	$R = \dfrac{N_{now}(\text{daughter Os})}{N_{now}(\text{surviving Re})}$ for age of meteorite, $t_{now} - t_{form}$	
		$2\cdot10^9$ a	$4\cdot10^9$ a
$5\cdot10^9$ a	$5.97\cdot10^{-11}$ a^{-1}	0.0419	0.0987
$10\cdot10^9$ a	$4.29\cdot10^{-11}$ a^{-1}	0.0384	0.0843
$20\cdot10^9$ a	$3.13\cdot10^{-11}$ a^{-1}	0.0365	→ 0.0773
$50\cdot10^9$ a	$2.18\cdot10^{-11}$ a^{-1}	0.0356	0.0731
∞ (never)	$1.73\cdot10^{-11}$ a^{-1}	0.0353	→ 0.0717

The calculated effect is so big in the case of the ^{187}Re-to-^{187}Os decay primarily because the relevant effective half-life, $40\cdot10^9$ a, is so long. For the α-decay of ^{238}U, where the apparent half-life is $4.51\cdot10^9$ a, the calculated accumulation ratio $R = N_{now}(\text{daughter } ^{234}\text{Th})/N_{now}(\text{surviving } ^{238}\text{U})$ in the same $4\cdot10^9$ a-old rock or meteorite (provided that it keeps its decay ^4He) is increased only 0.24 percent (from 0.8490 to 0.8510) by a turning of the tide that lies ahead in the future by the same $20\cdot10^9$ a. Forgetting this small correction, we can say that the ratio of daughter ^{234}Th to remaining ^{238}U tells the age of the mineral. This age once known, the past accumulation of ^{187}Os from ^{187}Re tests for a future turning of the tide.

The discussion given here for ^{187}Re ($T_{\frac{1}{2}} = 40\cdot10^9$ a) *vs.* ^{238}U ($T_{\frac{1}{2}} = 4.51\cdot10^9$ a) can be extended to other familiar long-lived radioactive substances, such as ^{40}K ($T_{\frac{1}{2}} = 1.3\cdot10^9$ a), ^{87}Rb ($T_{\frac{1}{2}} = 50\cdot10^9$ a) and ^{147}Sm ($T_{\frac{1}{2}} = 130\cdot10^9$ a).

The apparent ages of $\sim 4\cdot10^9$ a-old terrestrial rocks and meteorites, as deduced from accumulations from the radioactive decay of three substances, U, ^{40}K and ^{87}Rb, of very different apparent half-lives, have been found compatible by PEEBLES and DICKE [221]. Those ages would have been in observable discrepancy, one against the other, they conclude, if the fine-structure constant were changing at a rate more than 3 parts in 10^{13} per year. On the other hand, if we assume no change in the fine-structure constant, the same considerations will put an upper limit on the « turnabout effects » that we have been considering here. Otherwise stated, there is not the slightest evidence in the data cited sixteen years ago by PEEBLES and DICKE for anything in the way of an impeding reversal of statistics coming up at a cosmologically reasonable time in the future.

The great advances that have taken place in radiochemical age determinations in the meantime give room for a reexamination of this question. Even more needed is a consistent theory of « doubled-ended statistics » in a fully cosmological context. How can such varied physical processes as heat conduction, thermonuclear reactions, electromagnetic radiation and radioactive decay, with their very different characteristic times, couple together to give an orchestrated turning of the tide? Until one has an answer to this question of theory, one will not really understand the first thing about what it means observationally to test for a « turning of the tide ».

It is conceivable that one will someday understand the origin of initial-value data so well that one can say that statistics of necessity always runs in one direction. Today we are not in that happy situation. Therefore at the moment it cannot be excluded that statistical turnaround occurs. If so, and if it can be detected, it will at one stroke 1) give a cosmological foundation for statistical mechanics, 2) tell the scale of time from big bang to big stop and thus 3) provide evidence that the Universe is closed.

8. – Memory and the arrow of time.

It's a poor memory that only works backwards.

White Queen to Alice [222].

« If physics is four-dimensional, and if past, present and future are all laid out shiningly in one vast space-time diagram, why is there any 'now' in our apprehension of physics? Nothing has done more to suggest to some of us a way out of this mystery than some comments made in conversation by Hugh EVERETT. He compares the brain of the observer with a servomechanism, or—if I may go beyond EVERETT in explicitness—the computer of an aircraft gun. The radar unit mounted on the gun carriage sights on the enemy plane (fig. 10). Minute by minute it feeds information about the position of that plane into the computer. From this information the computer extrapolates the future position of the plane. It then fires a shell to intercept that plane an appropriate number of minutes later. The computer thus carries within it information about a few minutes of past history—and also information about a few minutes of forthcoming history. »

« It would be possible for the computer to remember more, perhaps the position of the enemy plane yesterday. But that outdated information would be of no use in the present crisis. Remembering it would only impose a more complicated burden on the electronics and increase the weight to be hauled along as the gun is moved from site to site. Similarly, the computer can be forced to extrapolate the flight of the enemy plane over a much greater reach of time, even to this hour tomorrow. However, that prediction would obviously have no value whatsoever. A few minutes of the future, like a few minutes of the past, are all that the computer memory will carry. The memory span

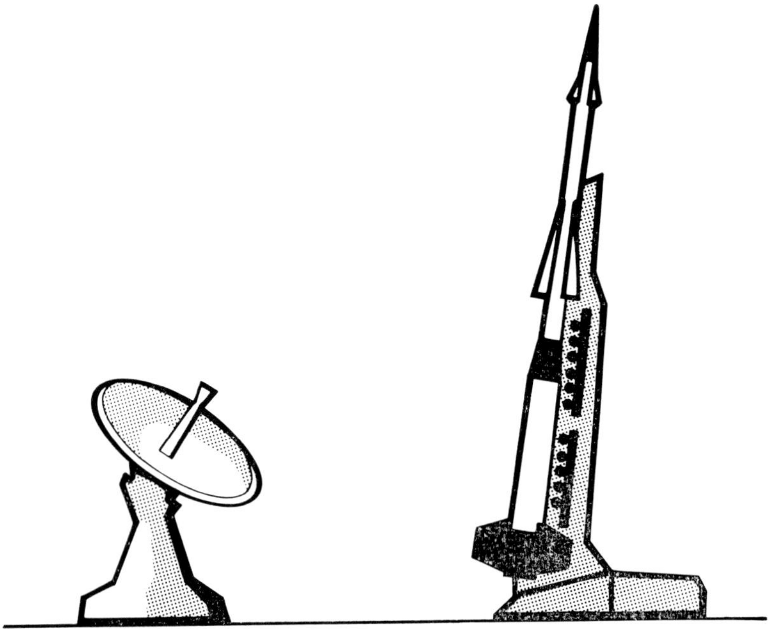

Fig. 10. – Role of the memory in linking observation with action symbolically represented. The radar has followed the enemy plane for some minutes past. Out of the electronic memory of this past a computer program projects the motion of the plane into the future and gives orders when and where to fire to intercept it.

can be no wider if the antiaircraft gun is to be as light and simple as possible. Otherwise it could not stand up in the competition with rival devices. Thus the struggle for survival trims the memory down to 'now'. This 'now' is remarkable. On it are vividly engraved not only a few minutes of the past, but also a few minutes of the future. Moreover, this 'memory' (or, more precisely, anticipation) of the immediate future is green, whereas the memory of yesterday has altogether withered away. So in the human species the struggle for survival—Everett's analogy would suggest—has built into our minds a type of 'now' in which the old past is remembered less well than the immediate future.

« Can one trace out Everett's 'servomechanism explanation of *now*' in quantitative—and even quantum-mechanical—detail, on the basis of one or another simple model? Of course, devices of the feed-back type have been studied quite thoroughly, but never from exactly the point of view of interest here. We all know that when we try to describe the behavior of such devices, we use the ideas of 'purpose', 'planning ahead', and so on—teleological ideas, all of which form a part of our consciousness. But to fit a description of such a system, with its resistances and 'dash pots', into the Hamiltonian formulation of the kind we now require is an 'analysis that has not yet been undertaken'. »

(The foregoing quoted from ref. [223]; for investigations on the mechanism of memory see for example ref. [54-56].)

9. – The gates of time [96].

... every substance ... can only begin by creation and end only by annihilation.

G. W. LEIBNIZ [224]

Given memory, uncover the machinery of memory: that was the challenge of sect. **8**. In **9** the concern is differerent. Given a machinery of memory—dynamics—explain how that machinery can ever stop remembering.

Not the slightest warrant does Einstein's equation give for thinking there can be any such thing as a « before » before the big bang or an « after » after the big crunch or after the collapse of a star to a black hole. These three processes mark three « gates of time ».

For time to come to an end is to say that time is not an ultimate category in the description of Nature. Therefore a deeper description of Nature must transcend the category of time: this is the conclusion suggested by a review of available evidence on cosmology, theoretical and observational; this is the theme of this final « frontier of time ».

The characteristic feature of a gate of time is collapse to a singularity, not only for matter but also for the space geometry that envelops this matter. Moreover, at a singularity Einstein's field equation loses its applicability. If the mathematics fails at the singularity, how can one argue consistently about the physics at the singularity? How then can there be any foundation for believing that time ends at a gate of time? The point is simple. Time does not today stand in splendid isolation, a concept with an independent existence of its own, free of entangling alliances with the rest of physics. General relativity has subdued the concept time to membership in a larger kingdom: space-time. There is nothing that we know about time, there is nothing we do with the concept of time, there is no meaningful attribute of time that is not subsumed, defined and given meaning through Einstein's 1915 and still standard geometrodynamics. Equations stop and time go on? That might once have seemed conceivable. Today it is not. Time has been robbed of the power to go off on a voyage of its own. There is no time today except the time of space-time. Where the one stops, so does the other. Time before space-time? That is a question, a proposal, a commentary to which one does not even know how to give the smallest shred of meaning.

Story though this is in brief of the gates of time, it is a story that can and must receive expansion in the rest of this section. Six topics will come into consideration: 1) the validity of general relativity, 2) evidence for the big bang, 3) do black holes exist? 4) is the Universe closed? 5) will it collapse? 6) what happens to a black hole when the Universe collapses?

First, how certain is one that the particular description of space-time that is given by Einstein's general relativity is the most reasonable one? On that point the available evidence is summarized in ref. [73]: experimental tests in Chapt. 38; analysis of alternative theories and their difficulties, Chapt. 39; solar-system experiments, Chapt. 40; gravitational waves and possibilities of detecting them to get new tests of relativity, Chapt. 35-37; cosmology and its relevance to general relativity, Chapt. 27-31; gravitational collapse and the theory of the black hole, Chapt. 32-34.

In brief « no theory more resembles Maxwell's electrodynamics in its simplicity, beauty and scope than Einstein's geometrodynamics. Few principles in physics are more firmly established than those on which it rests: the local validity of special relativity, the equivalence principle, the conservation of momentum and energy and the prevalence of second-order field equations throughout physics. Those principles and the demand for no 'extraneous fields' (*e.g.*, Dicke's scalar field) and 'no prior geometry' lead to the conclusion that the geometry of space-time must be Riemannian and the geometrodynamic law must be Einstein's. »

« To say that the geometry is Riemannian is to say that the interval between any two nearby events C and D, anywhere in space-time, stated in terms of the interval AB between two nearby fiducial events, at quite another point in space-time, has a value CD/AB independent of the route of intercomparison. There are a thousand routes. By this hydraheaded prediction, Einstein's theory thus exposes itself to destruction in a thousand ways. »

« Geometrodynamics lends itself to being disproven in other ways as well. The geometry has no option about the control it exerts on the dynamics of particles and fields. The theory makes predictions about the equilibrium configurations and pulsations of compact stars. It gives formulae for the deceleration of the expansion of the Universe, for the density of mass-energy and for the magnifying power of the curvature of space, the tests of which are not far off. It predicts gravitational collapse and the existence of black holes, and a wealth of physics associated with these objects. It predicts gravitational waves. In the appropriate approximation, it encompasses all the well-tested predictions of the Newtonian theory of gravity for the dynamics of the solar system, and predicts testable post-Newtonian corrections besides, including several already verified effects. »

« No inconsistency of principle has ever been found in Einstein's geometric theory of gravity. No purported observational evidence against the theory has ever stood the test of time. No other acceptable account of physics of comparable simplicity and scope has ever been put forward » [73].

Second, how certain are we of the initial gate of time, the big bang? No one has found any way to escape the big bang acceptable to even 10 % of the community of physics and astrophysics. The reason is simple. There is too much evidence that is correlated by the concept of a big bang that has not been brought into line by any other reasonable proposal: the Hubble expansion [225]; the primordial cosmic fireball radiation [226-229]; the time scale of the astrophysical evolution of stars [230] and star clusters [231]; and the physical conditions and the time required for the formation of the elements [232, 233].

Third, ever so much nearer in time to today than big bang or big crunch is the black hole, the second gate of time; but how certain are we that there is even one genuine black hole anywhere in the Universe? We can agree with the words of Laplace in 1795. It « would not, in consequence of its attraction, allow any of its rays to arrive at us; it is, therefore, possible that the largest luminous bodies in the Universe may, through this cause, be invisible » [234]. We can recognize that a neutron star, member of a double-star system, can receive almost unlimited amounts of matter via stellar wind from an appropriate companion in a double-star system [235, 236]. We can note the absolute inescapability, in theory, of collapse for such a system when it attains more than a critical mass, less than three solar masses according to the best available estimate [237, 238]. But has collapse to a black hole actually happened anywhere?

There is no absolutely compelling evidence for a black hole today. However, no one sees any other reasonable way to account for the unusual properties of the compact X-ray object Cyg X-1 discovered by R. GIACCONI and his collaborators in 1973 [239].

« The optical component of the pair moves back and forth a distance of $5.2 \cdot 10^6$ km in the line of sight with a 5.6 day period. Its mass, from two lines of evidence (spectral character and absolute luminosity), is concluded to be of the order of 25 solar masses. The invisible component, in order to swing by its gravitational pull so big a visible mass back and forth so great a distance and so quickly has to be of the order of 10 solar masses, and certainly greater than 5 $M_\odot$, one reasons. An ordinary star of this mass would be quite visible in the optical, quite invisible in the X-ray spectrum. This is not an ordinary star. Moreover, it is too heavy to be a white dwarf or a neutron star. No one sees any natural and reasonable interpretation for it except as a black hole. [Of course the] X-radiation does not come out of a black hole. It comes out of gas on its way towards the black hole from the normal star. Gas is drawn in towards the compact component by its powerful gravitational attraction.

In the ensuing 'traffic jam' it is compressed and heated ... to temperatures so high that the gas cannot avoid emitting X-rays before it reaches the horizon of the black hole » [240].

No hope to make compelling identification of a black hole is today the focus of more numerous and more active investigations than the signature of such an object: that combination of fluctuations in time and spectral characteristics of X-ray emission which will divide putative black holes unambiguously into true and counterfeit.

In parallel with the search for black holes of few solar masses goes the search for black holes of a million solar masses or more. The characteristic distance associated by general relativity with an object of mass m is its Schwarzschild « radius » $2Gm/c^2$, which amounts to 29.4 km for a black hole of 10 solar masses, but for such an object with $10^6\ M_\odot$ amounts to $2.94 \cdot 10^6$ km or about 0.02 times the distance from the Earth to the Sun (equals 0.02 A.U.).

Oort [241] gives evidence that it is reasonable to think of the center of the Milky Way containing a black hole of mass $\sim 4 \cdot 10^6 M_\odot$. The relevant region is too obscured by intervening dust to be seeable in the visible spectrum, but it can and has been investigated via radio waves and infrared. Stars are unmistakably present at distances of 100 A.U. from the center. Moreover their Doppler shifts can be measured. From the velocity and distance from the center one can deduce the amount of mass sufficient to curve such rapidly moving objects into orbits so great, $\sim 6 \cdot 10^6 M_\odot$. On the other hand, from the luminosity in this region one concludes that the amount of the mass in the form of stars may be only $\sim 2 \cdot 10^6 M_\odot$. Oort tentatively attributes the difference, $m \sim 4 \cdot 10^6 M_\odot$, to a single black hole. His original paper has to be read for a careful statement of caveats and consequences.

Tentative evidence for a still more massive black hole, $m \sim 5 \cdot 10^9 M_\odot$, at the center of the galaxy M 87 has been reported still more recently [242, 243]. The evidence comes from studying the distribution in luminosity with very high resolution very close to the center of M 87. The investigators looked at « slices » of the telescopic image of the galaxy M 87 at different distances from the center. They used Doppler measurements to tell how fast the stars are moving in each slice. The stars near the center are moving much faster than one would expect, and as if orbiting around a concentrated but invisible object of mass $\sim 5 \cdot 10^9 M_\odot$.

It is not possible to say that the present evidence incontrovertibly establishes the existence of black holes. However, the evidence is sufficiently impressive to make one comfortable about accepting two very general considerations: the possibility of such objects is an inescapable consequence of general relativity; and there are several very plausible astrophysical scenarios, the inevitable outcome of which is the formation of a black hole.

Fourth, any evidence that the Universe will some day collapse up against the third gate of time, dealing as it does with the future, necessarily contains a strong component of theory. Of this the most important ingredient is closure of the Universe. « Is the Universe open or closed? On no central issue of cosmology is there greater divergence of evidence today. Einstein's philosophical arguments speak for closure. [So does an appreciable body of physical evidence.] An appreciable body of astrophysical evidence speaks against it. » « To determine the so-called deceleration parameter q_0

$$q_0 = -\frac{\mathrm{d}^2(\text{radius of Universe})/\mathrm{d}t^2}{\text{radius}}\left[\frac{\text{radius}}{\mathrm{d}(\text{radius})/\mathrm{d}t}\right]^2 \tag{87}$$

from source counts is the goal of some of the greatest and most skilled observers of our times. This important measurement nevertheless requires such care in interpretation, demands so many corrections, and is afflicted with such uncertainties that the final number still today leaves the door open to either cosmology [244]. »

« The quickest way to see that the expansion may be slowing down is still the most elementary. One has only to compare the actual time back to the start of the expansion, a time of the order or $10 \cdot 10^9$ years, as judged from the rate of evolution of stars and clusters of stars, with the apparent, or extrapolated or Hubble time of $\sim 20 \cdot 10^9$ years. This is the time it would have taken galaxies to get to their present separations from us, moving with their present separation velocities, with no allowance for the greater velocity in times past. Of course, considerable uncertainties attend both numbers, uncertainties of the order of 30 percent or, conceivably, even more. Even so, it is difficult to find evidence more impressive anywhere else in cosmology for the predicted slowing down of the expansion. »

« If to fix ideas we take the two numbers, $10 \cdot 10^9$ years and $20 \cdot 10^9$ years, as 100 percent accurate and assume a homogeneous isotropic spherical universe and neglect the pressure and energy content of radiation in comparison to the mass-energy of inchoate material ('dust'), then Einstein's theory straightforwardly gives all the other illustrative numbers of table II. The 30-fold discrepancy between the density of the Universe today as called for by these calculations and the density estimated by Oort [245] gives rise to the well-known 'mystery of the missing mass' .... Of all the evidence for a low density cited by Gott, Gunn, Schramm and Tinsley [246] and by Gunn and Oke [247], none is more impressive than the abundance of primordial deuterium. The sensitivity of the deuterium abundance to density arises ... from the dependence of the expansion rate on density and from the fact that only a few minutes are required for primordial neutrons to decay to protons. Unhappily, less satisfactory than this theoretical side of the study is the observational evidence. Determinations of deuterium abundance are made by looking at the absorption of light in interstellar space on its way from a star to the telescope. Only a few such determinations have been made. No one knows how representative

TABLE II. – *Major features of the Universe according to Einstein's theory, as normalized by two key astrophysical data, each believed uncertain by an amount of the order of* 30%: 1) *the actual time,* $\sim 10 \cdot 10^9$ y, *back to the start of the expansion, as determined from the evolution of the stars and the elements, and* 2) *the «Hubble time», or time linearly extrapolated back to the start of the expansion,* $\sim 20 \cdot 10^9$ y, *that is, the time needed for galaxies to reach their present distances if they had always been receding from us with their present velocities* (*adapted from ref.* [73]).

Illustrative values all derived from	
Time from start to now	$10 \cdot 10^9$ y
Hubble time now	$20 \cdot 10^9$ y

Hubble expansion rate now	49.0 (km/s)/megaparsec
Rate of increase of radius now	0.66 ly/y
Radius now	$13.19 \cdot 10^9$ ly
Radius at maximum	$18.94 \cdot 10^9$ ly
Time, start to end	$59.52 \cdot 10^9$ ly
Density now	$14.8 \cdot 10^{-30}$ g/cm^3
Amount of matter	$5.68 \cdot 10^{56}$ g
Equivalent number of baryons	$3.39 \cdot 10^{80}$

are the samples of gas intervening nor how much they have been altered between primordial times and today by cosmic-ray impacts and contaminated by ejecta from stars and supernovae. New light on missing mass comes from the recent work of Ostriker and Peebles [248] and Ostriker, Peebles and Yahil [249]. They give arguments from [the gravitational theory of] galactic stability that the mass of the typical galaxy must be of the order of 3 to 20 times as great as one has previously estimated. They give reasons to believe that this matter is in the form of stars of modest mass and very low luminosity. Happily for the subject, the direct observational search for this 'halo' is now underway» [250].

Quite another way to get at the effective overall density of matter in the Universe has been developed by PEEBLES and his associates [251]. The focus of attention in this work is galaxy clustering and the correlation in space between galaxies.

What comes into play here is the force of gravitation, which one understands, and the density, which one wants to understand. Negligible by comparison are other factors such as radiation pressure, degree of ionization, opacity and nuclear reactions, important though they are in the internal machinery of individual stars and galaxies. This enormous simplification in the analysis opens the door to meaningful statistical analysis of the correlation in position between galaxies and its change in time. DAVIS, GROTH and PEEBLES [251] find that the logarithm of an accurately defined correlation function, plotted as a function of the logarithm of the angular separation between galaxy and galaxy, shows a sharp break in slope at a separation, $r \sim 9\,h^{-1}\,\mathrm{Mpc} = 2.8 \cdot 10^{25} h^{-1}$ cm. Here h is the ratio of the actual Hubble expansion rate today, whatever measurements of high precision may someday disclose it to be, to the working figure of 55 km $\mathrm{s}^{-1}\,\mathrm{Mpc}^{-1}$, adopted for convention's sake. This behavior is reproduced by gravitation theory when PEEBLES and his collaborators assume a simple power law spectrum of initial perturbations and density in the early universe and when they assume in addition that the actual magnitude of the density today is that called for by the condition of closure. In contrast when a substantially lower density is assumed the calculations based on gravitation theory fail to produce the observed break in the distribution function. They conclude, « the analysis presented here yielding $\Omega \gtrsim 0.3$ [a density greater than about 30 % of the requisite amount] conflicts with arguments based on other lines of evidence that have been taken to show $\Omega \lesssim 0.1$ [that the density is less than about 10 % of the requisite amount] (*e.g.*, GOTT *et al.* [246]). Our approach will require considerable further work before it can offer a definitive constraint on the cosmology. On the other hand, we consider that the same applies to the other methods of estimating Ω, so this discrepancy is an interesting indication that something is not well understood but, at the moment, it is hardly a serious problem for the gravitational instability picture. » (End of quote from ref. [251]; following is completion of quote from ref. [250].) « It is difficult to name any single issue in all of astrophysics which draws together a wider variety of important investigations than those going on today concerning in one way or another the mystery of the missing mass. »

« It has often been suggested that one should make a direct geometrical determination of the curvature of space in the large. In this way, the hope has been expressed, one could find out whether the Universe is closed or open even prior to a reliable determination of the average mass density of the Universe. More than one calculation has been made and reported [252] of the apparent angular diameter of an object of standard dimensions (if there be any such) as a function of distance (as defined by red-shift). In Euclidean space, a 'standard' object has an apparent angular diameter which decreases in inverse proportion to distance. However, when the object is far enough away in an ideal spherical space, it is magnified by a kind of lens effect. Then the apparent angular diameter, rather than decreasing, increases with distance. Moreover, the double radio sources associated with quasi-stellar objects offer a conspicuous 'ruler'. If anything, the length of this 'ruler' will be shortened in early double radio sources as compared to more recent ones by the greater density at early times of the matter through which the 'twin exhausts' [253, 254] have to plough their way. Thus if double radio sources of a sufficiently great red-shift were to begin to show an increase in apparent angular diameter, one could hardly do anything but regard this effect as evidence for the predicted lens effect. »

« A closer consideration shows that the situation is by no means as simple as would be indicated by these elementary considerations. It was already pointed out by ZEL'DOVICH [255] and by DASHEVSKY and ZEL'DOVICH [256] (references to this and the subsequent literature in PRESS and GUNN [257]) that the clustering of matter into galaxies, deviation from uniformity unimportant for the question of openness or closure, is vitally important for the focusing process. A spray of light rays that starts at a point, and spreads out as it goes, *continues* to spread out as it travels through matter-free interstellar space, even though the Universe itself is contracting. Nothing like the elementary focusing effect takes place. ... R. C. ROEDER [258] [stresses] the difficulties posed by this circumstance for any proposed cosmological test of closure, via measurement of apparent angular diameters as a function of red-shift. However, if one hope fades another brightens. PRESS and GUNN [257] show that [condensed objects present to] a cosmologically significant density [have a] high probability to cause a distant point source to be gravitationally imaged into two roughly equal images—an effect with testable consequences. » In spite of these difficulties hopes remain very much alive that someday an astrophysical means will be found to determine the large-scale curvature of space. Among these hopes the conceivable anomalies in radioactive decay rates cited in **7** may also be mentioned. Much astrophysical work, and justified astrophysical work, is underway to get a 'yes', 'no' answer to the simple question: Is the Universe closed? However much this issue belongs to science and however important general relativity is in dealing with it, one cannot forget that this science and this tool took their birth in philosophy. Therefore, it would be unbalanced not to quote Einstein's own considerations about closure, 'Thus we may present the following arguments against the conception of a space-infinite, and for the conception of a space-bounded, universe: 1) From the standpoint of the theory of relativity, the condition for a closed surface is very much simpler than the corresponding boundary condition at infinity of the quasi-Euclidean structure of the Universe. 2) The idea that MACH expressed, that inertia depends upon the mutual action of bodies, is contained, to a first approximation, in the equations of the theory of relativity; .... But this idea of Mach's corresponds only to a finite universe, bounded in space, and not to a quasi-Euclidean infinite universe' [259]. In another place EINSTEIN [260] states, 'In my opinion the general theory of relativity can only solve this problem statisfactorily if it regards the world as spatially self-enclosed'. »

In our own time a fresh consideration argues for closure: the difficulty of any alternative. (The following is quoted from ref. [250].) « [T]he 'initial-value data' are essential in formulating what general relativity is all about. There are alternatives to closure as part of the formulation of the initial-value data but no alternative so simple as closure. It is one alternative to postulate asymptotic flatness at infinity. It is another alternative to postulate more particularistic data on some closed 2-surface that bounds the 3-geometry embraced in the 'initial-value problem'. What kind of data should be given on such a 2-surface? Mathematical tools we have on hand to try to answer such a question, but no slightest hint of any physical consideration that would make this a reasonable route to follow. And asymptotic flatness (see, for example, the 'hierarchical cosmology' of Alfvén and Klein [261] and De Vaucouleurs [262]) makes double difficulties. First, it takes the geometry of faraway space out of physics and makes it part of theology, to be discovered by reading Euclid's bible. It puts us back to the days before RIEMANN, days when, as EINSTEIN [263] puts it, '... space was still, for them [physicists], a rigid, homogeneous something, susceptible of no change or conditions. Only the genius of RIEMANN, solitary and uncomprehended, had already won its way by the middle of the last century to a new conception of space, in which space was deprived of its rigidity, and in which its power to take part in physical events was recognized as possible'. »

« Why accept this advance for near space and undo it for faraway space? Moreover, 'asymptotic flatness' leaves one lost. How can anyone even define the idea of asymptotic flatness? According to the most elementary considerations of quantum theory there is no such thing as *the* geometry of space. Geometry is not deterministic, it is probabilistic. There is a probability amplitude $\psi({}^{(3)}\mathscr{G})$ for this, that, and the other 3-geometry that differs from the first by an amount of the order $\Delta g \sim L^*/L$ in a region of order L. Thus, no matter how 'far away' one goes, one can never arrive at a place where the fluctuations have less than standard strength. Difficult as it is under these circumstances to define 'far away', it is even more difficult to see where else one can turn for a satisfactorily sharp boundary condition compatible with quantum fluctuations, except to *closure.* »

Fifth: Will there be a big crunch? No factor bears so directly on this point as the question of closure. As simplest illustration, the distinction may be recalled between the Friedmann open model universe with the metric

$$ds^2 = -\,dt^2 + a^2[(d\chi)^2 + (\sinh\chi)^2(d\theta^2 + \sin^2\theta\, d\varphi^2)] \tag{88}$$

and the closed universe with the metric

$$ds^2 = -\,dt^2 + a^2[(d\chi)^2 + (\sin^2\chi)(d\theta^2 + \sin^2\theta\, d\varphi^2)]\,. \tag{89}$$

In one case the scale factor, a, and the time, t, are connected with each other by the relation

$$\begin{cases} a = (a_0/2)(\cosh\eta - 1)\,, \\ t = (a_0/2)(\sinh\eta - \eta)\,, \end{cases} \tag{90}$$

and both are ever growing quantities. In the case of the closed Friedmann model universe, things come to an end after a finite time and at that third gate of time the radius itself falls to zero,

$$(91)\qquad \begin{cases} a = (a_0/2)(1 - \cos\eta)\,, \\ t = (a_0/2)(\eta - \sin\eta)\,. \end{cases}$$

However, when the 3-sphere model universe is replaced by a 3-torus universe of repetition length $L(t)$, then the story is quite different. The dimension, $L(t)$, following the big bang increases forever [264]. This type of closed space has not been explored enough to know what its difficulties are. In default of the deeper analysis that is required we shall exclude it from attention. When we speak of a closed model universe, we shall mean a geometry that has in the large the qualitative character of a 3-sphere, however much it may be pocked in the small with multiple connectedness: wormholes or handles.

A model universe that has the topology of a 3-sphere, that obeys Einstein's geometrodynamic law, and that contains a nowhere negative density of mass-energy, almost inevitably develops a singularity according to reasoning traced out in successively greater detail by TOLMAN [265], AVEZ [266], GEROCH [267], HAWKING and PENROSE [268] and HAWKING and ELLIS [269]. Only in a set of cases of measure zero does the system escape big bang or big crunch or both, it is widely believed.

An illustration is provided by the Taub model universe [270]. This model universe is of exceptional theoretical interest and simple in this respect, that it is curved up into closure neither by matter nor by electromagnetic radiation, but by gravitational radiation alone, and this of the longest wavelength that will fit into a closed universe [217, 271]. Despite the fact that the volume of this system is zero at a time, t_1, and at another time, t_2, with one maximum in between, the geometry does not become singular at either t_1 or t_2. Instead it transforms itself smoothly and continuously [272] into another topology, one where there exist closed timelike lines. Such a space-time contradicts every normal idea of causality. In it past and future are inextricably confused.

ELLIS and KING have given other examples of such « whimper » model universes, that just barely escape the singularity of big bang or big crunch [273]. The transition from closed to open geometry had been investigated in detail only in the case of Taub universes [272]. It is found that the « continuity is achieved only at the cost of having certain classes of world-lines spiral round the Universe in the final states of its collapse to tighter and tighter packing. Thus the presence of the slightest 'real matter' builds up an ever-increasing density (in this connection, see also [274]). As it goes to infinity, this density destroys the relevance of the model with which one started. One returns to something closer to a Friedmann cosmology with a Friedmann singularity » [250]. Near both gates of time, it is reasoned in several interesting papers by members of the Moscow group [154, 275-282], the singularity in the generic case is characterized by a general « mixmaster oscillation » of the local geometry [283] with the phase, amplitude and orientation of the principal axes of this deformation of the geometry varying from point to point (see also [284]). If this is the characteristic behavior of the generic solution, then, it is suggested by BELINSKY, KHALATNIKOV and COLLINS [285], the « whimper » solutions form a set of measure zero among these generic solutions. (See also [286].) If these tentative conclusions are sustained by more detailed mathematical analysis, then one would seem justified to say that the big crunch is « almost inevitable ».

Sixth, the singularity of the black hole is not separate and distinct from the singularity of the big crunch in a model universe that collapses to a singularity. The icicle that hangs from the ceiling of a cave of ice is not separate and distinct from the ceiling. The one is part and parcel of the other. How best to bring into mathematical evidence this point, first suggested by PENROSE [287] (see also ref. [288]), is a point under active investigation. One proposal has it that space-time is best foliated by a 1-parameter family of spacelike hypersurfaces distinguished one from another by the value of the trace of the extrinsic curvature, constant on each hypersurface but differing from hypersurface to hypersurface. As successive members of this family are examined, each higher within the ice cave than the one before, none will touch the hanging icicle. Instead each will envelop it more closely than the one before, after the manner of a glove. The value of the trace, apart from a numerical constant, is identical with the so-called York time [289]. With respect to growing values of this time parameter one expects to see the mixmaster oscillations of Belinsky, Khalatnikov and Lifshitz played out. The black hole shows itself up, not as something new and strange, but as a special case of the mixmaster oscillations. Ahead though this description is of what the mathematics of the moment allows one to say, and afflicted though it is with some uncertainty as to the appropriate scheme of foliation, it nevertheless puts together the major features of the best thinking of today as to how the generic singularity is approached. When there are several black holes they coalesce: but the singularity of the individual black holes and of their coalescence is still described in terms of deformation oscillations of the geometry as the foliation parameter rises without limit.

This review of the three gates of time shows in what sense and with what caveats and with what degree of certainty one can say that the Universe begins with a singularity and ends with a singularity. Little escape is evident from these words: there is no « before » before the big bang and no « after » after the big crunch. Time ends with space-time. The Universe does not endure from everlasting to everlasting. Everything came from « nothing ». Of all the frontiers of time examined here, that one would seem to be most pregnant with the future.

REFERENCES

[1] C. LYELL: *Principles of Geology, being an attempt to explain the former changes of the earth's surfaces, by reference to cause now in operation*, Vol. II (London, 1830-1833).

[2] C. DARWIN: as quoted in H. WARD: *Darwin: The Man and His Warfare* (Indianapolis, Ind., 1927), p. 297.

[3] J. A. WHEELER: *From relativity to mutability*, in ref. [4], pp. 202-247.

[4] J. MEHRA, editor: *The Physicists' Conception of Nature* (Dordrecht, 1973).

[5] M. EIGEN: *The origin of biological information*, in ref. [4], pp. 594-632.

[6] M. EIGEN and R. WINKLER: *Das Spiel: Naturgesetze steuern den Zufall* (München, 1975).

[7] P. S. LAPLACE: *Essai philosophique sur les probabilités* (Paris, 1814), 2nd edition, pp. 3-4.

[8] N. BOHR: *Das Quantenpostulat und die neuere Entwicklung der Atomistik, Naturwiss.*, **16**, 245-257 (1928).

[9] W. HEISENBERG: *Über den auschaulichen Inhalt der quantentheoretischen Kinematik und Mechanik, Zeits. Phys.*, **43**, 172-198 (1927).

[10] J. BURCKHARDT: *Die Kultur der Renaissance in Italien* (Leipzig, 1860); authorized English translation from the 15th edition, by S. G. C. MIDDLEMORE: *The Civilization of the Renaissance in Italy* (New York, N. Y., 1929).

[11] P. GAY: *Style in History* (New York, N. Y., 1974).

[12] H. EVERETT III: *The theory of the universal wave function*, doctoral dissertation, Princeton University (1957); published in abbreviated form in ref. [13]; published in full for the first time in ref. [14].

[13] H. EVERETT III: *«Relative state» formulation of quantum mechanics, Rev. Mod. Phys.*, **29**, 454-462 (1957); reprinted in ref. [14].

[14] B. S. DEWITT and N. GRAHAM, editors: *The Many-Worlds Interpretation of Quantum Mechanics* (Princeton, N. J., 1973).

[15] J. VON NEUMANN: *Mathematischen Grundlagen der Quantenmechanik* (Berlin, 1932); translated as *Mathematical Foundations of Quantum Mechanics* (Princeton, N. J., 1955).

[16] J. A. WHEELER: *Assessment of Everett's «relative state» formulation of quantum mechanics, Rev. Mod. Phys.*, **29**, 463-465 (1974); reprinted in ref. [14].

[17] E. P. WIGNER: *Epistemological perspective on quantum theory*, in *Contemporary Research in the Foundations and Philosophy of Quantum Theory*, edited by C. A. HOOKER (Dordrecht, 1973), pp. 369-385.

[18] E. P. WIGNER: *Remarks on the mind-body question*, in *The Scientist Speculates*, edited by I. J. GOOD (London, 1962), pp. 284-302.

[19] E. P. WIGNER: *Are we machines?, Proc. Amer. Phil. Soc.*, **113**, 95-101 (1969).

[20] E. P. WIGNER: *Physics and the explanation of life, Found. Phys.*, **1**, 35-45 (1970).

[21] C. F. VON WEIZSÄCKER: *Classical and quantum decriptions*, in ref. [4], pp. 635-667.

[22] J. A. WHEELER: *Include the observer in the wave function?*, in *Quantum Mechanics, a Half Century Later*, edited by J. LEITE LOPES and M. PATY (Dordrecht, 1977).

[23] N. BOHR: *Atomic Theory and the Description of Nature* (Cambridge, 1934).

[24] N. BOHR: *Atomic Physics and Human Knowledge* (New York, N. Y., 1958).

[25] N. BOHR: *Essays 1958-1962 on Atomic Physics and Human Knovledge* (New York, N. Y., 1963).

[26] A. EINSTEIN: *Autobiological Notes*, in ref. [27], quotation from p. 81.

[27] P. A. SCHILPP, editor: *Albert Einstein: Philosopher-Scientist* (Evanston, Ill., 1949), and subsequent paperback editions elsewhere.
[28] N. BOHR: *Can quantum-mechanical description of physical reality be considered complete?*, *Phys. Rev.*, **48**, 696-702 (1935); the quotation comes from p. 697.
[29] T. MANN: *Freud, Goethe, Wagner* (New York, N. Y., 1937), p. 20; translated by H. T. LOWE-PORTER from *Freud und die Zukunft* (Vienna, 1936); the cited words were included in the lecture given at the 80th birthday celebration for Sigmund FREUD, 8 May 1936.
[30] C. S. PEIRCE: *The Philosophy of Peirce: Selected Writings*, edited by J. BUCHLER (London, 1940); paperback reprint, *Philosophical Writings of Peirce* (New York, N. Y., 1955); « fallibilist », p. 358.
[31] B. K. HARRISON, M. WAKANO and J. A. WHEELER: *Matter-energy at high density; end point of thermonuclear evolution*, in *La structure et l'évolution de l'univers, Onzième conseil de physique Solvay* (Bruxelles, 1958), pp. 124-146; terminology « crushing points », pp. 134-136.
[32] J. R. GOTT III, J. E. GUNN, D. N. SCHRAMM and B. M. TINSLEY: *Will the Universe expand forever*, *Sci. Amer.*, **234**, 62-79 (March 1976); terminology « big crunch », p. 69.
[33] J. A. WHEELER: *Our universe: the known and the unknown*, address before the American Association for the Advancement of Science, New York, December 29, 1967; *Amer. Scholar*, **37**, 248-274 (1968); terminology « black hole », pp. 258-262.
[34] J. A. WHEELER: *Superspace and the nature of quantum geometrodynamics*, in *Battelle Rencontres: 1967 Lectures in Mathematics and Physics*, edited by C. DEWITT and J. A. WHEELER (New York, N. Y., 1968), pp. 242-307; « no before, no after », p. 253.
[35] W. SHAKESPEARE: *The Tempest*, London, about 1610; Prospero in Act IV, Scene I, lines 148-158.
[36] J. A. WHEELER: ref. [3]; no magic particle, no magic field, p. 235.
[37] C. M. PATTON and J. A. WHEELER: *Is physics legislated by cosmogony?*, in *Quantum Gravity*, edited by C. J. ISHAM, R. PENROSE and D. W. SCIAMA (Oxford, 1975); no magic mathematics, pp. 589-591. © Oxford University Press 1975, by permission of Oxford University Press. Appreciation also expressedl to Charles PATTON for permission to quote cited passages.
[38] A. I. OPARIN, editor: *Evolutionary Biochemistry, Proceedings of the V International Congress on Biochemistry* (Moscow, 1961; London, 1963), pp. 12-51.
[39] G. W. LEIBNIZ: source of quotation not traced.
[40] J. A. WHEELER: *Genesis and Observership*, ref. [41], pp. 3-33; see p. 29.
[41] R. E. BUTTS and K. J. HINTIKKA: *Foundational Problems in the Special Sciences* (Dordrecht, 1977).
[42] N. BOHR: ref. [23]; Bohr's stick, p. 99.
[43] F. J. BELINFANTE: *Measurements and Time Reversal in Objective Quantum Theory* (Oxford, 1975); terminology « indelible », p. 39.
[44] E. P. WIGNER: *Are we machines?*, *Proc. Amer. Phil. Soc.*, **113**, 95-101 (1969); quotation from p. 97.
[45] E. P. WIGNER: *The philosophical problem*, in ref. [46], pp. 1-3; quotation from p. 3.
[46] B. D'ESPAGNAT, editor: *Foundations of Quantum Mechanics* (New York, N. Y., 1971).
[47] N. BOHR: ref. [24]; irreversible amplification, p. 88.
[48] N. BOHR: ref. [24]; closed by irreversible amplification, p. 73.
[49] N. BOHR: ref. [25]; plain language, p. 3.
[50] N. BOHR: ref. [25]; unambiguously communicable, pp. 5, 6.

[51] J. Kalckar: letter to J. A. Wheeler dated June 10, 1977. Appreciation is expressed here to Dr. J. Kalckar both for the letter and for subsequent permission to quote from it. He adds that he cannot guarantee to have correct wording; and that Bohr's reply, as so often, was a joke with a definite « point »—hence to be taken seriously, but not *quite* seriously.

[52] K. R. Popper and J. C. Eccles: *The Self and Its Brain* (Berlin, New York, N. Y., and London, 1977); see especially pp. 207-208, 438-440 and 515.

[53] J. von Neumann: *The Computer and the Brain* (New Haven, Conn., 1958); see especially pp. 60-61.

[54] L. N. Cooper: *A possible organization of animal memory and learning*, in *Nobel Symposium on the Collective Properties of Physical Systems*, edited by B. Lundqvist and S. Lundqvist (New York, N. Y., 1973), pp. 252-264.

[55] M. M. Nass and L. N. Cooper: *A theory for the development of feature-detecting cells in visual cortex*, *Biol. Cyber.*, **19**, 1-18 (1975).

[56] L. N. Cooper: *A theory for the acquisition of animal memory*, in *Lepton and Hadron Structure* (*1974 International School of Subnuclear Physics, Erice, Trapani, Sicily: July 14-31, 1974*), edited by A. Zichichi (New York, N. Y., 1975), pp. 808-839.

[57] G. E. Pugh: *On the Origin of Human Values* (New York, N. Y., 1976), chapter « Human values, free will, and the conscious mind »; preprinted in *Zygon*, **11**, 2-24 (1976).

[58] International Union of Pure and Applied Physics: *Report of the Commission on Symbols, Units and Nomenclature* (Amsterdam, 1948).

[59] J. A. Wheeler: *The « past » and the « delayed-choice » double-slit experiment*, in ref. [60].

[60] A. R. Marlow, editor: *Mathematical Foundations of Quantum Theory* (New York, N. Y., 1978).

[61] E. H. Gombrich: *Art and Illusion: A Study in the Psychology of Pictorial Representation* (Princeton, N. J., 1961), 2nd edition, revised, p. 394. Appreciation is expressed to Prof. E. H. Gombrich and Princeton University Press for permission to quote cited passages in the text.

[62] E. H. Gombrich: ref. [61], p. 273.

[63] E. H. Gombrich: ref. [61], p. 329.

[64] A. Einstein, B. Podolsky and N. Rosen: *Can quantum-mechanical description of physical reality be considered complete?*, *Phys. Rev.*, **47**, 777-780 (1935).

[65] R. M. F. Houtappel, H. Van Dam and E. P. Wigner: *The conceptual basis and use of the geometric invariance principles*, *Rev. Mod. Phys.*, **37**, 595-632 (1965); see especially §§ 4.1-4.5 on pp. 610-616.

[66] C. N. Yang and R. L. Mills: *Conservation of isotopic spin and isotopic gauge invariance*, *Phys. Rev.*, **96**, 191-195 (1954).

[67] S. Hojman, K. Kuchař and C. Teitelboim: *New approach to general relativity*, *Nature Phys. Sci.*, **245**, 97-98 (1973).

[68] S. A. Hojman, K. Kuchař and C. Teitelboim: *Geometrodynamics regained*, *Ann. of Phys.*, **76**, 88-135 (1976).

[69] C. Teitelboim: *Surface deformations, space-time structure and gauge invariance*, in *Relativity, Fields, Strings and Gravity: Proceedings of the Second Latin American Symposium on Relativity and Gravitation SILARG II* held in Caracas, December 1975, Universidad Simon Bolivar, edited by C. Aragone (Caracas, 1976).

[70] J. E. Nelson and C. Teitelboim: *Hamiltonian for the Einstein-Dirac field*, *Phys. Lett.*, **69** B, 81-84 (1977).

[71] J. A. Wheeler: ref. [3]; machinery hidden, pp. 236-240.

[72] J. A. Wheeler: *Beyond the end of time*, Marchon lecture, University of Newcastle Upon Tyne, May 18, 1971, and Nuffield lecture, Cambridge University,

July 19,1971; unpublished lectures, updated in chapter of same title in ref. [73]; see especially pp. 1206-1208.
[73] C. W. MISNER, K. S. THORNE and J. A. WHEELER: *Gravitation* (San Francisco, Cal., 1973). Appreciation is expressed to Charles MISNER, Kip THORNE and Freeman and Co. for permission to quote the cited passages in the text.
[74] D. Z. FREEMAN, P. VAN NIEUWENHUIZEN and S. FERRARA: *Progress toward a theory of supergravity*, *Phys. Rev. D*, **13**, 3214-3218 (1976).
[75] S. DESER and B. ZUMINO: *Consistent supergravity*, *Phys. Lett.*, **62** B, 335-337 (1976).
[76] C. TEITELBOIM: *Supergravity and square roots of constraints*, *Phys. Rev. Lett.*, **38**, 1106-1110 (1977).
[77] R. TABENSKY and C. TEITELBOIM: *The square root of general relativity*, *Phys. Lett.*, **69** B, 453-456 (1977).
[78] R. P. FEYNMAN: *The Principle of Least Action in Quantum Mechanics*, doctoral dissertation (Princeton University, 1942); unpublished; available from University Microfilms Inc. (Ann Arbor, Mich., 48106).
[79] R. P. FEYNMAN and A. R. HIBBS: *Quantum Mechanics and Path Integrals* (New York, N. Y., 1965).
[80] J. R. KLAUDER: *The action option and a Feynman quantization of spinor fields in terms of ordinary c-numbers*, *Ann. of Phys.*, **11**, 123-168 (1960).
[81] J. A. WHEELER: ref. [34]; space-time an approximation, pp. 252-253.
[82] C. W. MISNER: *Feynman quantization of general relativity*, *Rev. Mod. Phys.*, **29**, 497-509 (1957).
[83] L. D. FADDEEV: *Symplectic structure and quantization of the Einstein gravitation theory*, in *Actes du Congrès International des Mathématiciens, Nice, Septembre 1-10, 1970* (Paris, 1971), pp. 35-40.
[84] G. W. GIBBONS and S. W. HAWKING: *Action integrals and partition functions in quantum gravity*, *Phys. Rev. D*, **15**, 2752-2756 (1977).
[85] Ref. [73]: pp. 733-742 (Friedmann); pp. 1274-1275 (Schwarzschild); p. 940 (Taub).
[86] Ref. [73]: pp. 419-423, 713 ff. and 1181.
[87] Ref. [73]: pp. 511-516 and 520.
[88] Ref. [73]: pp. 493-498 and Chap. 21.
[89] J. A. WHEELER: ref. [34], p. 253.
[90] M. PLANCK: *Über irreversible Strahlungsvorgänge*, *Sitzber. Deut. Akad. Wiss. Berlin, Kl. Math.-Phys. Tech.*, 440-480 (1899).
[91] J. A. WHEELER: *On the nature of quantum geometrodynamics*, *Ann. of Phys.*, **2**, 604-614 (1957).
[92] Ref. [73]: sect. **1** and **2** of Chap. 21.
[93] J. A. WHEELER and R. P. FEYNMAN: *Interaction with the absorber as the mechanism of radiation*, *Rev. Mod. Phys.*, **17**, 157-181 (1945).
[94] J. A. WHEELER and R. P. FEYNMAN: *Classical electrodinamics in terms of direct interparticle action*, *Rev. Mod. Phys.*, **21**, 425-433 (1949). Appreciation is expressed here to Richard FEYNMAN and the Editor of *Reviews of Modern Physics* to quote the cited passages in the text.
[95] W. RITZ and A. EINSTEIN: *Zum gegenwartigen Stand des Strahlungsproblems*, *Phys. Zeits.*, **10**, 323-324 (1909).
[96] J. A. WHEELER: ref. [40], p. 11, terminology « gates of time ».
[97] N. BOHR: *Discussion with Einstein on epistemological problems in atomic physics*, in P. A. SCHILPP, editor: *Albert Einstein: Philosopher-Scientist* (Evanston, Ill., 1949), pp. 199-241.
[98] M. A. DUGUAY: *The ultrafast optical Kerr shutter*, in E. WOLF, editor: *Progress*

in Optics, Vol. 14 (Amsterdam, 1976), pp. 161-193, compares the new ultrafast *optical* Kerr shutter to the older and slower conventional and *electrically* operated Kerr cell.
[99] N. Bohr: ref. [97], p. 230.
[100] A. Petersen: *Quantum Mechanics and the Philosophical Tradition* (Cambridge, Mass., 1968), gives a preliminary account of the stages in the development of the concept of « phenomenon ».
[101] N. Bohr: *Atomic Physics and Human Knowledge* (New York, N. Y., 1958), p. 64.
[102] E. P. Wigner: *Epistemological perspective on quantum theory*, ref. [103], pp. 369-385.
[103] C. A. Hooker, editor: *Contemporary Research in the Foundations and Philosophy of Quantum Theory* (Dordrecht, 1973).
[104] A. Einstein, B. Podolsky and N. Rosen: *Can quantum-mechanical description of physical reality be considered complete?*, *Phys. Rev.*, **47**, 777-780 (1935). See [64].
[105] N. Bohr: ref. [28].
[106] D. Bohm: *Quantum Theory* (Englewood Cliffs, N. J., 1951), pp. 614-619.
[107] M. Born and E. Wolf: *Principles of Optics*, 4th edition (New York, N. Y., 1970).
[108] B. A. Kagali: *A note on the correlation of photon polarizations in positronium annihilation*, *Phys. Rev. D*, **17**, 3280 (1978).
[109] J. A. Wheeler: *Polyelectrons*, *Ann. New York Acad. Sci.*, **48**, 219-238 (1946). An error in one of the numbers in this paper is corrected in ref. [110].
[110] M. H. I. Pryce and J. C. Ward: *Angular correlation effects with annihilation radiation*, *Nature*, **160**, 435 (1947).
[111] H. S. Snyder, S. Pasternack and J. Hornbostel: *Angular correlation of scattered annihilation radiation*, *Phys. Rev.*, **73**, 440-448 (1948).
[112] E. Bleuler and H. L. Bradt: *Correlation between the states of polarization of the two quanta of annihilation radiation*, *Phys. Rev.*, **73**, 1398 (1948).
[113] R. C. Hanna: *Polarization of annihilation radiation*, *Nature*, **162**, 332 (1948).
[114] N. A. Vlasov and B. S. Dzeljepov: *Poljarizatzija annigiljatzionnikh gamma-kvantov*, *Dokl. Akad. Nauk SSSR*, **69**, 777-779 (1949).
[115] C. S. Wu and I. Shakhov: *The angular correlation of scattered annihilation radiation*, *Phys. Rev.*, **77**, 136 (1950).
[116] F. Hereford: *The angular correlation of photo-electrons ejected by annihilation quanta*, *Phys. Rev.*, **81**, 482 (1951), see also pp. 627-628.
[117] G. Bertolini, M. Bettoni and E. Lazzarini: *Angular correlation of scattered annihilation radiation*, *Nuovo Cimento*, **2**, 661-662 (1955).
[118] H. Langhoff: *Die Linearpolarisation der Vernichtungsstrahlung von Positronen*, *Zeits. Phys.*, **160**, 186-193 (1960).
[119] L. Kasday, J. Ullman and C. S. Wu: *The Einstein-Podolsky-Rosen argument: positron annihilation experiment*, *Bull. Amer. Phys. Soc.*, **15**, 586 (1970).
[120] L. Kasday: *Experimental test of quantum predictions for widely separated photons*, in B. d'Espagnat, editor: *Foundations of Quantum Mechanics, Proceedings of the International School of Physics « Enrico Fermi »*, Course IL (New York, N. Y., 1971), pp. 195-210.
[121] G. Faraci, D. Gutkowski, S. Notarrigo and A. R. Pennisi: *An experimental test of the EPR paradox*, *Lett. Nuovo Cimento*, **9**, 607-611 (1974).
[122] A. R. Wilson, J. Lowe and D. K. Butt: *Measurement of the relative planes of polarization of annihilation quanta as a function of separation distance*, *J. Phys. G*, **2**, 613-624 (1976).
[123] C. A. Kocher and E. D. Commins: *Polarization correlation of photons emitted in an atomic cascade*, *Phys. Rev. Lett.*, **18**, 575-577 (1967).

[124] S. J. FREEDMAN and J. F. CLAUSER: *Experimental test of local hidden-variable theories*, *Phys. Rev. Lett.*, **28**, 938-941 (1972).

[125] R. A. HOLT: *Atomic cascade experiments*, doctoral dissertation (Harvard University, Cambridge, Mass., 1973), unpublished; available from University Microfilms, Inc. (Ann Arbor, Mich. 48106).

[126] J. F. CLAUSER: *Experimental investigation of a polarization correlation anomaly*, *Phys. Rev. Lett.*, **36**, 1223-1226 (1976).

[127] S. J. FREEDMAN and R. A. HOLT: *Tests of local hidden-variable theories in atomic physics*, *Comm. Atom. Molec. Phys.*, **5**, 55-62 (1975).

[128] E. S. FRY and R. C. THOMPSON: *Experimental test of local hidden-variable theories*, *Phys. Rev. Lett.*, **37**, 465-468 (1976).

[129] G. W. LEIBNIZ: *On nature in itself; or on the force residing in created things, and their actions*, in G. M. DUNCAN: *The Philosophical Works of Leibniz*, 2nd (revised) edition (New Haven, Conn., 1908), pp. 119-134; translated from the original *De ipsa natura sive de vi insita creaturarum* of 1698.

[130] P. A. M. DIRAC, V. A. FOCK and B. PODOLSKY: *On quantum electrodynamics*, *Phys. Zeits. Sow.*, **2**, 468-479 (1932).

[131] S. TOMONAGA: *On a relativistically invariant formulation of the quantum theory of wave fields*, *Prog. Theor. Phys.*, **1**, 27-42 (1946).

[132] L. D. LANDAU and E. LIFSHITZ: *The Classical Theory of Fields*, 3rd revised English edition (Reading, Mass., and London, 1971).

[133] K. GÖDEL: *An example of a new type of cosmological solution of Einstein's field equations of gravitation*, *Rev. Mod. Phys.*, **21**, 447-450 (1949).

[134] D. HILBERT: *Die Grundlagen der Physik*, *Konigl. Gesell. Wiss. Gött.*, *Nachr.*, *Math.-Phys. Kl.*, 395-407 (1915).

[135] J. A. WHEELER: ref. [34], p. 273.

[136] C. TEITELBOIM: *How commutators of constraints reflect space-time structure*, *Ann. of Phys.*, **79**, 542-557 (1973).

[137] C. TEITELBOIM: *The Hamiltonian structure of space-time*, doctoral thesis (Princeton, 1973); available from University Microfilms, Inc. (Ann Arbor, Mich. 48106).

[138] K. KUCHAŘ: *Canonical quantization of gravity*, in W. ISRAEL, editor: *Relativity, Astrophysics and Cosmology* (Dordrecht, 1973), pp. 238-288.

[139] K. KUCHAŘ: *Geometrodynamics regained: a Lagrangian approach*, *Journ. Math. Phys.*, **15**, 708-715 (1974).

[140] G. W. LEIBNIZ: *Refutation of Spinoza*, written about 1708 as *Animadversiones ad Joh. George Wachteri librum de recondita Hebraeorum philosophia*; English translation in G. M. DUNCAN: *The Philosophical Works of Leibniz*, 2nd (revised) edition (New Haven, Conn., 1908), pp. 264-273.

[141] A. EINSTEIN: as quoted by Alyesa FORSEE in *Albert Einstein, theoretical physicist* (New York, N. Y., 1963), p. 81.

[142] B. RIEMANN: *Gesammelte Mathematische Werke*, 2nd edition, edited by H. WEBER, paperback reprint (New York, N. Y., 1953). The *Habilitationsvorlesung* of 1854 was translated into English by W. K. CLIFFORD: *Nature*, **8**, 14 (1873).

[143] A. EINSTEIN: *Essays in Science* (New York, N. Y., 1934), p. 68. Translated from *Mein Weltbild* (Amsterdam, 1933).

[144] E. CARTAN: *Sur les équations de la gravitation de Einstein*, *Journ. Math. Pures Appl.*, **1**, 141-203 (1922).

[145] E. CARTAN: *La théorie des groupes et les recherches récentes de géométrie différentielle*, in *Conference Proceedings of the International Congress of Mathematicians, Toronto, 1924*, *L'Enseign. math.*, **24**, 1-18 (1925).

[146] P. A. M. DIRAC: *Fixation of co-ordinates in the Hamiltonian theory of gravitation*, *Phys. Rev.*, **114**, 924-930 (1959).

[147] P. A. M. DIRAC: *Lectures on Quantum Mechanics*, Belfer Graduate School of Science Monograph Series Number Two (New York, N. Y., 1964).
[148] R. ARNOWITT, S. DESER and C. W. MISNER: *The dynamics of general relativity*, in L. WITTEN, editor: *Gravitation: An Introduction to Current Research* (New York, N. Y., 1962), pp. 227-265.
[149] C. W. MISNER, K. S. THORNE and J. A. WHEELER: ref. [73], pp. 10-12.
[150] J. A. WHEELER: *Superspace*, in R. P. GILBERT and R. G. NEWTON, editors: *Analytic Methods in Mathematical Physics* (New York, N. Y., 1970), pp. 335-378.
[151] A. E. FISCHER: *The theory of superspace*, in M. CARMELI, S. I. FICKLER and L. WITTEN, editors: *Relativity* (New York, N. Y., 1970), pp. 303-357.
[152] B. RIEMANN: *Gesammelte Mathematische Werke*, 2nd edition, edited by H. WEBER, paperback reprint (New York, N. Y., 1953).
[153] L. BERS: *Universal Teichmüller space*, in R. P. GILBERT and R. NEWTON, editors: *Analytic Methods in Mathematical Physics* (New York, N. Y., 1970), pp. 65-83.
[154] E. M. LIFSHITZ and I. M. KHALATNIKOV: *Investigations in relativistic cosmology*, *Adv. Phys.*, **12**, 185-249 (1963); translated from the earlier Russian version by J. L. BEEBY; results available in boiled down form in L. D. LANDAU and E. M. LIFSHITZ: ref. [132]; see especially pp. 350-355.
[155] T. REGGE: *Relativity without co-ordinates*, *Nuovo Cimento*, **19**, 558-571 (1961).
[156] C. W. MISNER, K. S. THORNE and J. A. WHEELER: ref. [73], Chap. 42.
[157] R. F. BAIERLEIN, D. H. SHARP and J. A. WHEELER: *Three-dimensional geometry as carrier of information about time*, *Phys. Rev.*, **126**, 1864-1865 (1962).
[158] H. LEUTWYLER: *Gravitational field: Equivalence of Feynman quantization and canonical quantization*, *Phys. Rev.*, **134**, B 1155-B 1182 (1964).
[159] B. DEWITT: *Quantum theory of gravity*. - I, *Phys. Rev.*, **160**, 1113-1148 (1967).
[160] B. DEWITT: *Quantum theory of gravity*. - II: *The manifestly covariant theory*, *Phys. Rev.*, **162**, 1195-1239 (1967).
[161] B. DEWITT: *Quantum theory of gravity*. - III: *Applications of the covariant theory*, *Phys. Rev.*, **162**, 1239-1256 (1967).
[162] E. P. WIGNER and H. SALECKAR: *Quantum limitations of the measurement of space-time distances*, *Phys. Rev.*, **109**, 571-577 (1958).
[163] R. F. MARZKE and J. A. WHEELER: *Gravitation as geometry*. - I: *The geometry of space-time and the geometrodynamical standard meter*, in H.-Y. CHIU and W. F. HOFFMAN, editors: *Gravitation and Relativity* (New York, N. Y., 1964), pp. 40-64.
[164] B. DEWITT: *Dynamical theory of groups and fields*, in C. DEWITT and B. DEWITT, editors: *Relativity, Groups and Topology* (New York, N. Y., 1964), pp. 587-822; see especially pp. 598-614.
[165] J. EHLERS, F. A. E. PIRANI and A. SCHILD: *The geometry of free fall and light propagation*, in L. O'RAIFEARTAIGH, editor: *General Relativity, Papers in Honor of J. L. Synge* (London, 1972), pp. 63-84.
[166] C. W. MISNER, K. S. THORNE and J. A. WHEELER: ref. [73], p. 72.
[167] L. LANDAU and R. PEIERLS: *Erweiterung des Unbestimmtheitsprinzips für die relativistische Quantentheorie*, *Zeits. Phys.*, **69**, 56-69 (1931).
[168] N. BOHR and L. ROSENFELD: *Zur Frage der Messbarkeit der electromagnetischen Feldgrössen*, *Kgl. Danske Videnskab. Selskab, Mat.-fys. Medd.*, **12**, No. 8 (1933).
[169] N. BOHR and L. ROSENFELD: *Field and charge measurements in quantum electrodynamics*, *Phys. Rev.*, **78**, 794-798 (1950).
[170] A. PERES: *On the Cauchy problem in general relativity*. - II, *Nuovo Cimento*, **26**, 53-62 (1962).
[171] U. GERLACH: *Derivation of the ten Einstein equations from the semi-classical approximation to classical geometrodynamics*, *Phys. Rev.*, **177**, 1929-1941 (1969).
[172] J. A. WHEELER: *Geometrodynamics* (New York, N. Y., 1962), pp. 71-83.

[173] K. KUCHAŘ: *Ground state functional of the linearized gravitational field, Journ. Math. Phys.*, **11**, 3322-3334 (1970).
[174] C. M. PATTON and J. A. WHEELER: ref. [37], pp. 577 ff.
[175] E. MACH: *Beiträge zur Analyse der Empfindungen* (Jena, 1886; 5th edition, 1906, entitled *Die Analyse der Empfindungen*).
[176] J. PIAGET: *The Construction of Reality in the Child,* translated from the French by M. COOK (New York, N. Y., 1954); also the summary of the lesson of this and his books *The Child's Conception of Movement and Speed* and *The Child's Conception of Time,* in J. PIAGET: *Le structuralisme* (Paris, 1968); translated by C. MASCHLER as *Structuralism* (New York, N. Y., 1971).
[177] R. J. E. CLAUSIUS: *Abhandlungen über die mechanische Wärmetheorie* (Braunschweig, 1864-1867).
[178] B. RUSSELL, as quoted by G. STEINER: *Has truth a future?—The first annual Bronowski Memorial Lecture, BBC Listener 99,* No. 2542, 42-46 (January 12, 1978).
[179] W. J. COCKE: *Statistical time symmetry and two-time boundary conditions in physics and cosmology, Phys. Rev.*, **160**, 1165-1170 (1967).
[180] V. I. ARNOLD and A. AVEZ: *Ergodic Problems of Classical Mechanics* (New York, N. Y., 1968).
[181] T. GOLD: *The arrow of time,* in *Onzième Conseil de Physique Solvay: La structure et l'évolution de l'univers* (Brussels, 1959), pp. 81-95.
[182] T. GOLD: *The arrow of time, Amer. Journ. Phys.*, **30**, 403-410 (1962).
[183] T. GOLD: *The arrow of time,* in *Recent Developments in General Relativity* (New York, N. Y., and Warsaw, 1962), pp. 225-234.
[184] H. BONDI: *Physics and cosmology, Observatory,* **82**, 133-143 (1962).
[185] P. G. BERGMANN, H. BONDI, T. GOLD, A. GRÜNBAUM, F. HOYLE, D. LAYZER, P. MORRISON, I. ROBINSON, L. ROSENFELD, L. SCHIFF, J. A. WHEELER and « X »: *The instability of the future,* pp. 138-148, and *General discussion,* pp. 223-243, in ref. [186].
[186] T. GOLD, editor: *The Nature of Time* (Ithaca, N. Y., 1967).
[187] H. SCHMIDT: *Model of an oscillating cosmos which rejuvenates during contraction, Journ. Math. Phys.*, **7**, 495-509 (1966).
[188] J. A. WHEELER: *At last—a sane look at the « arrow of time »,* a review of P. C. W. DAVIES: *The Physics of Time Asymmetry, Phys. Today,* **28**, 49-50 (June 1975).
[189] G. K. CHESTERTON: quoted in A. L. MACKAY, collector: *The Harvest of the Quiet Eye: A Selection of Scientific Quotations* (Bristol, 1977), as adapted for use here.
[190] S. G. BRUSH, editor: *Kinetic Theory.* - Vol. 2: *Irreversible Processes* (Oxford, 1976). The author thanks Prof. S. G. BRUSH and Pergamon Press for permission to make the quotes from this book indicated in ref. [192-194] and [196-198] and refers to Chapter 14 of a book by S. G. BRUSH: *The Kind of Motion We Call Heat* (Amsterdam and New York, N. Y., 1977).
[191] J. W. GIBBS: *On the equilibrium of heterogeneous substances, Trans. Conn. Acad.*, **3**, 108-248 (1875).
[192] L. BOLTZMANN: *Über die Beziehung eines allgemeine mechanischen Satzes zum zweiten Hauptsatze der Wärmetheorie, Sitzber. Akad. Wiss.*, part. II, **75**, 67-73 (1877), translated into English as *On the relation of a general mechanical theorem to the second law of thermodynamics,* in ref. [190], pp. 188-193; quotes from pp. 190, 192 and 193.
[193] H. POINCARÉ: *Le méchanisme et l'experience, Revue de Metaphysique et de Morale,* **1**, 534-537 (1893), translated into English as *Mechanism and experience,* in ref. [190], pp. 203-207; quote from p. 206.
[194] E. ZERMELO: *Über einen Satz der Dynamik and die mechanische Wärmetheorie, Ann. der Phys.*, **57**, 485-494 (1896), translated into English as *On a theorem of*

dynamics and the mechanical theory of heat, in ref. [190], pp. 208-217; quotes from pp. 208, 209, 214 and 215.

[195] H. Poincaré: *Sur le problème des trois corps et les équations de dynamique*, *Acta Math.*, **13**, 1-270 (1890).

[196] L. Boltzmann: *Entgegnung auf die wärmetheoretischen Betrachtungen der Hrn. E. Zermelo*, *Ann. der Phys.*, **57**, 773-784 (1896); translation, *Reply to Zermelo's remarks on the theory of heat*, in ref. [190], pp. 218-228; quotes from pp. 219-221 and 223-226.

[197] E. Zermelo: *Über mechanische Erklärungen irreversibler Vorgänge*, *Ann. der Phys.*, **59**, 793-801 (1896); translation, *On the mechanical explanation of irreversible processes*, in ref.[190], pp. 229-237; quote from p. 235.

[198] L. Boltzmann: *Zu Hrn. Zermelo's Abhandlung über die mechanische Erklärung irreversibler Vorgänge*, *Ann. der Phys.*, **60**, 392-398 (1897); translation, *On Zermelo's paper « On the mechanical explanation of irreversible processes »*, in ref. [190], pp. 238-245; quotes from pp. 238, 239, 241, 242.

[199] T. Ehrenfest-Afanassjewa: *Note*, appended to the 1959 reprinting of P. and T. Ehrenfest: *Begriffliche Grundlagen der statistischen Auffassung in der Mechanik*, in *Encyklopädie der mathematischen Wissenschaften*, IV 2, II, Heft 6 (Leipzig, 1912), pp. 1-90, and printed in ref. [200], pp. 301-302; quote from p. 302.

[200] M. J. Klein, editor: *Paul Ehrenfest: Collected Scientific Papers* (Amsterdam, 1959).

[201] G. E. Uhlenbeck: *An outline of statistical mechanics*, in ref. [202], pp. 1-29; quotes from p. 4 and p. 19. Appreciation is expressed to G. E. Uhlenbeck and E. G. D. Cohen for permission to quote cited passages.

[202] E. G. D. Cohen, editor: *Fundamental Problems in Statistical Mechanics. - II* (Amsterdam, 1968).

[203] N. N. Bogoliubov: *Problem dynamicheskoi teorii v statisticheskoi fizike* (Moscow, 1946); translation by E. K. Gora: *Problems of a dynamical theory in statistical physics*, in ref. [204], pp. 1-118.

[204] J. de Boer and G. E. Uhlenbeck: *Studies in Statistical Mechanics*, Vol. I (Amsterdam, 1962).

[205] E. G. D. Cohen: *Kinetic approach to non-equilibrium phenomena*, in ref. [4], pp. 548-560; the quote comes from p. 560. Appreciation is expressed to E. G. D. Cohen and J. Mehra for permission to quote cited passages.

[206] I. Prigogine: *Time, irreversibility and structure*, in ref. [4], pp. 561-591.

[207] M. J. Klein: *Paul Ehrenfest. - Vol. I: The Making of a Theoretical Physicist* (Amsterdam, 1970).

[208] H. Reichenbach: *The Direction of Time* (Berkeley, Cal., 1956).

[209] P. C. W. Davies: *The Physics of Time Asymmetry* (Berkeley and Los Angeles, Cal., 1974).

[210] P. and T. Ehrenfest: *Über eine Aufgabe aus der Wahrscheinlichkeitsrechnung, die mit der kinetischen Deutung der Entropievermehrung zusammenhängt*, *Math.-Naturw. Blätter*, **3**, No. 11 and 12 (1906); discussed and key passages translated in ref. [207], pp. 115-119.

[211] P. and T. Ehrenfest: *Über zwei bekannte Einwände gegen das Boltzmannsche H-Theorem*, *Phys. Zeits.*, **8**, 311-314 (1907); reprinted in ref. [200].

[212] E. Schrödinger and K. W. F. Kohlrausch: *Die Ehrenfestsche Modell der H-Kurve*, *Phys. Zeits.*, **27**, 306-313 (1926).

[213] M. Kac: *Probability and Related Topics in Physical Science* (New York, N. Y., 1959).

[214] A. H. Taub: *Empty space-times admitting a three-parameter group of motions*, *Ann. of Math.*, **53**, 472-490 (1951).

[215] C. W. MISNER: *Taub-NUT space as a counterexample to almost anything*, in J. EHLERS, editor: *Relativity Theory and Astrophysics*. - Vol. I: *Relativity and Cosmology* (Lectures in Applied Mathematics, Vol. **8**) (Providence, R. I., 1967), pp. 160-169.
[216] C. W. MISNER and A. H. TAUB: *A singularity-free empty universe*, *Sov. Phys. JETP*, **28**, 122-133 (1969). See [272].
[217] J. A. WHEELER: *The beam and the stay of the Taub universe*, preprint, University of Texas, Austin (July 1978).
[218] I. PRIGOGINE: this volume, p. 308.
[219] I. PRIGOGINE: *Time, structure and fluctuations*, Nobel Prize Lecture in Chemistry, December 1977, University of Texas at Austin preprint.
[220] K. L. HAINEBACH and D. N. SCHRAMM: *Comments on galactic evolution and nucleocosmochronology*, *Astrophys. Journ*, **212**, 347-359 (1977), give references to the literature on the ^{187}Re decay constant.
[221] R. H. DICKE and P. J. E. PEEBLES: *Cosmology and the radioactive decay ages of terrestrial rocks and meteorites*, *Phys. Rev.*, **128**, 2006-2011 (1962).
[222] C. L. DODGSON, *alias* Lewis CARROLL: *Through the Looking-Glass* (London, 1871); on p. 170 in this as reprinted in *Alice's Adventures in Wonderland and Through the Looking-Glass* (London, 1954, reprinted 1966).
[223] J. A. WHEELER: *General discussion*, in ref. [186], pp. 234-255.
[224] G. W. LEIBNIZ: *A new system of nature ...*, *Journal des Savans*, June 27, 1695; reprinted in P. P. WIENER, editor: *Leibniz Selections* (New York, N. Y., 1951), p. 108.
[225] E. P. HUBBLE: *A relation between distance and radial velocity among extragalactic nebulae*, *Proc. Nat. Acad. Sci. U.S.*, **15**, 169-173 (1929).
[226] R. A. ALPHER, H. A. BETHE and G. GAMOW: as cited in ref. [73], p. 760.
[227] R. H. DICKE: as cited in ref. [73], p. 760.
[228] A. A. PENZIAS and R. W. WILSON: as cited in ref. [73], p. 760.
[229] R. H. DICKE, P. J. E. PEEBLES, P. G. ROLL and D. T. WILKINSON: as cited in ref. [73], p. 760.
[230] M. SCHWARZSCHILD: *Structure and Evolution of the Stars* (Princeton, N. J., 1958).
[231] A. P. LIGHTMAN and S. L. SHAPIRO: *The dynamical evolution of globular clusters*, *Rev. Mod. Phys.*, **50**, 437-481 (1978); see especially table I on p. 442 for ages $\sim 10^{10}$ a for nine representative clusters in the Milky Way.
[232] W. FOWLER and F. HOYLE: *Nucleosyntheses in Massive Stars and Supernovae* (Chicago, Ill., 1964).
[233] R. V. WAGONER: *Bing-bang nucleosynthesis revisited*, *Astrophys. Journ.*, **179**, 343-360 (1973).
[234] P. S. LAPLACE: *Exposition du Système du monde*, Vol. **2** (Paris, 1796); the relevant page is reproduced photographically in ref. [73], p. 623.
[235] B. PACZYNSKI: *A model of accretion disks in close binaries*, *Astrophys. Journ.*, **216**, 822-826 (1977).
[236] E. VAN DEN HEUVEL: *Discussion of binary stars*, in *Astrophysics and Gravitation: Proceedings of the Sixteenth Conference in Physics of the University of Brussels, September 1973* (Bruxelles, 1974).
[237] J. WILSON: *Rapidly rotating neutron stars*, *Phys. Rev. Lett.*, **30**, 1082-1084 (1973).
[238] M. NAUENBERG and G. CHAPLINE jr.: *Determination of properties of cold stars in general relativity by a variational method*, *Astrophys. Journ.*, **179**, 277-287 (1973).
[239] R. GIACCONI: *Progress in X-ray astronomy*, *Phys. Today*, **26**, No. 5, 38-47 (1973).
[240] J. A. WHEELER: *The Universe as home for man*, in *The Nature of Scientific Discovery*, edited by O. GINGERICH (Washington, D.C., 1975), pp. 276-278.
[241] J. H. OORT: *The galactic center*, *Ann. Rev. Astron. Astrophys.*, **15**, 295-362 (1977).

[242] P. YOUNG, J. A. WESTPHAL, J. KRISTIAN, C. P. WILSON and F. P. LANDAUER: *Evidence for a supermassive object in the nucleus of the galaxy* M 87 *from* SIT *and* CCD *area photometry, Astrophys. Journ.*, **221**, 721-730 (1978).
[243] W. L. W. SARGENT, P. J. YOUNG, A. BOKSENBERG, K. SHORTRIDGE, C. R. LYNDS and F. D. A. HARTWICK: *Dynamical evidence for a central mass concentration in the galaxy* M 87, *Astrophys. Journ.*, **221**, 731-744 (1978).
[244] C. W. MISNER, K. S. THORNE and J. A. WHEELER: ref. [73], pp. 797-798.
[245] J. H. OORT: *Distribution of galaxies and the density of the universe*, in *Onzième Conseil de Physique Solvay: La structure et l'évolution de l'univers* (Brussels, 1958).
[246] J. F. GOTT, J. E. GUNN, D. N. SCHRAMM and B. M. TINSLEY: *An unbound universe?*, *Astrophys. Journ.*, **194**, 543-553 (1974).
[247] J. E. GUNN and J. B. OKE: *Spectrophotometry of faint cluster galaxies and the Hubble diagram: An approach to cosmology*, *Astrophys. Journ.*, **195**, 255-268 (1975).
[248] J. P. OSTRIKER and P. J. E. PEEBLES: *A numerical study of the stability of flattened galaxies; or, can cold galaxies survive?*, *Astrophys. Journ.*, **186**, 467-480 (1973).
[249] J. P. OSTRIKER, P. J. E. PEEBLES and A. YAHIL: *The size and mass of galaxies and the mass of the Universe*, *Astrophys. Journ. Lett.*, **193**, L1-4 (1974).
[250] J. A. WHEELER: *Conference summary: more results than ever in gravitation physics and relativity*, in *General Relativity and Gravitation, Proceedings of the Seventh International Conference* (GR7), *Tel-Aviv, June 1974*, G. SHAVIV and J. ROSEN, editors (New York, N. Y., 1975), pp. 299-344.
[251] M. DAVIS, E. J. GROTH and P. J. E. PEEBLES: *Study of galaxy correlations: Evidence for the gravitational instability picture in a dense universe*, *Astrophys. Journ.*, **212**, L107-111 (1977). Also see earlier references cited therein.
[252] C. W. MISNER, K. S. THORNE and J. A. WHEELER: ref. [73], pp. 795-796.
[253] R. D. BLANDFORD and M. J. REES: *Extragalactic double radio sources—the current observational and theoretical position*, *Contem. Phys.*, **16**, 1-16 (1975).
[254] R. D. BLANDFORD and M. J. REES: *A «twin-exhaust» model for double radio sources*, *Mont. Not. Roy. Astr. Soc.*, **169**, 395-415 (1974).
[255] YA. B. ZEL'DOVICH: *Observations in a universe homogeneous in the mean*, *Astron. Žurn.*, **41**, 19-24 (1964); English translation in *Sov. Astron.*, *Astron. Journ.*, **8**, 13-16 (1964).
[256] V. M. DASHEVSKY and YA. B. ZEL'DOVICH: *The propagation of light in a non-flat universe*. - II, *Astron. Žurn.*, **41**, 1071-1074 (1964); English translation in *Sov. Astron.*, *Astron. Journ.*, **8**, 854-856 (1965).
[257] W. H. PRESS and J. E. GUNN: *Method for detecting a cosmological density of condensed objects*, *Astrophys. Journ.*, **185**, 397-412 (1973).
[258] R. C. ROEDER: report presented at the *Seventh International Conference* (GR7) at Tel-Aviv University, June 23-28, 1974.
[259] A. EINSTEIN: *The Meaning of Relativity*, 3rd edition (Princeton, N. J., 1950), pp. 107-108.
[260] A. EINSTEIN: *Essays in Science* (New York, N. Y., 1934). Translated from *Mein Weltbild* (Amsterdam, 1933). The quote appears on p. 52 of the 1934 translation.
[261] H. ALFVÉN and O. KLEIN: *Matter-antimatter annihilation and cosmolovy*, *Ark. Fys.*, **23**, 187-194 (1962).
[262] G. DE VAUCOULEURS: *The large-scale distribution of galaxies and clusters of galaxies*, *Astron. Soc. Pacific Publ.*, **83**, 113-143 (1971).
[263] A. EINSTEIN: ref. [260], p. 68.

[264] D. R. BRILL and J. A. ISENBERG: *Dynamics of an open universe, Bull. Amer. Phys. Soc.*, **21**, 766 (1976).

[265] R. C. TOLMAN: *Relativity, Thermodynamics, and Cosmology* (Oxford, 1934).

[266] A. AVEZ: *Propriétés globales des espace-temps périodiques clos, Compt. Rend.*, **250**, 3585-3587 (1960).

[267] R. P. GEROCH: *Singularities in the space-time of general relativity: their definition, existence, and local characterization*, doctoral dissertation, Princeton University (1967).

[268] S. W. HAWKING and R. PENROSE: *The singularities of gravitational collapse and cosmology, Proc. Roy. Soc. London*, **314** A, 529-548 (1969).

[269] S. W. HAWKING and G. F. R. ELLIS: *The Large-Scale Structure of Space-time* (Cambridge, 1973).

[270] A. H. TAUB: *Empty space-times admitting a 3-parameter group of motions, Ann. of Math.*, **53**, 472-490 (1951).

[271] J. A. WHEELER: *Geometrodynamics and the issue of the final state*, in C. DEWITT and B. S. DEWITT, editors: *Relativity, Groups and Topology* (New York, N. Y., 1964), pp. 315-520, see especially p. 501 and 519.

[272] C. W. MISNER and A. H. TAUB: *A singularity-free empty universe, Žurn. Ėksp. Teor. Fiz.*, **55**, 233-255 (1968); English original in *Sov. Phys. JETP*, **28**, 122-133 (1969).

[273] G. F. R. ELLIS and A. R. KING: *Was the big bang a whimper?, Comm. Math. Phys.*, **38**, 119-156 (1974).

[274] R. PENROSE: *Internal instability in a Reissner-Nordstrøm black hole, Intern. Journ. Theor. Phys.*, **7**, 183-197 (1973).

[275] E. M. LIFSHITZ and I. M. KHALATNIKOV: *Problems of relativistic cosmology, Usp. Fiz. Nauk*, **80**, 391-438 (1963); English translation in *Sov. Phys. Usp.*, **6**, 495-522 (1964).

[276] E. M. LIFSHITZ and I. M. KHALATNIKOV: *Oscillatory approach to singular point in the open cosmological model, Žurn. Ėksp. Teor. Fiz. Pis'ma*, **11**, 200-203 (1970); English translation in *Sov. Phys. JETP Lett.*, **11**, 123-125 (1971).

[277] V. A. BELINSKY and I. M. KHALATNIKOV: *On the nature of the singularities in the general solution of gravitational equations, Žurn. Ėksp. Teor. Fiz.*, **56**, 1700-1712 (1969); English translation in *Sov. Phys. JETP*, **29**, 911-917 (1969).

[278] V. A. BELINSKY and I. M. KHALATNIKOV: *General solution of the gravitational equations with a physical singularity, Žurn. Ėksp. Teor. Fiz.*, **57**, 2163-2175 (1969); English translation in *Sov. Phys. JETP*, **30**, 1174-1180 (1970).

[279] V. A. BELINSKY and I. M. KHALATNIKOV: *General solution of the gravitational equations with a physical oscillatory singularity, Žurn. Ėksp. Teor. Fiz.*, **59**, 314-321 (1970); English translation in *Sov. Phys. JETP*, **32**, 169-172 (1971).

[280] I. M. KHALATNIKOV and E. M. LIFSHITZ: *General cosmological solutions of the gravitational equations with a singularity in time, Phys. Lett.*, **24**, 76-79 (1970).

[281] V. A. BELINSKY, I. M. KHALATNIKOV and E. M. LIFSHITZ: *Oscillatory approach to a singular point in the relativistic cosmology, Usp. Fiz. Nauk*, **102**, 463-500 (1970); English translation in *Adv. Phys.*, **19**, 525-573 (1970).

[282] V. A. BELINSKY, E. M. LIFSHITZ and I. M. KHALATNIKOV: *Oscillatory mode of approach to a singularity in homogeneous cosmological models with rotating axes, Žurn. Ėksp. Teor. Fiz.*, **60**, 1969-1979 (1971); English translation in *Sov. Phys. JETP*, **33**, 1061-1066 (1971).

[283] C. W. MISNER, K. S. THORNE and J. A. WHEELER: ref. [73], pp. 805-814.

[284] D. EARDLEY, E. LIANG and R. SACHS: *Velocity-dominated singularities in irrotational dust cosmologies, Journ. Math. Phys.*, **13**, 99-107 (1972).

[285] V. A. BELINSKY, I. M. KHALATNIKOV and B. COLLINS: 24 June 1976 draft pre-

print entitled *Dealing with the instability of special solutions of the field equations.* I thank Dr. V. A. BELINSKY for the opportunity to see this unpublished paper.

[286] J. A. WHEELER: *Singularity and unanimity*, *Gen. Rel. Grav.*, **8**, 713-715 (1977).

[287] R. PENROSE: *Singularities in cosmology*, in M. S. LONGAIR, editor: *Confrontation of Cosmological Theories with Observational Data* (Dordrecht, 1974).

[288] J. A. WHEELER: *The black hole*, in *Astrophysics and Gravitation*: *Proceedings of the XVI Conference on Physics at the University of Brussels, September 1973* (Bruxelles, 1974).

[289] C. W. MISNER, K. S. THORNE and J. A. WHEELER: ref. [73], pp. 539-543.

ALPHABETICAL LISTING OF REFERENCES

Author	Reference number(s)
ALFVÉN, H.	261
ALPHER, R. A.	226
ARAGONE, C.	69
ARNOLD, V. I.	180
ARNOWITT, R.	148
AVEZ, A.	180, 266
BAIERLEIN, R. F.	157
BELINFANTE, F. J.	43
BELINSKY, V. A.	277, 278, 279, 281, 282, 285
BERS, L.	153
BERGMANN, P. G.	185
BERTOLINI, G.	117
BETHE, H. A.	226
BETTONI, M.	117
BLANDFORD, R. D.	253, 254
BLEULER, E.	112
BOGOLIUBOV, N. N.	203
BOHM, D.	106
BOHR, N.	8, 23, 24, 25, 28, 42, 47, 48, 49, 50, 97, 99, 101, 105, 168, 169
BOKSENBERG, A.	243
BOLTZMANN, L.	192, 196, 198
BONDI, H.	184, 185
BORN, M.	107
BRADT, H. L.	112
BRILL, D. R.	264
BRUSH, S. G.	190
BURCKHARDT, J.	10
BUTT, D. K.	122
BUTTS, R. E.	41
CARMELI, M.	151
CARROLL, L.	222
CARTAN, E.	144, 145
CHESTERTON, G. K.	189
CHIU, H.-Y.	163
CLAUSER, J. F.	124, 126

PROCEEDINGS OF THE INTERNATIONAL SCHOOL OF PHYSICS
« ENRICO FERMI »

Course I
Questioni relative alla rivelazione delle particelle elementari, con particolare riguardo alla radiazione cosmica
edited by G. PUPPI

Course II
Questioni relative alla rivelazione delle particelle elementari, e alle loro interazioni con particolare riguardo alle particelle artificialmente prodotte ed accelerate
edited by G. PUPPI

Course III
Questioni di struttura nucleare e dei processi nucleari alle basse energie
edited by G. SALVETTI

Course IV
Proprietà magnetiche della materia
edited by L. GIULOTTO

Course V
Fisica dello stato solido
edited by F. FUMI

Course VI
Fisica del plasma e applicazioni astrofisiche
edited by G. RIGHINI

Course VII
Teoria della informazione
edited by E. R. CAIANIELLO

Course VIII
Problemi matematici della teoria quantistica delle particelle e dei campi
edited by A. BORSELLINO

Course IX
Fisica dei pioni
edited by B. TOUSCHEK

Course X
Thermodynamics of Irreversible Processes
edited by S. R. DE GROOT

Course XI
Weak Interactions
edited by L. A. RADICATI

Course XII
Solar Radioastronomy
edited by G. RIGHINI

Course XIII
Physics of Plasma: Experiments and Techniques
edited by H. ALFVÉN

Course XIV
Ergodic Theories
edited by P. CALDIROLA

Course XV
Nuclear Spectroscopy
edited by G. RACAH

Course XVI
Physicomathematical Aspects of Biology
edited by N. RASHEVSKY

Course XVII
Topics of Radiofrequency Spectroscopy
edited by A. GOZZINI

Course XVIII
Physics of Solids (Radiation Damage in Solids)
edited by D. S. BILLINGTON

Course XIX
Cosmic Rays, Solar Particles and Space Research
edited by B. PETERS

Course XX
Evidence for Gravitational Theories
edited by C. MØLLER

Course XXI
Liquid Helium
edited by G. CARERI

Course XXII
Semiconductors
edited by R. A. SMITH

Course XXIII
Nuclear Physics
edited by V. F. WEISSKOPF

Course LIII
Developments and Borderlines of Nuclear Physics
edited by H. Morinaga

Course LIV
Developments in High-Energy Physics
edited by R. R. Gatto

Course LV
Lattice Dynamics and Intermolecular Forces
edited by S. Califano

Course LVI
Experimental Gravitation
edited by B. Bertotti

Course LVII
Topics in the History of 20th Century Physics
edited by C. Weiner

Course LVIII
Dynamic Aspects of Surface Physics
edited by F. O. Goodman

Course LIX
Local Properties at Phase Transitions
edited by K. A. Müller

Course LX
C^*-Algebras and their Applications to Statistical Mechanics and Quantum Field Theory
edited by D. Kastler

Course LXI
Atomic Structure and Mechanical Properties of Metals
edited by G. Caglioti

Course LXII
Nuclear Spectroscopy and Nuclear Reactions with Heavy Ions
edited by H. Faraggi and R. A. Ricci

Course LXIII
New Directions in Physical Acoustics
edited by D. Sette

Course LXIV
Nonlinear Spectroscopy
edited by N. Bloembergen

Course LXV
Physics and Astrophysics of Neutron Stars and Black Holes
edited by R. Giacconi and R. Ruffini

Course LXVI
Health and Medical Physics
edited by J. Baarli

Course LXVII
Isolated Gravitating Systems in General Relativity
edited by J. Ehlers

Course LXVIII
Metrology and Fundamental Constants
edited by A. Ferro Milone, P. Giacomo and S. Leschiutta

Course LXIX
Elementary Modes of Excitation in Nuclei
edited by A. Bohr and R. A. Broglia

Course LXX
Physics of Magnetic Garnets
edited by A. Paoletti

Course LXXI
Weak Interactions
edited by M. Baldo Ceolin

Tipografia Compositori - Bologna - Italy

Course XXIV
Space Exploration and the Solar System
edited by B. Rossi

Course XXV
Advanced Plasma Theory
edited by M. N. Rosenbluth

Course XXVI
Selected Topics on Elementary Particle Physics
edited by M. Conversi

Course XXVII
Dispersion and Absorption of Sound by Molecular Processes
edited by D. Sette

Course XXVIII
Star Evolution
edited by L. Gratton

Course XXIX
Dispersion Relations and Their Connection with Causality
edited by E. P. Wigner

Course XXX
Radiation Dosimetry
edited by F. W. Spiers and G. W. Reed

Course XXXI
Quantum Electronics and Coherent Light
edited by C. H. Townes and P. A. Miles

Course XXXII
Weak Interactions and High-Energy Neutrino Physics
edited by T. D. Lee

Course XXXIII
Strong Interactions
edited by L. W. Alvarez

Course XXXIV
The Optical Properties of Solids
edited by J. Tauc

Course XXXV
High-Energy Astrophysics
edited by L. Gratton

Course XXXVI
Many-Body Description of Nuclear Structure and Reactions
edited by C. Bloch

Course XXXVII
Theory of Magnetism in Transition Metals
edited by W. Marshall

Course XXXVIII
Interaction of High-Energy Particles with Nuclei
edited by T. E. O. Ericson

Course XXXIX
Plasma Astrophysics
edited by P. A. Sturrock

Course XL
Nuclear Structure and Nuclear Reactions
edited by M. Jean

Course XLI
Selected Topics in Particle Physics
edited by J. Steinberger

Course XLII
Quantum Optics
edited by R. J. Glauber

Course XLIII
Processing of Optical Data by Organisms and by Machines
edited by W. Reichardt

Course XLIV
Molecular Beams and Reaction Kinetics
edited by Ch. Schlier

Course XLV
Local Quantum Theory
edited by R. Jost

Course XLVI
Physics with Storage Rings
edited by B. Touschek

Course XLVII
General Relativity and Cosmology
edited by R. K. Sachs

Course XLVIII
Physics of High Energy Density
edited by P. Caldirola and H. Knoepfel

Course IL
Foundations of Quantum Mechanics
edited by B. d'Espagnat

Course L
Mantle and Core in Planetary Physics
edited by J. Coulomb and M. Caputo

Course LI
Critical Phenomena
edited by M. S. Green

Course LII
Atomic Structure and Properties of Solids
edited by E. Burstein